W0193115

Fluid
Flow in
Pipes and
Channels

Fluid Flow in Pipes and Channels

G. L. Asawa
Professor of Civil Engineering
Indian Institute of Technology Roorkee
Roorkee 247 667
India

CBS Publishers & Distributors Pvt. Ltd.

New Delhi • Bengaluru • Chennai • Kochi • Kolkata • Mumbai
Hyderabad • Nagpur • Patna • Pune • Vijayawada

ISBN : 978-81-239-1723-8

First Edition : 2009
Reprint : 2010, 2012, 2014
Reprint:2017

Copyright © Author & Publisher

All rights reserved. No part of this book may be reproduced or transmitted in any form or by any means, electronic or mechanical, including photocopying, recording, or any information storage and retrieval system without permission, in writing, from the publisher.

Published by Satish Kumar Jain for
CBS Publishers & Distributors Pvt. Ltd.,
CBS Plaza, 4819/XI Prahlad Street, 24 Ansari Road, Daryaganj,
New Delhi - 110002, India.   • Website: www.cbspd.com
e-mail: delhi@cbspd.com, cbspubs@airtelmail.in
Ph.: 23289259, 23266861, 23266867   • Fax: 011-23243014

Branches:

• *Bengaluru:* Seema House, 2975, 17th Cross, K.R. Road,
  Bansankari 2nd Stage, Bengaluru - 560070
    • Ph.: +91-80-26771678/79 • Fax: +91-80-26771680
    • E-mail: cbsbng@gmail.com, bangalore@cbspd.com
• *Pune:* Bhuruk Prestige, Sr. No. 52/12/2+1+3/2,
  Narhe, Haveli (Near Katraj-Dehu Road By-pass), Pune - 411041
    • Ph.: +91-20-64704058/59, 020-32392277   • E-mail: pune@cbspd.com
• *Kochi:* 36/14, Kalluvilakam, Lissie Hospital Road,
  Kochi - 682018, Kerala • Ph.: +91-484-4059061-65
    • Fax: +91-484-4059065   • E-mail: cochin@cbspd.com
• *Chennai:* 20, West Park Road, Shenoy Nagar, Chennai - 600030
  Ph.: +91-44-26260666, 26208620   • Fax: +91-44-42032115
    • E-mail: chennai@cbspd.com
• *Mumbai:* 83-C, Ist Floor, Dr. E. Moses Road, Worli, Mumbai-400 018,
  Maharashtra   Ph.: +91-9833017933, 022-24902340/24902341
    • E-mail: mumbai@cbspd.com

Printed at :
India Binding House, Noida (UP)

to

all my family members

Preface

Most of the courses in hydraulic engineering (or hydraulics) have now been replaced with courses in the mechanics of fluids; the contents of the two courses, however, remain almost the same. An engineer dealing with fluids very often encounters problems related to the basics of fluid mechanics as well as flow of fluids in pipes and channels. Majority of the books on fluid mechanics cover open channel flow in only one chapter. This is rather inadequate from the point of view of the importance of open channel flow and, therefore, the reader has to invariably look for additional information dealing with open channel flow. The books on open channel flow, on the other hand, do not deal with the basics of fluid mechanics and hydraulics of pipe flow. This book includes the basics of fluid mechanics and flow of fluids in pipes as well as channels in enough detail. It is, therefore, hoped that the book will be useful to the undergraduate students of fluid mechanics and hydraulic engineering, and the practising engineers dealing with flow of fluids in pipes and channels.

This book is based on the teaching notes prepared by the author over a period of several years for the purpose of teaching undergraduate students at the Indian Institute of Technology Roorkee (formerly University of Roorkee). The author consulted several texts as listed in the bibliography and available literature on the subject as listed at the end of the relevant chapters for preparing the teaching notes. The author would like to sincerely acknowledge and appreciate the efforts of the learned authors as well as the publishers of the referenced and other literature related to the subject. The contents of Chapter 17 and parts of Chapters 12, 14 and 16 of the book are largely based on the text that appears in the author's another book *Irrigation and Water Resources Engineering* (2005).

The author has been immensely benefited by his association with his present and former colleagues at the Indian Institute of Technology Roorkee who have been associated with teaching, research, and consultancy in the fields of hydraulic engineering and water resources.

The author has been very fortunate in receiving lots of love, respect, and cooperation from all his family members, including his brothers and their family members, all through his life. Therefore, this book is being dedicated to all these family members. The forbearance of my wife Savi during the period I was busy writing the manuscript of this book is especially appreciated.

Suggestions from the readers and users of the book are welcome.

G. L. Asawa

Contents

CHAPTER 1

Fluid Properties

1.1 FLUID

Any matter exists in either a solid or fluid (i.e., liquid or gas) state. The distinction between the two states is obvious to anyone. However, the technical distinction between the two states of matter lies in their response (or reaction) to an applied shear (or tangential) stress[1]. A solid can resist an applied shear stress by a *static* deflection (or deformation). But, a fluid cannot resist shear stress, no matter how small, and starts moving *continuously* as long as the stress is applied. Thus, a *fluid* is a substance that deforms *continuously* under the application of a shear stress, irrespective of the magnitude of the applied shear stress. This, obviously, implies that if a fluid is at rest, no shear force can exist in it. A solid, however, can resist a shear force even while at rest; the shear force may cause some displacement of one layer over another, but the material does not continue to move indefinitely even if the shear force continues to exist. However, in case of a fluid, the material *continues* to move under the application of the shear force and comes to rest when the shear force is withdrawn.

Figure 1.1.1 (a) shows a solid mass abcd that has been bonded to the upper and lower plates. A constant shear force, F is applied to the solid mass through the upper plate. The solid gets deformed to abc'd' under the application of the shear stress, $\tau = F/A$, where A is the area of the surface in contact with the upper plate. Provided that the elastic limit for the solid material is not exceeded, the deformation is proportional to the shear stress τ.

If the same experiment is repeated with a fluid between the plates, Fig. 1.1.1 (b), one observes that the fluid element abcd continues to deform increasingly as long as the shear force F is applied. Figure 1.1.1 (b) shows the shape of the fluid element at successive instants of time t_o, t_1 and t_2. Since the fluid motion *continues* under the application of a shear stress, one concludes that a fluid cannot (i) sustain shear when at rest, and (ii) attain equilibrium deformation under the action of a shear force, no mater how small. Therefore, the fluid does not resist a shear force by acquiring an equilibrium deformation. But, the fluid does acquire an equilibrium velocity the value of which increases with the applied shear

---

[1] A shear force is the force component parallel to a surface. When one divides the force component by the area of the surface, one obtains the shear stress over the *area*. Shear stress at a *point* is the limiting value of the shear force per unit of the surface area when the surface area is reduced to a point.

force. Therefore, a fluid resists a shear force not by attaining an equilibrium deformation but by attaining an equilibrium *rate* of deformation. Hence, a fluid, when subjected to a shear force, deforms continuously at a finite rate that is determined by the magnitude of the applied shear force and the fluid properties.

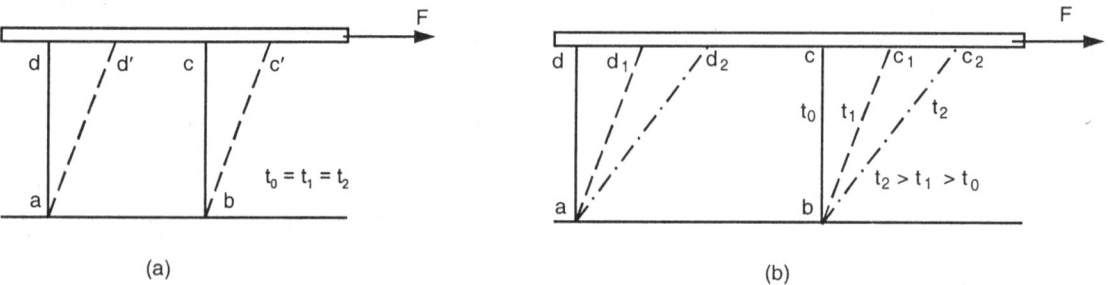

Fig. 1.1.1. Response of (a) solid and (b) fluid, when subjected to a constant shear force (2)

The distinction between the two classes of fluids, i.e., liquids and gases, is made on the basis of the effect of the cohesive forces between the molecules of the fluid. A liquid, composed of relatively close-packed molecules with strong cohesive forces, retains its volume and forms a free surface in a gravitational field, if unconfined from above. On the other hand, a gaseous substance is composed of wide-spaced molecules with negligible cohesive forces and is free to expand until it meets confining walls.

Fluid mechanics deals with the study of fluids either in motion (fluid dynamics) or at rest (fluid statics). Fluid mechanics finds its application in breathing, blood flow, swimming, windmills, pumps, pipes, channels, rivers, ships, automobiles, airplanes etc. Almost everything on this planet is either a fluid itself or surrounded by a fluid. Fluid mechanics is, therefore, a subject that is important in everyday life as well as in modern technology.

1.2 CONTINUUM

All fluids are conglomeration of separate molecules that are widely spaced in gases and closely spaced in liquids. The distance between molecules in both cases, however, is very large compared with their molecular diameter. Also, the molecules move freely relative to each other. Therefore, a fluid property such as the mass density (i.e., mass per unit volume) does not have precise meaning since the number of molecules occupying a given volume changes continuously due to enormous interchange of molecules across the boundaries of the chosen volume (6). This effect, however, would become insignificant if the chosen volume is relatively much larger compared with, say, the cube of the molecular spacing in which condition the number of molecules within the chosen volume will remain almost constant in spite of the enormous interchange of molecules across the boundaries.

An absolutely perfect analysis of the fluid behaviour would have to take into account the behaviour (or action) of each individual molecule. However, fortunately, the interest centres on the *average* conditions of the fluid properties (such as density, temperature etc.) and flow properties (such as velocity) in most engineering applications. Therefore, in fluid mechanics, one assumes, almost always, continuous distribution of matter with no empty space, i.e., the fluid is a *continuum*. As a result of this assumption, each fluid (and flow) property is assumed to have a definite value at every point in space at any given time. This means that the fluid properties such as mass density, temperature, velocity etc. are considered to be *continuous* functions of position and time. If mass density of fluid is ρ, the spatial coordinates are x, y, and z, and t represents the time, then one can write for a fluid, assumed to be continuum,

$$\rho = \rho\,(x\,,\,y,\,z,\,t) \qquad\qquad (1.2.1)$$

1.3 DIMENSIONS AND UNITS

Dimension can be defined as the measure by which a physical variable (such as mass, length, time, force etc.) is expressed quantitatively. *Unit* is a particular way by which a number is attached to the quantitative dimension. For example, length [L] is a dimension associated with variables such as distance, displacement, height, depth etc. Centimeters, inches etc. are the numerical units for quantifying the dimension of length. Dimension is a powerful concept around which technique of "dimensional analysis", used for obtaining solutions of several engineering problems, has been developed (Chapter-6).

The International System (SI) of units has been adopted in most countries and is expected to be adopted soon elsewhere too. In this system, the primary dimensions of mass [M], length [L], time [T] and force [F] have base units of kilogram (kg), metre (m), second (s) and newton (N) respectively.

There are seven base (or basic) units and their internationally agreed names and symbols are as in Table 1.3.1.

Table 1.3.1: Names and symbols of basic units of primary dimensions

Quantity	Unit	Symbol
Length	metre	m
Mass	kilogram	kg
Time interval	second	s
Temperature	Kelvin	K
Electric current	Ampere	A
Luminous intensity	candela	cd
Amount of substance	mole	mol

Units of all other quantities are derived from the base units. Some of these have been given internationally agreed special names, Table 1.3.2.

Temperature is conventionally expressed in "degree celcius", °C (formerly known as degree centigrade) such that $t°C = (273.15 + t)$ K. The unit of pressure, Pa is small for most applications and, therefore, multiples of Pa are used. Pressure is, many times, expressed in "*bar*" which equals 10^5 Pa and, approximately, the atmospheric pressure (written as "*atm*").

Table 1.3.2: Names and symbols of basic units of some other quantities

Quantity	Unit	Symbol	Equivalent combination of other units
Force	newton	N	$kg\ m/s^2$
Pressure and stress	pascal	Pa	N/m^2
Work and energy	joule	J	Nm
Power	watt	W	J/s or Nm/s
Frequency	hertz	Hz	s^{-1}

To avoid inconvenience while using too large or too small quantities, one may use prefixes (Table 1.3.3) with the unit names.

Table 1.3.3: Prefixes for units (3)

Sl. No.	Multiple	SI prefix	Abbreviation	Word equivalent	Sl. No.	Multiple	SI prefix	Abbreviation
1	10^{15}	peta	P	quadrillion	8	10^{-2}	centi	c
2	10^{12}	tera	T	trillion	9	10^{-3}	mille	m
3	10^{9}	giga	G	billion	10	10^{-6}	micro	μ
4	10^{6}	mega	M	million	11	10^{-9}	nano	n
5	10^{3}	kilo	k	thousand	12	10^{-12}	pico	p
6	10^{2}	hecto	h	hundred	13	10^{-15}	femto	f
7	10^{-1}	deci	d		14	10^{-18}	atto	a

1.4 FLUID PROPERTIES

1.4.1 Density Related Properties

The *mass density* of a fluid is its mass per unit volume. The *weight den*sity (also termed as *specific* [or *unit*] *weight*) of a fluid is its weight per unit volume. Mass density is denoted by ρ (rho) and weight density by γ (gamma) and they are related through the gravitational acceleration, g as $\gamma = \rho g$. The density of gases is highly variable and increases nearly proportionally to the pressure level. But, density for liquids is nearly constant. For example, the density of water ρ_w (= 998 kg/m^3 at atmospheric pressure and 20°C) would increase only by about 1 percent if the pressure is increased by a factor of 220(6). It is for this reason that most of the liquid flows are treated "incompressible" in their analysis. The weight density (or specific weight) of water, γ_w at 20°C and atmospheric pressure is (998 kg/m^3) (9.807 m/s^2) = 9790 N/m^3. The term specific weight would be often used in the applications of hydrostatic pressure. The mass density and weight density of air, at atmospheric pressure and 20°C are, respectively, 1.205 kg/m^3 and 11.8 N/m^3. In the present text, the densities for water and air (at 20°C and atmospheric pressure) for numerical problems have been approximated to the following values:

$$\rho_w = 1000 \text{ kg/m}^3 \qquad\qquad \rho_a = 1.2 \text{ kg/m}^3$$
$$\gamma_w = 9800 \text{ N/m}^3 \qquad\qquad \gamma_a = 11.8 \text{ N/m}^3$$

with
$$g = 9.8 \text{ m/s}^2$$

Specific volume, V_s is the reciprocal of the mass density and is, therefore, volume per unit mass. *Specific gravity* of a solid or liquid is the ratio of its mass (or weight) density and mass (or weight) density of water at 4°C. For gases, however, the reference (or standard) density may be that of air or hydrogen.

1.4.2 Pressure

Molecules of any fluid are in a continuous state of collision. Therefore, every part of the fluid, as well as the solid surface the fluid is in contact with, experiences a force exerted upon it by the surrounding fluid. This force may vary from location to location in magnitude and/or direction. Hence, it is more meaningful to work with pressure (or pressure intensity) that equals force per unit area. After velocity, the pressure (or pressure intensity) p is another important dynamic variable in fluid mechanics. *Pressure* is the stress (compression) at a point in a fluid and is the consequence of normal force acting on a plane in the fluid or a plane surface that the fluid is in contact with. The pressure p at a point in the fluid is the ratio of the normal force F to the area A of the plane as the area approaches a small value enclosing that point. This means, that

$$p = \lim_{A \to 0} \frac{F}{A} \tag{1.4.1}$$

Therefore, pressure has the units of N/m^2, also called pascal (Pa). The standard atmospheric pressure is often taken as 1 atm $\equiv$ 101, 300 Pa $\equiv$ 101.3 k Pa $\equiv$101.3 kN/m^2. Liquids can often sustain a considerable pressure without any noticeable change in their density and there does not exist any universal relationship between pressure and density of a liquid. But, gases respond to changes in pressure (or compressive force). For ideal gases, the pressure and density are related through the perfect-gas law,

$$p = \rho \, RT \tag{1.4.2}$$

where, T is the temperature in K and R the gas constant. Differences (or gradients) in pressure often cause a fluid flow, especially in conduits or ducts.

1.4.3 Viscosity

Viscosity of a fluid is an important property in the study of the fluid flow. *Viscosity* may be defined as the property of a fluid due to which the fluid is able to resist shear (or the movement of the fluid). In a way, viscosity of a fluid is a quantitative measure of the resistance to its flow that the flowing fluid offers. Moving in air, whose viscosity is low, is easy compared to moving in water that is about 50 times more viscous than air. Sliding one's hand would be more difficult in molasses than in glycerin whose viscosity is about one-fifth of the viscosity of molasses and about 1500 times the viscosity of water.

The resistance of any fluid to its movement depends upon cohesion and the rate of transfer of molecular momentum in the fluid. Because of intermolecular distance being much less, the liquids have much larger cohesion. Therefore, cohesion appears to be the main cause of viscosity in a liquid. Cohesion decreases with increase in temperature. Therefore, viscosity of a liquid also decreases with increase in temperature. On the other hand, gases have much lesser cohesive force. Therefore, transfer of molecular momentum is the prime cause of viscosity in a gas. The viscosity of a gas, therefore, increases with increase in temperature that results in the increased molecular activity and, hence, transfer of molecular momentum.

The viscosity of a fluid increases only marginally with pressure and, therefore, the variation of viscosity with pressure is generally neglected in most engineering applications. Temperature, however, has a strong effect on the variation of viscosity of a fluid.

Consider a fluid element abcd with its upper surface dc moving at a speed δu relative to the lower surface ab, Fig. 1.4.1 (a). The fluid element is therefore, subjected to a shear stress τ due to which shear strain angle $\delta\theta$ (at any time δt) continuously increases as long as the shear stress τ exists. Common fluids such as water and air exhibit a linear relation between the applied shear stress τ and the resulting shear strain (or deformation) rate $\delta\theta / \delta t$. This means,

$$\tau \propto \frac{\delta\theta}{\delta t}$$

From Fig. 1.4.1(a),

$$\tan \delta\theta = \frac{\delta u \; \delta t}{\delta y}$$

For infinitesimal change (i.e., $\delta\theta$ and δt both tending to be very small and approaching zero; therefore, $\tan \delta\theta \approx \delta\theta$), one can obtain relation between shear strain rate ($d\theta/dt$) and velocity gradient (du/dy) as

$$\frac{d\theta}{dt} = \frac{du}{dy} \qquad (1.4.3)$$

The applied shear stress τ is, therefore, proportional to the velocity gradient and introducing a coefficient of proportionality μ (mu, termed as *viscosity coefficient* or *dynamic viscosity* or *absolute viscosity* or, simply, *viscosity*), one may write

$$\tau = \mu \frac{du}{dy} \qquad (1.4.4)$$

Dimensions of μ would be $\{M/(LT)\}$ or $\{FT/L^2\}$. Dynamic viscosity, therefore, has the unit of kg/(ms), i.e., kilograms per metre-second or Ns/m^2, i.e., newton-second per squared metre. Equation (1.4.4) is known as Newton's law of viscosity and the fluids that follow this law are called *Newtonian fluids*, after Sir Isaac Newton, who first postulated this law in 1687(6).

Figure 1.4.1(b) illustrates a velocity profile of a moving fluid near a solid boundary. At the boundary, the velocity of the fluid, relative to the boundary, is zero. This is called the no-slip condition that would exist for all viscous fluid flows past a solid surface. The shear stress at any point is proportional to the slope of the velocity profile, du/dy at the point and is, therefore, the greatest at the solid boundary.

The ratio of the dynamic viscosity μ and the mass density ρ of a fluid is termed the *kinematic viscosity* ν (nu) of the fluid.

$$\nu = \frac{\mu}{\rho} \qquad (1.4.5)$$

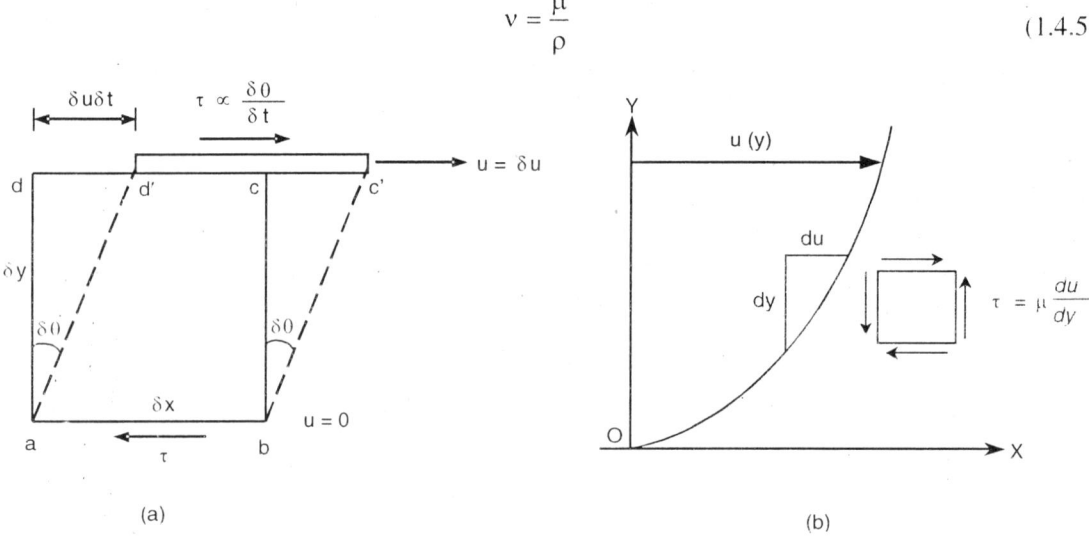

Fig. 1.4.1. Deformation of a fluid element due to shear applied on its surfaces

The dimensions of ν would be $\{L^2/T\}$ and, therefore, the corresponding unit is m^2/s. At 20°C, the values of μ and ν for air and water can be taken as follows:

$$\mu_{air} = 1.80 \times 10^{-5} \text{ Kg/ms} \qquad \mu_{water} = 10^{-3} \text{ Kg/ms}$$
$$\nu_{air} = 1.5 \times 10^{-5} \text{ m}^2/\text{s} \qquad \nu_{water} = 10^{-6} \text{ m}^2/\text{s}$$

A fluid having no viscosity (i.e., $\mu = 0$) is termed *non-viscous* or *inviscid* or *ideal fluid*. No real fluid is an inviscid fluid. But, in many situations, the assumption of inviscid fluid makes analysis of motion simpler and yields results that are not far from the true ones.

Mechanics of fluids (or simply, fluid mechanics) deals with Newtonian fluids. Fluids that do not follow Newton's law of viscosity, Eq. (1.4.4), are termed *non-Newtonian* fluids. Equation (1.4.4) may, alternatively, be written as

$$\tau = K \left(\frac{du}{dy} \right)^n \tag{1.4.6}$$

in which K is the consistency index and n is the flow behaviour index and, for Newtonian fluids, $K = \mu$ and $n = 1$. Equation (1.4.6) can also be written as

$$\tau = K \left| \frac{du}{dy} \right|^{n-1} \frac{du}{dy} = \eta \frac{du}{dy} \tag{1.4.7}$$

in which η is termed the *apparent viscosity* (2).

Many common fluids are non-Newtonian. Figure 1.4.2 compares four types of non-Newtonian fluids with a Newtonian fluid. A *dilatant* (or shear-thickening) fluid (example: concentrated solution of sugar in water) is the one for which resistance increases with increasing applied shear stress ($n > 1$). For *pseudoplastic* (or shear-thinning) fluids (examples: gelatine, milk, blood, liquid cement), resistance decreases with increasing applied shear stress ($n < 1$). If the thinning effect is very strong, the fluid is called *plastic*. Some plastic fluids require a finite yield stress, τ_0, before they begin to flow. Toothpaste will not flow out of its tube until a finite stress is applied to it by squeezing the tube. An *ideal plastic* or *Bingham plastic* (examples: clay suspension, drilling mud, toothpaste) exhibit linear relation between the applied shear stress and rate of deformation and for which (2)

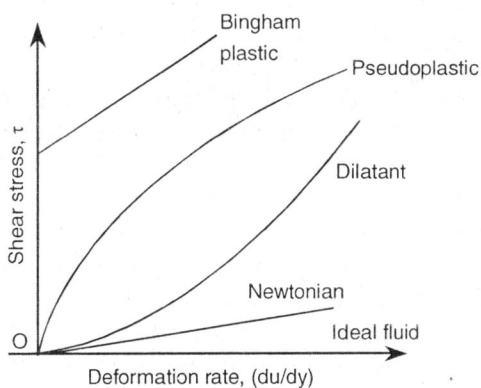

Fig. 1.4.2. Variation of shear stress and deformation rate for Newtonian and non-Newtonian fluids

$$\tau = \tau_0 + \mu_p \left(\frac{du}{dy} \right) \tag{1.4.8}$$

in which μ_p is the viscosity of the plastic fluid. *Rheology* deals with non-Newtonian fluids.

1.4.4 Surface Tension

A liquid will always form an interface with another liquid or gas unless the liquid is in a container that is covered and filled completely. Molecules deep inside the liquid, being surrounded by similar molecules, are subjected to inter-molecular forces of the same magnitude in all directions, Fig. 1.4.3(4).

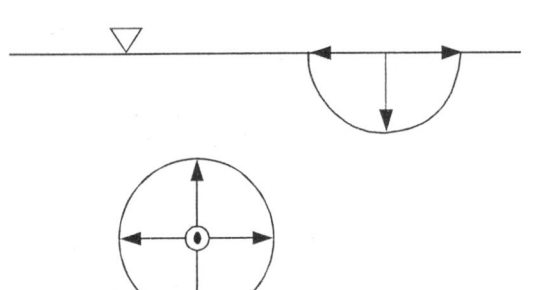

Fig. 1.4.3. Intermolecular forces near a liquid surface

The molecules on the free surface (or interface), however, are surrounded by the molecules of other liquid or gas over the surface. The apparent physical consequence of this unbalanced force along the interface is to create a hypothetical skin or membrane. This unbalanced tensile force, acting in the plane of the surface, is termed the surface tension that opposes any increase in the area of a liquid surface. Such an increase, therefore, is at the expense of some mechanical energy. Hence, the existence of a free surface implies the presence of free surface energy that is equal to the work done in forming the free surface(4). These surface effects, in fluid mechanics, are adequately accounted for with the concept of surface tension.

The effect of surface tension is to reduce the surface of a body of liquid to a minimum as increasing the surface area would require bringing the molecules to the surface from the body of the liquid against the unbalanced attraction that pulls the surface molecules inwards (1). It is for this reason that drops of liquid tend to attain spherical shape so as to minimize surface area.

If one imagines a line drawn on the surface of a liquid, then the liquid on one side of the line pulls the liquid on the other side. The magnitude (or coefficient) of surface tension or, simply, surface tension is the magnitude of the tensile force acting across and perpendicular to a unit length of straight elements of the line. Alternatively, the surface tension can be regarded as the surface energy (in N-m) per unit area (in m²) of the interface. The surface tension, σ (sigma) has dimensions of {F/L} or {M/T²}. The value of σ for water-air interface at 20°C is 0.073 N/m. Most organic liquids have surface tension between 0.020 and 0.030 N/m and mercury about 0.51 N/m when in contact with air. For water, the surface tension decreases when small quantities of organic solutes like soap or detergents are added. Detergents change the surface properties of oil, grease and dirty particles in contact with water making it easier to clean them away. Salinity in water increases the surface tension of water.

The surface tension of a liquid drop creates a higher pressure within the drop compared to that existing in the surrounding fluid. The pressure inside a drop (of radius R) of liquid can be determined using the free-body diagram of the drop that is cut in half, Fig. 1.4.4. The force developed around the edge due to surface tension is $2\pi R\sigma$ that must be balanced by the pressure difference Δp, between the internal pressure p_i, and the external pressure p_a, acting over the circular area πR^2. Thus,

$$2\pi R\sigma = \Delta p \pi R^2$$

$$\therefore \qquad \Delta p = p_i - p_a = \frac{2\sigma}{R} \qquad (1.4.6)$$

For the case of a soap bubble of the same radius *R*, the surface tension force would be

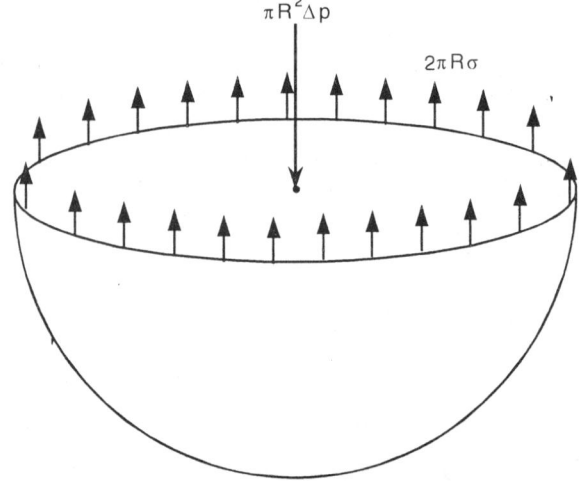

Fig. 1.4.4. Free-body diagram of a spherical drop for pressure and surface tension forces

$2 \times 2\,\pi R\sigma$ as there is an additional surface inside of the bubble on which too surface tension force exists. Thus, for a bubble of radius R,

$$4\pi R\sigma = \Delta p \pi R^2$$

$$\Delta p = \frac{4\sigma}{R} \qquad\qquad (1.4.7)$$

The surface of a cylindrical jet of water (radius R) too experiences tension. Consider a section of jet of unit length, that is, the ring between two planes normal to the jet axis and unit length apart. If one half of this ring were taken as free body, Fig. 1.4.5, the tensions per unit length at top and bottom would be σ_1 and σ_2. For high pressures in the jet, the pressure centre can be taken at the centre of the jet and then $\sigma_1 = \sigma_2 = \sigma$. The horizontal component of force (due to the pressure p - in the jet - exceeding the surrounding pressure) acts through the pressure centre of the projected area and, therefore, equals $2pR$. Therefore,

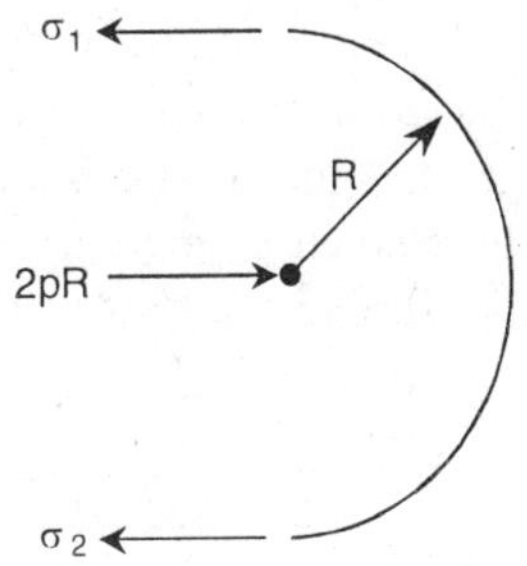

Fig. 1.4.5. Free-body diagram of a cylindrical jet

$$2pR = \sigma_1 + \sigma_2 = 2\sigma$$

$$\therefore \qquad\qquad p = \frac{\sigma}{R} \qquad\qquad (1.4.8)$$

Another common phenomenon associated with the surface tension is the rise (or fall) of a liquid in a capillary tube, i.e., a tube of very small diameter. Consider a capillary tube inserted into water, Fig. 1.4.6(a). The water level inside the tube would rise above the water level outside the tube. The rise (or fall) of liquid in the glass tube is decided by the adhesive forces (between the molecules of the liquid and the glass) and cohesive forces (between the molecules of the liquid). For the illustrated case, adhesion is stronger than cohesion and the molecules are, therefore, pulled up the tube wall. Hence, the liquid level inside the tube rises and *wets* the tube wall. If water is replaced with mercury, cohesion becomes stronger than adhesion and, therefore, mercury level inside the tube falls and mercury does not

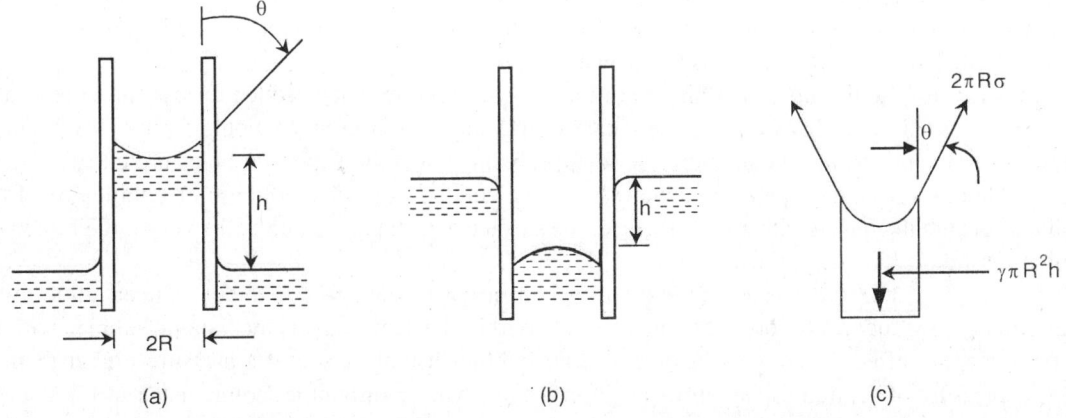

Fig. 1.4.6. Capillary action: (a) Rise of level of a wetting liquid (b) Depression of level of a non-wetting liquid (c) Free-body diagram of a liquid column

wet the surface, Fig.1.4.6(b). From the free-body diagram of Fig. 1.4.6(c), vertical force due to surface tension $2 \pi R \sigma \cos \theta$ must equal the weight of the liquid in the capillary tube, i.e., $\pi R^2 h \gamma$. Thus,

$$\pi R^2 h \gamma = 2 \pi R \sigma \cos \theta$$

$\therefore$
$$h = \frac{2\sigma \cos \theta}{\gamma R} \qquad (1.4.9)$$

When an interface between two fluids, such as water and air or mercury and air, meets a solid surface, the interface forms an angle with respect to the solid surface. This angle is termed the angle of contact θ that depends on the liquid, the surrounding fluid and the solid surface. For non-wetting liquids, such as mercury, the angle of contact is greater than 90° (for mercury in contact with clean glass, $\theta \approx 130°$). For water in contact with clean glass, $\theta \approx 0$ and, hence, Eq. (1.4.9) reduces to

$$h = \frac{2\sigma}{\gamma R} \qquad (1.4.10)$$

Surface tension plays an important role in several problems of fluid mechanics including the movement of liquids through soil and other porous media, flow of thin films (or sheets) of liquid, formation of drops and bubbles and breaking of liquid jets etc. In model studies too, sometimes, the surface tension becomes an important parameter.

1.4.5 Vapour Pressure

The molecules of any liquid are in a state of continuous motion. Some of these molecules in the surface layer of the liquid would have sufficient energy to enable them to overcome the cohesive attraction of the surrounding molecules and escape into the space above the layer. Some of these escaped molecules would return to the layer, but other molecules of the liquid would take their place. If the space above the liquid surface is confined, an equilibrium would reach when the number of molecules of the liquid in the space above the liquid surface is constant and the gas above the surface is saturated with vapour. These molecules exert partial pressure on the liquid surface, i.e., free surface. This pressure is known as the *vapour pressure* in the space(5).

With the increase in temperature, the molecular activity increases and, therefore, vapour pressure will also increase. When the pressure above the liquid equals the vapour pressure of the liquid, the liquid starts boiling. Boiling of water, therefore, can start at room temperature if the pressure in the room is reduced to the vapour pressure of water.

In a flowing liquid, under certain conditions, local pressures may reduce to less than the vapour pressure of the liquid. This would cause local boiling and vapour bubbles would form. This phenomenon is known as *cavitation* that causes serious problems when the vapour bubbles are carried, by the flowing liquid, into a high-pressure zone resulting in sudden implosive collapse of bubbles(6). If this collapse occurs in contact with a solid surface, the surface gets damaged due to the very large force with which the liquid hits the surface.

Cavitation can also occur when the dissolved air or gas is released due to the reduced solubility as the pressure is reduced. Gas or air bubbles are released in the same way as the vapour bubbles with the same damaging effects. The releases of gas or air bubbles usually occur at a pressure higher than the vapour pressure of the liquid. Therefore, cavitation due to gas or air bubbles commences earlier than the vapour cavitation.

Cavitation can affect performance of hydraulic machinery when cavitation has caused local erosion of metallic surfaces of the machine parts. Surface of a spillway can be damaged on account of cavitation.

1.4.6 Bulk Modulus of Elasticity

For most of the engineering applications, a liquid can be considered as incompressible. But, if the liquid is subjected to sudden or large changes in its pressure, the compressibility of the liquid becomes important. Also, when the velocity of a fluid (or that of a body in a fluid) is higher than the velocity of sound in that fluid, the compressibility effects become important. The compressibility of a liquid is expressed by its *bulk modulus of elasticity K* defined as the ratio of the change in the pressure dp and the resulting volumetric strain, $d\forall/\forall$, where $\forall$ is the volume of the liquid. This means,

$$K = -\frac{dp}{\dfrac{d\forall}{\forall}} = -\forall\frac{dp}{d\forall} \tag{1.4.11}$$

The dimensions of K is the same as that of pressure, i.e., $\{F/L^2\}$ or $\{M/LT^2\}$ and its unit is N/m^2. The negative sign in Eq. (1.4.11) appears because an increase in pressure causes a decrease in volume. Since a decrease in volume for a given mass results in an increase in the density ρ, one can also write,

$$K = \frac{dp}{\dfrac{d\rho}{\rho}} = \rho\frac{dp}{d\rho} \tag{1.4.12}$$

The values of K for common liquids are large (for water, $K = 2.0 \times 10^9$ N/m^2) indicating that a large pressure change is required for a small change in volume. When the liquid pressure is increased, the molecules come closer and K, therefore, becomes higher, i.e., the liquid is still less compressible. The value of K gets doubled if the pressure is increased from 1 to about 3000 atm.

Tables 1.4.1 and 1.4.2 list the values of some fluid properties of some common fluids.

Table 1.4.1: Viscosity and density of water at 1 atm

Temperature T, °C	Mass density ρ, Kg/m³	Dynamic viscosity 1000 μ, Ns/m²	Kinematic viscosity 10⁶ v, m²/s
0	1000	1.788	1.788
10	1000	1.307	1.307
20	998	1.003	1.005
30	996	0.799	0.802
40	992	0.657	0.622
50	988	0.548	0.555
60	983	0.467	0.475
70	978	0.405	0.414
80	972	0.355	0.365
90	965	0.316	0.327
100	958	0.283	0.295

Table 1.4.2: Approximate properties of common liquids at 1 atm and 20°C.

Liquid	Specific gravity, S	Bulk modulus of elasticity, K, Gpa	Vapour pressure Kpa	*Surface tension, σ, N/m	Viscosity μ Kg/m.s	Kinematic viscosity ν, m^2/s
Benzene	0.88	1.03	10.00	0.029	6.51×10^{-4}	7.40×10^{-7}
Carbon tetrachloride	1.59	1.10	13.10	0.027	9.67×10^{-4}	6.08×10^{-7}
Glycerin	1.26	4.34	14.00	0.063	1.49	1.18×10^{-3}
Mercury	13.56	26.20	0.00017	0.510	1.56×10^{-3}	1.15×10^{-7}
Water	1.00	2.20	2.45	0.073	1.00×10^{-3}	1.0×10^{-6}

* in contact with air

EXAMPLES

E 1.1: If the Kinematic viscosity of benzene is 7.4×10^{-3} stokes (i.e., cm^2/s) and its mass density is 880 kg/m^3, determine its dynamic viscosity in poise (i.e., gm/cm.s).

Solution: $\nu = 7.4 \times 10^{-3}$ stokes $= 7.4 \times 10^{-3}$ cm^2/s

$\qquad = 7.4 \times 10^{-3} \times 10^{-4}$ m^2/s $= 7.4 \times 10^{-7}$ m^2/s

$\qquad \mu = \rho\nu = 880 \times 7.4 \times 10^{-7}$ kg/ms (or Ns/m^2)

$\qquad = 6.512 \times 10^{-4}$ kg/ms

$\qquad = 6.512 \times 10^{-3}$ g/cm.s $= 6.512 \times 10^{-3}$ poise

E 1.2: A rectangular plate of 0.4 m × 0.4 m dimensions weighing 400 N slides down a plane inclined at an angle of 25° with the horizontal. The velocity of the plate is 1.8 m/s and the space between the plate and plane is 2 mm. The space is filled with an oil. Find the viscosity of the oil.

Solution: From Newton's law of viscosity, Eq.(1.4.4), the shear stress along the surface of the plate,

$$\tau = \mu \frac{du}{dy} \cong \mu \frac{\Delta U}{\Delta y}$$

$$= \mu \frac{1.75}{2 \times 10^{-3}}$$

Therefore, the shear force,

$$= (0.4 \times 0.4)\mu \frac{1.75}{2 \times 10^{-3}}$$

$$= 140 \, \mu \, N$$

For equilibrium, this shear force must be equal to the component of the weight of the plate along the inclined plane.

$\qquad \therefore \qquad 400 \sin 25° = 140 \, \mu$

$\qquad \therefore \qquad \mu = 1.21$ Ns/m^2 = 1.21 kg/ms

E 1.3:   A needle 40 mm long rests on a water surface at 20°C. Determine the force, over and above the needle's own weight, that would be required to lift the needle from contact with the water surface.

Solution: The required force is to overcome the surface tension force (0.073 N/m) that holds the needle on the water surface. The total length over which the surface tension force is acting equals twice the length of the needle. Thus, the force due to surface tension, i.e., the required force,

F = 2 × 0.04 × 0.073 = 5.84 × 10⁻³ N

E 1.4:   Compute the approximate depression of mercury at 20°C (σ = 0.51 N/m and γ = 135.6 kN/m³) in a capillary glass tube of radius 1.5 mm when the tube is dipped in mercury. The contact angle with glass surface for mercury is 140°.

Solution: From Eq. (1.4.9), $h = \dfrac{2\sigma\cos\theta}{\gamma R}$

$$= \frac{2\times0.51\times\cos140°}{(135.6\times10^{3})(1.5\times10^{-3})}$$

$$= -3.84\times10^{-3}\text{ m}$$

$$= -3.84\text{ mm}$$

The negative sign indicates depression with respect to the surrounding mercury level.

E 1.5:   At some large depth in an ocean, the pressure is 90 Mpa with reference to the atmospheric pressure. Assuming that the specific weight at the surface of the ocean is 10 kN/m³ and the average bulk modulus of elasticity is 2.4 Gpa, find the specific volume and specific weight at that large depth.

Solution: At the surface of the ocean, the specific volume

$$V_{s1} = \frac{1}{\rho_1} = \frac{g}{\gamma_1} = \frac{9.8}{10\times10^{3}} = 9.8\times10^{-4}\text{ m}^3/\text{kg}$$

Since, $E = -\dfrac{dp}{\dfrac{dV_s}{V_s}}$

$$2.4\times10^{9} = (-)\frac{90\times10^{6}}{\dfrac{dV_s}{9.8\times10^{-4}}}$$

$\therefore$ $dV_s = (-)\dfrac{90\times10^{6}\times9.8\times10^{-4}}{2.4\times10^{9}} = (-)\,0.3675\times10^{-4}\text{ m}^3/\text{kg}$

Therefore, the specific volume at the large depth,

$$V_{s2} = V_{s1} + dV_s = (9.8 - 0.3675)\times10^{-4} = 9.4325\times10^{-4}$$

$$= 9.43\times10^{-4}\text{ m}^3/\text{kg}$$

and $\gamma_2 = \dfrac{g}{V_{s2}} = \dfrac{9.80}{9.43 \times 10^{-4}} = 1.04 \times 10^4 = 10.4 \text{ kN/m}^3$.

E 1.6:   Eight kilometers below the surface of an ocean, the pressure is 82 Mpa. Determine the density of seawater at this depth if the density at the surface of the ocean is 1020 kg/m³ and the average bulk modulus of elasticity is 2.34 GPa.

Solution: From Eq. (1.4.12)

$$K = \rho \frac{dp}{d\rho}$$

$$\therefore \quad \frac{d\rho}{\rho} = \frac{1}{K} dp$$

or $\log_e \dfrac{\rho_1}{\rho_2} = \dfrac{1}{K}(p_1 - p_2)$

or $\log_e \dfrac{\rho_1}{1020} = \dfrac{82 \times 10^6 - 101.3 \times 10^3}{2.34 \times 10^9}$

$\therefore \quad \rho_1 = 1056.33 \text{ kg/m}^3$

PROBLEMS

P1.1   A block of weight W slides down an inclined plane (inclination with horizontal, θ) that has been lubricated by a thin film of oil (viscosity $= \mu$). The thickness of the film is h and contact area of the film with the block is A. Assuming a linear velocity distribution in the film, determine the terminal (zero acceleration) velocity V of the block.

P1.2   A thin plate is separated from two fixed plates by viscous liquids of viscosity μ_1 and μ_2 respectively. All three plates are parallel to each other and the plate spacings b_1 and b_2 are unequal. The surface area of the central plate is A. Assuming a linear velocity distribution in each fluid, determine the force F required to pull the central plate with velocity V.

P1.3   A solid cylinder of diameter D, length L, and density ρ_s falls due to gravity inside a tube of diameter D_0. The clearance between the cylinder and tube, filled with a fluid (density $= \rho$ and viscosity $= \mu$), is much smaller than the diameter of the cylinder. Derive an expression for the terminal fall velocity of the cylinder and, thus, obtain the terminal fall velocity of a steel cylinder with $D = 2$ cm, $D_0 = 2.03$ cm, $L = 12$ cm and the fluid is oil ($\mu = 0.3$ kg/ms). Neglect buoyancy effects.

P1.4   Rotating cylinder viscometer consisting of two concentric cylinders is often used to measure the viscosity of liquids. In this device, the outer cylinder of radius R_0 is fixed and the inner cylinder of radius R_i is rotated with an angular velocity, ω, the torque T required to develop ω is measured and the viscosity is calculated from these measurements. Assuming that the velocity distribution in the space (filled with a liquid of viscosity μ) between the two cylinders is linear and neglecting the end effect, obtain suitable expression for the estimation of viscosity.

P1.5   A uniform film of oil 0.16 mm thick separates two discs, each of 220 mm diameter, mounted coaxially. Ignoring edge effects, calculate the torque necessary to rotate one disc relative to the other at a speed of 10 revolutions per second. Assume that the oil has a viscosity of 0.15 Pa s.

P1.6 Estimate the excess pressure inside a raindrop of diameter equal to 2.5 mm.

P1.7 If the hydrogen-water interface is comparable to air-water interface, estimate the excess pressure within a hydrogen bubble (of diameter 0.01 mm) moving in a body of water whose surface tension is 0.073 N/m.

P1.8 A 10-mm diameter jet of water discharges vertically into the atmosphere. Owing to surface tension, the pressure inside the jet will be slightly higher than the surrounding atmospheric pressure. Determine this difference in pressure.

P1.9 An open, clean glass tube having a diameter of 3mm is inserted vertically into a dish of mercury at 20°C. How far will the column of mercury in the tube be depressed?

P1.10 What diameter of glass tube is required, if the capillary effects on the water within the tube are not to exceed 0.4 mm?

P1.11 The bulk modulus of elasticity of water is 2.2 GPa. What pressure change is required to reduce its volume by 0.5 percent?

REFERENCES

(1) Douglas, JF, Gasiorek, JM, and Swaffield, JA: *Fluid Mechanics*, 4th Edition, Pearsonn Education (Singapore) Pte. Ltd., 2002.

(2) Fox, RW, and McDonald, AT; *Introduction to Fluid Mechanics*, 4th Edition, John Wiley and Sons, Inc., USA, 1994.

(3) Massey, BS (revised by John Ward-Smith): *Mechanics of Fluids*, 7th Edition, Chennai Micro Print Pvt. Ltd., Chennai, India, 1998.

(4) Rouse, H.:, *Elementary Mechanics of Fluids*, Wiley Eastern Private Limited, New Delhi, India, 1970.

(5) Streeter, VL, Wylie, EB, and Bedford, KW: *Fluid Mechanics*, 9th Edition, McGraw-Hill Companies, Singapore, 1998.

(6) White, FM: *Fluid Mechanics*, 5th Edition, McGraw-Hill Companies, USA, 2003.

CHAPTER 2

Fluid Statics

2.0 GENERAL

In a static homogeneous fluid (such as water stored in a reservoir) or in a fluid undergoing rigid body motion (such as a tank filled with a liquid that has been spinning for a long time), a fluid particle retains its identity, with respect to its surrounding fluid particles, for all times, and fluid elements do not deform. This absence of angular deformation implies the absence of shear stresses. Hence, fluids either at rest or in rigid-body motion experience only normal stresses. The analysis of such fluid problems is relatively simpler in comparison to that for fluids undergoing angular deformation.

2.1 SURFACE AND BODY FORCES

Surface and body forces are always encountered in the study of fluid mechanics. *Surface forces* exist by virtue of direct contact between various fluid particles and, thus, act on the boundaries of a medium through direct physical contact (3), Fig. 2.1.1(a). If one considers a central fluid particle surrounded by several fluid particles, then the effect of the surrounding particles (that are in contact with the central fluid particle) around the central particle can be considered in terms of an equivalent system of resultant force ΔF (due to the individual forces ΔF_n, ΔF_{s_1}, and ΔF_{s_2}) acting on the area of contact between the surrounding particles and the central particle, Fig. 2.1.1(b). Since the surface areas, in general, are arbitrarily curved, it is more convenient to work with stress, which is a scalar quantity being the ratio of force and area both of which are vector quantities. These stresses are generally classed into two categories: normal stress, σ_n and shear stress, τ, Fig. 2.1.1(c). The stresses are defined in a limit sense (i.e., the contact area approaches zero) that is,

$$\sigma_n = \lim_{\Delta A \to 0} \frac{\Delta F_n}{\Delta A} \qquad \qquad ...(2.1.1)$$

$$\tau_{ss_1} = \lim_{\Delta A \to 0} \frac{\Delta F_{s_1}}{\Delta A} \qquad \qquad ...(2.1.2)$$

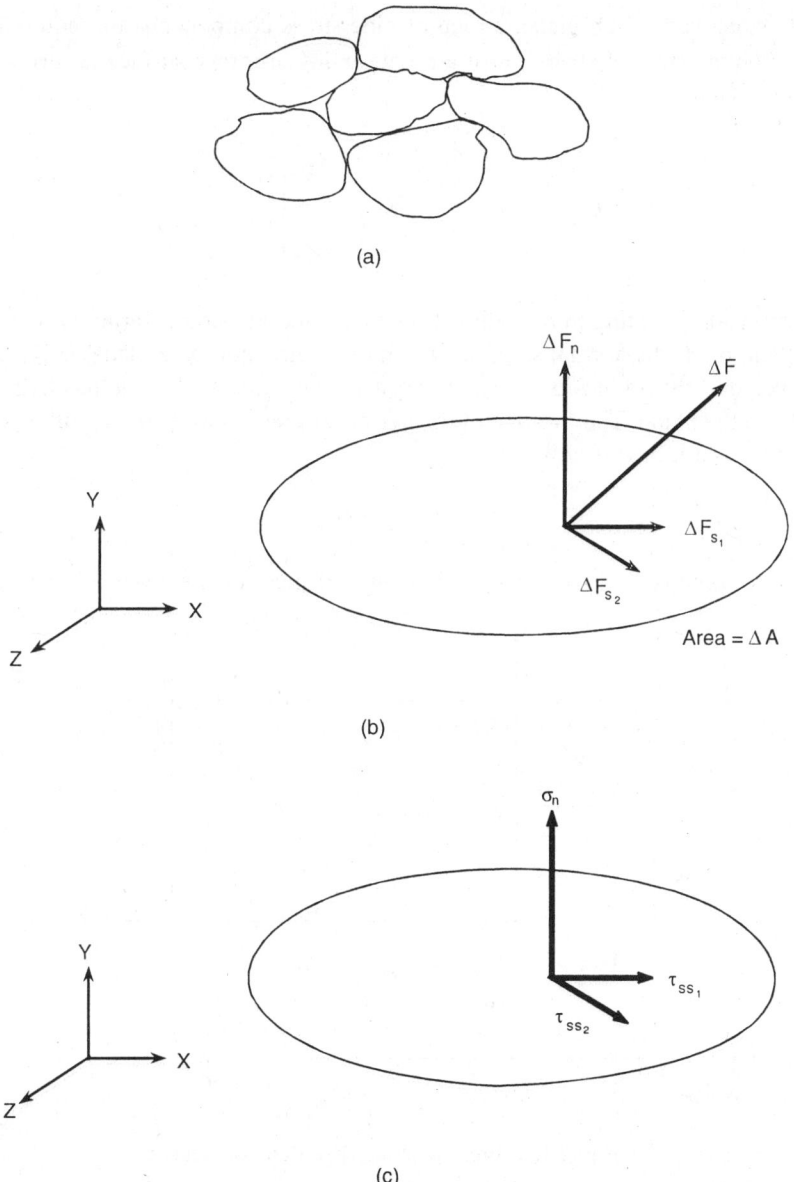

Fig. 2.1.1. Forces and stresses acting on the contact surface of a fluid element(3)

$$\tau_{ss_2} = \lim_{\Delta A \to 0} \frac{\Delta F_{s_2}}{\Delta A} \qquad \qquad ...(2.1.3)$$

Again, there can be many planes (i.e., areas) that can arise at a point in a given fluid. Therefore, it is necessary to define the stress in terms of an orthogonal system of forces referenced to global coordinates and an accompanying series of orthogonal planes passing through the origin so that the state of stress can be unambiguously defined with the minimum number of components. For three orthogonal planes and three stresses on each plane, a total of nine stress components are required, in general, to completely describe the state of stress, $\bar{\tau}$ at a point on any arbitrary surface in terms of fixed global coordinate system. Thus,

$$\bar{\tau} = \begin{pmatrix} \sigma_{xx} & \tau_{yx} & \tau_{zx} \\ \tau_{xy} & \sigma_{yy} & \tau_{zy} \\ \tau_{xz} & \tau_{yz} & \sigma_{zz} \end{pmatrix} \qquad \qquad ...(2.1.4)$$

Here, τ_{yx} is the shear stress acting in the x-direction on a plane perpendicular to the y-axis (i.e., parallel to the xz plane) and σ_{yy} is the normal stress acting in the y-direction on a plane perpendicular to the y-axis. All stress vectors, shown in Fig. 2.1.2, are positive. This means that the positive normal stress is directed away from the plane. The average of the normal stresses is termed the bulk stress σ, which, in turn, defines the pressure, p as follows (3):

$$p = -\sigma = -\frac{1}{3}\left(\sigma_{xx} + \sigma_{yy} + \sigma_{zz}\right) \qquad \qquad ...(2.1.5)$$

Therefore, pressure force is positive towards the centre of mass of the surface it acts upon.

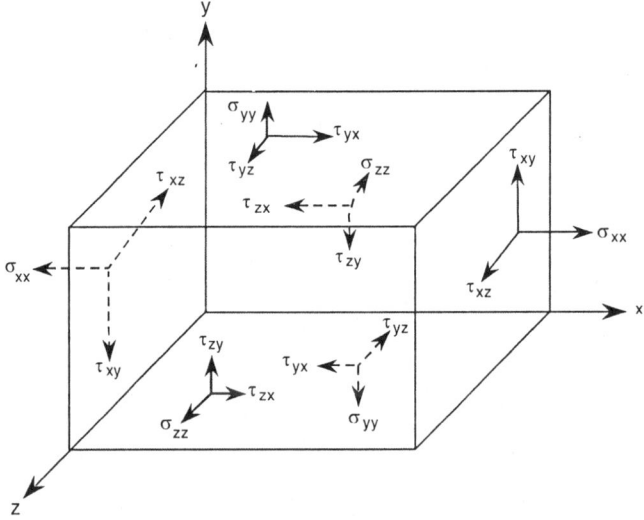

Fig. 2.1.2. Stresses acting on a fluid element

Forces that develop without physical contact but are distributed over the volume of the fluid under consideration are known as *body forces* (1). Gravitational and electromagnetic forces are examples of body forces. The weight of a fluid is on account of the gravitational force and is the only body force that has been considered in this text. The gravitational body force acting on a fluid element of volume $d\forall$ is given by $\rho g\, d\forall$.

2.2 PRESSURE AT A POINT

A fluid at rest cannot support the shear stress and, hence, a fluid element at rest will have only normal stress on any of its planes. This normal stress is, therefore, a point property called the *fluid pressure p* that is taken positive for compression.

Consider a free-body diagram, Fig. 2.2.1 obtained by removing a small triangular wedge of fluid from some arbitrary location within a fluid element that is either at rest or moves as a rigid body (with no acceleration) so that there is no shear on any of its planes. The chosen fluid wedge, Fig. 2.2.1, is therefore, in equilibrium under the action of pressure and gravity forces that have been marked in Fig. 2.2.1. The pressures p_x (acting on the surface of the wedge parallel to the y-z plane), p_y, p_z, and p_s are the average pressures on the corresponding faces. The summation of forces, in both x and y directions, must be zero as there is no acceleration. Hence,

$$p_x \, \delta_y \, \delta_z - p_s \, \delta_y \, \delta_s \sin \theta = 0 \qquad \qquad ...(2.2.1)$$

and

$$p_z \, \delta_x \, \delta_y - p_s \, \delta_y \, \delta_s \cos \theta - \gamma \frac{\delta_x \delta_y \delta_z}{2} = 0 \qquad \qquad ...(2.2.2)$$

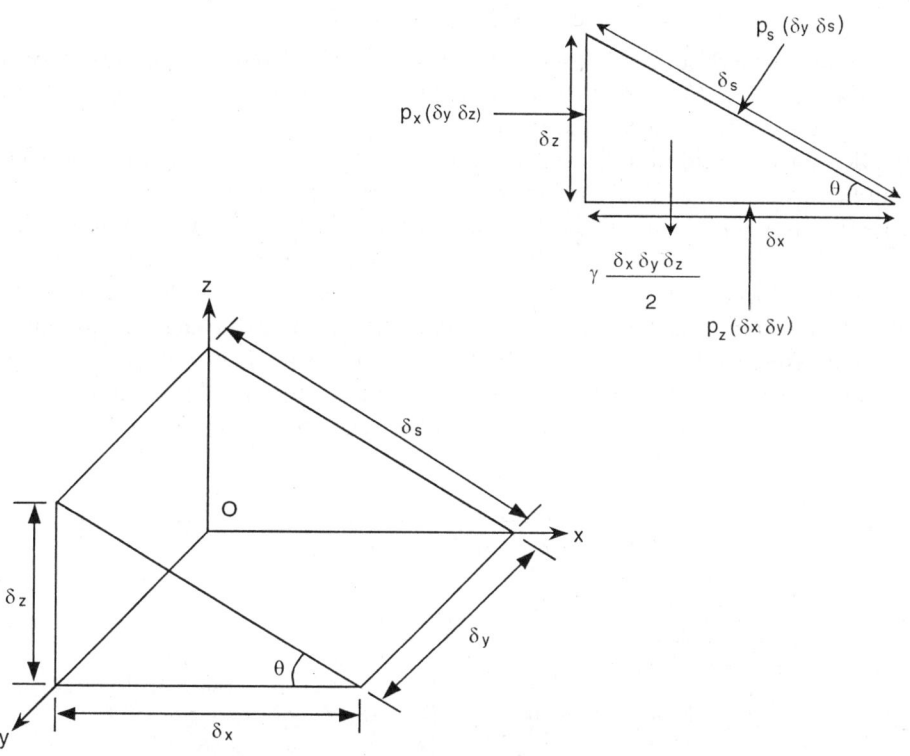

Fig. 2.2.1. Free-body diagram of a wedge-shaped fluid element

From the geometry of the chosen wedge, Fig. 2.2.1,

$$\delta_x = \delta_s \cos \theta \text{ and } \delta_z = \delta_s \sin \theta$$

Therefore, Eqs. (2.2.1 and 2.2.2) yield,

$$p_x = p_s \qquad \qquad ...(2.2.3)$$

and

$$p_z = p_s + \tfrac{1}{2} \gamma \delta_z \qquad \qquad ...(2.2.4)$$

In the limit, the wedge reduces to a point, i.e., δ_x, δ_y, and δ_z tend to zero. Therefore,

$$p_x = p_z = p_s$$

Similarly, by taking a wedge where one of the faces is parallel to the x-z plane, one can obtain

$$p_y = p_z = p_s$$

Therefore,

$$p_x = p_y = p_z \qquad \qquad ...(2.2.5)$$

Since the wedge and angle θ were arbitrarily chosen, one can conclude that the pressure at a point in a fluid is independent of direction as long as there are no shearing stresses, i.e., the fluid is either at rest or in rigid-body motion with no acceleration. This important conclusion is known as Pascal's law named in honour of Blaise Pascal (1623–1662), a French mathematician who made significant contributions in hydrostatics (2).

For moving fluids in which the shear stresses exist, the normal stresses at a point, which corresponds to pressure in fluids at rest, are not necessarily the same in all directions. In such situations, the pressure p is taken as the *average* of any three mutually perpendicular normal stresses on the element.

$$p = -\,1/3\,(\sigma_{xx} + \sigma_{yy} + \sigma_{zz}) \qquad \qquad ...(2.2.6)$$

The minus sign occurs because a compressive stress is considered to be negative whereas pressure p is positive. Equation (2.2.6) is, however, subtle and rarely required because a large majority of viscous flows have negligible viscous normal stresses.

For the fictitious inviscid fluid, no shear stress can exist for any motion of the fluid and, hence, at any point in the inviscid fluid, the pressure is the same in all directions.

Pressure at any point in any given fluid mass is specified as either the *absolute* (i.e., the total magnitude) pressure or the *gauge* (i.e., relative to the *local* atmospheric pressure) pressure. The need of gauge pressure arises as many pressure-measuring instruments are of differential type and measure, not the absolute magnitude, but, the difference between the fluid pressure and the atmospheric pressure. The measured pressure p (absolute), greater than the local atmospheric pressure p_a (absolute), is termed gauge pressure, p (gauge) that equals $p - p_a$. If p is less than p_a then $p_a - p$ is termed either negative or suction or vacuum gauge pressure. Figure 2.2.2 illustrates the concept of the gauge and absolute pressures.

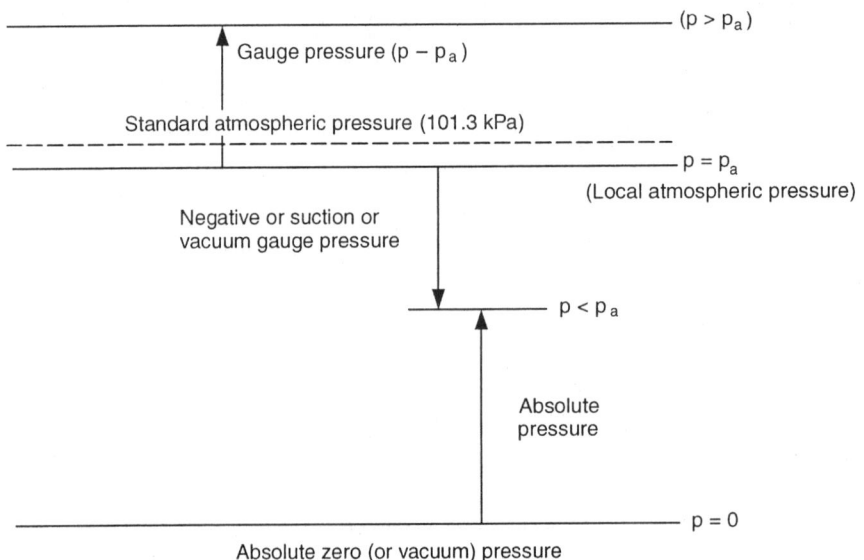

Fig. 2.2.2. Absolute and gauge pressures

In Fig. 2.2.2, standard atmospheric pressure may be considered as an idealized representation of year-round mean conditions of the earth's atmosphere.

The Bourdon gauge is a typical of the instruments used for measuring gauge pressures. The pressure element is a curved, hollow, flat metallic tube of elliptical cross-section closed at one end, Fig. 2.2.3. Other end of the pressure element is connected to the pressure to be measured. When the pressure inside the tube increases, the tube tends to straighten and thus moves a pointer, connected to the tube through a link, on a dial. The dial can be calibrated in suitable units such that the pointer reads zero when pressure inside the tube is the same as that outside. Other pressure (or pressure difference) measuring devices have been described in Sec. 2.4.

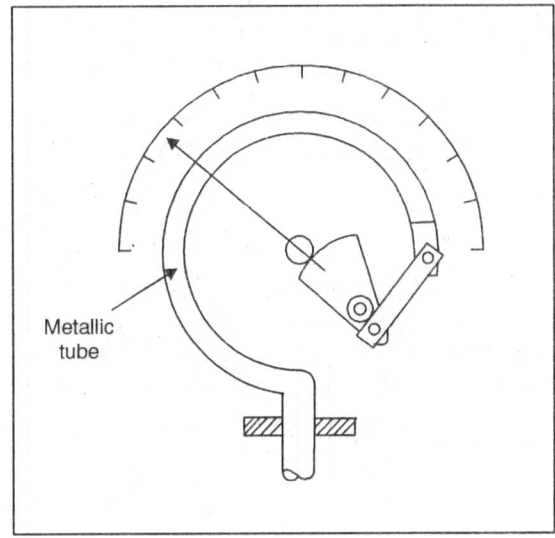

Fig. 2.2.3. Bourdon gauge

2.3 BASIC EQUATION FOR PRESSURE FIELD

Consider a differential element of fluid within the mass of a fluid. The fluid element is stationary with respect to the stationary rectangular coordinate system shown in Fig. 2.3.1. The forces acting on this element are only surface forces and body forces. Let pressure at the centre of the element be p. Therefore, average pressure forces on the two faces (parallel to the y-z plane) of the element are as shown in Fig. 2.3.1 and, therefore, the resultant surface force δF_x acting on the element in the x-direction is given as

$$\delta F_x = \left(p - \frac{\partial p}{\partial x} \cdot \frac{\delta x}{2} \right) \delta y\, \delta z - \left(p + \frac{\partial p}{\partial x} \cdot \frac{\delta x}{2} \right) \delta y\, \delta z$$

$$\therefore \qquad \delta F_x = -\frac{\partial p}{\partial x} \delta x\, \delta y\, \delta z \qquad\qquad ...(2.3.1)$$

Similarly, one can write expressions for the resultant surface forces δF_y and δF_z acting on the element in the y- and z-directions, respectively, as follows:

$$\therefore \qquad \delta F_y = -\frac{\partial p}{\partial y} \delta x\, \delta y\, \delta z \qquad\qquad ...(2.3.2)$$

$$\text{and} \qquad \delta F_z = -\frac{\partial p}{\partial z} \delta x\, \delta y\, \delta z \qquad\qquad ...(2.3.3)$$

The resultant surface force δF_s, acting on the element, may now be written in vector form as

$$\delta F_s = \hat{i}\, \delta F_x + \hat{j}\, \delta F_y + \hat{k}\, \delta F_z$$

$$= -\left(\hat{i} \frac{\partial p}{\partial x} + \hat{j} \frac{\partial p}{\partial y} + \hat{k} \frac{\partial p}{\partial z} \right) \delta x\, \delta y\, \delta z$$

where, $\hat{i}$, $\hat{j}$, and $\hat{k}$ are unit vectors.

$$\therefore \qquad \delta F_s = - (\nabla p)\, \delta x\, \delta y\, \delta z \qquad\qquad ...(2.3.4)$$

where, the symbol ∇ is the vector operator and termed 'grad' or 'gradient' or 'del' and ∇p is termed the pressure gradient. Therefore, the resultant surface force per unit volume is

$$\frac{\delta Fs}{\delta x \delta y \delta z} = -\nabla p \qquad\qquad ...(2.3.5)$$

Thus, the pressure gradient (i.e., the rate at which pressure changes with distance) is equal to the negative of the surface force per unit volume. The gradient of a scalar field (i.e., pressure p) gives a vector field (i.e., δF_s). One should note that it is the pressure gradient, rather than the magnitude of pressure, that matters in evaluating the pressure force and is very useful in the study of fluid mechanics. Only other force that is acting on the differential fluid element of Fig. 2.3.1 is the body force δF_B, on account of gravity, that equals $\rho g \delta x \delta y \delta z$. Since z-axis is vertical (with +ve upwards), one can write

$$\delta F_B = - \rho g \delta x \delta y \delta z\, \hat{k} \qquad\qquad ...(2.3.6)$$

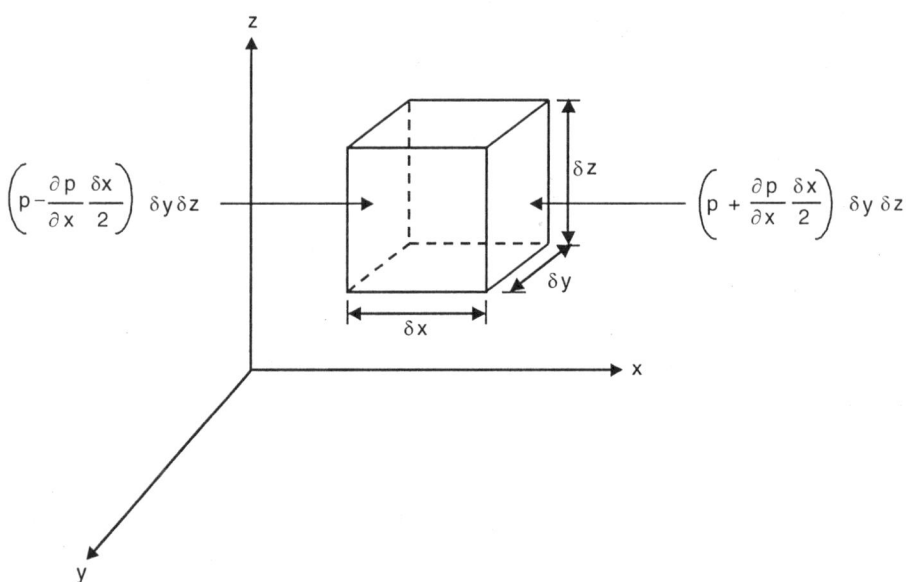

Fig. 2.3.1. Pressure forces (acting in the x-direction) on a fluid element

Thus, the total force δF, acting on the fluid element, is

$$\delta F = \delta F_s + \delta F_B = - (\nabla p)\, \delta x \delta y \delta z - \rho g\, \delta x \delta y \delta z\, \hat{k}$$

If the fluid particle is in motion (without angular deformation, i.e., rigid-body motion) with acceleration $\bar{a} = \left(\hat{i} a_x + \hat{j} a_y + \hat{k} a_z \right)$, then Newton's second law of motion gives

$$\delta F = - (\nabla p)\, \delta x \delta y \delta z - \rho g\, \delta x \delta y \delta z\, \hat{k} = \rho\, \delta x \delta y \delta z\, \bar{a}$$

$$\therefore \qquad - \nabla p - \rho g\, \hat{k} = \rho\, \bar{a} \qquad\qquad ...(2.3.7)$$

This equation is the general equation of motion for a fluid in rigid-body motion when there is no shear or angular deformation within the fluid. For a static fluid, however, $\bar{a} = 0$. Thus,

$$\nabla p + \hat{k}\gamma = 0$$

i.e.,

$$-\nabla p = \hat{k}\gamma$$

or

$$-\left(\hat{i}\frac{\partial p}{\partial x} + \hat{j}\frac{\partial p}{\partial y} + \hat{k}\frac{\partial p}{\partial z} \right) = \hat{k}\gamma$$

$\therefore$

$$\frac{\partial p}{\partial x} = 0 \qquad \qquad ...(2.3.8)$$

$$\frac{\partial p}{\partial y} = 0 \qquad \qquad ...(2.3.9)$$

and

$$\frac{\partial p}{\partial z} = -\gamma \qquad \qquad ...(2.3.10)$$

This means that p (in a static fluid) does not vary either in the x-direction or in the y-direction, i.e., p does not vary in a horizontal plane. Since p does not vary in a horizontal plane in a static fluid, the free surface of a liquid is always horizontal (more precisely, spherical i.e., perpendicular to gravity) as the free surface of the liquid is subjected to uniform atmospheric pressure everywhere. Equations (2.3.8 – 2.3.10) indicate that in a continuously distributed homogeneous, static, and incompressible fluid, the pressure (i) is the same at all points on any horizontal plane, (ii) increases with the depth of the fluid, and (iii) remains unaffected by the size and shape of the container holding the fluid(4).

One can replace $\dfrac{\partial p}{\partial z}$ with dp/dz in Eq. (2.3.10) as $p \neq f(x,y)$.

$\therefore$

$$\frac{dp}{dz} = -\gamma = -\rho g \qquad \qquad ...(2.3.11)$$

This is the basic equation for any static fluid and is true regardless of whether γ is constant (as in liquids) or is not constant (as in gases). Equation (2.3.11) indicates that the pressure gradient in the positive vertical direction is negative. This means that the pressure decreases in the upward direction and increases in the downward direction. Equation (2.3.11) assumes that (i) the fluid is static, (ii) the gravity is the only body force, and (iii) the z-axis is vertical and upward. To determine the pressure distribution in a static fluid, one needs to integrate Eq. (2.3.11) and apply the boundary conditions suitably.

2.3.1 Pressure Variation in a Static Incompressible Fluid

Equation (2.3.11) can be integrated to obtain the variation of pressure p in a fluid mass with z, provided one knows the variation of the mass density ρ of the fluid and the gravitational acceleration g. The variation in g is negligible for most of the engineering applications. Therefore, one needs to consider variation of only the mass density. For liquids, again, the mass density does not vary significantly even over large vertical distances. Therefore, one can assume specific weight γ to be constant while dealing with liquids. Hence, for incompressible fluids, ρ = constant and integration of Eq. (2.3.11) with appropriate boundary condition ($p = p_0$ at $z = z_0$) yields

$$\int_{p_0}^{p} dp = -\int_{z_0}^{z} \rho g\, dz$$

$\therefore$
$$p - p_0 = -\rho g\,(z - z_0) = \rho g\,(z_0 - z) = \gamma\,(z_0 - z) \qquad \qquad ...(2.3.12)$$

For liquids, reference level (i.e., $z = z_0$) is often taken at the free surface (where $p_0 = p_a$ = the atmospheric pressure) and distances in the vertical direction are measured in terms of the depth h below the free surface. This means,

$$p - p_a = p\ (gauge) = \rho g h = \gamma h \qquad \qquad ...(2.3.13)$$

For points 1 and 2 in Fig. 2.3.2, Eq. (2.3.12) attains the form

$$p_1 - p_2 = -\rho g\,(z_1 - z_2) = \rho g\,(z_2 - z_1) = \rho g h = \gamma h$$

or
$$p_1 = p_2 + \gamma\, h \qquad \qquad ...(2.3.14)$$

where, h is the depth measured below the location of point 2 to point 1. This type of pressure distribution is termed the *hydrostatic* distribution in which pressure increases linearly with the increase in depth, as per Eq. (2.3.14), irrespective of the shape of the container. The hydrostatic pressure distribution holds for moving fluids too provided that there is no acceleration in the vertical direction. If the flow of a fluid is curved in vertical plane, the pressure distribution is non-hydrostatic.

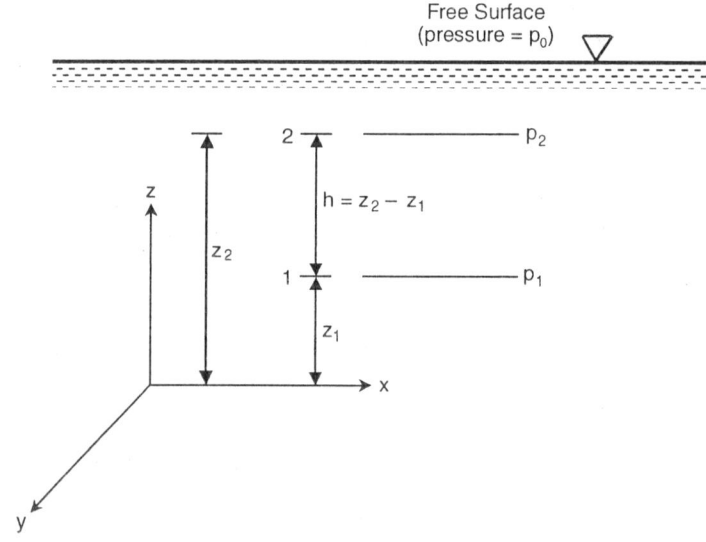

Fig. 2.3.2. Pressure variation in an incompressible liquid with a free surface

Equation (2.3.13) indicates that the gauge pressure p is directly proportional to the distance (or depth) h. Therefore, pressure can also be visualized in terms of the vertical distance $h = p/\rho g$ that is termed as the *pressure head* corresponding to the pressure p. Thus, pressures are sometimes expressed in terms of millimetres of mercury or metres of water. The concept of pressure head has been found very useful and is employed even when free surface does not exist above the point under consideration. For example, in a pipe flow, Fig. 2.3.3, which is a pressure flow and without free surface, $p/\rho g$ corresponds to the height above the centre line of the pipe to which liquid level would rise in a small diameter vertical tube of sufficient length that is open to the atmosphere – known as *piezometer tube* – and connected to the pipe.

On integration, Equation (2.3.11) also yields

$$p = -\rho g z + \text{constant}$$

$$p + \rho g z = \text{constant} \qquad ...(2.3.15)$$

or

$$\frac{p}{\rho g} + z = \text{constant} \qquad ...(2.3.16)$$

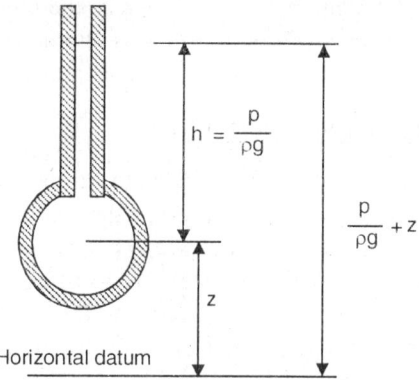

Fig. 2.3.3. Pressure and Piezometric heads

This means that for an incompressible fluid, the sum of the pressure head and elevation above the chosen horizontal datum plane is constant. This constant is known as the *piezometeric head* and corresponds to the height of the free surface above the chosen datum plane. The quantity, $(p + \rho g z)$, is, accordingly, termed the *piezometeric pressure*.

2.3.2 Pressure Variation in a Static Compressible Fluid

Specific weights (γ) of common gaseous fluids are relatively very small compared to those of common liquids. Equation (2.3.11), therefore, suggests that the pressure gradient in the vertical direction is very small and, therefore, the pressure for a gas will remain constant even for large vertical distances. Hence, one can neglect the effect of elevation changes on the pressure in gases in tanks, pipes etc. for which the elevation changes involved are small.

For larger variations in elevation, however, one must consider variation in pressure. Combining the equation of state for an ideal (or perfect) gas, Eq. (1.4.2), with Eq. (2.3.11), one obtains

$$\frac{dp}{dz} = -\frac{gp}{RT} \qquad ...(2.3.17)$$

On separating the variables, and assuming g and R constant over the elevation change from z_1 to z_2, Eq. (2.3.17) reduces to

$$\int_{p_1}^{p_2} \frac{dp}{p} = -\frac{g}{R} \int_{z_1}^{z_2} \frac{dz}{T} \qquad ...(2.3.18)$$

For isothermal conditions over the elevation change, the temperature T has a constant value T_0 and Eq. (2.3.18) gives

$$\ln \frac{p_2}{p_1} = -\frac{g(z_2 - z_1)}{RT_0}$$

or

$$p_2 = p_1 \exp\left[-\frac{g(z_2 - z_1)}{RT_0}\right] \qquad ...(2.3.19)$$

Equation (2.3.19) provides the pressure-elevation relationship for a compressible fluid under iso-thermal condition. For non-isothermal conditions, one can, similarly, derive pressure-elevation relation for known temperature – elevation relation.

2.4 MANOMETERS

Equation (2.3.12) indicates that a change in elevation $z_0 - z$ within a liquid is equal to change in pressure head $(p_0 - p)/\gamma$. Therefore, a static column of one or more liquids (or gases) can be employed to measure difference of pressure at two locations. *Manometers* are devices that employ columns of a suitable liquid (or gas) for measuring difference in pressures. This difference could be either between the pressure at a certain point and the atmospheric pressure or between the pressures at two points which may be in the same or different fluid systems and neither of which is necessarily at atmospheric pressure. Figure 2.4.1 illustrates the use of Eq. (2.3.12) for a column of multiple fluids. Pressure at $z = z_1$ is p_1 and is known. Pressures at z_2, z_3, and z_4 are obtained using Eq. (2.3.12) as follows:

$$p_2 - p_1 = -\rho_o g \, (z_2 - z_1)$$

$$p_3 - p_2 = -\rho_w g \, (z_3 - z_2)$$

$$p_4 - p_3 = -\rho_m g \, (z_4 - z_3)$$

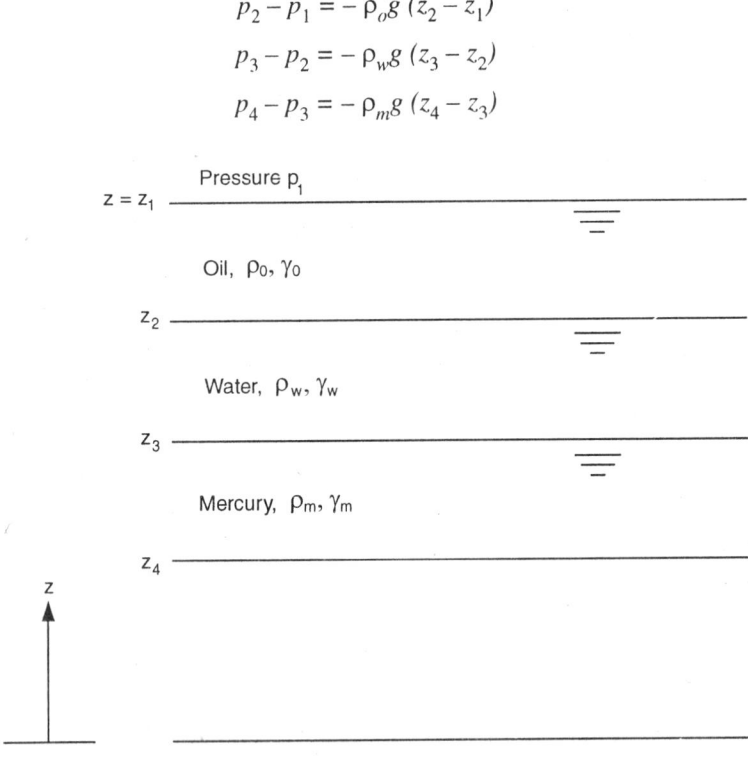

Fig. 2.4.1. Pressure changes through a column of multiple liquids

Thus, for obtaining $p_4 - p_1$ one may add the successive changes $p_2 - p_1$, $p_3 - p_2$ and $p_4 - p_3$ to get

$$p_4 - p_1 = -\rho_o g(z_2 - z_1) - \rho_w g \, (z_3 - z_2) - \rho_m g \, (z_4 - z_3) \qquad \text{...(2.4.1)}$$

Equation (2.3.12) combines two negative signs to have the pressure increase downward. This equation is easy to remember and use if written as

$$p_{down} = p^{up} + \gamma |\Delta z| \qquad \text{...(2.4.2)}$$

This means that the pressure increases if one moves downward and the pressure decreases when one moves upward. Using this memory device (4), Eq. (2.4.1) could be rewritten as

$$p_4 = p_1 + \gamma_0 |z_1 - z_2| + \gamma_w |z_2 - z_3| + \gamma_m |z_3 - z_4| \qquad ...(2.4.3)$$

Equation (2.4.2) would prove to be very helpful in manometeric computations.

2.4.1 Piezometers

The most elementary form of the manometers, usually called piezometer, is shown in Fig. 2.3.3. The gauge pressure at the centre of the pipe is, obviously, $\rho g h$. Such piezometers would not work for negative gauge pressure, as atmospheric air would be sucked into the fluid system. For large pressure, the piezometer tube would have to be very long. Therefore, it would be impractical to use piezometer for measuring large pressure. Also, piezometer can measure pressures only in liquids.

2.4.2 Open U-tube Manometer

Figure 2.4.2 shows a simple open U-tube manometer that can be used to overcome the difficulties noted for the piezometers. The fluid in the manometer is heavier than the fluid in the chamber A to keep z_2 small. The following simpler procedure may be followed in all manometeric computations:

Start from A where the pressure is, say, p_A. Apply Eq. (2.4.2) to obtain p_1 at B ("down") and jump across to C through the manometeric fluid to the same pressure p_1 and then use Eq. (2.4.2) to obtain pressure p_2 ($=p_a$) at D ("up"). Therefore,

$$p_A + \gamma_1 |z_A - z_1| - \gamma_m |z_2 - z_1| = p_2 = p_a \qquad ...(2.4.4)$$

One can jump across (4) from B to C as pressures at any two points at the same elevation in a continuous mass of the same static fluid will be the same.

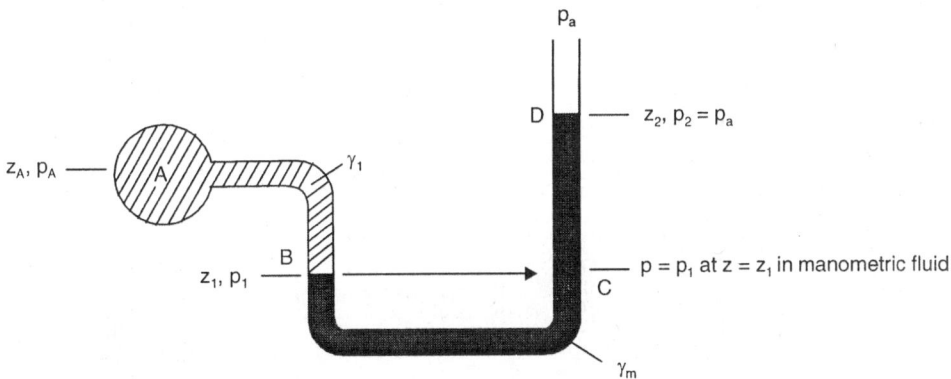

Fig. 2.4.2. Open U-tube manometer

2.4.3 Multi-fluid Manometer

Figure 2.4.3 illustrates a multi-fluid manometer for finding $p_A - p_B$ as follows:

$$p_A + \gamma_1 |z_A - z_1| - \gamma_2 |z_2 - z_1| + \gamma_3 |z_2 - z_3| - \gamma_4 |z_B - z_3| = p_B$$

or

$$p_A - p_B = -\gamma_1 |z_A - z_1| + \gamma_2 |z_2 - z_1| - \gamma_3 |z_2 - z_3| + \gamma_4 |z_B - z_3| \qquad ...(2.4.5)$$

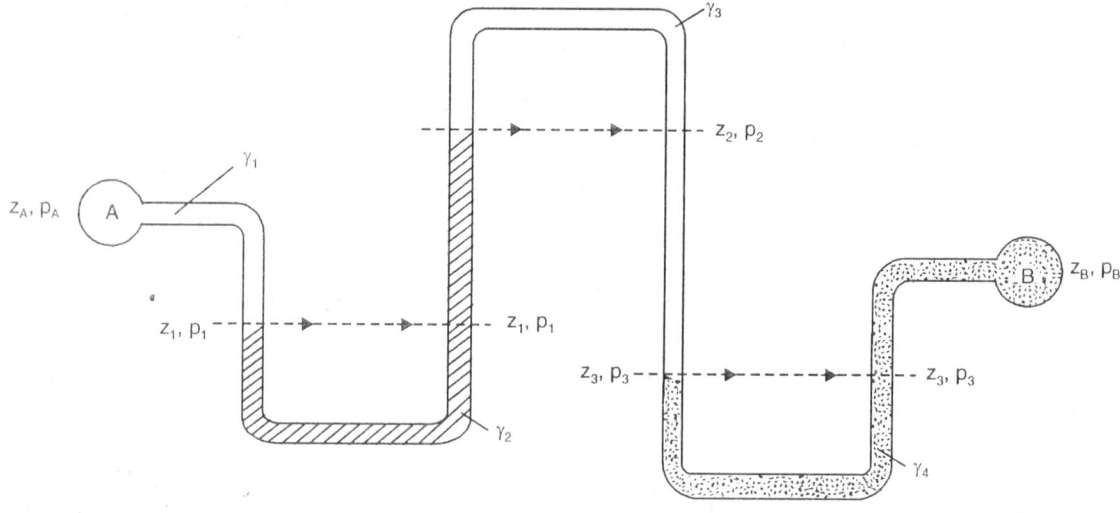

Fig. 2.4.3. Multi-fluid manometer

2.4.4 Inclined Manometer

Accuracy of pressure measurements, while measuring small pressure differences, can be increased by either choosing a lighter manometer fluid or by inclining the manometer limbs, or by attempting both, that would result in magnified reading which will be converted to the desired level difference in the vertical direction for use in the manometeric equation, Eq. (2.4.4 or 2.4.5).

2.4.5 Micromanometer

Another type of manometer that measures the difference in elevation of liquid levels in the two tubes of the manometer very accurately is known as micromanometer, Fig. 2.4.4. Since the difference in the liquid levels in the two reservoirs, $2\Delta h$ is very small, one uses a small telescope for measurement of Δh and, hence, pressure difference $p_c - p_d$ is obtained by using the following manometeric equation:

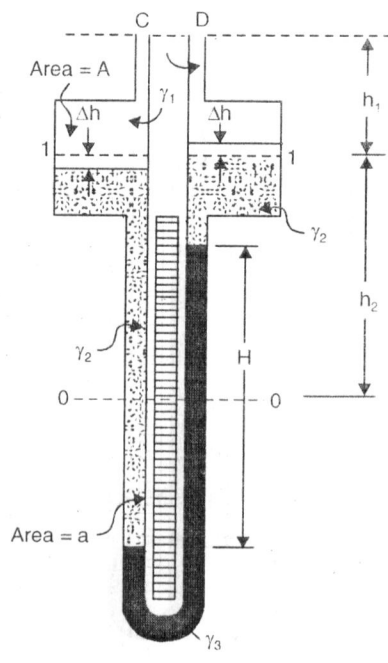

$$p_C + (h_1 + \Delta h)\gamma_1 + \left(h_2 + \frac{H}{2} - \Delta h\right)\gamma_2 - H\gamma_3$$

$$- \left(h_2 - \frac{H}{2} + \Delta h\right)\gamma_2 - (h_1 - \Delta h)\gamma_1 = p_D$$

$$\therefore \qquad p_c - p_D = -2\Delta h\gamma_1 - H\gamma_2 + 2\Delta h\gamma_2 + H\gamma_3$$

since

$$\Delta h A = \frac{H}{2}a$$

$$\therefore \qquad \Delta h = \frac{H}{2}\frac{a}{A}$$

Fig. 2.4.4. Micromanometer

Hence,
$$p_C - p_D = -H\frac{a}{A}\gamma_1 - H\gamma_2 + H\frac{a}{A}\gamma_2 + H\gamma_3$$

$$= H\left[\gamma_3 - \gamma_2\left(1 - \frac{a}{A}\right) - \gamma_1\frac{a}{A}\right] \ .$$

2.4.6 Comments on use of Manometers

The manometer in its various forms is an extremely useful pressure gauge requiring no calibration against any standard as the pressure difference is calculated using the first principles. While it can be adopted for measuring small pressure differences, it is inconvenient for extremely large pressure differences. Surface tension effects, resulting in capillary rise, can be minimized by selecting U-tubes of diameter larger than 15 mm. Liquids that do not form well-defined meniscus should not be used. A major disadvantage of the manometer is its slow response, which makes it unsuitable for measuring fluctuating pressures. There should be no air bubble in the manometer liquid and the liquid whose pressure is being measured.

2.5 FORCES ON SURFACES SUBMERGED IN STATIC FLUID

Any surface submerged in a static fluid experiences fluid force that is distributed on the submerged surface. A resultant force can conveniently replace the distributed forces. This resultant force, termed as *hydrostatic force* or *fluid thrust*, and its line of action can be determined by (i) integration, (ii) formula, and (iii) using the pressure prism. Computation of such force is required for the design of tanks, gates, and other structures.

2.5.1 Horizontal Plane Surfaces

A horizontal plane surface submerged in a static liquid, Fig. 2.5.1, is subjected to a constant pressure p all over the surface. On a small elemental area δA, the force due to the pressure p is $p\delta A$. The elemental forces $p\delta A$ acting on the entire surface of area A are all parallel and have the same direction normal to the surface and act towards the surface. Therefore, a scalar summation of all such elemental forces would give the desired magnitude of the resulting force F acting on one side of the surface as follows:

$$F = \int p\,dA = p\int dA = pA = \gamma hA \qquad \qquad ...(2.5.1)$$

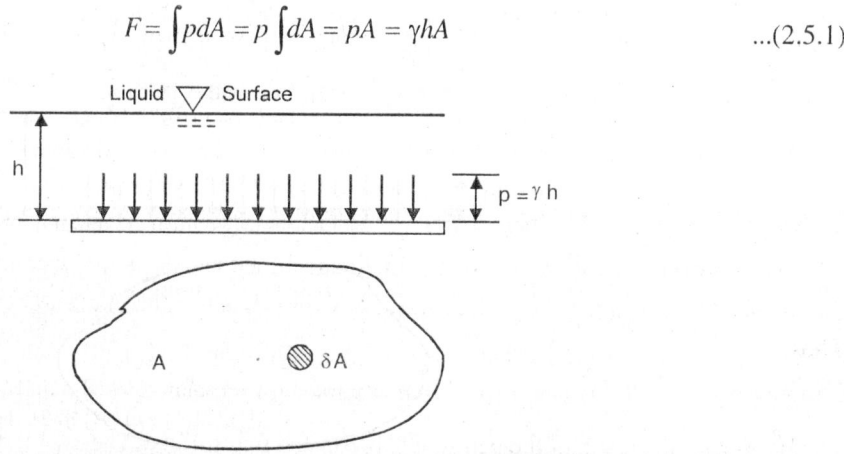

Fig. 2.5.1. Force on a horizontal plane surface in a static liquid

The line of action of the resultant force would, obviously, pass through the centroid of the surface.

2.5.2 Inclined Plane Surfaces

Figure 2.5.2 shows a plane surface of an arbitrary shape and is inclined at an angle θ with the horizontal. The surface is completely submerged in a liquid of weight density γ. The intersection of this plane with the plane of the free surface is taken as the x-axis while the y-axis is taken along the inclined plane. For an elemental area δA of the plane, which is at a depth h below the free surface, the elemental force is

$$\delta F = p\,\delta A = \gamma h\,\delta A = \gamma y \sin\theta\,\delta A$$

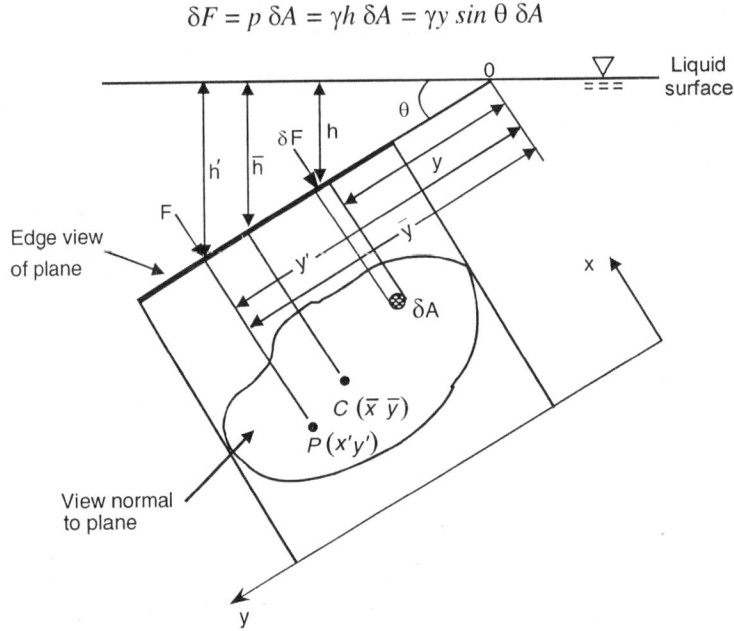

Fig. 2.5.2. Force on an inclined surface in a static liquid

Since the fluid is not moving, there are no shear stresses and, hence, δF is perpendicular to the element. Since the surface is plane, all the elemental forces would, therefore, be parallel. Therefore, the desired magnitude of the resultant force F acting on one side of the inclined surface is

$$F = \int_A \gamma y \sin\theta\,dA = \gamma \sin\theta \int_A y\,dA \qquad \qquad ...(2.5.2)$$

Since $\int_A y\,dA$ is the first moment of the area about the x-axis and may, therefore, be represented by $A\,\bar{y}$ where A represents the total area of the inclined surface and $\bar{y}$ is the distance of the centroid C from the x-axis. The coordinates of the centroid C with respect to the chosen coordinate axes are ($\bar{x}$, $\bar{y}$). Thus,

$$F = \gamma \sin\theta\,A\,\bar{y} = \gamma A\,\bar{h} \qquad \qquad ...(2.5.3)$$

Here, $\bar{h}$ is the depth of the centroid C below the free surface. Therefore, the magnitude of the force exerted on one side of a plane surface submerged in a liquid is the product of its area A and and the pressure $\gamma \bar{h}$ at its centroid and is independent of the shape of the surface and also the angle θ at which it is inclined with respect to the horizontal. Since all the elemental forces are normal to the surface, the resultant force F would also be normal to the surface. One can also conclude that the resulting force F would remain unchanged even if the surface were rotated about any axis through its centroid provided that the surface remains submerged in the static liquid. It should also be noted that the presence of free surface is not essential. One only needs to know the pressure at the centroid for which one can use any suitable means or method. Therefore, Eq. (2.5.3) is valid for any static fluid of uniform density that is in equilibrium, no matter how the pressure is produced.

It still remains to determine the point at which line of action of the resultant force F meets the plane surface. This point is known as the *centre of pressure* P, with coordinates (x', y'), although *centre of thrust* would have been more appropriate term. It should be noted that unlike the centre of pressure for the horizontal surface, the center of pressure for an inclined surface is not at the centroid but below it toward the high-pressure side. To obtain the location of the centre of pressure, the moments of the resultant force (i.e., F) about y-axis (= Fx') and x-axis (= Fy') are equated to the corresponding moments of the elemental forces pdA about y-axis and x-axis, respectively. Thus,

$$x'F = \int_A xpdA = \int_Z x\gamma y\sin\theta\,dA = \gamma\sin\theta\int_A xydA \qquad \qquad ...(2.5.4)$$

and

$$y'F = \int_A ypdA = \int_Z \gamma y^2\sin\theta\,dA = \gamma\sin\theta\int_A y^2dA \qquad \qquad ...(2.5.5)$$

Substituting the value of $F = \gamma A \bar{y}\ sin\ \theta$,

$$x' = \frac{\displaystyle\int_A xy\,dA}{A\bar{y}} = \frac{I_{xy}}{A\bar{y}} \qquad \qquad ...(2.5.6)$$

$$y' = \frac{\displaystyle\int_A y^2\,dA}{A\bar{y}} = \frac{I_x}{A\bar{y}} \qquad \qquad ...(2.5.7)$$

Here, I_x is the second moment of area about the x-axis from which y is measured. When the surface has an axis of symmetry in the y-direction, this axis may be chosen as the y-axis and then $\int_A xy\,dA$ becomes zero. This means that the centre of pressure lies on the axis of symmetry at a distance y' (measured down the plane surface) from the x-axis that equals the ratio of I_x (i.e., the second moment of area about the intersection of its plane with that of the free surface) and $A\bar{y}$. From the parallel axes theorem,

$$I_x = I_c + A\bar{y}^2$$

so that Eq. (2.5.7) is rewritten as

$$y' = \bar{y} + \frac{I_c}{A\bar{y}} \qquad \qquad ...(2.5.8)$$

Here, I_c is the second moment of area about its horizontal centroidal axis and is always positive. Therefore, $y' > \bar{y}$. This means that the centre of pressure for any submerged plane surface is always lower than the centroid of the surface except when the surface is horizontal. Also, when the surface is submerged very deep (i.e, $\bar{y}$ is very large), the contribution of the second term on the right hand side of Eq. (2.5.8) becomes small and, therefore, the centre of pressure is closer to the centroid of the surface. This is so because at larger depths the pressure becomes greater and its variation becomes relatively smaller so that the distribution of pressure is almost uniform. Therefore, the centroid of a submerged plane surface may be considered as the centre of pressure, if the variation of pressure is negligible. This is true for gases (as, in gases, the pressure changes very little with depth) and for liquids, only if the depth is very large. One should also note that whereas the total hydrostatic force (or fluid thrust) F on a submerged plane surface acts at the centre of pressure, its magnitude is given by the product of the area and pressure at the centroid. Since $y' = h'/\sin\theta$, the distance y' will increase when the depth of submergence h' increases, or, for a given h', the surface is rotated so that angle θ decreases.

It might have been noticed that the pressure at any point in the fluid has been computed as γh (and not $p_a + \gamma h$) with h as the vertical distance of the point below the free surface. Therefore, the reference pressure is the local atmospheric pressure p_a. When the opposite side of the surface is open to the atmosphere, the force exerted by the atmosphere on the surface is $p_a A$ based on absolute zero as datum. The force on the liquid side, with absolute zero as datum, would be $(p_a + \gamma h)A$. Therefore, the force due to the atmospheric pressure $p_a A$ acts on both sides and, hence, does not affect the resultant force or its location.

Another method for determining the resultant fluid thrust and line of action of the thrust on a plane surface is based on the concept of a pressure prism. The surface forms the base of the prism and height (or altitude) of the prism at any point (h below the free surface) of the surface is given by γh, Fig. 2.5.3. One can imagine a free surface to define h in the absence of real free surface. The force acting on an elemental area δA is

$$\delta F = \gamma h\, \delta A = \delta \forall \qquad \qquad ...(2.5.9)$$

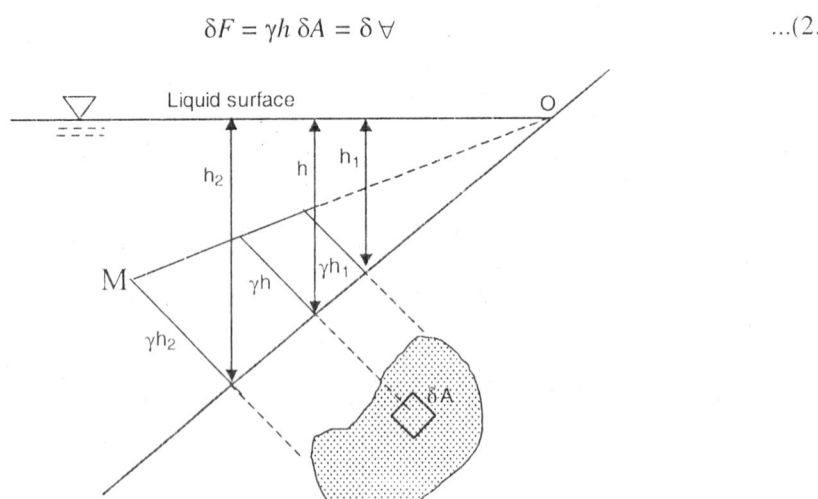

Fig. 2.5.3. Pressure prism

Here, $\delta \forall$ is an elemental volume (i.e., $\gamma h \delta A$) of the pressure prism. Hence, the desired fluid thrust F, obtained by integration of Eq. (2.5.9), equals $\forall$ that is the volume of the pressure prism. Using Eqs.

(2.5.4) and (2.5.5)

$$x'F = \int xp\,dA = \int x\,d\forall$$

$$y'F = \int yp\,dA = \int y\,d\forall$$

∴

$$x' = \frac{1}{\forall}\int_\forall x\,d\forall$$

and

$$y' = \frac{1}{\forall}\int_\forall y\,d\forall$$

This means that x' and y' are distances, along the axes, from O to the centroid of the pressure prism. Therefore, the line of action of the fluid thrust passes through the centroid of the pressure prism and is normal to the surface.

2.5.3 Curved Surfaces

On a curved surface submerged in a fluid, the pressure forces $p\delta A$ on individual area elements, being normal to the element, will differ in direction along the surface and, therefore, cannot be added numerically. They have to be added vectorially. A simple way to determine the fluid thrust on a curved surface is to calculate horizontal and vertical components of the total thrust and then combine these components vectorially to obtain the total thrust.

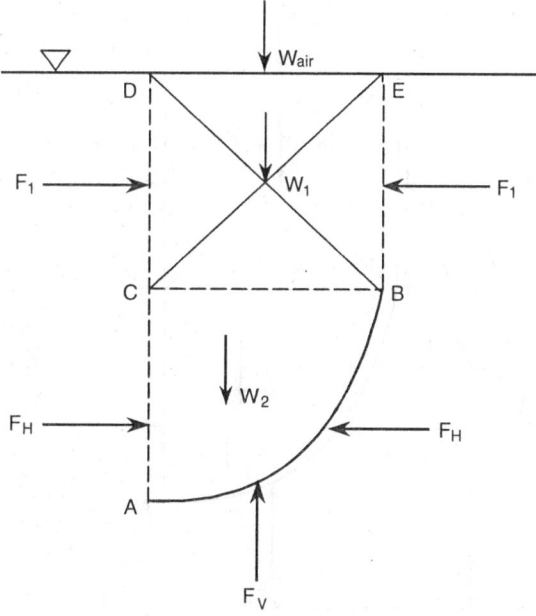

Fig. 2.5.4. Free-body diagram of the fluid over a curved surface (4)

Consider an arbitrary curved surface AB as shown in Fig. 2.5.4, which also shows a free-body diagram of the column of fluid (liquid) contained within the curved surface, its vertical projections from its edges, and the free surface. Consider the equilibrium of the body of the liquid ABEDC. The curved surface exerts a force on the liquid ABC in response to the fluid thrust on the surface. Let F_H and F_V be the horizontal and vertical components of the force exerted by the surface on the fluid. For the upper part of the fluid column BCDE, the horizontal components F_1 balance each other. Therefore, the desired force F_H due to the curved surface must be equal (in magnitude) to the force F_H on the vertical face AC and this can be computed using Eq. (2.5.3). The two forces would, obviously, be collinear and acting in the opposite direction. The line of action of the force F_H on the face AC would pass through the centre of pressure for the plane AC. The vertical force F_V, which is equal to the vertical component of the fluid thrust on the curved surface, is obtained from

$$F_V = W_1 + W_2 + W_{air} \approx W_1 + W_2 \qquad \text{...(2.5.6)}$$

The line of action of F_V is, obviously, through the centroid of the fluid volume ABEDC.

Thus, the horizontal component F_H of the total fluid thrust, exerted by the fluid on a curved surface immersed in a fluid, is equal to the fluid thrust on the projection of the surface on a vertical plane and acts through the centre of pressure of the projection (i.e., a plane surface). The vertical component F_V of the total fluid thrust on the curved surface will be due to the weight W of the liquid enclosed by the curved surface, free surface, and the vertical planes through the edges of the surface and the line of action of F_V would pass through the centroid of the volume ABEDC.

Sometimes, it is the underside, and not the upper side, of a curved surface that is subjected to the fluid pressure. In such a case, the vertical component of the fluid thrust on the surface acts in the upward direction. The magnitude of this vertical component equals the weight of an imaginary volume of the fluid extending from the curved surface to the free surface. This is so because if the imaginary fluid were actually present, fluid pressures on the two sides of the surface would be identical in magnitude and opposite in direction and, therefore, balance each other. This method is also applicable to such fluids which do not have a free surface. In such situations, the free surface is assumed or imagined to be at a height p/γ above a point at which the pressure p is known.

2.5.4 Tensile Stress in a Circular Pipe

When a fluid flows under pressure in a circular pipe, the pipe, under the action of an internal pressure, is in tension around its periphery, as shown in Fig. 2.5.5 which also shows the pipe's half cross-section as the free body. Assuming that there is no longitudinal stress in pipe walls, the tensions per unit length of pipe are marked as T_1 and T_2 as shown in the figure. The horizontal component of the pressure force is $p(2r)$ and acts through the centre of pressure of the projected area. Here, p is the pressure at the centre

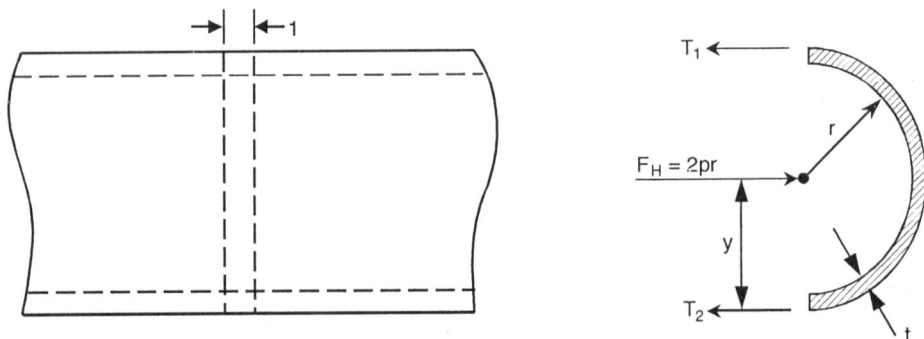

Fig. 2.5.5. Tensile stress in a pipe

line and r is the radius of the pipe. For high pressures, the centre of pressure can be taken as the centre of the pipe. Then,

$$T_1 + T_2 = 2pr \text{ and } T_1 = T_2$$

$$\therefore \qquad T_1 = T_2 = T = pr$$

in which T is the tensile force per unit length of the pipe whose wall thickness is t and, therefore, the tensile stress in the pipe wall is

$$\sigma = \frac{T}{t} = \frac{pr}{t} \qquad \qquad ...(2.5.7)$$

For larger variations in pressure between top and bottom of the pipe (when the centre of pipe cannot be taken as the centre of pressure), the location of the centre of pressure is computed and then

$$T_1 + T_2 = 2pr$$

And by taking moments about the lower end of the pipe,

$$2rT_1 - 2pry = 0$$

Thus, $\qquad \qquad T_1 = py \text{ and } T_2 = p\,(2R - y)$

2.5.5 Tensile Stress in a Shell

One may consider a thin spherical shell subjected to an internal pressure p. The free body of the sphere would be a hemisphere cut from the sphere by a vertical plane, Fig. 2.5.6. One may neglect the weight of the fluid within the sphere. The component of the fluid force normal to the plane acting on the inside of the sphere is $p\pi r^2$ where r is the radius of the sphere. If the thickness of the sphere wall is t, and the tensile stress in the wall is σ, then

$$\sigma\left(2\pi rt\right) = p\pi r^2$$

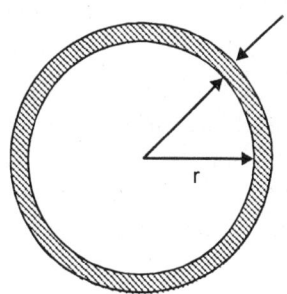

$$\therefore \qquad \sigma = \frac{pr}{2t}$$

Fig. 2.5.6. Free-body diagram of a spherical shell

2.6 BUOYANCY AND STABILITY

2.6.1 Buoyancy

The *buoyant force* is the resultant force exerted on a body by a static fluid in which the body is either submerged or floating. Consider any vertical plane through a submerged (or floating) body, Fig. 2.6.1(a). The projected area of each of the two sides of the body on the chosen vertical plane will be equal and at the same depth below the free surface. Therefore, there can be no horizontal component of the resultant force on the submerged or floating body due to the pressure of the surrounding fluid. The buoyant force, therefore, always acts vertically upward as the fluid thrust on the lower part of the body is larger than that on the upper part of the body.

The method of calculating fluid thrust on a curved surface (Art. 2.5.3) is applicable to all shapes of the surface and, hence, to the surface of a submerged body too. Thus, the surface ADC of the submerged body ABCD, Fig. 2.6.1(a), experiences an upward force F_{v2} equal to the weight of the liquid ADCFE. The downward force F_{v1} on the surface ABC, likewise, equals the weight of the liquid ABCFE.

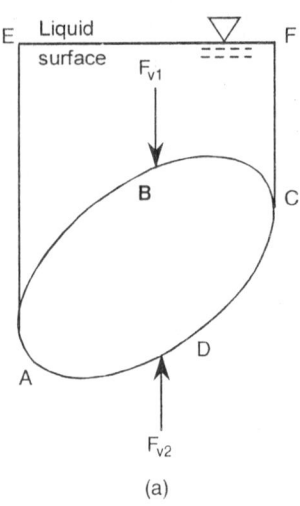

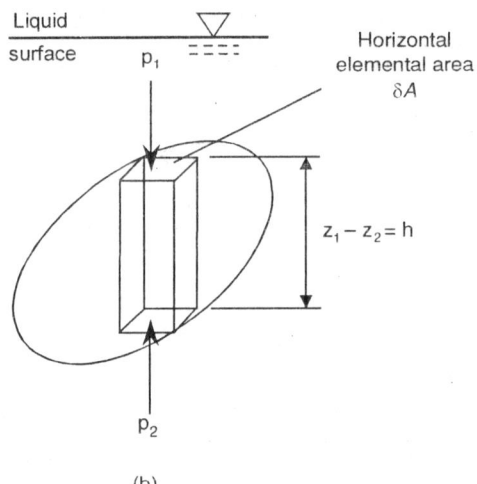

(a)

(b)

Fig. 2.6.1. Buoyant force on a submerged body (4)

The net force on the body, i.e., the buoyant force F_B is the difference between the two forces that would be equal to the weight of liquid ABCD that is *displaced* by the submerged body. This is Archimedes' first law of buoyancy. In equation form,

$$F_B = \gamma \forall \qquad \qquad ...(2.6.1)$$

in which $\forall$ is the volume of the fluid (or liquid) displaced and γ is the specific weight of the displaced fluid (or liquid). The same formula holds for floating bodies too in which case only a part of the body is submerged, with the remaining part poking up out of the free surface. Equation (2.6.1), for the floating bodies is rewritten as

$$F_B = \gamma \text{ (displaced volume)} = \text{weight of the floating body} \qquad ...(2.6.2)$$

Obviously, the buoyant force and weight of the floating body have to be collinear so that the net moment is zero and the body remains in static equilibrium. This means that *a floating body displaces its own weight in the fluid in which it floats*. This is Archimedes' second law of buoyancy and Eq. (2.6.2) is its mathematical form.

Alternatively, consider an elemental vertical prism, Fig. 2.6.1(b), of cross-section δA and volume $\delta \forall$ $(=h\delta A)$. The vertical upward force exerted on this element is

$$\delta F_B = (p_2 - p_1)\,\delta A$$
$$= \gamma h \delta A$$
$$= \gamma \delta \forall$$

Integrating over the entire body volume, one obtains

$$\int_\forall dF_B = \int_\forall \gamma d\forall$$

Assuming that γ is constant throughout the volume,

$$\therefore \qquad \qquad F_B = \gamma \forall$$

Thus, *a body immersed in a fluid experiences a vertical buoyant force equal to the weight of the fluid it displaces*. Archimedes discovered this law of buoyancy in the third century B.C.(4).

Since an iceberg (specific gravity of ice = 0.9) floats in sea water (specific gravity of sea water = 1.025), the volume of displaced sea water would be 900/1025, i.e., about 88% the volume of the iceberg. This means only about 12% of the iceberg (tip of the iceberg) is outside the sea surface.

The line of action of the buoyant force can be found by taking moments of elemental buoyant forces about any convenient axis, and equating these to the moment of the resultant F_B. Thus,

$$\gamma \int_\forall x d\forall = \gamma \forall \bar{x}$$

or

$$\bar{x} = \frac{1}{\forall} \int_\forall x d\forall \qquad\qquad ...(2.6.3)$$

in which $\bar{x}$ is the distance from the chosen axis to the line of action. This equation, therefore, yields the distance to the centroid of the volume. This means, the buoyant force acts through the centroid of the displaced volume of fluid which is termed the *centre of buoyancy*.

2.6.2 Stability

A body that floats in a static liquid has vertical stability. If one gives a small upward displacement to this floating body, the volume of the displaced liquid decreases. Therefore, there is an unbalanced downward force, due to the weight of the body remaining the same, which tends to bring back the body to its original position when weight of the body equals the weight of the displaced liquid. Likewise, a small downward displacement to the floating body results in increased buoyant force causing an unbalanced upward force that restores the body to its original position.

If a floating body is given a small angular displacement, a couple is set up since the weight of the body and the buoyant force are no longer collinear. If the couple so formed further increases the angular displacement, the body is in *unstable* equilibrium. On the other hand, if the couple restores the body to its original position, the body is in *stable* equilibrium. If no couple is set up due to the angular displacement, the body is in *neutral* equilibrium. This means that the floating body has rotational stability when a restoring couple is set up by any small angular displacement. The three types of the equilibrium are illustrated in Fig. 2.6.2. A light piece of wooden rod with a metal sphere at its lower end, Fig. 2.6.2(a), has stable equilibrium. When the metal sphere is at the top, Fig. 2.6.2(b), the body may float, but a small angular displacement would cause the body to attain the position shown in Fig. 2.6.2(a). In Fig. 2.6.2(c)

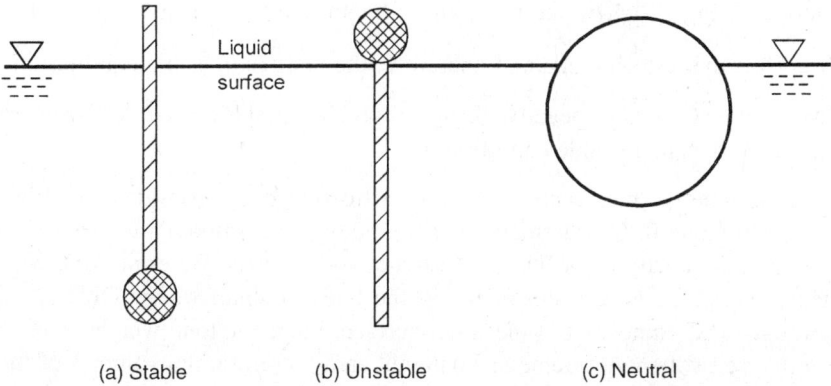

(a) Stable   (b) Unstable   (c) Neutral

Fig. 2.6.2. Equilibrium of floating bodies

is shown a homogeneous sphere or a cylinder that remains in equilibrium for any angular position and no couple results due to any angular displacement.

A completely submerged body, Fig. 2.6.3, has rotational stability when its centre of gravity G is below its centre of buoyancy B as only then a restoring couple would result following any angular displacement. Normally, when a body is too heavy to float, it submerges and goes down until it rests on the bottom.

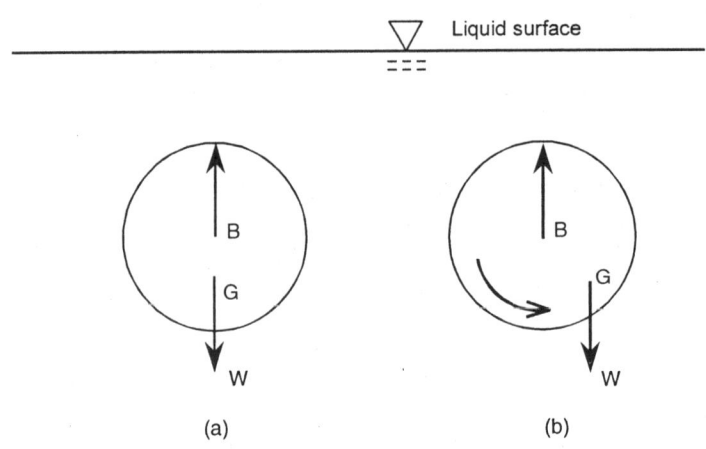

Fig. 2.6.3. Rotational stability of a completely submerged body

Naval architects have perfected the art of designing a floating body, like ship, so as to avoid floating instability. Only the basic method for determining the static stability of a floating body has been explained in the following (4):

1. Determine the body's centre of mass G and the centre of buoyancy B. The weight of the body is, say, W and the buoyant force is F_B, Fig. 2.6.4(a).

2. On tilting the body by a small angle θ, a new waterline (on the body) is set-up and new position B′ of the centre of buoyancy is computed. A vertical line drawn upward from B′ intersects the line of symmetry GB at a point M, called the *metacentre*, which is independent of θ for small angles of tilt. The distance MG is known as the *metacentric height*.

3. If M is above G, Fig. 2.6.4(b), the metacentric height $\overline{MG}$ is positive and the resulting couple ($W.\overline{MG} \sin θ$) is the restoring one and, therefore, the body is in stable equilibrium. On the other hand, if M is below G ($\overline{MG}$ is negative), Fig. 2.6.4(c), the resulting couple is overturning one and, therefore, the body is in unstable equilibrium.

In order to determine the metacentric height of a floating body, consider one, Fig. 2.6.5, that is tilted through a small angle θ. It is assumed that the body has a smooth variation of shape near the waterline. The line of symmetry of the body is assumed to be y-axis. Waterline CD, when the body is not tilted, is taken as x-axis. As a result of tilting the body, a small wedge OFD is submerged and another small wedge OAC comes out of the water surface. Since the total weight of the body remains unaltered, the immersed volume too remains unaltered and, therefore, the volumes of the two wedges are equal. However, the centre of buoyancy is now at its new position B′, the vertical through which intersects y-axis at the metacentre M. The centre of buoyancy was at B when the body had not been

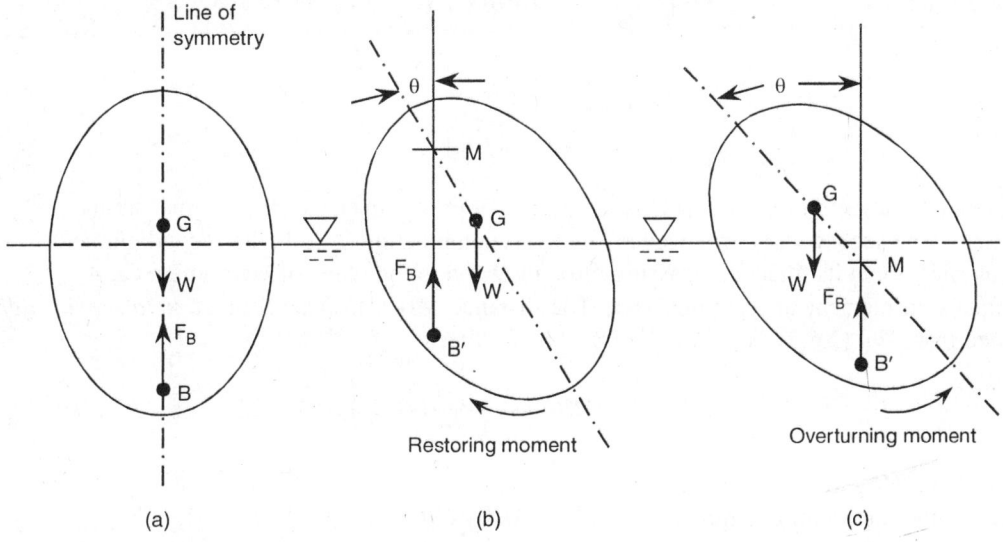

Fig. 2.6.4. Restoring and overturning moments for a tilted body (4)

tilted. The distance between the cente of gravity G and the metacentre M, i.e., MG is the *metacentric height*. So as to be able to estimate MG, one needs to determine the new position of the centre of buoyancy B′ (i.e., the centroid of the submerged volume AOFDEA of the body). Let the new centre of buoyancy be $\bar{x}$ away from the original position of the centre of buoyancy B. Therefore, taking the moment of the volumes (considering length of the body perpendicular to the plane of the diagram as L) about B, one gets

$$\bar{x}\, LA_{AOFDEA} = \int_{CODEAC} xL\, dA + \int_{OFDO} xL\, dA - \int_{AOCA} xL\, dA \qquad \qquad ...(2.6.4)$$

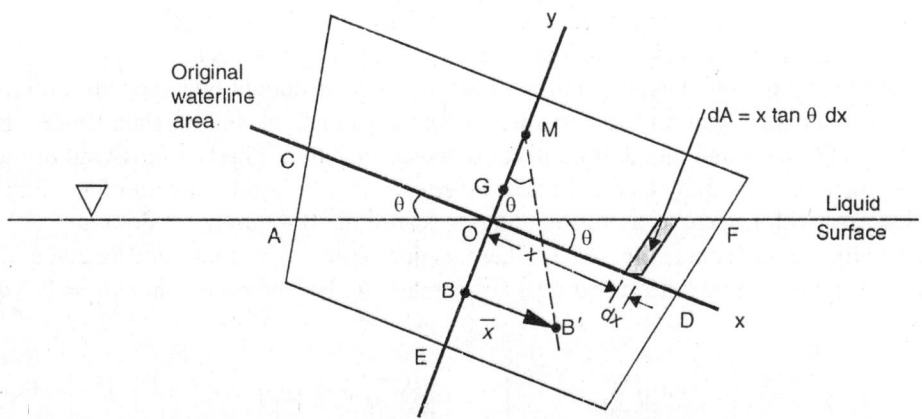

Fig. 2.6.5. A floating body tilted through a small angle θ

$$= 0 + \int_{OFDO} xL(x\tan\theta)\,dx - \int_{AOCA} xL(-x\tan\theta)\,dx$$

$$= \tan\theta \int_{waterline} x^2 \, dA_{waterline}$$

$$\therefore \qquad\qquad \overline{x}\, LA_{AOFDEA} = I_0 \tan\theta \qquad\qquad ...(2.6.5)$$

The first integral of Eq. (2.6.4) is zero because of the symmetry of the original submerged portion CODEAC. The second and third "wedge" integrals combine into I_0 (i.e., the second moment of area of the waterline about its tilted axis passing through O and in the plane of waterline area, $A_{waterline}$) since Ldx equals an element of waterline area. The distance MB also (also termed *metacentric radius*) is obtained from Eq. (2.6.5) as

$$MB = \frac{\overline{x}}{\tan\theta} = \frac{I_0}{\forall_{sub}} \qquad\qquad ...(2.6.6)$$

Since $MB = MG + GB$

Therefore, the metacentric height, $\qquad MG = MB - GB$

or $\qquad\qquad\qquad\qquad MG = \dfrac{I_0}{\forall_{sub}} - GB \qquad\qquad ...(2.6.7)$

Here, $\forall_{sub}$ is the immersed (or submerged) volume of the body.

2.7 FLUIDS IN RIGID-BODY MOTION

2.7.1 Uniform Linear Acceleration

It is interesting to note that in certain types of fluid motion, the methods of fluid statistics are applicable. The moving fluid must, however, satisfy the condition of no deformation. For example, if a fluid moves uniformly (i.e., there is no variation in its velocity and all the fluid particles are moving with the same velocity) in a straight line, there is no acceleration and there are no shear forces either. Therefore, no force acts on the fluid on account of its motion and the only surface force that acts on each element is due to pressure. Hence, the principles of fluid statics are applicable as such.

If the fluid in motion is undergoing uniform acceleration in a straight line, then also, no fluid layer moves relative to another layer and, hence, there are no deformations and no shear forces. However, there is additional force acting on the fluid to cause the acceleration. Fluids in this kind of motion are said to be in either the *rigid-body motion* or *relative equilibrium*. In rigid-body motion, a fluid particle retains its identity, with respect to its surroundings, because the fluid does not deform.

Consider an open container, Fig. 2.7.1. The container contains a liquid and is given a uniform horizontal acceleration a_x and uniform vertical acceleration a_z. For the acceleration $a_y = 0$, Eq. (2.3.7) reduces to

$$\left(i\frac{\partial p}{\partial x} + j\frac{\partial p}{\partial y} + k\frac{\partial p}{\partial z} \right) = -\rho g k - \rho\left(ia_x + ka_z \right) \qquad\qquad ...(2.7.1)$$

Therefore, $\qquad\qquad\qquad\qquad \dfrac{\partial p}{\partial x} = -\rho a_x \qquad\qquad ...(2.7.2)$

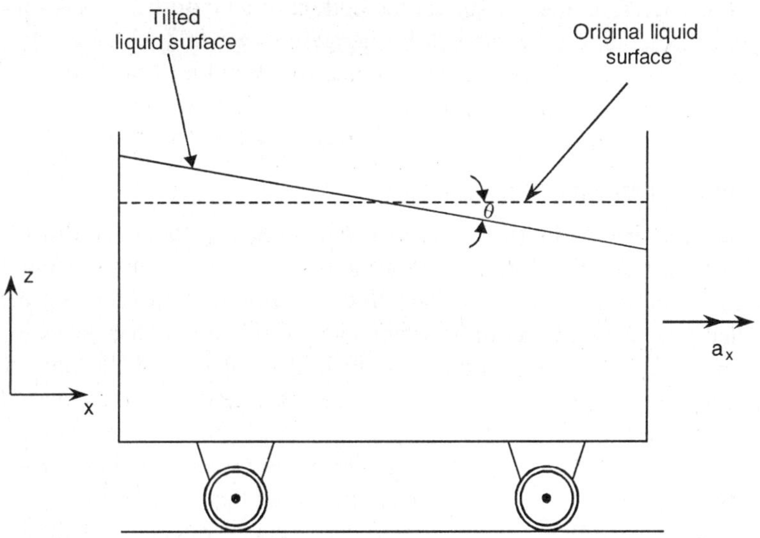

Fig. 2.7.1. Rigid-body motion of a liquid

$$\frac{\partial p}{\partial y} = 0 \qquad \qquad ...(2.7.3)$$

and

$$\frac{\partial p}{\partial z} = -\rho \left(g + a_z \right) \qquad \qquad ...(2.7.4)$$

Equation (2.7.3) shows that the pressure does not vary in y-direction. Therefore, the change in pressure between two points located at (x, z) and $(x + dx, z + dz)$ is

$$dp = \frac{\partial p}{\partial x}.dx + \frac{\partial p}{\partial z}.dz$$

Substituting values of $\frac{\partial p}{\partial x}$ and $\frac{\partial p}{\partial z}$ from Eqs. (2.7.2) and (2.7.4), one gets

$$dp = -\rho a_x \, dx - \rho \left(g + a_z \right) dz \qquad \qquad ...(2.7.5)$$

Along a line of constant pressure (such as free surface) $dp = 0$ and, therefore, Eq (2.7.5) yields the slope (*tan* θ) of this line as

$$\tan \theta = \frac{dz}{dx} = -\frac{a_x}{g + a_z} \qquad \qquad ...(2.7.6)$$

This means that the free surface (i.e., constant pressure line) for the accelerating liquid of Fig. 2.7.1 will be inclined if $a_x \neq 0$. Further, all lines of constant pressure will be parallel to the free surface. When $a_x = 0$ and $a_z \neq 0$, i.e., the liquid in the container of Fig. 2.7.1 is accelerating only in the vertical direction, the free surface will be horizontal as $dz/dx = 0$ from Eq. (2.7.6). But, the pressure distribution is not hydrostatic and is given by Eq. (2.7.4). This means that for fluids of constant density, pressure varies linearly with depth, but the variation is due to the combined effects of gravitational and externally induced accelerations. Therefore, the pressure at the bottom of a liquid-filled tank that is resting on the floor of an elevator (accelerating upward) will be increased over that which exists when the tank is either at rest or moving with a constant velocity. For a freely falling fluid mass (i.e., $a_z = -g$), the pressure gradients in all three directions are zero. This means that pressure is the same everywhere in the fluid.

2.7.2 Uniform Rotation about a Vertical Axis

A body of fluid, contained in a cylindrical container that is rotating about a vertical axis with uniform angular velocity, eventually attains relative equilibrium (i.e., rigid-body motion) and, therefore, rotates with the same angular speed ω as the container. Such rotation of fluid, moving as a solid, about a vertical axis, is called *forced-vortex* motion in which every fluid particle has the same angular velocity (3). On the other hand, in a *free-vortex* motion (as liquid only being rotated in a kitchen mixing jar) each particle of the fluid moves in a circular path with a speed that varies inversely with the distance from the centre.

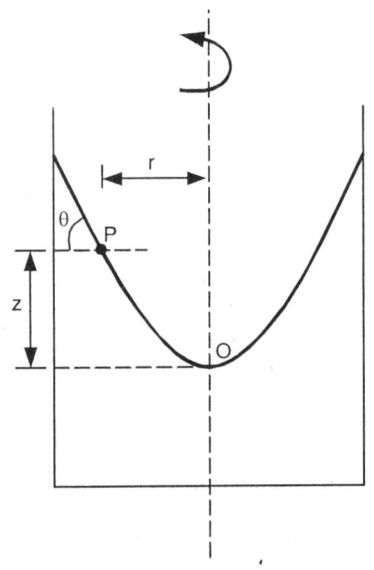

Fig. 2.7.2. Forced vortex

In forced-vortex motion, Fig. 2.7.2, no shear stresses exist in the liquid that is being rotated and the only acceleration that occurs is directed radially inward towards the axis of rotation. Therefore, the acceleration of any particle of liquid at radius r due to rotation will be $-\omega^2 r$ in radial direction with direction of r being positive outward from the axis.

Since p is a function of r and z only, the total differential dp is

$$dp = \frac{\partial p}{\partial r} dr + \frac{\partial p}{\partial z} dz$$

Using Eqs. (2.7.2) and (2.7.4), and replacing x with r

$$dp = \rho \omega^2 r dr - \gamma dz \qquad \text{...(2.7.7)}$$

For a liquid of constant density, integration of Eq. (2.7.7) gives

$$p = \rho \frac{\omega^2 r^2}{2} - \gamma z + C$$

in which C is the constant of integration and can be obtained using the boundary condition that the value of pressure, p at the origin O ($r = 0$, $z = 0$) is p_0. Thus,

$$p = p_0 + \rho \frac{\omega^2 r^2}{2} - \gamma z \qquad \text{...(2.7.8)}$$

If one selects a horizontal plane ($z = 0$) for which $p_0 = 0$ everywhere on the plane and Eq. (2.7.8) is divided by γ ($=\rho g$), then

$$h = \frac{p}{\gamma} = \frac{\omega^2 r^2}{2g} \qquad \text{...(2.7.9)}$$

Equation (2.7.9) gives the surface profile of the forced vortex and shows that the head, or vertical depth (above $z = 0$ plane), varies as the square of the radius. The surfaces of equal pressure (including the free surface) are, therefore, paraboloids of revolution.

At any point P on the free surface of the forced vortex, the inclination θ of the free surface, obtained from Eq. (2.7.6), is

$$\tan \theta = -\frac{a_r}{g + a_z} = \frac{\omega^2 r}{g} \qquad \qquad ...(2.7.10)$$

Therefore, inclination of the free surface varies linearly with r.

EXAMPLES

E 2.1.   A horizontal cylindrical fuel oil storage tank, Fig. E2.1, of internal diameter 3m is half full of liquid consisting of a layer of fuel oil (specific gravity = 0.87) above a layer of water whose thickness is 0.2m. The upper half of the tank is vented to the atmosphere. Calculate the gauge pressure at the bottom of the tank.

Solution:

$$P_B = P_a - g \int_0^{-1.3} \rho_{oil} \, dz - g \int_0^{-0.2} \rho_w \, dz$$

$\therefore$

$$P_B - P_a = \rho_{oil} \, g (1.3) + \rho_w \, g (0.2)$$

$$= \rho_w g (1.3 \times 0.87 + 0.2)$$

$$= 1000 \times 9.81 \, (1.131 + 0.2)$$

$$= 13057 \text{ N/m}^2$$

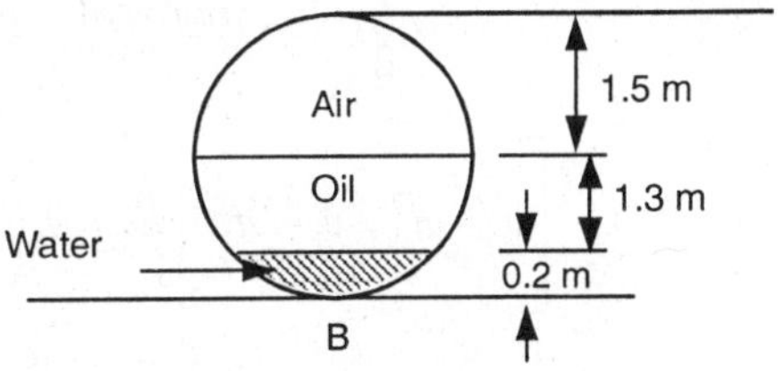

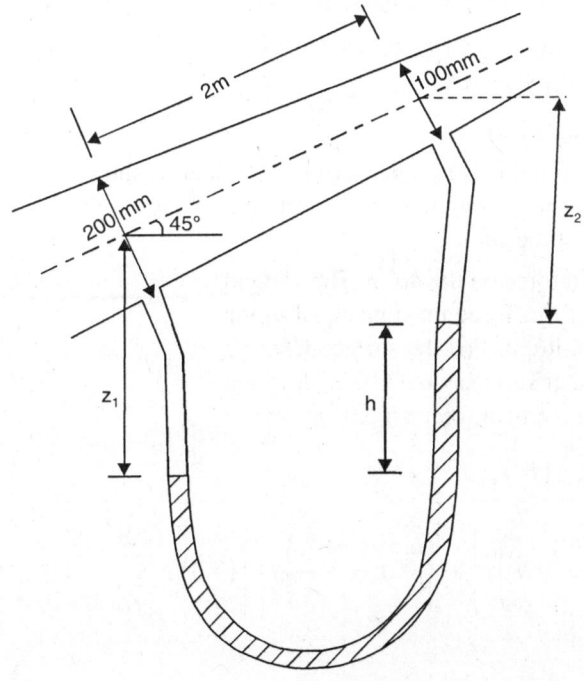

Fig. E2.1

E 2.2.   A pipe inclined at 45° to the horizontal converges over a length of 2 m from a diameter d_1 of 200 mm to a diameter d_2 of 100 mm at the upper end, Fig. E2.2. Oil of relative density 0.9

Fig. E2.2

flows through the pipe at a mean velocity v_1 (at the lower end) of 2 m/s. Find the pressure difference across the 2m length ignoring any loss of energy. Also find the difference in level that would be shown on a mercury (specific gravity = 13.6) manometer connected across this length.

Solution: Using Bernoulli's equation (Chapter 4)

$$\frac{p_1}{\rho_{oil}g}+\frac{v_1^2}{2g}+z_1=\frac{p_2}{\rho_{oil}g}+\frac{v_2^2}{2g}+z_2$$

$\therefore$

$$p_1-p_2=\frac{1}{2}\rho_{oil}\left(v_2^2-v_1^2\right)+\rho_{oil}g(z_2-z_1)$$

$$=\frac{1}{2}(0.9\times1000)(8^2-2^2)+900(2\sin45°)\times9.81$$

$$\left[v_2\frac{\pi}{4}d_2^2=v_1\frac{\pi}{4}d_1^2\quad\therefore v_2=v_1\left(\frac{d_1}{d_2}\right)^2=2\left(\frac{200}{100}\right)^2=8\,m/s\right]$$

$\therefore$

$$p_1-p_2=39486\text{ N/m}^2$$

For the manometer reading,

$$p_1+\rho_{oil}\,g\,z_1=p_2+\rho_{oil}\,g(z_2-h)+\rho_{Hg}\,gh$$

$\therefore$

$$p_1-p_2=\rho_{oil}\cdot g(z_2-z_1)+gh(\rho_{Hg}-\rho_{oil})$$

$\therefore$

$$39486=900\times9.81\,(2\sin45°)+9.81\,h\,(13.6-0.9)\,1000$$

$\therefore$

$$h=0.2167\text{ m}$$

$$=21.67\text{ cm}$$

E 2.3. A cylindrical drum acts as a barrier to hold water as shown in Fig. E 2.3. The radius of the drum is 2.5m and its length is 3m. Determine (a) the weight of the cylinder and (b) the force exerted against the wall.

Solution: Water exerts a force on the drum. The vertical component of this force must be equal to the weight of the drum. For the surface CD, the imaginary water surface would be at elevation A. Hence, the vertical force on ABCD is

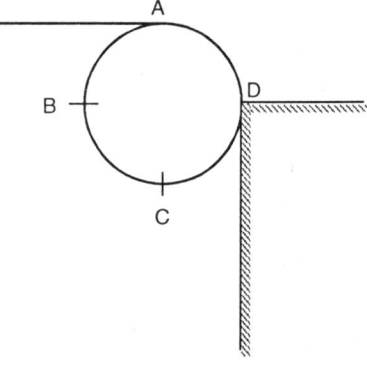

$$F_v=\left(F_v\right)_{BCD}-\left(F_v\right)_{AB}$$

$$=\left[\left\{\left(\frac{\pi r^2}{2}+2r.r\right)\gamma\right\}-\left\{\left(r.r-\frac{\pi r^2}{4}\right)\gamma\right\}\right](3.0)$$

$$=\left(\frac{3\pi r^2}{4}+r^2\right)(3.0\gamma)$$

Fig. E2.3

$$= \left(\frac{3\pi}{4} + 1\right)\left(3.0\gamma r^2\right)$$

$$= \left(\frac{3\pi}{4} + 1\right)\left[3.0 \times 9800 \times (2.5)^2\right]$$

∴   The weight of the drum = 616700.7 N

The force exerted by the drum against the wall is, obviously, the horizontal force on ABC minus the horizontal force on CD. Since horizontal forces on BC and CD are equal in magnitude and opposite to each other, the required force [Eq. (2.5.3)] is

$$F_H = \left(F_H\right)_{ABC} - \left(F_H\right)_{CD} = \left(F_H\right)_{AB} = \gamma r l \left(r/2\right) = 9800 \times 2.5 \times 3 \times \left(2.5/2\right) = 91875 \text{ N}$$

E 2.4.   The inner diameter of a steel pipe is 0.1m and wall thickness of the pipe is 5 mm. What is the allowable maximum fluid pressure in the pipe, if the permissible tensile stress is 70 MPa.

Solution: From Eq. (2.5.7),

$$p = \frac{\sigma t}{r}$$

∴

$$p = \frac{\left(70 \times 10^6\right)\left(5 \times 10^{-3}\right)}{0.05}$$

$$= 7 \times 10^6 \text{ Pa}$$

$$= 7 \text{ MPa}$$

E 2.5.   A rectangular pontoon of width 5 m and length 10 m has a draught (i.e,, the depth of its submergence, D)of 1.2 m in fresh water. Compute (i) the weight of the pontoon W, (ii) its draught in sea water ($\rho = 1025 \text{ kg/m}^3$) and, (iii) the load W_L that can be supported by the pontoon in sea water if the maximum permissible draught is 2.0 m.

Solution: In fresh water and unloaded condition,

Weight of the pontoon W = Total upthrust = weight of the water displaced

$$= 9.80 \times 1000 \times 5 \times 10 \times 1.2$$

∴ $$W_p = 588 \text{ kN}$$

In sea water and unloaded condition,

W = Total upthrust =weight of the sea water displaced

or $$588 \times 10^3 = 9.80 \times 1025 \times 5 \times 10 \times D$$

∴ $$D = 1.17 \text{ m}$$

For the loaded condition of pontoon in sea water,

Total upthrust, $W_T = 9.80 \times 1025 \times 5 \times 10 \times 2$

$$W_T = 1004.5 \text{ kN} = W_p + W_L$$

Therefore, $W_L = 1004.5 - 588$

$= 416.5$ kN

E 2.6.   A wooden (specific gravity = 0.6) cylinder of uniform diameter (=2m) is required to float (in water) upright under stable and neutral equilibrium. Compute the maximum permissible height of the cylinder.

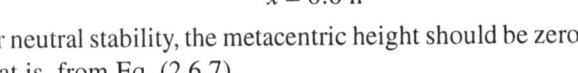

Fig. E 2.6

Solution: A floating body displaces its own weight in the fluid in which it floats. Therefore, if the cylinder, Fig. E2.6, is of diameter D and height h, and if x is the depth of its submergence, then

$$\frac{\pi}{4}D^2 h(0.6\times1000\times9.8)=\frac{\pi}{4}D^2(x)(1000\times9.8)$$

∴ $x = 0.6$ h

For neutral stability, the metacentric height should be zero. That is, from Eq. (2.6.7),

$$\frac{I_0}{\forall_{sub}} - BG = 0$$

where, I_0 is the second moment of area of the plane of floatation (i.e., waterline) about centroidal axis perpendicular to plane of rotation and is equal to $\dfrac{\pi D^4}{64}$, $\forall_{sub}$ is the immersed volume $\left(\dfrac{\pi}{4}D^2 x\right)$, and BG is the distance between the centre of gravity of the cylinder, G and the centre of buoyancy, B, i.e., $\left(\dfrac{h}{2}-\dfrac{x}{2}=0.2h\right)$. Therefore,

$$\frac{\pi D^4}{64\left(\dfrac{\pi D^2}{4}x\right)}=0.2h$$

∴ $D^2 = 16 \times 0.2 \times 0.6 \ h^2 = 1.92 \ h^2$ as $x = 0.6 \ h$

∴ $D = 1.386 \ h$

∴ $h = 1.443$ m

E 2.7.   Figure E2.7 shows a pressure tank with hemispherical dome of diameter equal to 2 m and vertical walls of height 2 m. The vessel is filled with an oil of mass density equal to 800 kg/m³. The mercury manometer connected to the pressure tank monitors the pressure in the tank and the mercury level in the manometer is as shown in Fig. E2.7. Compute the net hydrostatic force acting on the dome and walls of the tank. The length of the tank is 5 m.

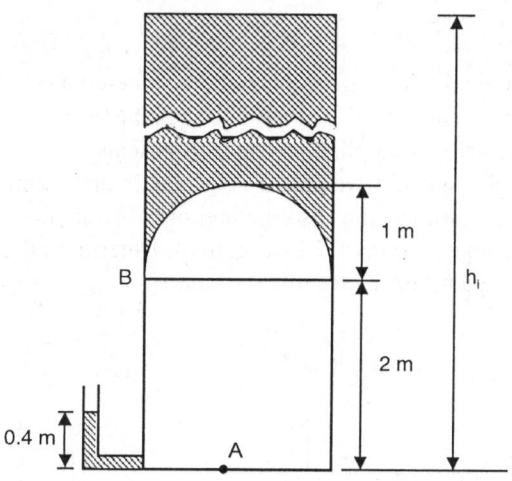

Fig. E2.7

Solution: Pressure force acts normal to a surface. Therefore, the dome would have net resultant force only in the vertical direction and this force would be equal to the weight of the imaginary volume of the fluid extending from the dome surface to the free surface corresponding to the pressure at A (Art. 2.5.3).

Gauge pressure at $A = \rho g h$

$$= 13.6 \times 1000 \times 9.8 \times 0.4$$

$$= 53312 \text{ Pa}$$

Let the imaginary free surface of oil be h_i above A.

$\therefore$ $\qquad (\rho g)_{oil} \ h_i = 53312$

or $\qquad 0.8 \times 1000 \times 9.8 \ h_i = 53312$

$\therefore$ $\qquad h_i = 6.8 \text{ m}$

Therefore, weight of oil between the imaginary free surface and the dome surface

$$= \left[2 \times 5(6.8-2) - \pi(1)^2 \times 5 \right] (0.8 \times 1000 \times 9.8)$$

$$= 253169.6 \text{ N}$$

$$= 253.17 \text{ kN}$$

which is the desired hydrostatic force acting on the dome in the vertical direction.

Pressure at B = $0.8 \times 1000 \times 9.8 \times (6.8 - 2) = 37632$ Pa.

Hence, pressure force on vertical walls in horizontal direction

$$= \frac{1}{2}(53312 + 37632)(2 \times 5)$$

$$= 454.72 \text{kN}$$

PROBLEMS

P2.1 Inside diameter of a vertical, clean, glass piezometer tube connected to a water tank is 1 mm. When pressure is applied, water (at 20°C) rises into the tube to a height of 20 cm. After correcting for surface tension, estimate the applied pressure in pascals.

P2.2 At a depth of 800 m, the specific weight of seawater is approximately 10,520 N/m³. At the surface, $\gamma \approx 10,050$ N/m³. Estimate the absolute pressure, in atm, at the depth of 800 m in the sea.

P2.3 In Fig. P2.3, pressure gauge A reads 1.5 kPa (gauge). Determine the elevations z, in meters, of the liquid levels in the open piezometer tubes B and C.

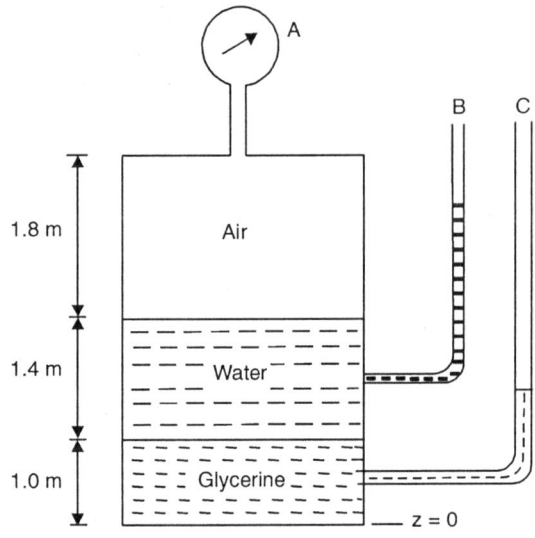

Fig. P2.3

P2.4 What is the density of the oil in Fig. P2.4, in kg/m³? Neglect surface tension effects.

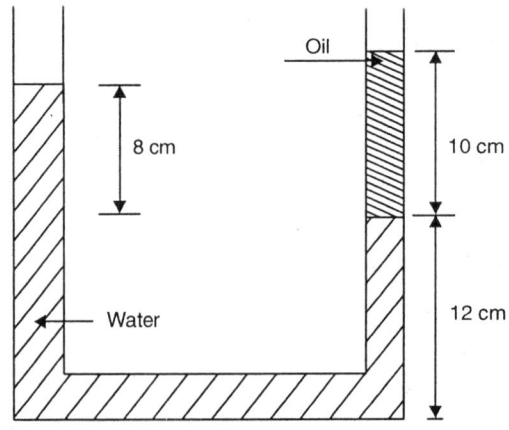

Fig. P2.4

P2.5 Water flows upward in a pipe inclined at 30°, as shown in Fig. P2.5. What is the pressure difference $p_1 - p_2$ in the pipe?

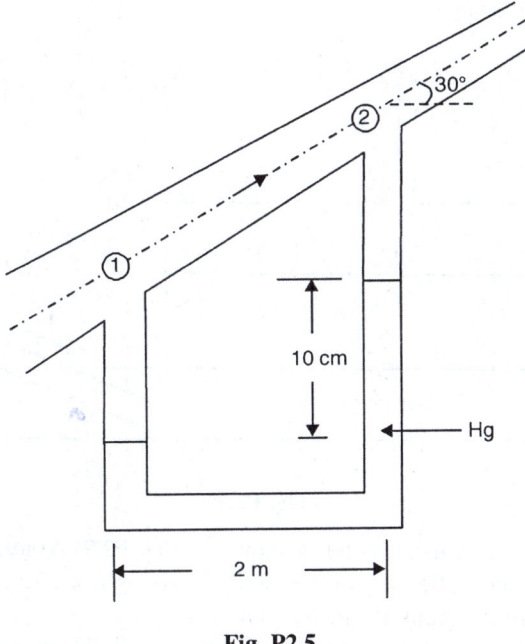

10 cm

Hg

2 m

Fig. P2.5

P2.6 In Fig. P2.6, determine the gauge pressure at point A in Pa.

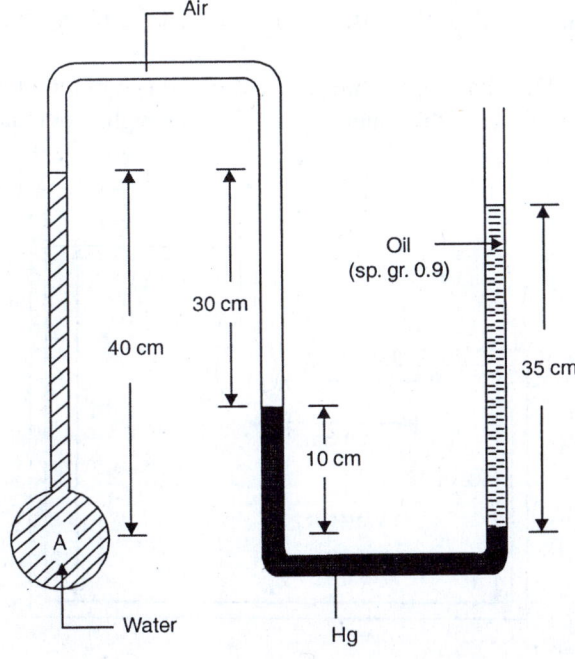

Air

Oil
(sp. gr. 0.9)

30 cm

40 cm

35 cm

10 cm

A

Water

Hg

Fig. P2.6

P2.7 For the inclined-tube reservoir manometer shown in Fig. P 2.7, obtain a general expression for the liquid deflection with respect to the original level, *L*, in the inclined leg, in terms of the applied pressure difference, Δ*p*.

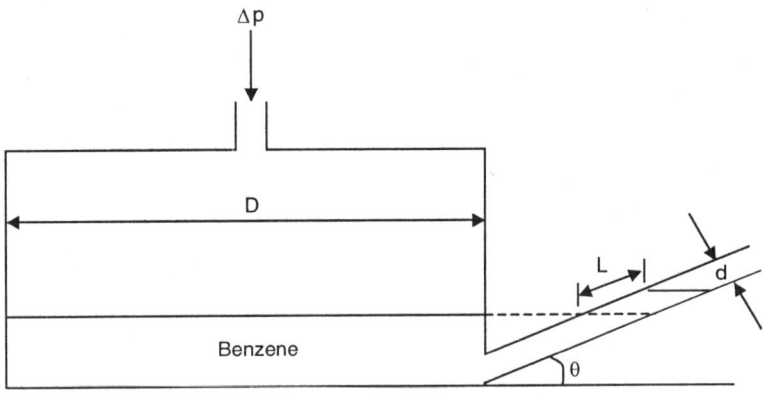

Fig. P2.7

P2.8 An inclined-tube reservoir manometer is shown in Fig. P2.7. Assume θ = 15.0°, *D* = 70.0mm, *d* = 6.30mm. The liquid deflection in the inclined tube is *L* = 230 mm when the manometer is connected to pressure differential Δ*p*. Evaluate Δ*p*.

P2.9 A vertical lock gate is 4m wide and separates water levels of 2m and 3m, respectively. Find the moment about the bottom of the gate required to keep the gate stationary.

P2.10 A closed container contains water at a depth of 4m. The absolute pressure above the water surface is 0.3 atm. Calculate the absolute pressure on the inside surface of the bottom of the container.

P2.11 Gate ABC in Fig. P2.11 has a fixed hinge line at B and is 3 m long into the paper. The gate will open at *A* to release water if the water depth is high enough. Compute the depth *h* at which the gate will begin to open.

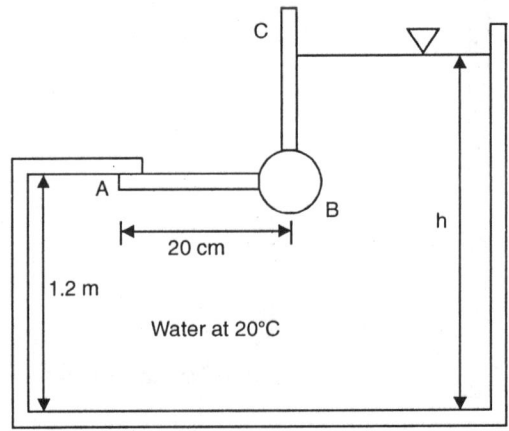

Fig. P2.11

P2.12 The tank in Fig. P2.12 contains benzene and is pressurized to 200 kPa (gauge) in the air gap. Determine the vertical hydrostatic force on circular arc section AB and its line of action.

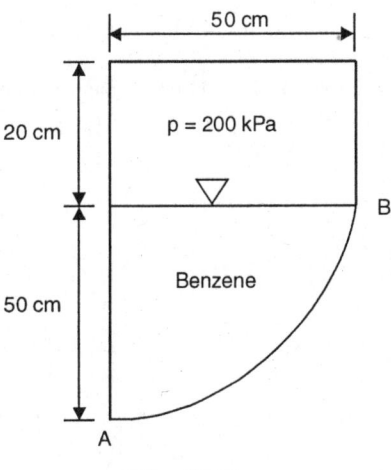

Fig. P2.12

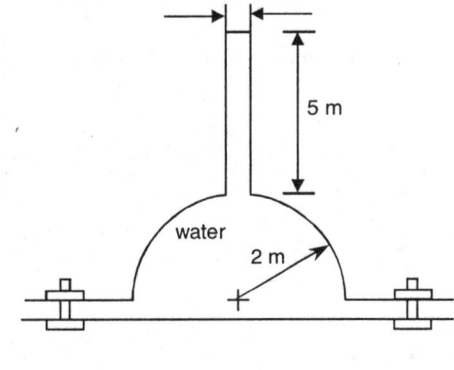

Fig. P2.13

P2.13 The hemispherical dome in Fig. P2.13 weighs 25 kN and is filled with water and attached to the floor by four equally spaced bolts on each side. What is the force in each bolt required to hold down the dome?

P2.14 An equilateral triangle with one edge in a water surface extends downward at a 30° angle. Locate the centre of pressure in terms of the length of a side of the triangle.

P2.15 (a) Determine the horizontal component of the force acting on the radial gate ABC (Fig. P2.15) and its line of action.

(b) Determine the vertical component of force and its line of action.

(c) Neglecting the weight of the gate, determine the force F required to open the gate?

(d) What is the moment about an axis normal to the paper and through the point O?

P2.16 A pipeline has an internal diameter of 1.00 m. (wall thickness of 10 mm) Lengths of pipe were capped and tested hydrostatically to a pressure of 12 Mpa. Calculate the maximum tensile stress in the pipe wall. Will the direction of the maximum stress in the pipe wall be axial or circumferential?

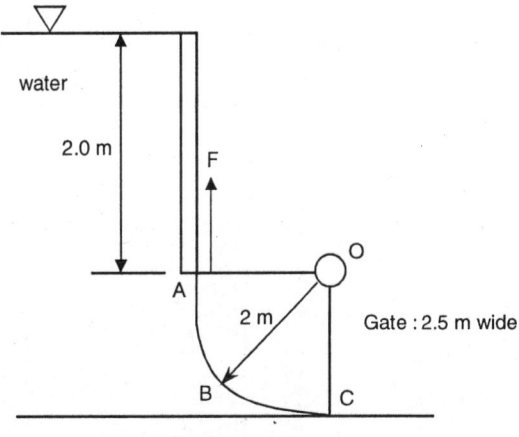

Fig. P2.15

P2.17 A door located in a plane vertical wall of a water tank is 1 m wide and 1.4 m high. The door is hinged along its upper edge which is 1 m below the water surface. Atmospheric pressure acts on the outer surface of the door and also at the water surface. What force must be applied at the lower edge of the door in order to keep the door closed?

P2.18 Justify the statement, "Only the tip of an iceberg shows (in seawater)."

REFERENCES

(1) Fox, RW, and McDonald, AT: *Introduction to Fluid Mechanics,* 4th Edition, John Wiley and Sons, Inc., USA, 1994.

(2) Munson, BR, Young, DF, and Okiishi, TH: *Fundamentals of Fluid Mechanics,* 4th Edition, John Wiley and Sons, Inc., Singapore, 2002.

(3) Streeter, VL, Wylie, EB, and Bedford, KW: *Fluid Mechanics*, 9th Edition, McGraw-Hill Companies, Singapore, 1998.

(4) White, FM: *Fluid Mechanics*, 5th Edition, McGraw-Hill Companies, USA, 2003.

Chapter 3

Fluid Kinematics

3.0 GENERAL

Fluids have a tendency to move or flow even if there exists a very small shear stress in the fluids. The study of velocity and acceleration of a flowing fluid and the description and visualization of the fluid motion are dealt with in *fluid kinematics*. The analysis of forces necessary to cause the fluid motion form the subject matter of *fluid dynamics*.

3.1 METHODS OF DESCRIBING FLUID MOTION

There are two different ways to analyze problems related to fluid motion. These are known as *Eulerian* and *Lagrangian* methods. The Eulerian method is concerned with the field of flow. The properties (pressure, velocity, density etc.) of fluid motion are expressed as functions of time and space. This means that in Eulerian method, one computes the pressure field $p(x, y, z, t)$ or velocity field $v(x,y,z,t)$ of the flow pattern and, not the pressure changes $p(t)$ or velocity changes $v(t)$ that a fluid particle experiences as it moves through the flow field. The Lagrangian method follows an individual particle (or a given mass) moving through the flow field and, thus, determines the fluid properties of individual particles (or masses) as a function of time, like $p(t)$ or $v(t)$.

A fluid is composed of very large number of particles. Therefore, it is convenient to use Eulerian method of describing fluid motion. The Eulerian method is suitable for fluid dynamic measurements too. For example, a pressure probe, fixed at a specific position (x, y, z) in a fluid flow in a laboratory, measures Eulerian pressure field $p(x, y, z, t)$. For Lagrangian description, the probe would be required to move downstream with the fluid particle.

3.2 SYSTEM AND CONTROL VOLUME

Use of free-body diagram in Chapter-2 for estimating forces on some arbitrary fixed mass was the Lagrangian concept. This is an example of *system* that refers to a fixed, identifiable quantity of mass. The system distinguishes this mass from all other matter called *surroundings*. The system boundaries, forming a closed surface, separate the system from the surroundings. The system boundaries may be fixed or movable. But, there is no mass transfer across the system boundaries. Gas contained in a familiar piston-cylinder assembly, Fig. 3.2.1, is an example of the system. If the assembly is heated from outside, the gas will expand and the piston moves. Alternatively, the gas in the cylinder could be compressed by moving the piston. Although heat and work have, thus, crossed the boundaries of the system, there is no mass transfer across the boundaries of the system, as the mass of the gas within the system boundaries has remained unchanged.

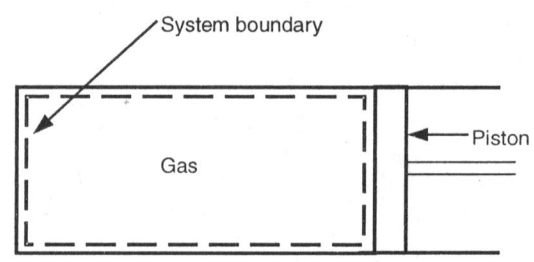

Fig. 3.2.1. Piston – cylinder system

In cases of fluid flow in pipes, channels, nozzles etc., it would be difficult to focus attention on a fixed identifiable quantity of mass. Instead, it would be rather more convenient to focus attention on a volume, in a flow field, through which the fluid flows. This volume is called *control volume* defined as an arbitrary volume in a flow field through which fluid flows. The geometric boundary of the control volume is called the *control surface* that may be real or imaginary; or may be at rest or in motion. A possible control volume for fluid flow in a pipe is shown in Fig. 3.2.2. The inside surface of the pipe forms real physical control surface of the control volume. But, the vertical portions of the control surface are imaginary. The size and shape of the control volume can be chosen in any arbitrary manner. The location of the control surface, however, affects the computational procedure. Therefore, the control surface should be chosen carefully. Generally, the control surfaces are made to coincide with the solid boundaries of the flow field or they are drawn normal to the flow direction.

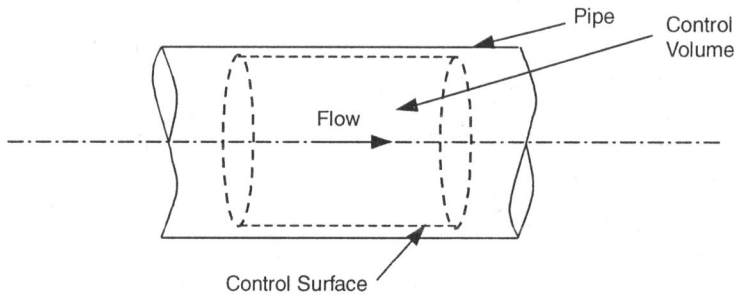

Fig. 3.2.2. Control volume for fluid flow in a pipe

3.3 TYPES OF FLUID FLOW

In general, a fluid flow is three-dimensional which means that the flow parameters, viz., velocity, pressure etc., vary in all three coordinate axes. For simplification of analysis of a fluid flow, it is, however, common to assume that the significant variation of the flow parameters occur in only two directions, or, even in only one.

One-dimensional (1D) flow is the one for which all the flow parameters may be expressed as functions of time and only one space coordinate that is usually the distance measured along the main flow direction. One-dimensional flow neglects variations of flow parameters in a plane transverse to the main flow direction (4). Flow of an ideal fluid in a pipe may be considered one-dimensional, Fig. 3.3.1(a), so that variations of pressure, velocity etc. may occur only along the length of the pipe and variations of these quantities over the cross-section are assumed insignificant. In reality, however, no flow is truly one-dimensional. Many practical problems can be analyzed with the assumption of one-dimensional flow.

In *two-dimensional flow*, all fluid particles are assumed to flow in parallel planes along identical paths (i.e., streamline pattern) in each of these planes. Hence, there are no variations in flow parameters normal to these planes and also the z-axis. This means that, the flow parameters are functions of time and only two rectangular space coordinates.

The flow in Fig. 3.3.1(b) is one-directional as the flow occurs in x-direction only, but it is two-dimensional as the pressure varies with x and the velocity varies with y. In the flow of Fig. 3.3.1(c), the velocity varies with both x and y coordinates and pressure varies with x when the streamlines are straight and with x and y when the streamlines are curved. Axi-symmetric flow, although not two-dimensional, may be analyzed as two-dimensional flow by making use of two cylindrical coordinates x and r.

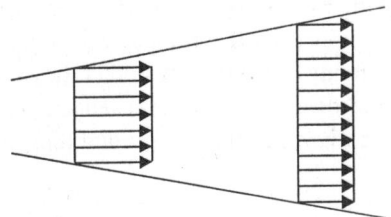

Fig. 3.3.1. (a) Hypothetical velocity profiles in one-dimensional flow

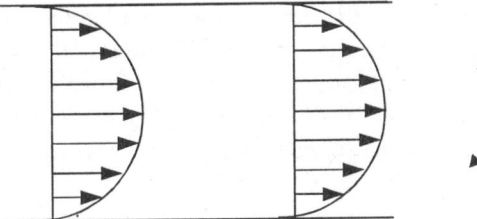

Fig. 3.3.1. (b) Flow between parallel plates

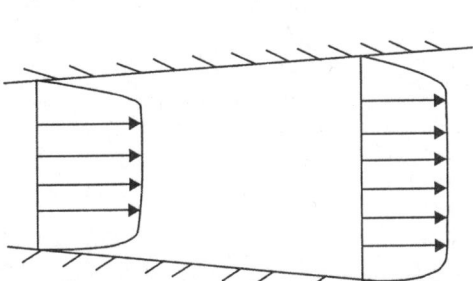

Fig. 3.3.1. (c) Flow between diverging planes

Three-dimensional flow is the most general and complex flow in which flow parameters vary with time and all three space coordinates x, y and z. River flow, though three-dimensional, is often assumed as two- or one-dimensional.

Flows are also classified as *steady* (when flow parameters are invariant with time) and *unsteady* (when flow parameters vary with time). Flows are, again, classified as *inviscid* or *non-viscous* (viscosity of the flowing fluid = 0) and *viscous* (viscosity of the flowing fluid ≠ 0) flows. Viscous flows are further grouped into two categories: (1) *Laminar flow* and (2) *Turbulent flow*. Flows in which density variation is negligible, are named as *incompressible flows*. Otherwise, the flow is called *compressible flow*. Liquid flows are frequently treated as incompressible whereas gas flows are examples of compressible flows.

Flows completely surrounded by solid surfaces are called *internal* or *duct flows*. Flows around bodies immersed in an unbounded fluid are termed *external flows*. Both internal and external flows may be laminar or turbulent, compressible or incompressible, steady or unsteady.

Flows may be uniform if the flow pattern does not vary from location to location (along the flow direction) at any instant. Otherwise, the flow is non-uniform. In addition, the fluid flows may be sonic or sub-sonic or supersonic depending upon whether the flow velocity is, respectively, equal to, or less than, or greater than the velocity of sound in the flowing fluid. In open channels, flows are classified as critical, supercritical, and subcritical depending upon whether the velocity of flow is, respectively, equal to, or greater than, or less than $\sqrt{gh}$ in which, h is the hydraulic depth.

3.4 VISUAL DESCRIPTION OF FLOW PATTERN

While dealing with fluid flow problems, it is often advantageous to obtain a visual representation of a flow field. The patterns of flow can be visualized in many different ways such as photographs, coloured dyes, flakes, hydrogen bubbles etc. Such visual representation can also be provided by streamlines, pathlines, and streaklines.

By looking at series of photographs of smoke leaving a chimney, one might approximately indicate on these photographs the speed and direction of air velocity by drawing arrows suitably at number of typical points. For a complete representation of the velocity, one would need to draw arrows at each and every possible point, making the result unsuitable for any meaningful analysis. However, a very satis-factory representation of the flow as a whole at any instant can be had by drawing a series of curves such that the velocity vectors for all points lying upon the curves would meet the curves tangentially. Such curves are known as *streamlines* that are drawn in a flow field at any given instant so that they are tangent to the direction of flow at every point in the flow field for that instant, Fig. 3.4.1(a). This means that a streamline is a line (or curve) that is everywhere tangent to the velocity field. Provided that the

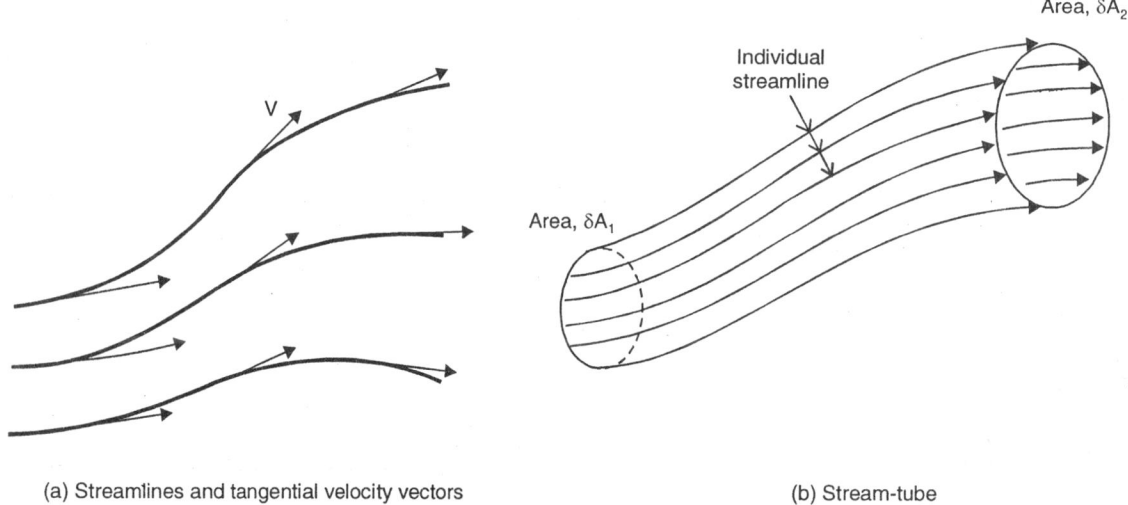

(a) Streamlines and tangential velocity vectors   (b) Stream-tube

Fig. 3.4.1. Streamline method of flow representation

flow is continuous, all streamlines of the flow would be continuous lines - either extending to the upstream and downstream limits of the flow or forming a closed curve. There can be no flow across a streamline. The boundaries of the flow are, therefore, always composed of streamlines. Streamlines can be determined analytically by integrating the equations defining lines tangent to the velocity field. An arbitrary velocity vector is shown in Fig. 3.4.2. The elemental length parallel to the velocity vector $\vec{V}$ is dr. Obviously, the components of $\vec{V}$ and dr in x-, y-, and z-directions must, respectively, be in proportion. Therefore,

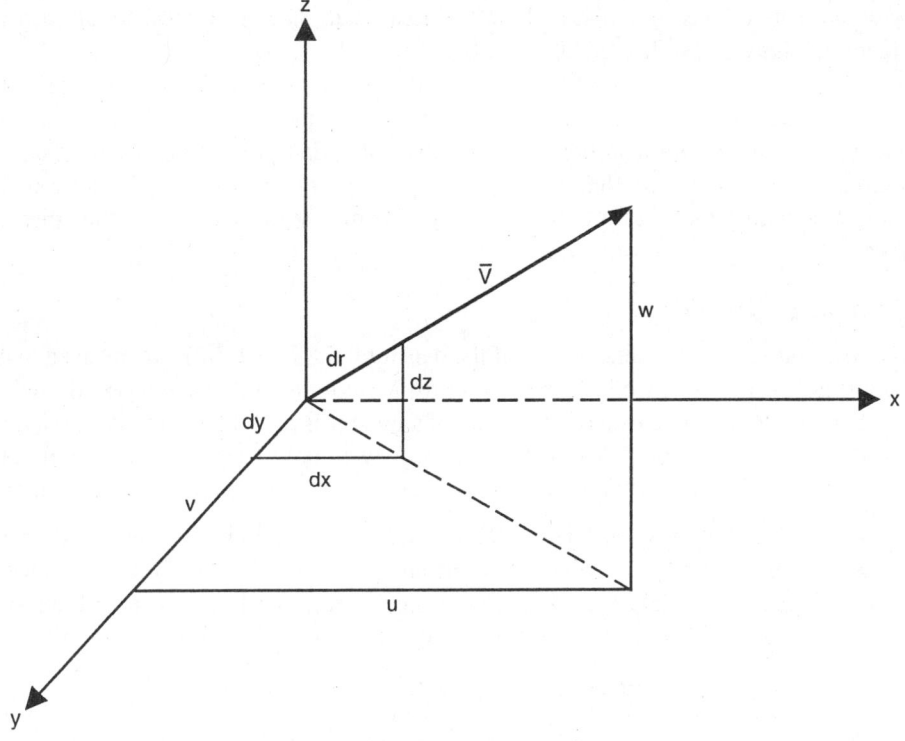

Fig. 3.4.2. Components of velocity vector $\vec{V}$

$$\frac{dx}{u} = \frac{dy}{v} = \frac{dz}{w} = \frac{dr}{\vec{V}} \qquad \qquad ...(3.4.1)$$

This means that if u, v and w are known in terms of space coordinates x, y, and z (and time t, if the flow is unsteady) one can obtain equation of the streamline. For two-dimensional flow, Eq. (3.4.1) is written as

$$\frac{dy}{dx} = \frac{v}{u} \qquad \qquad ...(3.4.2)$$

$$\therefore \qquad u\,dy - v\,dx = 0 \qquad \qquad ...(3.4.3)$$

If one follows a marked fluid particle and traces the path followed by the particle, one obtains *pathline*. The pathline is, therefore, the path or trajectory traced by a moving fluid particle in a flow field. A pathline is thus the actual path traversed by a given fluid particle in a flow field.

If one focuses one's attention on a fixed location of a flow field and joins all the fluid particles that passed through the identified location (point), the line, so obtained, is known as *streakline*. A streak line is, therefore, the locus of particles that have earlier passed through a specified point in the flow field.

The streamline can be calculated mathematically. But, other two lines, viz., pathline and streakline can be obtained experimentally with smoke, dye or bubble releases.

In a steady flow, the velocity at each point in the flow remains invariant with time. Therefore, streamlines of a steady flow do not vary with time. This means that a fluid particle on a given streamline will remain on the same streamline. Further, consecutive particles passing a fixed point on the identified streamline will remain on this streamline. Therefore, in a steady flow, pathlines, streaklines, and stream-lines are identical lines in the flow field.

Figure 3.4.1(b) shows a typical set of neighboring streamlines that form a passage through which the fluid flows. This passage (not necessarily circular) is termed *stream-tube*. *Stream filament* is a stream-tube with its cross-section sufficiently small so that variation of velocity over it may be considered negligible. By definition, the fluid within a stream-tube is confined there because it cannot cross the streamlines bounding the stream-tube. Thus, the stream-tube walls need not be solid and may be fluid surfaces.

3.5 VELOCITY FIELD

Continuum assumption that led to the notion of the density field, Eq. (1.2.1), can be used to define other fluid or fluid flow properties as well in terms of fields. Most important of the properties of a fluid flow is the velocity. Consider a fluid particle, consisting of small mass of fluid, which has a fixed identity and is at a point (x, y, z) in a flow field. The velocity at any given point (x,y,z) in a flow field is, therefore, the instantaneous velocity of the fluid particle that is passing through the point (x,y,z) at a given instant.

At any given instant, the velocity in a flow field, i.e., the velocity field $\vec{V}$ is a function of the space coordinates x, y, and z. Also, the velocity field at any point in the flow field may vary with time. Therefore, velocity in a flow field (i.e., the velocity field), in general, is a vector function of position and time and can be expressed in terms of three scalar components, u, v and w as follows:

$$\vec{V} = \vec{V}(x, y, z, t) = \hat{i}\, u(x,y,z,t) + \hat{j}\, v(x,y,z,t) + \hat{k} w(x,y,z,t) \qquad \qquad ...(3.5.1)$$

Here, u, v and w are three scalar components of the velocity $\vec{V}$ in x-, y-, and z- directions, respectively.

Thus, using the continuum hypothesis, it is possible to express properties of a flow field in terms of continuous functions of time and space. Obtaining the velocity vector field for a given fluid flow problem is almost equivalent to solving the problem completely. In this approach, the coordinates are fixed in space and one, therefore, observes the flow as it passes by. This is, therefore, Eulerian approach. This is in contrast with the Lagrangian approach in which one has to follow the moving position of individual particles (5). The field description of velocity (or any other property) is simple but powerful as it provides information for the entire flow field through one equation.

3.6 ACCELERATION OF FLUID PARTICLE IN A VELOCITY FIELD

To be able to apply Newton's second law of motion to a fluid particle moving along its streamline, one must know the particle acceleration in terms of the streamline coordinates. For two-dimensional flow, the acceleration, defined as the time rate of change of the velocity of the particle, has two components - one along the streamline a_s (i.e., the stream-wise or tangential acceleration) and other one normal to the streamline, a_n (i.e., the normal acceleration). For each of these two accelerations of a fluid particle, the change in velocity δv is the sum of the change due to the particle's change of position δs and the change due to the passing of time interval δt:

$$\delta v = \frac{\partial v}{\partial s}\delta s + \frac{\partial v}{\partial t}\delta t$$

and, hence, in the limit as $\delta t \to 0$

$$a_s = \lim_{\delta t \to 0}\frac{\delta v}{\delta t} = \frac{dv}{dt} = \frac{\partial v}{\partial s}\frac{ds}{dt} + \frac{\partial v}{\partial t}$$

Since

$$\frac{ds}{dt} = v,$$

$\therefore$

$$a_s = v\frac{\partial v}{\partial s} + \frac{\partial v}{\partial t} \qquad \qquad ...(3.6.1)$$

The term $\frac{\partial v}{\partial t}$ represents only the *local* or *temporal acceleration*, i.e., the rate of change of velocity of the particle with respect to time at a particular location of flow. The term $v\frac{\partial v}{\partial s}$ is termed the *convective acceleration*, i.e., the rate of change of velocity due to change of position of the particle. It is obvious that for steady flow, $\frac{\partial v}{\partial t} = 0$ and for uniform flow, $\frac{\partial v}{\partial s} = 0$. When the fluid particle moves along a curved path (and not straight as was assumed in deriving Eq. (3.6.1)), the velocity will be changing in direction (even if there is no change in its magnitude) and, hence, there will be acceleration in a direction perpendicular to its path. This acceleration is *normal* or *centrifugal* acceleration. Figure 3.6.1 shows a fluid particle moving from A to B along a curved path (i.e., streamline) of length δs that subtends a small angle $\delta\theta$ at the centre of curvature of the streamline. The change of velocity δv_n will be in a direction perpendicular to the direction of motion. From the diagrams of Fig. 3.6.1,

$$\delta v_n = v\delta\theta = v\frac{\delta s}{R}$$

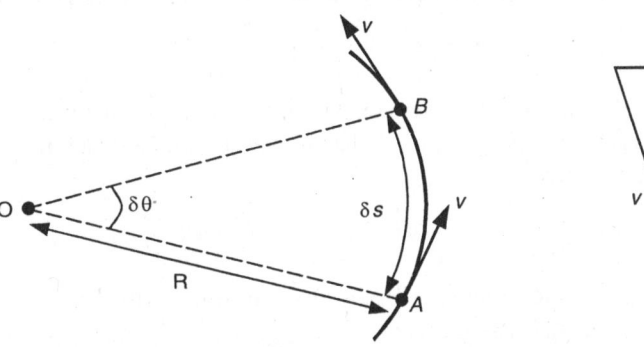

Fig. 3.6.1. Change of velocity during a circular path

Dividing by δt, the time in which the change occurs, and taking the limit $\delta t \rightarrow 0$,

$$\underset{\delta t \to 0}{\text{Lim}} \frac{\delta v_n}{\delta t} = \frac{dv_n}{dt} = \frac{v}{R}\frac{ds}{dt} = \frac{v^2}{R}$$

as

$$\frac{ds}{dt} = v$$

This is the convective term of the normal acceleration a_n to which must be added the temporal term $\frac{\partial v_n}{\partial t}$ in order to obtain the total normal acceleration. Thus,

$$a_n = \frac{v^2}{R} + \frac{\partial v_n}{\partial t} \qquad \qquad ...(3.6.2)$$

Figure 3.6.2 shows the presence or absence of tangential and normal accelerations for different stream-line patterns.

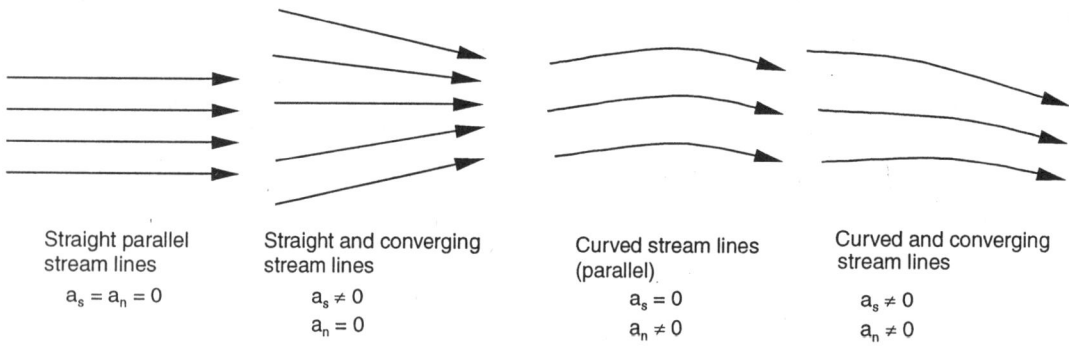

Straight parallel stream lines	Straight and converging stream lines	Curved stream lines (parallel)	Curved and converging stream lines
$a_s = a_n = 0$	$a_s \neq 0$	$a_s = 0$	$a_s \neq 0$
	$a_n = 0$	$a_n \neq 0$	$a_n \neq 0$

Fig. 3.6.2. Stream line patterns and the corresponding convective accelerations (2)

In general, the velocity of a fluid particle in a flow field is three-dimensional and depends on its position as well as time. If one were to describe particle-wise changes in velocity in a flow field, it would, obviously, be a cumbersome task. Therefore, a more general description is obtained when one considers a particle moving in a flow field. At any time t, the fluid particle is at a point (x, y, z) and has a velocity (corresponding to the velocity at that point in the flow field at time t)

$$\left. \overrightarrow{V_p} \right]_t = \overrightarrow{V}(x, y, z, t) \qquad \qquad ...(3.6.3)$$

Here, the subscript p denotes that the velocity refers to the fluid particle (when it is at point P and at time t) moving in a flow field. At time $t + dt$, the particle has moved to a new location Q $(x + dx, y + dy, z + dy)$ and, therefore, has a velocity

$$\left. \overrightarrow{V_p} \right]_{t+dt} = \overrightarrow{V}(x+dx, y+dy, z+dz, t+dt) \qquad \qquad ...(3.6.4)$$

The change in velocity of the particle in moving from P to Q in time dt is

$$d\overrightarrow{V_p} = \frac{\partial \overrightarrow{V}}{\partial x} dx + \frac{\partial \overrightarrow{V}}{\partial y} dy + \frac{\partial \overrightarrow{V}}{\partial z} dz + \frac{\partial \overrightarrow{V}}{\partial t}$$

Since, $\qquad \dfrac{dx}{dt} = u, \quad \dfrac{dy}{dt} = v \text{ and } \dfrac{dz}{dt} = w$

Therefore, $\qquad \overrightarrow{a_p} = \dfrac{d\overrightarrow{V_p}}{dt} = \dfrac{D\overrightarrow{V_p}}{Dt} = \left(u\dfrac{\partial \overrightarrow{V}}{\partial x} + v\dfrac{\partial \overrightarrow{V}}{\partial y} + w\dfrac{\partial \overrightarrow{V}}{\partial z} \right) + \dfrac{\partial \overrightarrow{V}}{\partial t} \qquad ...(3.6.5)$

The term $\dfrac{d\overrightarrow{V_p}}{dt}$ has been replaced with $\dfrac{D\overrightarrow{V_p}}{Dt}$ only to emphasize that the calculation of the acceleration of a fluid particle in a flow field requires a special derivative, commonly called the *substantial* or *material derivative* since it is computed for a particle of "substance" or "material"(1). As the particle moves from a given point P to another given point Q, its velocity may change for two reasons. First, during the time the given particle moves from P to Q, the velocity at Q changes. Second, fluid particles at Q may have a different velocity than that at P even at the same instant of time.

The term $\dfrac{\partial \overline{V}}{\partial t}$ is called the *local acceleration* that accounts for the change in velocity at the same location at different times. This term would be zero when the flow is steady, i.e., independent of time. The three terms in parentheses are called convective acceleration that arises when particle is moving in regions of spatially varying velocity. The sum of local and convective accelerations gives total acceleration and, hence, the derivative $\dfrac{D\overline{V}}{Dt}$ is also termed *total derivative*.

While considering an infinitesimal fluid system for analysis of fluid flow, one may be required to compute the acceleration vector $\vec{a}\left(=\dfrac{d\vec{v}}{dt}=\dfrac{D\overline{V}}{Dt}\right)$ of the flow at a point at a given time. Thus,

$$\vec{a}=\frac{D\overline{V}}{Dt}=\vec{i}\,\frac{du}{dt}+\vec{j}\,\frac{dv}{dt}+\vec{k}\,\frac{dw}{dt} \qquad ...(3.6.6)$$

Each of the three components u, v, and w in Eqs. (3.6.5 and 3.6.6) is dependent on four variables x, y, z and t. Hence, one may use the chain rule to obtain the time derivative of each of three terms u, v, and w. For example,

$$\frac{du(x,y,z,t)}{dt}=\frac{\partial u}{\partial t}+\frac{\partial u}{\partial x}\frac{dx}{dt}+\frac{\partial u}{\partial y}\frac{dy}{dt}+\frac{\partial u}{\partial z}\frac{dz}{dt}$$

But, by definition, $\dfrac{dx}{dt}$ is the local velocity component in x-direction, i.e., u and $\dfrac{dy}{dt}=v$ and $\dfrac{dz}{dt}=w$.

Therefore,

$$\frac{du}{dt}=\frac{Du}{Dt}=\frac{\partial u}{\partial t}+u\frac{\partial u}{\partial x}+v\frac{\partial u}{\partial y}+w\frac{\partial u}{\partial z}=(\overline{V}.\nabla)u \qquad ...(3.6.7)$$

Likewise,

$$\frac{dv}{dt}=\frac{Dv}{Dt}=\frac{\partial v}{\partial t}+u\frac{\partial v}{\partial x}+v\frac{\partial v}{\partial y}+w\frac{\partial v}{\partial z}=(\overline{V}.\nabla)v \qquad ...(3.6.8)$$

and

$$\frac{dw}{dt}=\frac{Dw}{Dt}=\frac{\partial w}{\partial t}+u\frac{\partial w}{\partial x}+v\frac{\partial w}{\partial y}+w\frac{\partial w}{\partial z}=(\overline{V}.\nabla)w \qquad ...(3.6.9)$$

Combining, Eqs. (3.6.6) to (3.6.9), one obtains the total acceleration at a point in the flow field as

$$\vec{a}=\frac{D\overline{V}}{Dt}=\frac{\partial \overline{V}}{\partial t}+u\frac{\partial \overline{V}}{\partial x}+v\frac{\partial \overline{V}}{\partial y}+w\frac{\partial \overline{V}}{\partial z}=\frac{\partial \overline{V}}{\partial t}+(\overline{V}.\nabla)\overline{V} \qquad ...(3.6.10)$$

It should be noted that the convective terms make the differential equations nonlinear and, therefore, the flows involving convective effects are more difficult to analyze than flows that do not involve convective changes. It is emphasized that Eq. (3.6.10) gives total time derivate of the velocity of a particle of fixed identity (in terms of location) in a flow field and is, therefore, useful for Eulerian description of the flow field(5).

3.7 MASS CONSERVATION (or CONTINUITY) EQUATION ALONG A STREAM-TUBE

Except in nuclear processes, matter is neither created nor destroyed. This principle of mass conservation is valid for a flowing fluid too. This means that in a fixed region of flow constituting a control volume, mass of fluid entering the volume per unit time must be equal to the mass of fluid leaving the control volume per unit time plus (or minus) the increase (or decrease) of mass of the fluid in the control volume per unit time. For steady flow, however, there cannot be any change in the mass within the control volume. Hence, for steady flow, mass of fluid entering a control volume per unit time equals the mass of fluid leaving the control volume per unit time. Consider a stream-tube of varying cross-section, Fig. 3.4.1(b). Obviously, no fluid can pass through the tube wall as the stream-tube is bounded by streamlines. This means that the mass of fluid passing one cross-section per unit time must equal, simultaneously, the mass of the fluid passing every other cross-section of the tube per unit time. In mathematical form, for steady flow,

$$\rho_1 v_1 \delta A_1 = \rho_2 v_2 \delta A_2 \qquad \qquad ...(3.7.1)$$

in which the velocities v_1 and v_2 are, respectively, measured at right angles to the cross-sectional areas δA_1 and δA_2. If the mass density does not vary, then

$$v_1 \delta A_1 = v_2 \delta A_2 \qquad \qquad ...(3.7.2)$$

i.e., the volume of fluid passing every section of a given stream-tube per unit time must be the same for all times when the flow is steady and for a given instant if the flow is unsteady. Equations (3.7.1) and (3.7.2) are basic continuity equations that are applicable to stream-tubes of finite cross-sectional areas as well so long as the velocity variation across the section is not excessive. For the flow of real fluid through a pipe or in an open channel, the velocity variation may be considerable. Using the average velocity, say V, the equation of continuity for steady flow in pipes and channels can be written as

$$\rho_1 V_1 A_1 = \rho_2 V_2 A_2 = \dot{m} \qquad \qquad ...(3.7.3)$$

where, $\dot{m}$ is the mass rate of flow. If the fluid flowing can be considered as incompressible, $\rho_1 = \rho_2$, and Eq. (3.7.3) reduces to

$$A_1 V_1 = A_2 V_2 = Q \qquad \qquad ...(3.7.4)$$

in which, Q is the volumetric rate of flow that is often termed as *discharge*.

3.8 DIFFERENTIAL EQUATION OF MASS CONSERVATION (CONTINUITY EQUATION)

Considering either an elemental control volume or an elemental system, one can derive all the basic differential equations governing fluid motion. Figure 3.8.1 shows, in Cartesian coordinates, the chosen control volume in the form of an infinitesimal rectangular parallelopiped whose sides are of length δx, δy, and δz. On this figure are also marked rates of mass inflow and mass outflow through the faces that are parallel to the y-z plane and have been calculated as follows:

Let u, v, and w represent, respectively, the velocities in the x-, y-, and z-directions and ρ be the mass density of the fluid. For the left face of the parallelopiped, the rate of the mass inflow entering the parallelopiped is $\rho u \, \delta y \, \delta z$. Therefore, the rate of mass outflow through the right face of the parallelopiped is

$$\rho u \, \delta y \, \delta z + \frac{\partial}{\partial x}(\rho u \, \delta y \, \delta z) \, \delta x$$

Hence, the net rate of mass inflow into the parallelopiped through left and right faces that are parallel to the y-z plane is

$$\rho u \, \delta y \, \delta z - \left\{ \rho u \, \delta y \, \delta z + \frac{\partial}{\partial x}\left(\rho u \, \delta y \, \delta z\right) \delta x \right\}$$

$$= -\frac{\partial}{\partial x}(\rho u) \, \delta x \, \delta y \, \delta z$$

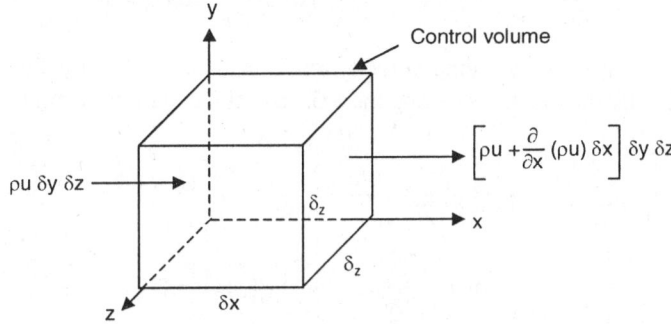

Fig. 3.8.1. Elemental control volume

Likewise, the net rate of mass inflow into the parallelopiped through back and front faces that are parallel to the x-y plane is

$$-\frac{\partial}{\partial z}(\rho w) \, \delta x \, \delta y \, \delta z$$

and that through bottom and top faces that are parallel to the x-z plane is

$$-\frac{\partial}{\partial y}(\rho v) \, \delta x \, \delta y \, \delta z$$

Since the mass within the parallelopiped is $\rho \, \delta x \, \delta y \, \delta z$, the rate of change of this mass on account of

inflows and outflows through the faces of the parallelopiped is $\dfrac{\partial \rho}{\partial t} \, \delta x \, \delta y \, \delta z$. In order to satisfy the law

of conservation of mass, the sum of the three mass inflow rates must equal the rate of change of mass of the fluid within the parallelopiped. Thus,

$$-\frac{\partial}{\partial x}(\rho u) \, \delta x \, \delta y \, \delta z - \frac{\partial}{\partial y}(\rho v) \, \delta x \, \delta y \, \delta z - \frac{\partial}{\partial z}(\rho w) \, \delta x \, \delta y \, \delta z = \frac{\partial \rho}{\partial t} \, \delta x \, \delta y \, \delta z$$

Dividing by the volume $\delta x \, \delta y \, \delta z$ of the parallelopiped, one gets

$$\frac{\partial \rho}{\partial t} + \frac{\partial}{\partial x}(\rho u) + \frac{\partial}{\partial y}(\rho v) + \frac{\partial}{\partial z}(\rho w) = 0 \qquad \qquad ...(3.8.1)$$

Equation (3.8.1) is often called the *equation of continuity* (or *continuity equation*) and requires no assumptions except that the density and velocity are continuum (i.e., continuous) functions. That is, Eq. (3.8.1) is applicable to all types of fluid flows that may be steady or unsteady, viscous or inviscid, compressible or incompressible.

The general continuity equation, Eq. (3.8.1), in vector form, is rewritten as

$$\frac{\partial \rho}{\partial t} + \nabla \cdot \left(\rho \overline{V} \right) = 0 \qquad \qquad ...(3.8.2)$$

In cylindrical polar coordinates, Eq. (3.8.1) becomes

$$\frac{\partial \rho}{\partial t} + \frac{1}{r} \frac{\partial}{\partial r} \left(\rho r v_r \right) + \frac{1}{r} \frac{\partial}{\partial \theta} \left(\rho v_\theta \right) + \frac{\partial}{\partial z} \left(\rho v_z \right) = 0 \qquad \qquad ...(3.8.3)$$

Here, v_r, v_θ, and v_z are, respectively, radial velocity, circumferential velocity, and axial velocity at an arbitrary point P defined by the distance z along the axis, a radial distance r from the axis, and a rotation angle θ about the axis.

For steady flow, $\dfrac{\partial}{\partial t} = 0$ and Eq. (3.8.1) reduces to

$$\frac{\partial}{\partial x} \left(\rho u \right) + \frac{\partial}{\partial y} \left(\rho v \right) + \frac{\partial}{\partial z} \left(\rho w \right) = 0 \qquad \qquad ...(3.8.4)$$

Equation (3.8.4) is applicable to both compressible and incompressible flows that are steady. For incompressible flow, the mass density ρ = constant. Therefore, Eq. (3.8.1) becomes

$$\frac{\partial u}{\partial x} + \frac{\partial v}{\partial y} + \frac{\partial w}{\partial z} = 0 \qquad \qquad ...(3.8.5)$$

Equation (3.8.5) is valid for both steady and unsteady incompressible flows.

3.9 STREAM FUNCTION

Consider a transparent plane and assume it to be parallel to the paper and a unit distance from the paper. Let the point A be a fixed point on the chosen plane and P a movable point, Fig. 3.9.1(a). The points A and P are joined by the arbitrary lines ABP and ACP. These lines, obviously, represent surfaces between the transparent plane and the plane of the paper. For an incompressible fluid (i.e., its density is constant), the volume rate of flow across any two lines ABP and ACP must be the same provided that no fluid is added to or withdrawn from the region ABPCA. This means that the rate of flow across ABP (or ACP) depends only on the end points A and P. Since A is fixed, one concludes, the rate of flow across ABP (or ACP) is a function only of the position of P. If this function is ψ (=$\psi(x,y)$), then ψ represents the volume flow rate across any line joining P (whose coordinates are (x,y)) to A at which point ψ is arbitrarily taken as zero. ψ is taken as positive if the flow is from right to left when one views the line from A looking toward P. The function ψ is known as the *stream function* (3). If ψ_1 and ψ_2 are the values of the stream function at points P_1 and P_2 (Fig. 3.9.1(b)), respectively, then the flow across $P_1 P_2$ is $\psi_2 - \psi_1$ and is independent of the location of A. If one takes another point Q in place of A, the values of ψ_1 and ψ_2 will be changed by the amount of the flow across AQ.

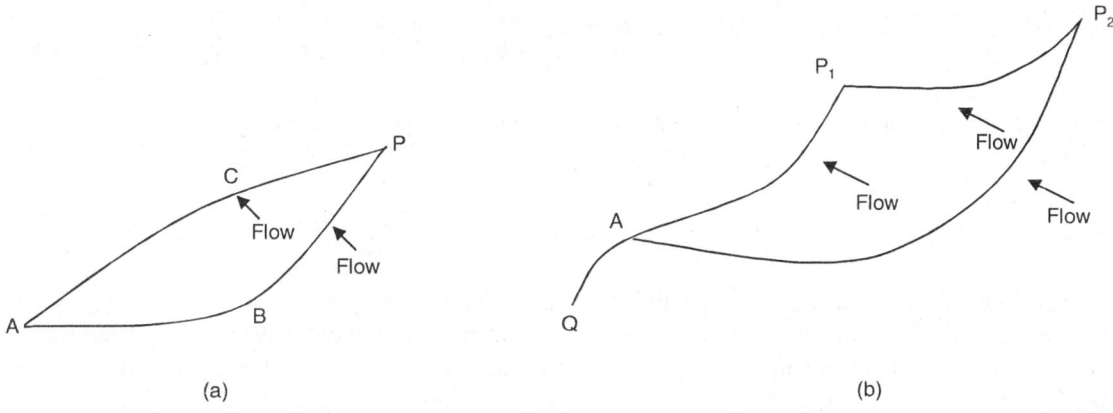

Fig. 3.9.1. (a) Fluid region for the definition of stream function
(b) Flow between two points in a fluid region

One can obtain expressions for the velocity components u and v in terms of ψ. The flow $\delta\psi$ across AP (Fig. 3.9.2(a)) due to the velocity component u is $u\delta y$ from left to right, i.e., $-u\delta y$ as per the sign convention. Therefore,

$$\delta\psi = -u\,\delta y$$

or
$$u = -\frac{\delta\psi}{\delta y}$$

or
$$u = -\frac{\partial\psi}{\partial y} \quad \text{as} \quad \delta y \to 0 \qquad \qquad ...(3.9.1)$$

and similarly,
$$\delta\psi = v\,\delta x$$

or
$$v = \frac{\partial\psi}{\partial x} \quad \text{as} \quad \delta x \to 0 \qquad \qquad ...(3.9.2)$$

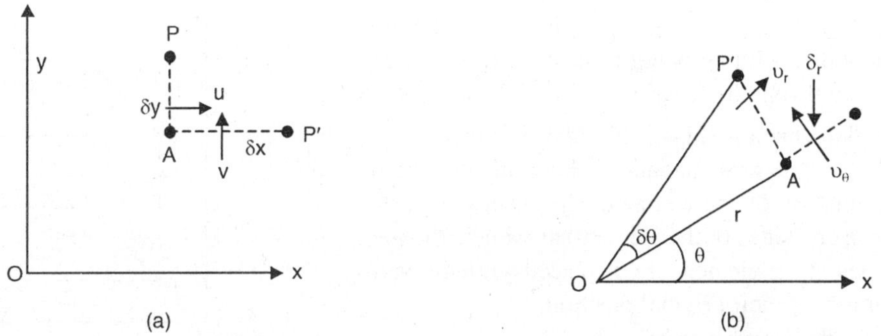

Fig. 3.9.2. Figure for relation between velocity components and stream function

This means that the partial derivative of the stream function with respect to any direction gives the velocity component at 90° anti-clockwise to that direction. The stream function is defined only for two-dimensional incompressible flow for which the continuity equation, Eq. (3.8.5) reduces to

$$\frac{\partial u}{\partial x} + \frac{\partial v}{\partial y} = 0 \qquad\qquad ...(3.9.3)$$

On substituting the values of u and v in terms of the stream function ψ, the left hand side of Eq. (3.9.3) becomes

$$\frac{\partial}{\partial x}\left(-\frac{\partial \psi}{\partial y}\right) + \frac{\partial}{\partial x}\left(\frac{\partial \psi}{\partial x}\right) = -\frac{\partial^2 \psi}{\partial x \partial y} + \frac{\partial^2 \psi}{\partial y \partial x} = 0$$

This means that the stream function always satisfies the continuity equation. The stream function, therefore, is a single mathematical function $\psi(x, y, t)$, that replaces the two velocity components, $u\ (x, y, t)$ and $v\ (x, y, t)$. The stream function ψ thus simplifies the problem of determining two unknown functions $u\ (x, y, t)$ and $v\ (x, y, t)$ to determining only one unknown function $\psi(x, y, t)$. Stream function can also be defined for three-dimensional flow with axial symmetry when all flow lines are in planes intersecting the same line (or axis) and the flow is identical in each of these planes, i.e., the flow is independent of the axial coordinate. From Fig. 3.9.2(b),

$$\delta\psi = -v_r\,(r\delta\theta)$$

$$\therefore \qquad\qquad v_r = -\frac{\partial \psi}{r\partial \theta} \text{ as } \delta\theta \to 0 \qquad\qquad ...(3.9.4)$$

and

$$\delta\psi = v_\theta \delta r$$

$$\therefore \qquad\qquad v_\theta = \frac{\partial \psi}{\partial r} \text{ as } \delta r \to 0 \qquad\qquad ...(3.9.5)$$

3.10 FLUID DISPLACEMENTS

In the study of fluid kinematics, the angular velocity of the fluid elements about their mass centres is also important. An infinitesimal element of a fluid, Fig. 3.10.1, may move in the fluid such that the element undergoes the following basic displacements (1, 2):

(i) Translation (Fig. 3.10.2(a)) in which the element moves bodily without being rotated (Fig. 3.10.2(b)) or deformed (Fig. 3.10.2(c)).

(ii) Linear deformation (Fig. 3.10.2(d)) in which shape of the elements gets changed without any change in the orientation of the element. The planes (i.e., the top, base and sides of the element) as well as the median lines of the element are displaced while remaining parallel to their original position.

(iii) Rotation (Fig. 3.10.2(b)) in which planes (i.e., top, base and sides of the element), and medians as well as diagonals of the element may change as a result of their rotating about any one (or all three) of the coordinate axes.

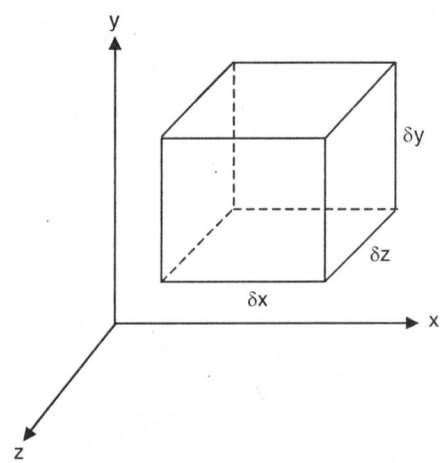

Fig. 3.10.1. Infinitesimal element of a fluid

(iv) Angular deformation (Fig. 3.10.2(c)) which involves a distortion of the fluid element in which planes that were initially perpendicular to each other are no longer so.

These four components of fluid motion, illustrated in Fig. 3.10.2, are for the motion in the *x-y* plane. Similar motions of the particle may occur in the *y-z* and *z-x* planes too in three-dimensional flows. It should be noted that for pure translation or rotation, the fluid retains its original shape and there is no deformation and, hence, no shear stress either.

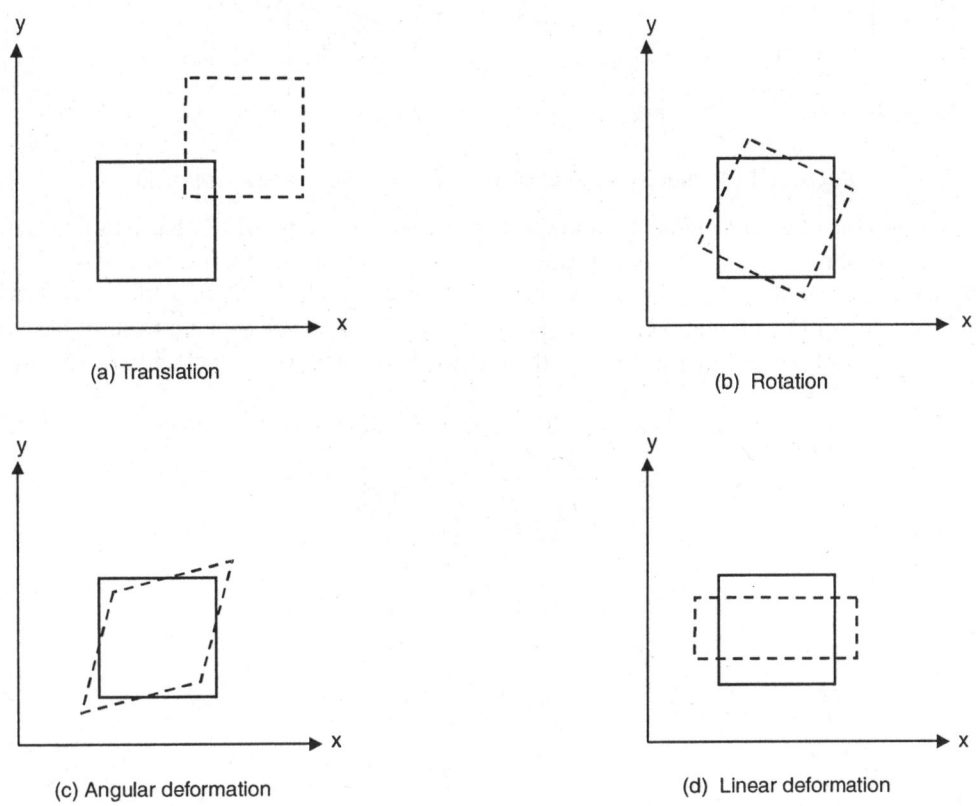

(a) Translation

(b) Rotation

(c) Angular deformation

(d) Linear deformation

Fig. 3.10.2 Fluid displacements

3.11 FLUID ROTATION

The rotation, $\bar{\omega}$, of a fluid particle is a vector quantity and is defined as the average angular velocity of any two mutually perpendicular line elements of the particle in each of the three orthogonal planes (1). Therefore, a particle, moving in general three-dimensional flow, may rotate about all three coordinate axes. Therefore, in general,

$$\varpi = \hat{i}\omega_x + \hat{j}\omega_y + \hat{k}\omega_z \qquad ...(3.11.1)$$

where, ω_x, ω_y, and ω_z are, respectively, the rotations about the x-, y-, and z-axes. Counter-clockwise rotation is considered positive. If the fluid particles in a flow region have rotation about any axis, the flow is called *rotational flow*, or *vortex flow*. When the fluid particles in a flow region have no rotation, the flow is called *irrotational flow*.

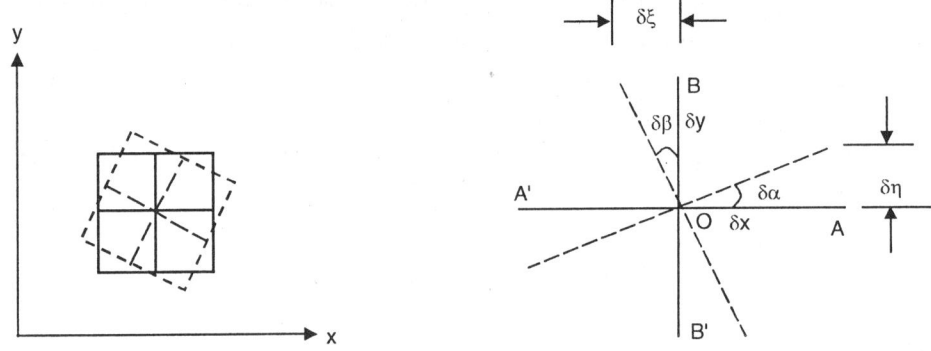

Fig. 3.11.1. Rotation of a fluid element in a two-dimensional flow field

Consider motion of a fluid element in the *x-y* plane as shown in Fig. 3.11.1 that also shows rotation of the element in the *x-y* plane. If velocities at points A and B are different than that at O, the two mutually perpendicular lines OA and OB would rotate by an angle $\delta\alpha$ during the time interval δt. Obviously, rotation of OA of length δx is on account of variations of the y-component of the velocity. If the y-component of the velocity at O is v_0, then, the y-component of the velocity at A is v that can

be written as $v = v_0 + \dfrac{\partial v}{\partial x}\,\delta x$ and, therefore, the angular velocity of the line OA is

$$\omega_{OA} = \lim_{\delta t \to 0} \frac{\delta\alpha}{\delta t} = \lim_{\delta t \to 0} \frac{\delta\eta/\delta x}{\delta t}$$

Since,

$$\delta\eta = \delta v\,\delta t = \frac{\partial v}{\partial x}\,\delta x\,\delta t$$

∴.

$$\omega_{OA} = \frac{\partial v}{\partial x}$$

Similarly,

$$\omega_{OB} = -\frac{\partial u}{\partial y}$$

As ω_{OB} is in the clockwise direction and rotation is positive in the counter-clockwise direction. The rotation of the fluid element about the z-axis, as per the definition of rotation, is, therefore,

$$\omega_z = \frac{1}{2}\left(\frac{\partial v}{\partial x} - \frac{\partial u}{\partial y}\right) \qquad \qquad ...(3.11.2)$$

Similarly, by considering the rotation of mutually perpendicular pair of lines in the *y-z* and *x-z* planes, one can obtain,

$$\omega_x = \frac{1}{2}\left(\frac{\partial w}{\partial y} - \frac{\partial v}{\partial z}\right) \qquad \qquad ...(3.11.3)$$

and

$$\omega_y = \frac{1}{2}\left(\frac{\partial u}{\partial z} - \frac{\partial w}{\partial x}\right) \qquad \qquad ...(3.11.4)$$

Hence, from Eq. (3.11.1),

$$\varpi = \frac{1}{2}\left[\hat{i}\left(\frac{\partial w}{\partial y} - \frac{\partial v}{\partial z}\right) + \hat{j}\left(\frac{\partial u}{\partial z} - \frac{\partial w}{\partial x}\right) + \hat{k}\left(\frac{\partial v}{\partial x} - \frac{\partial u}{\partial y}\right)\right] \qquad \text{...(3.11.5)}$$

In vector notation, Eq. (3.11.5) is

$$\varpi = \frac{1}{2}\nabla \times \vec{V} = \frac{1}{2}\text{ curl }\vec{V} \qquad \text{...(3.11.6)}$$

Another measure of the rotation of a fluid element is the *vorticity* $\vec{\xi}$ that is equal to twice the rotation. Thus,

$$\xi_x = \left(\frac{\partial w}{\partial y} - \frac{\partial v}{\partial z}\right) \qquad \text{...(3.11.7)}$$

$$\xi_y = \left(\frac{\partial u}{\partial z} - \frac{\partial w}{\partial x}\right) \qquad \text{...(3.11.8)}$$

$$\xi_z = \left(\frac{\partial v}{\partial x} - \frac{\partial u}{\partial y}\right) \qquad \text{...(3.11.9)}$$

$$\vec{\xi} = 2\varpi = \hat{i}\left(\frac{\partial w}{\partial y} - \frac{\partial v}{\partial z}\right) + \hat{j}\left(\frac{\partial u}{\partial z} - \frac{\partial w}{\partial x}\right) + \hat{k}\left(\frac{\partial v}{\partial x} - \frac{\partial u}{\partial y}\right) \qquad \text{...(3.11.10)}$$

A fluid particle moving, without rotation, in a flow field cannot rotate under the action of a body force (i.e., weight) and normal surface force (i.e., pressure force). A fluid particle, initially with no rotation, therefore, requires the action of shear stress on its surface for its rotation. Since shear stress is related to the rate of angular deformation through viscosity, one can conclude that the presence of viscosity causes the flow to become rotational.

The flow of inviscid or non-viscous fluid is irrotational. However, for all real fluids, shear stresses develop due to the presence of boundaries or other interactions between fluid processes and result in rotational flow. Nevertheless, one can assume flow in some regions of real fluid flow to be irrotational when, in those regions, the viscous forces (or effects) are negligible.

3.11.1 Angular Deformation

Angular deformation of a fluid element results when the angle between two mutually perpendicular lines of the element changes, Fig. 3.11.2. Thus, the rate of angular deformation of the fluid element, shown in Fig. 3.11.2, is the rate of decrease of the angle γ between lines OA and OB and is, therefore,

$$-\frac{d\gamma}{dt} = \frac{d\alpha}{dt} + \frac{d\beta}{dt}$$

Here, $\qquad \dfrac{d\alpha}{dt} = \underset{\delta t \to 0}{\text{Lim}} \dfrac{\delta\alpha}{\delta t} = \underset{\delta t \to 0}{\text{Lim}} \dfrac{\delta\eta/\delta x}{\delta t} = \underset{\delta t \to 0}{\text{Lim}} \dfrac{(\partial v/\partial x)\,\delta x\,\delta t/\delta x}{\delta t} = \dfrac{\partial v}{\partial x}$

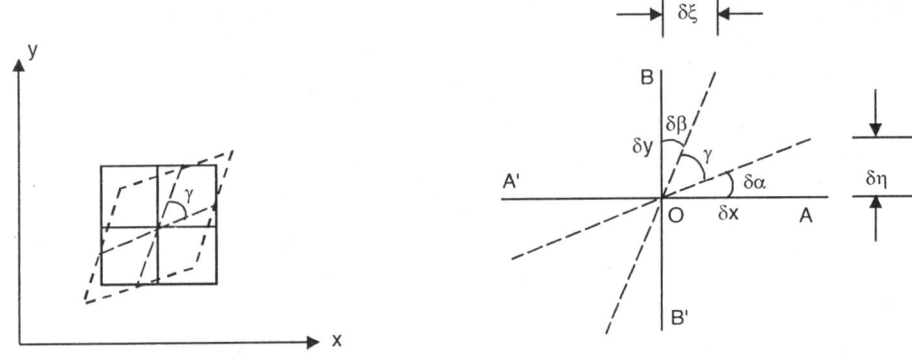

Fig. 3.11.2. Angular deformation of a fluid element in a two-dimensional flow field

and
$$\frac{d\beta}{dt} = \lim_{\delta t \to 0} \frac{\delta\beta}{\delta t} = \lim_{\delta t \to 0} \frac{\delta\xi/\delta y}{\delta t} = \lim_{\delta t \to 0} \frac{(\partial u/\partial y)\,\delta y\,\delta t/\delta y}{\delta t} - \frac{\partial u}{\partial y}$$

Hence, the rate of angular deformation in the *x-y* plane is

$$-\frac{d\gamma}{dt} = \frac{d\alpha}{dt} + \frac{d\beta}{dt} = \frac{\partial v}{\partial x} + \frac{\partial u}{\partial y} \qquad\qquad ...(3.11.11)$$

In any viscous flow, it is very unlikely that $\dfrac{\partial v}{\partial x}$ will be equal and opposite to $\dfrac{\partial u}{\partial y}$ throughout the

flow region(1). Therefore, the presence of viscous forces gives rise to angular deformation and, hence, the flow becomes rotational.

3.11.2 Circulation

The term *circulation*, Γ is defined as the line integral of the tangential velocity component around a closed curve in the flow region. That is

$$\Gamma = \oint_c \vec{V}\,d\vec{s} \qquad\qquad ...(3.11.12)$$

in which, $d\vec{s}$ is an elemental vector, of length ds, tangent to the curve. Γ is taken as positive for anti-clockwise path of integration around the curve.

One can develop a relationship between Γ and vorticity (or rotation) by considering a fluid element's boundary OACB of area $\delta A = \delta x \times \delta y$ as shown in Fig. 3.11.3. For the closed curve OACB,

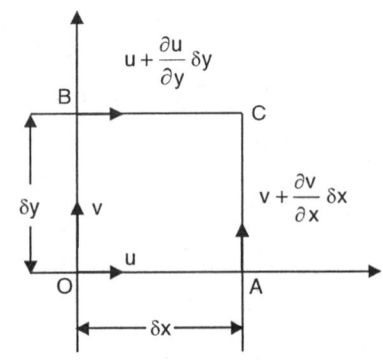

$$\Gamma = u\,\delta x + \left(v + \frac{\partial v}{\partial x}\delta x\right)\delta y - \left(u + \frac{\partial u}{\partial y}\delta y\right)\delta x - v\,\delta y$$

$$= \left(\frac{\partial v}{\partial x} - \frac{\partial u}{\partial y}\right)\delta x\,\delta y$$

Fig. 3.11.3 Velocity components on the boundaries of a fluid element

$$\therefore \qquad \frac{\Gamma}{\delta x\,\delta y} = \frac{\partial v}{\partial x} - \frac{\partial u}{\partial x} = 2\omega_z = \xi_z \qquad\qquad ...(3.11.13)$$

or
$$\Gamma = \oint_c \vec{V}\, d\vec{s} = \int_A 2\omega_z\, dA = \int_A \left(\nabla\times\vec{V}\right)_z dA \qquad ...(3.11.14)$$

This means that the circulation per unit area is equal to vorticity or twice the rotation about an axis perpendicular to the plane of the area.

3.12 VELOCITY POTENTIAL

The circulation round the closed curve ABPCA, Fig. 3.12.1, wholly in a flowing fluid is

$$\Gamma = \int_{ABP} v_s\, ds + \int_{PCA} v_s\, ds$$

$$\therefore \qquad \Gamma = \int_{ABP} v_s\, ds - \int_{ACP} v_s\, ds$$

in which v_s is the component of velocity along an element ds of the curve ABP or ACP. For irrotational flow, $\Gamma = 0$, so that

$$\int_{ABP} v_s\, ds = \int_{ACP} v_s\, ds \qquad\qquad ...(3.12.1)$$

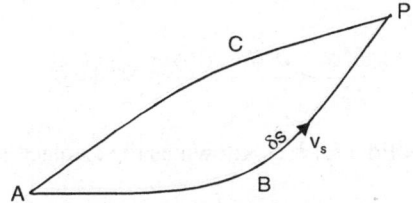

Fig. 3.12.1 Fluid region for definition of ϕ

This means that the value of the integral $\int_A^P v_s ds$ depends only on the position of P relative to A. On equating this integral to $-\phi$, one gets

$$-\phi = \int_A^P v_s\, ds \qquad\qquad ...(3.12.2)$$

$$\therefore \qquad \delta\phi = -v_s\,\delta s$$

or
$$v_s = \operatorname*{Lim}_{\delta_s \to 0}\left(-\frac{\delta\phi}{\delta s}\right)$$

i.e.,
$$v_s = -\frac{\partial\phi}{\partial s} \qquad\qquad ...(3.12.3)$$

The function ϕ is termed the *velocity potential* that decreases in the direction of flow just as the electric potential decreases in the direction of current flow. If the line element, of length δs, is perpendicular to a streamline then, obviously, $v_s = 0$ and, hence, $\delta\phi = 0$. This means, the velocity potential is constant along lines that are perpendicular to streamlines. The lines of constant potential are known as *equipotential lines*. Equation (3.12.3) expresses the velocity component v_s as the negative gradient of the velocity potential with respect to s. Hence, the velocity potential provides an alternative (with respect to the stream function, ψ) means to express the velocity components parallel to the coordinate axes in irrotational flow:

$$u = -\frac{\partial\phi}{\partial x} \qquad\qquad ...(3.12.4)$$

$$v = -\frac{\partial\phi}{\partial y} \qquad\qquad ...(3.12.5)$$

$$w = -\frac{\partial\phi}{\partial z} \qquad\qquad ...(3.12.6)$$

It should be noted that the velocity potential ϕ is not a measurable physical quantity, but a mathematical quantity defined by Eq. (3.12.2). A reference equipotential line of $\phi = 0$, like reference streamline with $\psi = 0$, may be chosen arbitrarily. One should, however, note that whereas the stream function applies to both irrotational and rotational flows, the velocity potential applies to *only* irrotational flow (also termed as *potential flow*). This is so because Eq. (3.12.1) is true only for irrotational flow.

Substitution of the values of u, v and w from Eqs. (3.12.4 – 3.12.6) into the continuity equation, Eq. (3.8.5) yields,

$$\frac{\partial^2\phi}{\partial x^2} + \frac{\partial^2\phi}{\partial y^2} + \frac{\partial^2\phi}{\partial z^2} \equiv \nabla^2\phi = 0 \qquad\qquad ...(3.12.7)$$

Thus, any function ϕ that satisfies Eq. (3.12.7), known as the Laplace's equation, is a velocity potential of a possible irrotational flow.

Substitution of ϕ in Eqs. (3.11.7 – 3.11.9) yields

$$\xi_x = \frac{\partial}{\partial y}\left(-\frac{\partial\phi}{\partial z}\right) - \frac{\partial}{\partial z}\left(-\frac{\partial\phi}{\partial y}\right) = -\frac{\partial^2\phi}{\partial y\partial z} + \frac{\partial^2\phi}{\partial z\partial y}$$

$$\xi_y = \frac{\partial}{\partial z}\left(-\frac{\partial\phi}{\partial x}\right) - \frac{\partial}{\partial x}\left(-\frac{\partial\phi}{\partial z}\right) = -\frac{\partial^2\phi}{\partial z\partial x} + \frac{\partial^2\phi}{\partial x\partial z}$$

and $$\xi_z = \frac{\partial}{\partial x}\left(-\frac{\partial\phi}{\partial y}\right) - \frac{\partial}{\partial y}\left(-\frac{\partial\phi}{\partial x}\right) = -\frac{\partial^2\phi}{\partial x\partial y} + \frac{\partial^2\phi}{\partial y\partial x}$$

Since $\dfrac{\partial^2\phi}{\partial y\partial z} = \dfrac{\partial^2\phi}{\partial z\partial y}$, $\dfrac{\partial^2\phi}{\partial z\partial x} = \dfrac{\partial^2\phi}{\partial x\partial z}$ and $\dfrac{\partial^2\phi}{\partial x\partial y} = \dfrac{\partial^2\phi}{\partial y\partial x}$ the vorticity must always be zero when velocity potential ϕ exists. This means that the velocity potential ϕ exists only for irrotational flow.

From Eq. (3.11.9), for two-dimensional irrotational flow in *x-y* plane,

$$\xi_z = \left(\frac{\partial v}{\partial x} - \frac{\partial u}{\partial y} \right) = 0 \qquad \qquad ...(3.12.8)$$

If one substitutes the values of *u* and *v*, in Eq. (3.12.8), in terms of the stream function ψ, one obtains

$$\frac{\partial}{\partial x}\left(\frac{\partial \psi}{\partial x} \right) - \frac{\partial}{\partial y}\left(-\frac{\partial \psi}{\partial y} \right) = 0$$

or
$$\frac{\partial^2 \psi}{\partial x^2} + \frac{\partial^2 \psi}{\partial y^2} = 0$$

This means that the stream function ψ too satisfies Laplace equation, if the flow is irrotational. Flows that do not satisfy Laplace's equation in ψ are, therefore, rotational flows for which velocity potential ϕ does not exist.

Substitution of ψ in the continuity equation for two-dimensional flow yields

$$\frac{\partial}{\partial x}\left(-\frac{\partial \psi}{\partial y} \right) + \frac{\partial}{\partial y}\left(\frac{\partial \psi}{\partial x} \right) = 0$$

or
$$-\frac{\partial^2 \psi}{\partial x \partial y} + \frac{\partial^2 \psi}{\partial y \partial x} = 0$$

Since $\dfrac{\partial^2 \psi}{\partial x \partial y} = \dfrac{\partial^2 \psi}{\partial y \partial x}$, the stream function, ψ, satisfies the continuity equation for incompressible flow.

The stream function, however, is not subject to the restriction of irrotational flow.

3.13 FLOWNETS

For any two-dimensional irrotational flow, one can draw two sets of lines or curves: (i) curves along which ψ is constant, i.e., streamlines, and (ii) curves along which ϕ is constant, i.e., equipotential lines that are, as mentioned earlier, perpendicular to lines of constant ψ. The plot of streamlines and equipotential lines together would form a grid of quadrilaterals having 90° corners. It is customary to draw the streamlines at equal increments of ψ and the equipotential lines at equal increments of ϕ. Considering

s (along the streamlines) and *n* (perpendicular to streamlines) coordinates, $v_s = \dfrac{\partial \psi}{\partial n} = -\dfrac{\partial \phi}{\partial s}$. This means, higher velocity would result in closer streamlines and equipotential lines for chosen increments of ψ and ϕ. When Δs and Δn, Fig. 3.13.1, are made equal and in the limit $\Delta s = \Delta n \rightarrow 0$, the quadrilaterals become perfect squares. It is customary to choose increments of ψ and ϕ such that $\Delta s \cong \Delta n$ and the resulting grid is called the *flownet*. To check the squareness of the grid (or flownet), one can draw the diagonals for all the squares. These diagonals should also result into a grid of squares, Fig. 3.13.2. Alternatively, one may use the fact that in each square of the flownet, the sides are tangent to a common circle.

For any given set of boundary conditions, there would be only one possible pattern of the irrotational flow (i.e., the flow of an inviscid fluid). Hence, a correctly drawn flownet satisfying these conditions provide a graphical solution of the given fluid motion as one can compute velocities at any point by the spacing of the streamlines or equipotential lines and, thereafter, pressure variations may be obtained by using the Bernoulli's equation (Chapter 4).

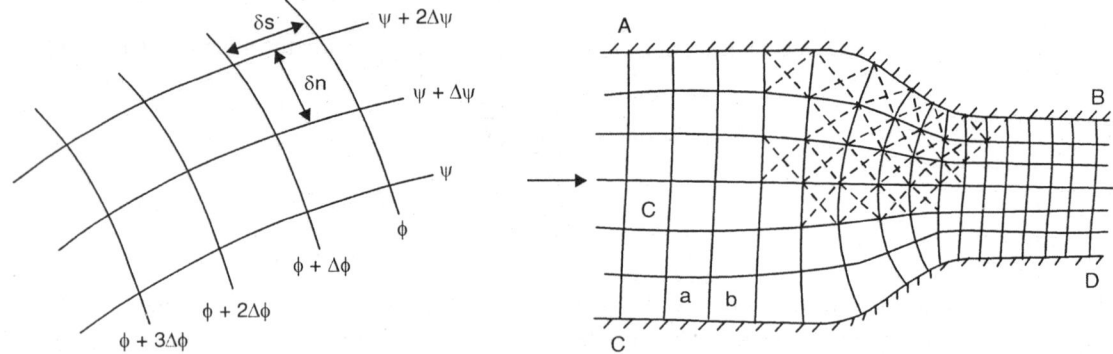

Fig. 3.13.1. Flownet   **Fig. 3.13.2.** Checking of flownet

3.14 COMBINATION OF FLOW PATTERNS

If one flow pattern described by $\psi = \psi_1$ is superimposed over another flow pattern $\psi = \psi_2$, the resulting flow pattern is described by a stream function $\psi = \psi_1 + \psi_2$ such that value of ψ for the resultant flow, at any point in the flow field is the algebraic sum of the stream function of the constituent flows at that point, Fig. 3.14.1. Therefore, using this principle, complicated motions can be analyzed by considering

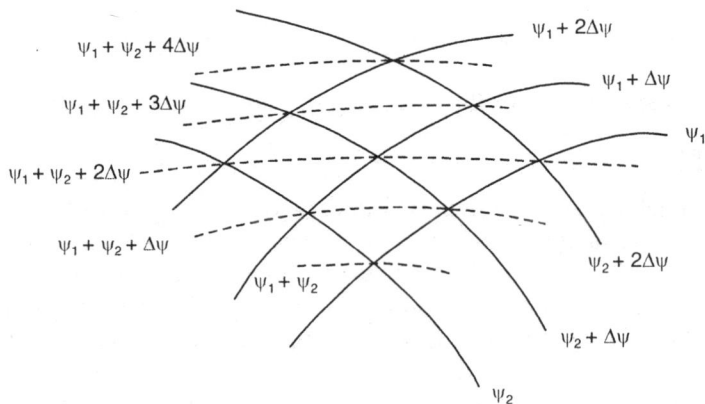

Fig. 3.14.1. Superimposition of flow patterns

them as combination of two or more simpler motions. The magnitude or velocity components of the resultant motion are, obviously, given by the algebraic sums of those for the constituent motions. That is,

$$u = -\frac{\partial}{\partial y}\left(\psi_1 + \psi_2\right) = -\frac{\partial \psi_1}{\partial y} - \frac{\partial \psi_2}{\partial y} = u_1 + u_2$$

and, similarly, for the *y*-direction component *v*.

For irrotational flows, the velocity potentials of constituent flows are, similarly, added to obtain the velocity potential of the resultant flow pattern. Laplace equation, Eq. (3.12.7) is linear in ϕ. Therefore, if ϕ_1 and ϕ_2, individually, were solutions of the Laplace equation, $\phi_1 + \phi_2$ would also be satisfying the Laplace equation.

EXAMPLES

E3.1: A discharge q (m³/s/m) is occurring between two parallel plates held apart at a distance d. Assuming the relation between the velocity v and the distance from one of the plates y as $v = ky\,(d - y)$, obtain the value of k.

Solution: From Fig. E3.1,

$$dq = v\,dy = ky\,(d - y)\,dy$$

$\therefore$

$$q = \int dq = \int_0^d ky\,(d - y)\,dy$$

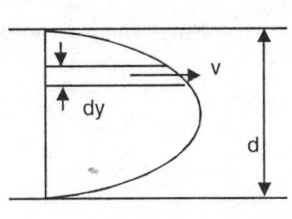

$$= k\left[\frac{dy^2}{2} - \frac{y^3}{3}\right]_0^d$$

Fig. E 3.1

$$= k\left[\frac{d^3}{2} - \frac{d^3}{3}\right]$$

$$= \frac{k}{6}d^3$$

$\therefore$

$$k = \frac{6q}{d^3}$$

E3.2: A 1 m diameter pipe is reduced to 50 cm diameter linearly in a distance of 2m. Initially (i.e., at $t = 0$), a discharge of 2 m³/s establishes through the pipe and it increases at the rate of 0.1 m³/s per hour. Compute the total acceleration at the beginning, the mid-section, and the end of the transition at the beginning and at ten minutes after the flow is established.

Solution: In Fig. E3.2, at any $x, d = 1 - \dfrac{1 - 0.5}{2} \cdot x$

$$d = 1 - 0.25\,x$$

at any time t, $\qquad Q = 2 + \dfrac{0.1}{3600}\,t$

$\therefore$

$$Q = 2 + 2.78 \times 10^{-5}\,t$$

Therefore, the velocity $V(x, t) = \dfrac{Q}{\dfrac{\pi}{4}d^2} = \dfrac{4Q}{\pi d^2}$

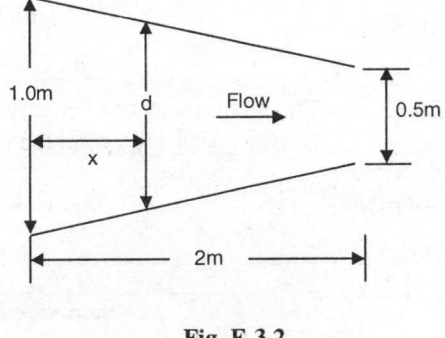

Fig. E 3.2

or
$$V(x,t)=\frac{4\left(2+\dot{2}.78\times10^{-5}t\right)}{\pi\left(1-0.25\,x\right)^{2}}$$

and
$$\frac{\partial V}{\partial x}=\frac{4Q}{\pi}(-2)\left(1-0.25x\right)^{-3}(-0.25)=\frac{2Q}{\pi}\left(1-0.25\,x\right)^{3}$$

and
$$\frac{\partial V}{\partial t}=\frac{4\left(2.78\times10^{-5}t\right)}{\pi\left(1-0.25\,x\right)^{2}}$$

Since there is no other flow component,

$$a(x,\,t)=\frac{DV}{Dt}=\frac{\partial V}{\partial t}+V\frac{\partial V}{\partial x}$$

$$=\frac{4\left(2.78\times10^{-5}\right)}{\pi\left(1-0.25\,x\right)^{2}}+\frac{8\left(2+2.78\times10^{-5}\,t\right)^{2}}{\pi^{2}\left(1-0.25\,x\right)^{5}}$$

or
$$a\!\left(x,t\right)=\frac{3.54\times10^{-5}}{\left(1-0.25\,x\right)^{2}}+\frac{8\left(2+2.78\times10^{-5}\,t\right)^{2}}{\pi^{2}\left(1-0.25\,x\right)^{5}}$$

∴
$$a\!\left(0,t\right)=3.54\times10^{-5}+0.81\!\left(2+2.78\times10^{-5}\,t\right)^{2}$$

Hence,
$$a\,(0,0) = 3.24 \text{ m/s}^2$$
$$a\,(0,600) = 3.297 \text{ m/s}^2$$

Now,
$$a\,(1,\,t) = 6.29 \times 10^{-5} + 3.41(2 + 2.78 \times 10^{-5}\,t)^2$$

∴
$$a\,(1,0) = 13.64 \text{ m/s}^2$$
$$a\,(1,\,600) = 13.87 \text{ m/s}^2$$

Similarly,
$$a\,(2,\,t) = 14.16 \times 10^{-5} + 25.92\,(2 + 2.78 \times 10^{-5}t)^2$$

∴
$$a\,(2,\,0) = 103.68 \text{ m/s}^2$$
$$a\,(2,\,600) = 105.42 \text{ m/s}^2$$

E3.3:   The x-component of velocity in an incompressible flow is $u = x^2 + z^2 + 5$ and the y-component is $v = y^2 + z^2$. Find the z-component of velocity w for the flow.

Solution:
$$u = x^2 + z^2 + 5 \therefore \frac{\partial u}{\partial x} = 2x$$

$$v = y^2 + z^2 \quad \therefore \frac{\partial v}{\partial y} = 2y$$

∴   Using Eq. (3.8.5), $\dfrac{\partial w}{\partial z} = -\left[2x + 2y\right] = -2(x + y)$

∴ $w = -2(x + y)z + f(x, y)$

E3.4:   The velocity field for a fluid flow is given as

$$\vec{V} = (3x + 2y)\hat{i} + \left(3z + 2x^2\right)\hat{j} + (2t - 3z)\hat{k}$$

Determine (a) whether the flow is incompressible, (b) the velocity components at a point (1,2,3) and the resultant speed at the point at time t.

Solution: (a) For the flow to be incompressible, Eq. (3.8.5) must be satisfied, that is,

$$\frac{\partial u}{\partial x} + \frac{\partial v}{\partial y} + \frac{\partial w}{\partial z} = 0$$

For the given problem, $u = 3x + 2y$ ∴ $\dfrac{\partial u}{\partial x} = 3$

$v = 3z + 2x^2$ ∴ $\dfrac{\partial v}{\partial y} = 0$

$w = 2t - 3z$ ∴ $\dfrac{\partial w}{\partial z} = -3$

Hence, $\dfrac{\partial u}{\partial x} + \dfrac{\partial v}{\partial y} + \dfrac{\partial w}{\partial z} = 3 + 0 + (-3) = 0$

Therefore, the velocity field represents incompressible flow.

(b) At point (1, 2, 3),

$$u = 3x + 2y = 3(1) + 2\,(2) = 7$$
$$v = 3z + 2x^2 = 3(3) + 2(1)^2 = 11$$
$$w = 2t - 3z = 2t - 3(3) = 2t - 9$$

Therefore, the speed V at (1, 2, 3) is $\sqrt{u^2 + v^2 + w^2}$

$$= \sqrt{49 + 121 + (2t - 9)^2}$$

$$= \sqrt{170 + (2t - 9)^2}$$

E3.5:   The velocity components for a flow are as follows:

$$u = a_1 x + b_1 y + c_1 z$$
$$v = a_2 x + b_2 y + c_2 z$$

and $w = a_3 x + b_3 y + c_3 z$

Determine the condition under which the flow would be incompressible.

Solution: Incompressible flow satisfies the continuity equation, Eq. (3.8.5).
Therefore,

$$\frac{\partial u}{\partial x} + \frac{\partial v}{\partial y} + \frac{\partial w}{\partial z} = 0$$

or
$$a_1 + b_2 + c_3 = 0$$

which is the required condition. Note that the requirement of continuity does not include the constants b_1, c_1, a_2, c_2, a_3, and b_3 and, therefore, there is no restriction on their values.

E3.6: The velocity components for a two-dimensional flow are
$$u = Ax \text{ and } v = -Ay$$

Determine the stream function ψ and velocity potential ϕ.

Solution: The given velocity field $\vec{V} = Ax\hat{i} - Ay\hat{j}$ satisfies the continuity equation, Eq. (3.8.5), for the incompressible flow. Therefore, from Eq. (3.8.1),

$$u = -\frac{\partial \psi}{\partial y} = Ax$$

Integrating with respect to y, one gets
$$\psi = -Axy + f(x) \qquad \qquad ...(1)$$

The function $f(x)$ is arbitrary and can be evaluated using Eq. (3.9.2) for v and Eq. (1). Thus

$$v = -Ay = \frac{\partial \psi}{\partial x} = -Ay + \frac{df(x)}{dx}$$

Therefore, $$\frac{df(x)}{dx} = 0$$

Hence, $f(x) =$ constant, say, C_1
Therefore, $$\psi = -Axy + C_1 \qquad \qquad ...(2)$$

Assigning the desired value to A and choosing any arbitrary value of C_1 (including zero), one can plot the streamlines on x-y plane for different values of ψ. Similarly, one can find the velocity potential ϕ as follows:

$$u = -\frac{\partial \phi}{\partial x} = Ax$$

$$\phi = -\frac{Ax^2}{2} + f(y)$$

Also,
$$v = -\frac{\partial \phi}{\partial y} = -Ay$$

$$\therefore \qquad \frac{\partial \phi}{\partial y} = Ay = \frac{df(y)}{dy}$$

Hence,
$$f(y)=A\frac{y^2}{2}+C_2$$

Here, C_2 is a constant of integration

$$\therefore \qquad \phi =\frac{A}{2}\left(y^2-x^2\right)+C_2$$

The constants C_1 and C_2 appearing in the expressions for ψ and ϕ can be assigned any arbitrary value (including zero).

E3.7:   The velocity components for a two-dimensional flow are

$$u = x - 4y \text{ and } v = -y - 4x$$

Do the velocity components represent incompressible flow? If yes, determine the stream function ψ for the flow. It the flow is irrotational too, determine the velocity potential ϕ.

Solution: Continuity equation for two-dimensional incompressible flow ($w = 0$) is obtained from Eq. (3.8.5) as

$$\frac{\partial u}{\partial x}+\frac{\partial v}{\partial y}=0$$

For the given flow,
$$\frac{\partial u}{\partial x}=1 \text{ and } \frac{\partial v}{\partial y}=-1$$

Therefore,
$$\frac{\partial u}{\partial x}+\frac{\partial v}{\partial y}=1-1=0$$

Hence, the given flow is incompressible.

Now,
$$u=-\frac{\partial \psi}{\partial y}=x-4y$$

$$\therefore \qquad \frac{\partial \psi}{\partial y}=-x+4y$$

$$\therefore \qquad \psi = -xy + 2y^2 + f(x)$$

Hence,
$$\frac{\partial \psi}{\partial x}=-y+\frac{\partial f(x)}{\partial x} \qquad \qquad ...(1)$$

Also,
$$v=\frac{\partial \psi}{\partial x}=-y-4x$$

$$\therefore \qquad \psi = -yx - 2x^2 + f(y)$$

$$\therefore \qquad \frac{\partial \psi}{\partial x}=-y-4x \qquad \qquad ...(2)$$

Equating Eqs. (1) and (2)

$$\frac{\partial f(x)}{\partial x} = -4x$$

$$\therefore \qquad f(x) = -2x^2 + C_1$$

where C_1 is a constant.
Hence,

$$\psi = -xy + 2y^2 - 2x^2 + C_1$$

If the flow is irrotational, $\qquad w_z = \frac{1}{2}\left(\frac{\partial v}{\partial x} - \frac{\partial u}{\partial y}\right) = 0$

i.e., $\qquad \frac{\partial v}{\partial x} - \frac{\partial u}{\partial y} = -4 - (-4) = 0$

$\therefore$ The flow is irrotational.

Alternatively, ψ should satisfy the Laplace's equation if the flow is irrotational
Since $\psi = -xy + 2y^2 - 2x^2 + C_1$

$$\therefore \qquad \frac{\partial \psi}{\partial x} = -y - 4x \ \text{ and } \ \frac{\partial^2 \psi}{\partial x^2} = -4$$

and

$$\frac{\partial \psi}{\partial y} = -x + 4y, \text{ hence, } \ \frac{\partial^2 \psi}{\partial y^2} = 4$$

$$\therefore \qquad \frac{\partial^2 \psi}{\partial x^2} + \frac{\partial^2 \psi}{\partial y^2} = 0$$

i.e., ψ satisfies the Laplace's equation.
Hence, the flow is irrotational.

Now $\qquad u = -\frac{\partial \phi}{\partial x} = x - 4y$

$$\therefore \qquad \frac{\partial \phi}{\partial x} = -x + 4y$$

$$\therefore \qquad \phi = -\frac{x^2}{2} + 4yx + f(y)$$

Hence, $\qquad \frac{\partial \phi}{\partial y} = 4x + \frac{\partial f(y)}{\partial y} = -v = y + 4x$

$$\therefore \qquad \frac{\partial f(y)}{\partial y} = y$$

$$\therefore \qquad f(y) = \frac{y^2}{2} + C_2$$

$$\therefore \qquad \phi = -\frac{x^2}{2} + 4yx + \frac{y^2}{2} + C_2$$

The constants C_1 and C_2 appearing in the expressions for ψ and ϕ can be assigned any arbitrary value (including zero).

E3.8: The velocity potential in a 2-D flow is $\phi = y + x^2 - y^2$. Find the stream function ψ for this flow.

Solution: $\phi = y + x^2 - y^2$

$$\therefore \qquad \frac{\partial \phi}{\partial x} = 2x = \frac{\partial \psi}{\partial y} \quad \therefore \ \psi = 2xy + f(x) \quad \therefore \ \frac{\partial \psi}{\partial x} = 2y + f'(x)$$

$$\frac{\partial \phi}{\partial y} = 1 - 2y = -\frac{\partial \psi}{\partial x} \quad \therefore \ \frac{\partial \psi}{\partial x} = 2y - 1$$

Hence $\qquad f'(x) = -1$

$$f(x) = -x + C$$

where C is a constant of integration.

$$\therefore \qquad \psi = 2xy - x + C$$

E3.9: The two-dimensional stream function for a flow is $\psi = 9 + 6x - 4y + 7xy$. Find the velocity potential.

Solution: $\qquad \psi = 9 + 6x - 4y + 7xy$

$$\frac{\partial \psi}{\partial y} = \frac{\partial \phi}{\partial x} = -4 + 7x \qquad \therefore \ \phi = -4x + \frac{7}{2}x^2 + f(y) \qquad \therefore \ \frac{\partial \phi}{\partial y} = f'(y)$$

$$\frac{\partial \psi}{\partial x} = -\frac{\partial \phi}{\partial y} = 6 + 7y$$

$$f'(y) = -6 - 7y$$

$$\therefore \qquad f(y) = -6y - \frac{7}{2}y^2 + C$$

where C is a constant of integration.

$$\therefore \qquad \phi = -4x + \frac{7}{2}x^2 - 6y - \frac{7}{2}y^2 + C$$

$$= -4x - 6y + \frac{7}{2}(x^2 - y^2) + C$$

E3.10: The velocity field in a fluid flow is

$$\bar{V} = \left(x^2 - y^2\right)\hat{i} + \left(y^2 - z^2\right)\hat{j} + \left(2xz + 2yz\right)\hat{k}$$

Compute the rotation vector, vorticity, and the acceleration at point (1,2,3).

Solution: From the velocity field

$$u=x^2-y^2 \qquad \therefore \frac{\partial u}{\partial y}=-2y \text{ and } \frac{\partial u}{\partial z}=0 \text{ and } \frac{\partial u}{\partial x}=2x$$

$$v=y^2-z^2 \qquad \therefore \frac{\partial v}{\partial x}=0 \text{ and } \frac{\partial v}{\partial z}=-2z \text{ and } \frac{\partial v}{\partial y}=2y$$

$$w=-2xz-2yz \qquad \therefore \frac{\partial w}{\partial x}=-2z \text{ and } \frac{\partial w}{\partial y}=-2z \text{ and } \frac{\partial w}{\partial z}=-2x-2y$$

$$w_x=\frac{1}{2}\left(\frac{\partial w}{\partial y}-\frac{\partial v}{\partial z}\right)=\frac{1}{2}(-2z+2z)=0$$

$$w_y=\frac{1}{2}\left(\frac{\partial u}{\partial z}-\frac{\partial w}{\partial x}\right)=\frac{1}{2}(0+2z)=z$$

$$w_z=\frac{1}{2}\left(\frac{\partial v}{\partial x}-\frac{\partial u}{\partial y}\right)=\frac{1}{2}(0+2y)=y$$

Hence, the rotation vector, $\vec{w}=z\hat{j}+y\hat{k}$

And, the vorticity vector, $\bar{\xi}=2z\hat{j}+2y\hat{k}$

Acceleration is $\bar{a}=\hat{i}\,a_x+\hat{j}a_y+\hat{k}a_z$

$$a_x=\frac{\partial u}{\partial t}+u\frac{\partial u}{\partial x}+v\frac{\partial u}{\partial y}+w\frac{\partial u}{\partial z}$$

$$= 0 + (x^2 - y^2)\,(2x) + (y^2 - z^2)\,(-2y) + (-2xz - 2yz)\,(0)$$
$$= 2x^3 - 2xy^2 - 2y^3 + 2yz^2$$
$$= 2(1)^3 - 2(1)\,(2)^2 - 2(2)^3 - 2(2)(3)^2 \text{ for point } (1, 2, 3)$$
$$= 2 - 8 - 16 + 36$$
$$= 14$$

Similarly,

$$a_y=\frac{\partial v}{\partial t}+u\frac{\partial v}{\partial x}+v\frac{\partial v}{\partial y}+w\frac{\partial v}{\partial z}$$

$$= 0 + (x^2 - y^2)\,(0) + (y^2 - z^2)\,(2y) + (-2xz - 2yz)\,(-2z)$$
$$= 2y^3 - 2yz^2 + 4xz^2 + 4yz^2$$
$$= 2y^3 + 2z^2\,(y + 2x)$$
$$= 2(2)^3 + 2(3)^2\,(2 + 2) \text{ for point } (1, 2, 3)$$

$$= 16 + 72$$
$$= 88$$

Likewise,

$$a_z = \frac{\partial w}{\partial t} + u\frac{\partial w}{\partial x} + v\frac{\partial w}{\partial y} + w\frac{\partial w}{\partial z}$$

$$= 0 + (x^2 - y^2)(-2z) + (y^2 - z^2)(-2z) + (-2xz - 2yz)(-2x - 2y)$$

$$= -2x^2z + 2y^2z - 2y^2z + 2z^3 + 4z(x + y)^2$$

$$= -2(1)^2(3) + 2(3)^3 + 4(3)(3)^2 \text{ for point } (1, 2, 3)$$

$$= -6 + 54 + 108$$

$$= 156$$

Hence, $\vec{a} = 14\hat{i} + 88\hat{j} + 156\hat{k}$

E3.11: If in a flow domain, $\phi = 3xy$ and $\psi = \dfrac{3\left(y^2 - x^2\right)}{2}$, compute the velocities at points (1, 3) and (3,4) and, hence, compute the discharge between the streamlines passing through these points. Also, compute the discharge on the basis of the stream functions at the two points.

Solution:

$$u = -\frac{\partial \phi}{\partial x} = -\frac{\partial \psi}{\partial y} = -3y$$

$$v = -\frac{\partial \phi}{\partial y} = \frac{\partial \psi}{\partial x} = -3x$$

$$u(1,3) = -9 \qquad \text{and} \qquad u(3,4) = -12$$

and $v(1,3) = -3$ and $v(3,4) = -9$

$\therefore$ Average value of $u = -10.5$

and average value of $v = -6$

Discharge between two streamlines is

$$q = d\psi = \frac{\partial \psi}{\partial x}\delta x + \frac{\partial \psi}{\partial y}\delta y$$

$$= v\,\delta x - u\,\delta y$$

Hence, discharge between the streamlines passing through the points (1, 3) and (3, 4) is

$$q = (-6)(2) - (-10.5)(1)$$

$$= -12 + 10.5$$

$$= -1.5 \text{ units}$$

The values of the stream function at the points (1, 3) and (3, 4) are

$$\psi\,(1,\,3) = 12 \text{ and } \psi\,(3,\,4) = 10.5$$

∴ $q = d\psi\ = \psi\,(3,\,4) - \psi\,(1,\,3) = 10.5 - 12 = -1.5$ units.

PROBLEMS

P3.1 A pipeline carries oil of specific gravity 0.9 at an average velocity of flow equal to 2.5 m/s through a 200 mm diameter pipe. At another section of the pipeline, the diameter of the pipe is 80 mm. Find the average velocity of the flow at this section and also the mass as well as volumetric rate of flow in the pipeline.

P3.2 Consider a cube with its edges, having length equal to 1 m, parallel to the coordinate axes and located in the first quadrant with one corner at the origin. The cube is a fluid element in a flow field,

$$\vec{V} = \left(3x\hat{i}\right) + \left(4y\hat{j}\right) + \left(-7z\hat{k}\right)$$

Find the flow through each face and, thus, determine if any mass is being accumulated within the cube if the fluid is of constant density.

P3.3 For a low Reynolds number (laminar) steady flow through a long tube, the axial velocity distribution is $v = C\,(R^2 - r^2)$ in which R is the radius of the tube and $r \le R$. Determine the total volumetric rate of flow (i.e., discharge) through the tube.

P3.4 Velocity $v\,(r)$ for turbulent flow of air in a 4.0 cm diameter tube is as follows:

r, cm	0	0.5	0.75	1.0	1.25	1.5	1.75	2.0
v, m/s	6.00	5.88	5.72	5.51	5.23	4.89	4.43	0.00

Estimate the value of the discharge through the given tube.

P3.5 An incompressible flow has the following three-dimensional velocity distribution

$$\vec{V} = 4xy^2\hat{i} + f\left(y\right)\hat{j} - zy^2\hat{k}$$

Determine f (y) that satisfies the continuity equation.

P3.6 Determine the unknown velocity u or v (ignoring constants of integration) that satisfy the continuity equation of two-dimensional incompressible flow for

    (a) $u = x^2y$ (b) $v = x^2y$

    (c) $u = x^2 - xy$ (d) $v = y^2 - xy$

P3.7 Consider the two-dimensional incompressible velocity potential $\phi = xy + x^2 - y^2$. Is it true that $\nabla^2\phi = 0$, and, if so, what does this mean? Determine the stream function $\psi(x,y)$ of this flow, and also find the equation of the streamline that passes through the point (2,1).

P3.8 Do the velocity distributions

    (i) $v = 3x\hat{i} + 4y\hat{j} - 10z\hat{k}$ and (ii) $v = 4x\hat{i} + 4y\hat{j} - 8z\hat{k}$

satisfy the mass conservation principle for incompressible flow?

P3.9 The potential function of a flow is $\phi\,(x,y) = x$. Determine the velocities $u(x,y)$ and $v(x,y)$ and plot the velocity field of the flow.

P3.10 The x and y components of fluid velocity in a two-dimensional flow field are, respectively, $u = x$ and $v = -y$. Determine the stream function and plot the streamlines for $\psi = 1$, $\psi = 2$, and $\psi = 3$. If a uniform flow described by $\psi = y$ is superimposed on the given flow, obtain the streamline pattern i.e., plot the streamlines for the resulting flow. Also, determine the stream function and velocity potential for the resulting flow.

P3.11 Determine the unknown velocity w or v (ignoring constants of integration) that satisfy the equation of three-dimensional incompressible continuity for

(a) $u = x^2yz$ $v = -y^2x$

(b) $u = 3z^2x$ $w = -z^3 + y^2$

Obtain the rotation and vorticity vector and also the conditions for this flow to be irrotational.

REFERENCES

(1) Fox, RW, and McDonald, AT: *Introduction to Fluid Mechanics*, 4th Edition, John Wiley and Sons, Inc., USA, 1994.

(2) Garde, RJ and Mirajgaokar, AG: *Engineering Fluid Mechanics*, Scitech Publications, Chennai, 2003.

(3) Massey, BS (revised by John Ward-Smith): *Mechanics of Fluids*, 7th Edition, Chennai Micro Print Pvt. Ltd., Chennai, India, 1998.

(4) Streeter, VL, Wylie, EB, and Bedford, KW: *Fluid Mechanics*, 9th Edition, McGraw-Hill Companies, Singapore, 1998.

(5) White, FM: *Fluid Mechanics*, 5th Edition, McGraw-Hill Companies, USA, 2003.

CHAPTER 4

Fluid Dynamics

4.1 NEWTON'S SECOND LAW OF MOTION FOR FLUIDS

The analysis of any fluid motion (as also the estimation of the forces exerted on a solid body by a fluid flowing around the body) primarily uses the Newton's second law of motion. This law, in its most general form, states that the net force acting on a body in any given direction equals the rate of change of momentum of the body in that direction(4). The specification of direction is essential since both momentum and force are vector quantities. When one is concerned with a collection of bodies of a system, such as a continuum of fluid particles, the law may be used for each of the bodies of the system individually. The resulting equations, when added together, yield total force in the chosen direction the same as the net force acting in the same direction at the boundaries of the system. All internal forces between the separate bodies occur in pairs of action and reaction and, hence, get cancelled and only the external boundary forces appear in the resulting final equation. For a fluid, the Newton's second law of motion (i.e., the momentum principle) can, therefore, be stated as follows:

The net external force acting on a fluid body in a certain direction is equal to the total rate of change of momentum of the fluid body in that very direction.

Momentum of an object or fluid particle is the product of its mass m and velocity v. The particles of a flowing fluid will possess momentum. When the velocity of the flowing fluid changes either in magnitude or in direction or in both, the momentum of the fluid particles would also change. As per the Newton's second law of motion, a force is needed to cause change in momentum. This force would be equal to the rate of change of momentum of the fluid particles. This force may be provided by the contact between the flowing fluid and a solid boundary or by one part of the flowing fluid acting on another part of the fluid. By the Newton's third law of motion, the fluid will exert an equal and opposite force on the solid boundary or the body of the fluid. Such dynamic forces are known as fluid dynamic forces or hydrodynamic (when fluid is water) forces.

4.2 MOMENTUM EQUATION FOR STEADY FLOW

Consider a stream-tube ABCD that constitutes a control volume, Fig. 4.2.1. Assume the flow to be steady so that

$$\rho_1 A_1 v_1 = \rho_2 A_2 v_2 = \dot{m}$$

in which $\dot{m}$ is the mass flow rate. The rate at which momentum enters the control volume ABCD through the control surface AB is $\rho_1 A_1 v_1 v_1$ or $\dot{m} v_1$ in the direction of motion. Likewise, the rate at which momentum leaves the control volume ABCD through the control surface CD is $\rho_2 A_2 v_2 v_2$ or $\dot{m} v_2$. Hence, the rate of change of momentum across the control volume and in the flow direction is $\dot{m} (v_2 - v_1)$. Hence, by the Newton's second law of motion, the resultant force F acting on the fluid element ABCD in the direction of motion is

$$F = \dot{m} (v_2 - v_1) \qquad \qquad ...(4.2.1)$$

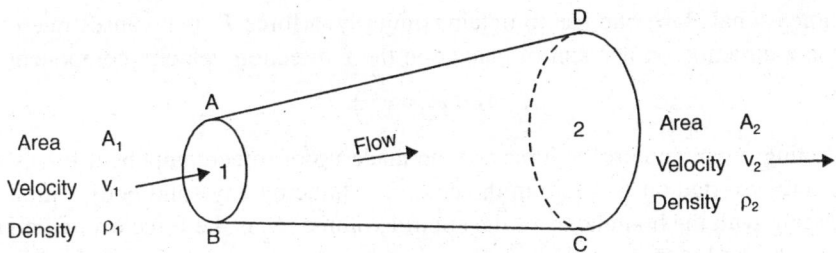

Fig. 4.2.1. Control volume for the momentum equation in a 1-D flow

By the Newton's third law of motion, the fluid ABCD will exert an equal and opposite reaction on its surroundings. Equation (4.2.1) is applicable to one-dimensional flow in which the velocities v_1 and v_2 were in the same direction. Consider a two-dimensional flow, Fig. 4.2.2, in which the velocities at the inlet and outlet sections are, respectively, v_1 (making an angle θ with the x-axis) and v_2 (making an angle ϕ with the x-axis). For such flow, it would be convenient to consider rate of change of momentum (or the external forces) in x- and y-directions. Thus,

$$F_x = \dot{m}(v_2 \cos\phi - v_1 \cos\theta) = \dot{m}(v_{x2} - v_{x1}) \qquad \qquad ...(4.2.2)$$

and

$$F_y = \dot{m}(v_2 \sin\phi - v_1 \sin\theta) = \dot{m}(v_{y2} - v_{y1}) \qquad \qquad ...(4.2.2)$$

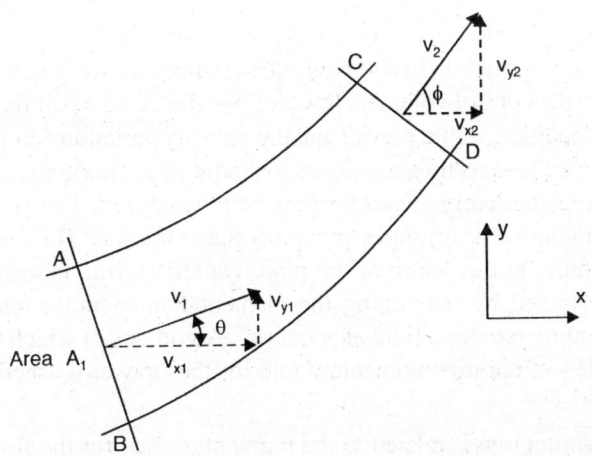

Fig. 4.2.2. Control volume for the momentum equation in a 2-D flow

Therefore, the resulting force, F (exerted by the surroundings on the fluid) is

$$F = \sqrt{F_x^2 + F_y^2} \qquad \qquad ...(4.2.4)$$

which acts in a direction that makes an angle α with the x-axis such that

$$\alpha = \tan^{-1}\left(\frac{F_y}{F_x}\right) \qquad \qquad ...(4.2.5)$$

Obviously, the fluid would exert a force on the surroundings that will be equal to F but in the opposite direction.

For three-dimensional flow, one has to obtain, similarly, a force F_z that causes rate of change of momentum in the z-direction on account of change in the z-direction velocity component. That is,

$$F_z = \dot{m}\left(v_{z2} - v_{z1}\right) \qquad \qquad ...(4.2.6)$$

The forces acting on any control volume will be made up of three component forces F_1, F_2, and F_3 (2). F_1 is the force exerted on the fluid in the control volume by any solid body within the control volume or coinciding with the boundaries of the control volume. F_2 is the force exerted on the fluid in the control volume by body forces such as gravity. F_3 is the force exerted on the fluid in the control volume by the fluid surrounding the control volume. Thus, for any given direction,

$$F = F_1 + F_2 + F_3 = \dot{m}\left(v_2 - v_1\right) \qquad \qquad ...(4.2.7)$$

Here, forces and velocities are, obviously, in the same direction. If R is the force exerted by the fluid on a solid body within the fluid or coinciding with the control surface, then, obviously,

$$R = -F_1 \qquad \qquad ...(4.2.8)$$

It is worth noting that the application of the momentum equation requires the details of flow only at the inlet and outlet sections of the chosen control volume and does not require the details of the flow inside the control volume. This feature of the momentum equation is, therefore, very useful in handling fluid flow problems involving considerable energy loss that is unknown to start with.

4.3 MOMENTUM CORRECTION FACTOR FOR ONE-DIMENSIONAL ANALYSIS

Application of the momentum principle to a stream-tube assumes uniform velocity distributions at the inlet and outlet sections of the control volume. However, for the flows of all real fluids in contact with a solid boundary, no-slip condition would prevail and the velocity variation across a flow section is non-uniform. This means that the flow section comprises a bundle of adjacent stream-tubes, each of cross-sectional area δA, which together carry the entire flow being analyzed. The velocity, in general varies from one stream-tube to another. For example, in a pipe, the velocity of flow would vary from zero at the pipe wall to the maximum at the centre of the pipe. Therefore, true momentum rate for any flow section can only be determined by integrating the elemental momentum rate over the entire flow section. That is, the momentum rate for a fluid element (of area δA and at which the velocity is v) would be $\rho(v\delta A)v$, i.e., $\rho v^2 \delta A$. Hence, the true momentum rate for the flow cross-section of area A would be

$\int_A \rho v^2 \, dA$. This true momentum rate is related to the momentum rate for the flow section, $\rho V^2 A$ calculated on the basis of the mean velocity V through a momentum correction factor β as follows:

True momentum rate = $\beta \times$ Mass per unit time $\times$ Mean velocity

i.e.,

$$\int_A \rho v^2 \, dA = \beta(\rho Q)V = \beta \rho A V^2$$

Thus,

$$\beta = \frac{\int_A \rho v^2 dA}{\rho A V^2} \qquad \qquad ...(4.3.1)$$

Equation (4.3.1) takes into account the variation of the mass density ρ as well. Therefore, it is applicable to the compressible fluids as well. For incompressible fluids, however,

$$\beta = \frac{1}{A} \int \left(\frac{v}{V}\right)^2 dA \qquad \qquad ...(4.3.2)$$

The average velocity can be computed from

$$V = \frac{\int_A v dA}{A} = \frac{1}{A} \int_A v \, dA \qquad \qquad ...(4.3.3)$$

The value of β would depend upon the shape of the flow cross-section and the velocity distribution. It should be noted that β would always be greater than 1. Further, for a velocity distribution that is closer to uniform velocity variation, β would only be slightly greater than unity. But, for more non-uniform velocity distribution, the values of β would be relatively higher.

4.4 DIFFERENTIAL MOMENTUM EQUATION

To derive the differential form of the momentum equation, one needs to apply the Newton's second law of motion to an infinitesimal fluid particle, Fig. 4.4.1, of mass dm and volume $\delta \forall = \delta x \, \delta y \, \delta z$. The forces acting on this fluid particle (or element of fluid) are body forces and surface forces. For example, stresses acting in the x-direction would cause surface forces in the x-direction. If the stresses, in the x-direction, at the centre of the fluid element are σ_{xx}, τ_{yx}, and τ_{zx}, then the stresses acting in the x-direction on all faces of the element, obtained by using Taylor series expansion about the centre of the fluid element, are as shown in Fig. 4.4.1. One can obtain the net surface force in the x-direction,

$$\delta F_{sx} = \left(\sigma_{xx} + \frac{\partial \sigma_{xx}}{\partial x} \frac{\delta x}{2}\right)\delta_y \delta_z - \left(\sigma_{xx} - \frac{\partial \sigma_{xx}}{\partial x} \frac{\delta x}{2}\right)\delta_y \delta_z$$

$$+ \left(\tau_{yx} + \frac{\partial \tau_{yx}}{\partial y} \frac{\delta y}{2}\right)\delta_x \delta_z - \left(\tau_{yx} - \frac{\partial \tau_{yx}}{\partial y} \frac{\delta y}{2}\right)\delta_x \delta_z$$

$$+ \left(\tau_{zx} + \frac{\partial \tau_{zx}}{\partial z} \frac{\delta z}{2}\right)\delta_x \delta_y - \left(\tau_{zx} - \frac{\partial \tau_{zx}}{\partial z} \frac{\delta z}{2}\right)\delta_x \delta_y$$

Thus,

$$\delta F_{sx} = \left(\frac{\partial \sigma_{xx}}{\partial x} + \frac{\partial \tau_{yx}}{\partial y} + \frac{\partial \tau_{zx}}{\partial z}\right)\delta x \, \delta y \, \delta z$$

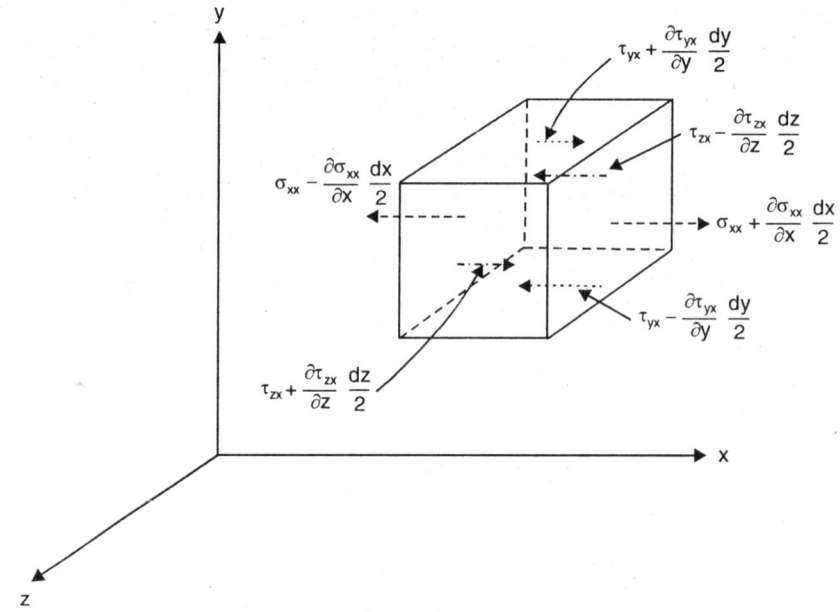

Fig. 4.4.1. Stresses in the x-direction on a parallelopiped element of fluid

On adding the body force due to gravity, the net force in the x-direction becomes

$$\delta F_x = \delta F_{xx} + \delta F_{Bx} = \left(\rho g_x + \frac{\partial \sigma_{xx}}{\partial x} + \frac{\partial \tau_{yx}}{\partial y} + \frac{\partial \tau_{zx}}{\partial z} \right) \delta x \, \delta y \, \delta z$$

Similarly, the net forces in the y- and z-directions would be

$$\delta F_y = \left(\rho g_y + \frac{\partial \tau_{xy}}{\partial x} + \frac{\partial \sigma_{yy}}{\partial y} + \frac{\partial \tau_{zy}}{\partial z} \right) \delta x \delta y \delta z$$

and

$$\delta F_z = \left(\rho g_z + \frac{\partial \tau_{xz}}{\partial x} + \frac{\partial \tau_{yz}}{\partial y} + \frac{\partial \sigma_{zz}}{\partial z} \right) \delta x \, \delta y \, \delta z$$

The terms in the bracket on the right hand side of the above three equations represent the force per unit volume in the x-, y-, and z-directions. The force per unit volume in a direction should equal the product of the mass density and acceleration in the corresponding direction (Eqs. (3.6.7 to 3.6.9)). Thus, the differential equations of motion are (3)

$$\rho \left(\frac{\partial u}{\partial t} + u \frac{\partial u}{\partial x} + v \frac{\partial u}{\partial y} + w \frac{\partial u}{\partial z} \right) = \left(\rho g_x + \frac{\partial \sigma_{xx}}{\partial x} + \frac{\partial \tau_{yx}}{\partial y} + \frac{\partial \tau_{zx}}{\partial z} \right) \qquad \text{...(4.4.1)}$$

$$\rho \left(\frac{\partial v}{\partial t} + u \frac{\partial v}{\partial x} + v \frac{\partial v}{\partial y} + w \frac{\partial v}{\partial z} \right) = \left(\rho g_y + \frac{\partial \tau_{xy}}{\partial x} + \frac{\partial \sigma_{yy}}{\partial y} + \frac{\partial \tau_{zy}}{\partial z} \right) \qquad \text{...(4.4.2)}$$

$$\rho \left(\frac{\partial w}{\partial t} + u \frac{\partial w}{\partial x} + v \frac{\partial w}{\partial y} + w \frac{\partial w}{\partial z} \right) = \left(\rho g_z + \frac{\partial \tau_{xz}}{\partial x} + \frac{\partial \tau_{yz}}{\partial y} + \frac{\partial \sigma_{zz}}{\partial z} \right) \qquad \text{...(4.4.3)}$$

Equations (4.4.1 to 4.4.3) are the differential equations of motion, obtained by applying the Newton's second law of motion, for any fluid (including non-Newtonian fluids) satisfying the continuum assumption. One can, however, use these equations for solving a fluid flow problem only after the stress terms are expressed in terms of velocity and pressure fields. For a Newtonian fluid, the viscous stress is proportional to the rate of shearing strain i.e., angular deformation rate. On substituting appropriate values of the stresses and rearranging the terms suitably, Eqs. (4.4.1 to 4.4.3) reduce to (5)

$$\rho\left(\frac{\partial u}{\partial t}+u\frac{\partial u}{\partial x}+v\frac{\partial u}{\partial y}+w\frac{\partial u}{\partial z}\right)=-\frac{\partial p}{\partial x}+\mu\left(\frac{\partial^2 u}{\partial x^2}+\frac{\partial^2 u}{\partial y^2}+\frac{\partial^2 u}{\partial z^2}\right) \qquad ...(4.4.4)$$

$$\rho\left(\frac{\partial v}{\partial t}+u\frac{\partial v}{\partial x}+v\frac{\partial v}{\partial y}+w\frac{\partial v}{\partial z}\right)=-\frac{\partial p}{\partial y}+\mu\left(\frac{\partial^2 v}{\partial x^2}+\frac{\partial^2 v}{\partial y^2}+\frac{\partial^2 v}{\partial z^2}\right) \qquad ...(4.4.5)$$

$$\rho\left(\frac{\partial w}{\partial t}+u\frac{\partial w}{\partial x}+v\frac{\partial w}{\partial y}+w\frac{\partial w}{\partial z}\right)=\rho g_z-\frac{\partial p}{\partial z}+\mu\left(\frac{\partial^2 w}{\partial x^2}+\frac{\partial^2 w}{\partial y^2}+\frac{\partial^2 w}{\partial z^2}\right) \qquad ...(4.4.6)$$

provided that the viscosity μ is constant, fluid flow is incompressible, z-axis is vertical, and the only forces that are acting on the fluid are due to gravity, pressure, and viscosity. Equations (4.4.4 to 4.4.6) are the equations of motion for Newtonian fluids and are known as the Navier-Stokes equations, named after C.L.M.H. Navier (1785–1836) and G.G. Stokes (1819–1903).

4.5 EULER'S EQUATIONS OF MOTION FOR THREE-DIMENSIONAL FLOW

Although all real fluids possess finite viscosity, there are many real flow conditions in which one may ignore the effects of viscosity. This is similar to neglecting friction in the analysis of some solid systems. Therefore, it is useful to study the dynamics of an ideal fluid that is both incompressible as well as inviscid. Since the shear stresses are absent in the flow of an ideal fluid, the analysis of the flow becomes simpler.

For inviscid or non-viscous or frictionless flow, $\mu = 0$ and the Navier-Stokes equations, Eqs. (4.4.4 to 4.4.6) reduce to the following Euler's equations of motion for three-dimensional flow:

$$\rho\left(\frac{\partial u}{\partial t}+u\frac{\partial u}{\partial x}+v\frac{\partial u}{\partial y}+w\frac{\partial u}{\partial z}\right)=-\frac{\partial p}{\partial x} \qquad ...(4.5.1)$$

$$\rho\left(\frac{\partial v}{\partial t}+u\frac{\partial v}{\partial x}+v\frac{\partial v}{\partial y}+w\frac{\partial v}{\partial z}\right)=-\frac{\partial p}{\partial y} \qquad ...(4.5.2)$$

$$\rho\left(\frac{\partial w}{\partial t}+u\frac{\partial w}{\partial x}+v\frac{\partial w}{\partial y}+w\frac{\partial w}{\partial z}\right)=\rho g_z-\frac{\partial p}{\partial z} \qquad ...(4.5.3)$$

4.5.1 Euler's Equations in Streamline Coordinates

Streamlines provide visual and graphical representation of a fluid flow. Even for unsteady flow, they provide a graphical representation of the instantaneous velocity field. Therefore, one may logically choose the distance along a streamline, say *s,* as one of the coordinates while deriving equations of motion. Other coordinate, say n, is taken as the distance normal to the streamline. Consider a small fluid element, Fig. 4.5.1, on a streamline in the x-z plane, at a point where the velocity vector is $\vec{V}$ (s,t) and the pressure is p. On applying the Newton's second law of motion in the stream-wise s-direction, one obtains,

$$\left(p-\frac{\partial p}{\partial s}\frac{\delta s}{2}\right)\delta n\delta y -\left(p+\frac{\partial p}{\partial s}\frac{\delta s}{2}\right)\delta n\delta y - \rho g\left(\delta s\delta n\delta y\right)\sin\theta = \rho\left(\delta s\delta n\delta y\right)a_s \qquad ...(4.5.4)$$

in which θ is the angle between the velocity vector and the horizontal (i.e., x-y plane) and a_s is the acceleration of the fluid element along the streamline, i.e., s-direction. Equation (4.5.4) simplifies to

$$-\frac{\partial p}{\partial s}-\rho g\sin\theta = \rho\, a_s \qquad ...(4.5.5)$$

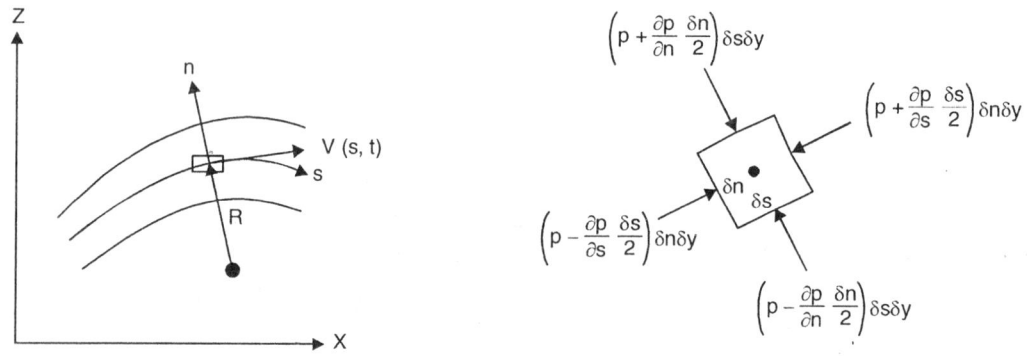

Fig. 4.5.1. Fluid element on a streamline

Since $\sin\theta=\dfrac{\partial z}{\partial s}$ and $a_s =\dfrac{\partial V}{\partial t}+V\dfrac{\partial V}{\partial s}$ from Eq. (3.6.1), Eq. (4.5.5) becomes

$$-\frac{1}{\rho}\frac{\partial p}{\partial s}-g\frac{\partial z}{\partial s} =\frac{\partial V}{\partial t} + V\frac{\partial V}{\partial s} \qquad ...(4.5.6)$$

Equation (4.5.6) is the Euler's equation of motion in the stream-wise direction with the z-axis directed vertically upward. Considering steady flow (i.e., $\dfrac{\partial V}{\partial t}=0$), Eq. (4.5.6) reduces to

$$-\frac{1}{\rho}\frac{\partial p}{\partial s} -g\frac{\partial z}{\partial s} = V\frac{\partial V}{\partial s} \qquad ...(4.5.7)$$

This means that, if one neglects body force term $\left(g\dfrac{\partial z}{\partial s}\right)$, decrease in velocity results in an increase in pressure along the streamline, i.e., in the flow direction.

Similarly, on applying the Newton's second law of motion in the normal direction, one obtains for the fluid element,

$$\left(p-\frac{\partial p}{\partial n}\frac{\delta n}{2}\right)\delta s\delta y -\left(p+\frac{\partial p}{\partial n}\frac{\delta n}{2}\right)\delta s\delta y - \rho g\left(\delta s\delta n\delta y\right)\cos\theta = \rho\left(\delta s\delta n\delta y\right)a_n \qquad ...(4.5.8)$$

in which a_n is the familiar centripetal acceleration of the fluid element in the n-direction towards the centre of curvature of the streamline. Equation (4.5.8) simplifies to

$$-\frac{1}{\rho}\frac{\partial p}{\partial n} - g\cos\theta = a_n$$

Since $\cos\theta = \dfrac{\partial z}{\partial n}$ and $a_n = -\dfrac{V^2}{R} + \dfrac{\partial v_n}{\partial t}$ from Eq. (3.6.2), one gets

$$-\frac{1}{\rho}\frac{\partial p}{\partial n} - g\frac{\partial z}{\partial n} = \frac{\partial v_n}{\partial t} - \frac{V^2}{R} \qquad \qquad ...(4.5.9)$$

in which v_n is the velocity component of V in the n-direction. For steady flow $\left(i.e., \dfrac{\partial v_n}{\partial t} = 0\right)$ in a

horizontal plane $\left(i.e., \dfrac{\partial z}{\partial n} = 0\right)$, Eq. (4.5.9) reduces to

$$-\frac{1}{\rho}\frac{\partial p}{\partial n} = \frac{V^2}{R} \qquad \qquad ...(4.5.10)$$

Therefore, pressure increases in the direction outward from the centre of curvature. When streamlines are straight $(R \approx \infty)$, there is no pressure variation normal to the streamlines.

4.5.2 Euler's Equation of Motion for Flow in a Stream-tube

Consider a fluid element so that it occupies part of a stream-tube of any shape but of small cross-section, Fig. 4.5.2. The ends of the element are, however, plane and perpendicular to the central streamline of the stream-tube. Assuming that the flow is steady and non-viscous, the only forces that remain to be included in the analysis are those due to (4): (i) pressure of the fluid on all surfaces of the chosen element, and (ii) gravity. The length δs of the element is sufficiently small so that curvature of the streamlines over the distance δs can be neglected. Here, s represents the distance measured along the

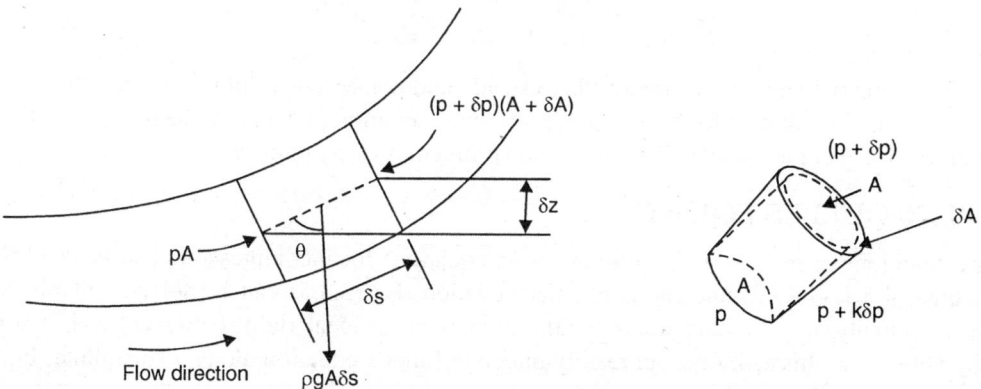

Fig. 4.5.2. Fluid element in a stream-tube (4)

steam-tube in the direction of flow. The flow is assumed to be steady and, hence, the variables such as pressure, velocity etc. may be regarded as dependent on s only. Pressures and pressure forces acting on the ends of the element are quite obvious and are as shown in Fig. 4.5.2. Sides of the element would be subjected to a varying pressure as the pressure varies along the length of the stream-tube. Since pressures at the ends of the element are p and $p + \delta p$, one can adopt a mean value $p + k\delta p$ as the pressure acting on the sides of the element. Here, k is less than unity. This pressure would cause a force on the sides of the element. This force will have a component in the flow direction as $(p+k\delta p)\delta A$ as the force in any direction is given by the product of the pressure and the projected area perpendicular to that direction. Hence, the total force acting on the fluid element, Fig. 4.5.2, in the direction of flow is

$$pA - (p+\delta p)(A + \delta A) + (p+k\delta p)\,\delta A - \rho g A \delta s \cos\theta$$

On neglecting the products of small quantities, this force is

$$-A\delta p - \rho g A \delta s \, Cos\,\theta$$
$$= -A\delta p - \rho g A \delta z$$

From the Newton's second law of motion, this net force must be equal to the mass of the element, $\rho A \delta s$ multiplied by the acceleration in the direction of force, i.e.,

$$\rho A \delta s \left(\frac{dV}{dt} \right)$$

$$= \rho A \delta s \left(\frac{\partial V}{\partial t} + V \frac{\partial V}{\partial s} \right)$$

$$= \rho A \delta s \, V \frac{dV}{ds}$$

as the flow is assumed to be steady and V is a function of s only for the stream-tube. Hence,

$$-A\delta p - \rho g A \delta z = \rho A \delta s \, V \frac{dV}{ds}$$

On dividing by $\rho A \delta s$ and taking the limit $\delta s \to 0$, one obtains

$$\frac{1}{\rho}\frac{dp}{ds} + V\frac{dV}{ds} + g\frac{dz}{ds} = 0 \qquad \qquad ...(4.5.11)$$

This differential equation is applicable to steady and non-viscous fluid flows in a stream-tube and referred to as Euler equation for flow in a stream- tube. Equation (4.5.11) can be integrated if ρ is either constant (as for incompressible fluid) or a known function of pressure p.

4.6 BERNOULLI'S EQUATION

Euler's equations of motion (i.e., the momentum equations for incompressible and inviscid fluid motion), Eqs. (4.5.1 – 4.5.3) and the continuity equation, Eq (3.8.5) can be solved suitably to obtain complete velocity and pressure fields for the motion of an ideal fluid. However, such a solution is usually not simple. Instead, one can readily integrate Euler's equation along a streamline, Eq. (4.5.7), for the steady conditions. Equation (4.5.7) is

$$-\frac{1}{\rho}\frac{\partial p}{\partial s} - g\frac{\partial z}{\partial s} = V\frac{\partial V}{\partial s} \qquad \qquad ...(4.6.1)$$

On multiplying Eq. (4.6.1) by δs (i.e., the distance travelled by a fluid particle along a streamline), one gets

$$-\frac{1}{\rho}\frac{\partial p}{\partial s}\delta s - g\frac{\partial z}{\partial s}\delta s = V\frac{\partial V}{\partial s}\delta s \qquad \text{...(4.6.2)}$$

If a fluid particle moves a small distance δs ($\delta s \to 0$) along a streamline, then changes in pressure dp, elevation dz, and velocity dV are, respectively, $\frac{\partial p}{\partial s}\delta s$, $\frac{\partial z}{\partial s}\delta s$, and $\frac{\partial V}{\partial s}\delta s$. Thus, when $\delta s \to 0$, Eq. (4.6.2) reduces to

$$-\frac{dp}{\rho} - g\,dz = V dV$$

or

$$\frac{dp}{\rho} + V dV + g\,dz = 0 \qquad \text{...(4.6.3)}$$

along a streamline. Integration of Eq. (4.6.3) yields

$$\int\frac{dp}{\rho} + \frac{V^2}{2} + gz = \text{constant} \qquad \text{...(4.6.4)}$$

Equation (4.6.4) is valid along a streamline and can be applied only if the relation between the pressure p, and the density ρ is known. For the case of incompressible flow, however, ρ is constant. Therefore, Eq. (4.6.4) yields

$$\frac{p}{\rho} + \frac{V^2}{2} + gz = \text{constant} \qquad \text{...(4.6.5)}$$

Likewise, Euler's equation of motion for flow in a stream-tube, Eq. (4.5.11), can also be integrated to yield, once again,

$$\frac{p}{\rho} + \frac{V^2}{2} + gz = \text{constant} \qquad \text{...(4.6.5)}$$

provided that ρ is constant.

Equation (4.6.5) is named as Bernoulli's equation in honour of a Swiss mathematician, Daniel Bernoulli (1700 – 1782). The second term, $\frac{V^2}{2}$, in Eq. (4.6.5) represents the ratio of the kinetic energy of a small fluid element to the mass of the element and is, therefore, the kinetic energy per unit mass of the flowing fluid. The third term gz, similarly, represents energy per unit mass and corresponds to the energy given to unit mass by raising its elevation by an amount z. Likewise, the term p/ρ represents the ability of unit mass of the fluid to do work by virtue of its pressure p. In Art. 4.5.2, it was observed that the contribution of the pressure force to the total force acting on the fluid element, Fig. 4.5.2, is $(-A\delta p)$ in the direction of motion. This force does the work on the fluid element when it moves the element by a distance δs, i.e., from a point where the pressure is p to another point where it is $p + \delta p$. The work done by the force equals $(-A\delta p)\,\delta s$. Since the mass of the element is $\rho A\delta s$, the work done by the force per unit mass of the fluid is

$$-\frac{A\delta p \delta s}{\rho A \delta s} = -\frac{\delta p}{\rho}$$

If the element were to move from a point where the pressure is p_1 to another point where the pressure is p_2, then the work done by the pressure force per unit mass of incompressible fluid (i.e., ρ is constant) is

$$\int_{p_1}^{p_2} -\frac{dp}{\rho} = \frac{p_1 - p_2}{\rho}$$

The term p/ρ in Bernoulli's equation, Eq. (4.6.5), can, therefore, be interpreted as the work that would be done on unit mass of the fluid by pressure force in moving the fluid from a point where the pressure is p_1 to another point where the pressure is p_2 (4). The work is done simply because of the flow of fluid.

Bernoulli's equation, Eq. (4.6.5) is a very useful equation for fluid flow analysis as it relates pressure changes to velocity and elevation changes along a streamline. However, while using the Bernoulli's equation, one must always keep in mind that the flow is: (i) along a streamline or along a stream-tube across which the flow parameters do not vary, (ii) steady, (iii) frictionless, and (iv) incompressible. In general, the constant appearing in the Bernoulli's equation would have different values along different streamlines.

Euler's equations for three-dimensional flow can also be integrated along a streamline to yield Bernoulli's equation subjected to the same restrictions of steady, incompressible and frictionless flow.

The Euler equation, Eq. (4.5.11) or (4.6.3) can also be integrated between any two sections (or points) 1 and 2 on a stream-tube (or streamline) to yield Bernoulli's equation of the form

$$\frac{p_2 - p_1}{\rho} + \frac{V_2^2 - V_1^2}{2} + g(z_2 - z_1) = 0 \qquad \text{...(4.6.6)}$$

or

$$\frac{p_1 - p_2}{\rho} + g(z_1 - z_2) = \frac{V_2^2 - V_1^2}{2} \qquad \text{...(4.6.7)}$$

This means that between two sections (or points) 1 and 2 in a frictionless flow, the increase of kinetic energy per unit mass is equal to the work done on the fluid per unit mass by the pressure and gravity forces.

Equation (4.6.6) can, alternatively, be written as

$$\frac{p_1}{\rho g} + \frac{V_1^2}{2g} + z_1 = \frac{p_2}{\rho g} + \frac{V_2^2}{2g} + z_2 \qquad \text{...(4.6.8)}$$

Equation (4.6.8) states that in a fluid flow, the sum of the pressure (or static) head, kinetic (or velocity) head, and datum (or elevation or gravity) head remains the same at all points along a streamline (or stream-tube) provided that the flow is incompressible, frictionless, and steady (i.e., steady flow of an ideal fluid). Hence, these three quantities are of the same kind.

The term p_1/ρ in the Bernoulli's equation is sometimes termed pressure energy since other two terms of the equation are energy terms. This is misleading as the fluid, in fact, does not even possess the pressure energy as it possesses the kinetic energy. The terms in the Bernoulli's equation, therefore, do not represent energy *stored* in the fluid but rather the total energy *transmitted* by the fluid(4).

Bernoullis' equation is, strictly speaking, applicable for a steady, incompressible, and frictionless flow along a streamline (or stream-tube). With care, however, these assumptions can be relaxed and the equation can be used for real fluid flows. For example, for an outflow from a reservoir, where the energy content everywhere is the same, one can choose two points 1 and 2 (for application of Bernoullis' equation) on different streamlines as the constant of integration does not change from one streamline to another. Likewise, when the change in pressure is only a small fraction of the absolute pressure, the flow can be considered incompressible even if the flowing fluid is gas. By adding a suitable energy loss (or gain) term in the Bernoulli's equation, it can be made applicable to real fluid flows as well. For example, if there occurs a head loss h_L or gain in head h_g between sections 1 and 2, Eq. (4.6.8) can be modified to

$$\frac{p_1}{\rho g} + \frac{V_1^2}{2g} + z_1 + h_g = \frac{p_2}{\rho g} + \frac{V_2^2}{2g} + z_2 + h_L \qquad ...(4.6.9)$$

Bernoulli's equation can also be applied, without any appreciable error, to unsteady flows with gradually changing conditions such as emptying of a reservoir.

It should be noted that integration of Eq. (4.5.6) would have yielded

$$\int_1^2 \frac{\partial V}{\partial t} ds + \int_1^2 \frac{dp}{\rho} + \frac{1}{2}\left(V_2^2 - V_1^2\right) + g(z_2 - z_1) = 0 \qquad ...(4.6.10)$$

Equation (4.6.10) would be valid for unsteady as well as compressible fluid flows that are friction-less. In order to make Eq. (4.6.10) applicable to such flows in which there may be either head loss h_L or gain in head h_g, Eq. (4.6.10) is modified to

$$\int_1^2 \frac{\partial V}{\partial t} ds + \int_1^2 \frac{dp}{\rho} + \frac{1}{2}\left(V_2^2 - V_1^2\right) + g(z_2 - z_1) + h_L - h_g = 0 \qquad ...(4.6.11)$$

Continuity, energy, and momentum equations play important role in the analysis of any fluid flow problem. Energy and momentum equations are complimentary to each other in the sense that complete information may not be obtainable by using only one of the two equations. One of the most common applications of the momentum equation is for flows (such as flow in an expanding pipe, or hydraulic jump etc.) for which energy loss is neither known and nor is negligible. In such cases, the momentum equation yields information that enables use of the energy equation for obtaining the energy loss.

4.7 ENERGY CORRECTION FACTOR FOR ONE-DIMENSIONAL ANALYSIS

In analysing several fluid flow problems, one often assumes that the flow is one-dimensional. This assumption implies that the entire flow section is treated as the cross-section of a large single stream-tube (comprising several small individual stream-tubes) with uniform velocity distribution over the stream-tube cross-section. Accordingly, the kinetic energy per unit mass is computed as $V^2/2$ where, V is the average velocity over the cross-section and equals the total discharge Q divided by the cross-sectional area of the flow. This would be completely justified only if

$$(\Sigma m)V^2 = (\Sigma m v^2) \qquad ...(4.7.1)$$

in which m and v represent, respectively, the mass and velocity of the flowing fluid in a short length of small individual stream-tube and V is the average velocity over the entire cross-section of the flow or large single stream-tube. Equation (4.7.1) is satisfied only if the velocity distribution over the cross-section of the flow is uniform. That is, at all points in the cross-section of the flow, the velocity remains the same. In real fluid flow, this will never be so due to the 'no-slip' condition at the boundaries on account of non-zero viscosity. The true kinetic energy per unit mass is, therefore, obtained by multiplying the computed value (on the basis of average velocity V) of the kinetic energy with a correction factor α that is termed as energy correction factor which is estimated as follows:

Consider a small fluid element of area δA across which there is no variation of the velocity v. The mass rate of flow through this element is, therefore, $\rho v \delta A$ and the kinetic energy passing through the element in unit time is $\frac{1}{2}(\rho v \delta A)v^2$. Hence, the true total kinetic energy passing through the entire flow cross-section in unit time is $\int_A \frac{1}{2}\rho v^3 dA$. The kinetic energy passing through the entire flow cross-section in unit time is $\frac{1}{2}(\rho A V)V^2$, if computed on the basis of the average velocity V over the entire flow cross-section. The energy correction factor α is, therefore,

$$\alpha \frac{1}{2}\rho A V^3 = \int_A \frac{1}{2}\rho v^3 dA$$

or

$$\alpha = \frac{1}{A}\int_A \left(\frac{v}{V}\right)^3 dA \qquad ...(4.7.2)$$

The average velocity can be computed from Eq.(4.3.3).

The energy correction factor α, like the momentum correction factor β, depends on the shape of the flow section and the velocity variation over the section and is always greater than unity. For a velocity distribution that is closer to uniform velocity distribution, the value of α would be marginally higher than unity.

The energy and momentum correction factors can also be approximated as(1)

$$\alpha = 1 + 3\varepsilon^2 - 2\varepsilon^3$$

and

$$\beta = 1 + \varepsilon^2$$

Here,

$$\varepsilon = \frac{U_m}{U} - 1$$

in which U_m and U are, respectively, the maximum and average velocities.

EXAMPLES

E4.1.   In a fluid flow for which $\vec{V} = 112.5\left(y^2\,\hat{i} + x^2\,\hat{j}\right)$ (in m/s), determine the pressure gradient at a point (1.25, 2). Assume specific gravity of the fluid as 1.4 and neglect viscous effects.

Solution: Given $\vec{V} = 112.5\left(y^2\,\hat{i} + x^2\,\hat{j}\right)$

$\therefore$ $\qquad\qquad u = 112.5\ y^2,\ v = 112.5\ x^2$ and $w = 0$

For these values, Eqs. (4.5.1) to (4.5.3) give

$$\rho\left(\frac{\partial u}{\partial t}+u\frac{\partial u}{\partial x}+v\frac{\partial u}{\partial y}+w\frac{\partial u}{\partial z}\right)=\rho v\frac{\partial u}{\partial y}=-\frac{\partial p}{\partial x}$$

$$\rho\left(\frac{\partial v}{\partial t}+u\frac{\partial v}{\partial x}+v\frac{\partial v}{\partial y}+w\frac{\partial v}{\partial z}\right)=\rho u\frac{\partial v}{\partial x}=-\frac{\partial p}{\partial y}$$

$$\rho\left(\frac{\partial w}{\partial t}+u\frac{\partial w}{\partial x}+v\frac{\partial w}{\partial y}+w\frac{\partial w}{\partial z}\right)=0=-\rho g-\frac{\partial p}{\partial z}$$

Hence, for the given velocity field and at the given point (1.25, 2)

$$(1.4 \times 1000)\ (112.5x^2)\ (225\ y) = -\frac{\partial p}{\partial x}$$

$\therefore\qquad \dfrac{\partial p}{\partial x}=(-)354.375\times10^5\ yx^2 =-1.11\times10^8$

Similarly, $(1.4 \times 1000)\ (112.5y^2)\ (225\ x) = -\dfrac{\partial p}{\partial y}$

$\therefore\qquad \dfrac{\partial p}{\partial y}=(-)354.375\times10^5\ xy^2 =-1.77\times10^8$

Similarly, $0 = -1.4\ (1000)\ (9.8)-\dfrac{\partial p}{\partial z}$

$\therefore\qquad \dfrac{\partial p}{\partial z}=-13720 =-0.14\times10^5$

$\therefore$ Pressure gradient vector is

$$- 1.11 \times 10^8\ \hat{i} - 1.77 \times 10^8\ \hat{j} - 0.14 \times 10^5\ \hat{k}$$

E4.2. Calculate the average velocity, momentum correction factor, and energy correction factor for the following velocity distribution in a pipe:

$$\frac{v}{v_{max}}=\left[1-\left(\frac{r}{R}\right)^2\right]$$

in which r is the radial distance and R is the radius of the pipe and $v = v_{max}$ at $r = 0$.

Solution: From Eq. (4.3.3), the average velocity of flow,

$$V=\frac{1}{A}\int_A v\,dA$$

$$= \frac{1}{\pi R^2} \int_0^R v_{max} \left[1 - \left(\frac{r}{R} \right)^2 \right] 2\pi r \, dr$$

$$= \frac{2}{R^2} v_{max} \int_0^R \left[r - \left(\frac{r^3}{R^2} \right) \right] dr$$

$$= \frac{2}{R^2} v_{max} \left[\frac{R^2}{2} - \frac{R^2}{4} \right]$$

$$= \frac{v_{max}}{2}$$

Momentum correction factor from Eq. (4.3.1)

$$\beta = \frac{1}{A V^2} \int_A v^2 \, dA$$

Now

$$\int_A v^2 \, dA = v_{max}^2 \int_0^R \left[1 - \left(\frac{r}{R} \right)^2 \right]^2 2\pi r \, dr$$

$$= 2\pi v_{max}^2 \int_0^R \left(r + \frac{r^5}{R^4} - 2\frac{r^3}{R^2} \right) dr$$

$$= 2\pi v_{max}^2 \left(\frac{R^2}{2} + \frac{R^2}{6} - \frac{R^2}{2} \right)$$

$$= \frac{1}{3} \pi R^2 v_{max}^2$$

$$= \frac{4}{3} A V^2$$

∴

$$\beta = \frac{1}{A V^2} \int_A v^2 \, dA = \frac{4}{3} = 1.333$$

Likewise, energy correction factor from Eq. (4.7.2)

$$\alpha = \frac{1}{A V^3} \int_A v^3 \, dA$$

$$= \frac{v_{max}^3 (2\pi)}{\pi R^2 \left(\dfrac{v_{max}}{2} \right)^3} \int_0^R \left[1 - \left(\frac{r}{R} \right)^2 \right]^3 r \, dr$$

$$= \frac{16}{R^2} \int_0^R \left[1 - 3\left(\frac{r}{R}\right)^2 + 3\left(\frac{r}{R}\right)^4 - \left(\frac{r}{R}\right)^6 \right] r \, dr$$

$$= \frac{16}{R^2} \left[\frac{R^2}{2} - \frac{3}{4}R^2 + \frac{1}{2}R^2 - \frac{1}{8}R^2 \right]$$

$$= \frac{16}{R^2} \left[\frac{1}{8}R^2 \right]$$

$$= 2$$

PROBLEMS

P4.1 Using the Navier–Stokes equations, derive the equations for one-dimensional steady, viscous, incompressible flow.

P4.2 Oil flows through a 0.8 m diameter pipeline. The flow in the pipeline is laminar and the velocity at any radius r is given by $u = (0.5 - 15r^2)$ m/s. Calculate (a) the volumetric rate of flow, (b) the mean velocity, (c) the momentum correction factor, and (d) the energy correction factor.

P4.3 A liquid flows through a circular pipe whose diameter is 0.6 m. Measurements of velocity taken at intervals along a diameter are:

Distance from wall, m	0	0.05	0.1	0.2	0.3
Velocity, m/s	0	2.0	3.8	4.6	5.0

Distance from wall, m	0.4	0.5	0.55	0.6
Velocity, m/s	4.5	3.7	1.6	0

Draw the velocity profile and calculate the
(a) mean velocity,
(b) momentum correction factor, and
(c) energy correction factor.

P4.4 Calculate the mean velocity and the momentum and energy correction factors for a velocity distribution in a circular pipe given by $\left(\dfrac{v}{v_0}\right) = \left(\dfrac{y}{R}\right)^{1/n}$, where v is the velocity at a distance y from the wall of the pipe, v_0 is the centreline velocity, R the radius of the pipe, and n a suitable number.

REFERENCES

(1) Chow, VT, Open-*Channel Hydraulics,* McGraw-Hill Book Co., 1959.
(2) Douglas, JF, Gasiorek, J.M., and Swaffield, J.A., *Fluid Mechanics* (4th Ed.), Pearson Education (Singapore) Pvt. Ltd., 2002.
(3) Fox, RW, and McDonald A.T., *Introduction to Fluid Mechanics*, (4th Ed.), John Wiley and Sons, Singapore.
(4) Massey, BS (revised by John Ward-Smith): *Mechanics of Fluids*, 7th Edition, Chennai Micro Print Pvt. Ltd., Chennai, India, 1998.
(5) White, FM, "*Fluid Mechanics*", 5th Edition, McGraw-Hill Companies, USA, 2003.

CHAPTER 5

Applications of Steady Flow Equations

5.1 STEADY FLOW EQUATIONS

Equations governing the motion of a fluid have been derived in Chapters 3 and 4. Of these equations, Eqs. (3.7.3, 3.7.4, 4.2.1, 4.6.8 and 4.6.9) governing the steady flow of a fluid are as follows:

Continuity equation: $$\rho_1 V_1 A_1 = \rho_2 V_2 A_2 = \dot{m} \qquad \qquad ...(5.1.1)$$

Continuity equation for incompressible fluids: $A_1 V_1 = A_2 V_2 = Q$ ...(5.1.2)

Momentum equation: $$F = \dot{m}(V_2 - V_1) \qquad \qquad ...(5.1.3)$$

Bernoulli's equation: $$\frac{p_1}{\rho g} + \frac{V_1^2}{2g} + z_1 = \frac{p_2}{\rho g} + \frac{V_2^2}{2g} + z_2 \qquad \qquad ...(5.1.4)$$

Modified Bernoulli's Equation: $$\frac{p_1}{\rho g} + \frac{V_1^2}{2g} + z_1 + h_g = \frac{p_2}{\rho g} + \frac{V_2^2}{2g} + z_2 + h_L \qquad \qquad ...(5.1.5)$$

Each of the terms in these equations is identified by the corresponding quantity at the inlet and outlet ends of the control volume.

5.2 STATIC, DYNAMIC, AND STAGNATION PRESSURES

The Bernoulli's equation, Eq. (5.1.4), can also be written as

$$p_1 + \frac{1}{2}\rho V_1^2 + \rho g z_1 = p_2 + \frac{1}{2}\rho V_2^2 + \rho g z_2 \qquad \qquad ...(5.2.1)$$

Each term of Eq. (5.2.1) has the dimensions of pressure. The first term p is termed the *static pressure* that can be measured by a piezometer or manometer or some other suitable device. The second term in Eq. (5.2.1) is termed the *dynamic pressure*. In Fig. 5.2.1, the piezometer at section 1 measures the static

102

pressure head $h\left(=\dfrac{p}{\rho g}\right)$ corresponding to the static pressure p at section 1. This is also the piezometeric

head with the axis of the pipe as the chosen datum. The fluid at point 2, i.e., the tip of an L-shaped tube (known as total head tube) pointing upstream, and also in the tube, will be stationary. This means that the point 2 is a stagnation point and H is the *stagnation pressure head H*, which is the sum of the static

pressure head $h\left(=\dfrac{p}{\rho g}\right)$ and the dynamic pressure head$\left(=\dfrac{V^2}{2g}\right)$.

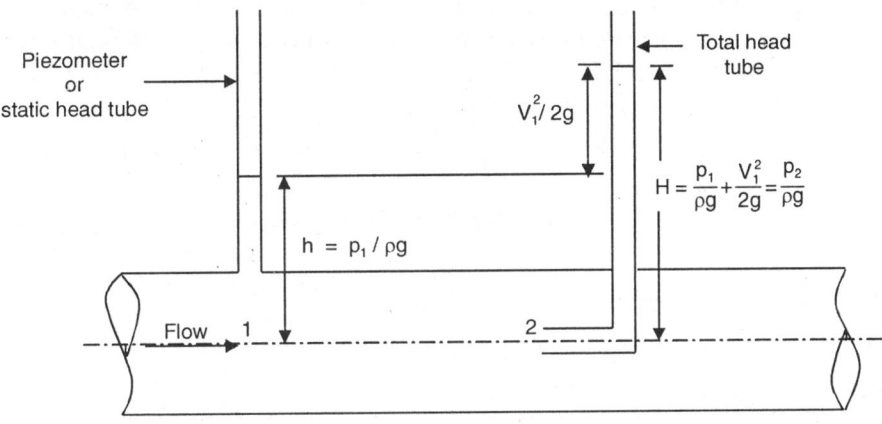

Fig. 5.2.1. Static and stagnation pressures

On applying Bernoulli's equation between points 1 and 2 and assuming that $z_1 = z_2$ and using $V_2 = 0$, one gets

$$p_2 = p_1 + \frac{1}{2}\rho V_1^2 = \rho g H \qquad \qquad ...(5.2.2)$$

This means that the stagnation pressure p_2 is greater than the static pressure p_1 by an amount equal to

the dynamic pressure $\dfrac{1}{2}\rho V_1^2$.

The third term on both sides of Eq. (5.2.1), $\rho g z$ or γz represents the change in pressure due to variations of potential energy of the fluid on account of elevation changes and may be termed hydrostatic pressure due to its similarity with regard to hydrostatic pressure variation. The sum of the static pressure, dynamic pressure and hydrostatic (or potential or datum) pressure is termed the total pressure. The Bernoulli's equation, therefore, states that the total pressure (or the total head) remains constant along a streamline if the flow is steady, incompressible and inviscid (i.e., non-viscous).

5.3 PITOT- STATIC TUBE

From Eq. (5.2.2), it is seen that the difference of the stagnation pressure and static pressure at a given location in a fluid flow is the dynamic pressure. Pitot-static tube uses this fact in order to enable one to measure the velocity at a point. From Eq. (5.2.2), one obtains

$$V_1 = \sqrt{\frac{2(p_2 - p_1)}{\rho}} \qquad\qquad ...(5.3.1)$$

or

$$V_1 = \sqrt{2g(H - h)} \qquad\qquad ...(5.3.2)$$

This is the Pitot formula, named after the French engineer, Henri de Pitot (1695-1771), who designed the device in 1732 for measuring velocities in river Seine (2). Pitot-static tube (or simply Pitot tube), Fig. 5.3.1, consists of two concentric tubes that are attached to two pressure gauges (or manometer) so that the difference of the stagnation pressure and static pressure at a given point can be measured. The inner tube is the total head tube and measures the stagnation pressure head while the outer tube is the static head tube and measures the static pressure head. Therefore, using Eq. (5.3.1 or 5.3.2), one can compute the velocity at any location in the flow field.

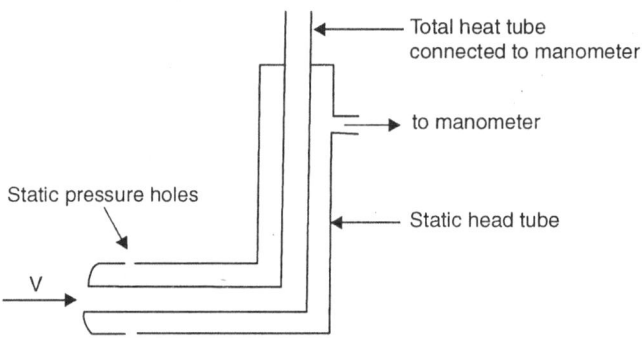

Fig. 5.3.1. Line sketch of a Pitot–static tube

Equation (5.3.2) assumes negligible elevation head difference between the total head tube and static pressure head tube as also the negligible friction loss that is true when the Reynolds number (i.e., Vd/v; Chapter-6) based on Pitot tube diameter is less than 1000. For error-free observations, the Pitot tube must be aligned with the flow direction. Very low flow velocity would develop small pressure difference that may be beyond the resolution of most pressure gauges. Therefore, Pitot tube is not suitable for measuring low velocity. Further, because of the slow response of the fluid-filled tubes that connect the Pitot tube with pressure sensors, like manometer, the Pitot tube is not suitable for fluctuating flow measurements.

5.4 HYDRAULIC AND ENERGY GRADE LINES

A useful visual and graphical representation of the Bernoulli's equation can be obtained by sketching,

along the flow, lines of piezometric pressure head $\left(= z + \dfrac{p_0}{\rho g} \right)$ and total head $\left(= z + \dfrac{p_0}{\rho g} + \dfrac{V^2}{2g} \right)$ that are,

respectively, known as *hydraulic grade line* (HGL) and *energy line* or *energy grade line* (EGL). The elevation of the EGL can be obtained by measuring the stagnation pressure with a total head tube, Fig. 5.2.1. The piezometer or static pressure tube, Fig. 5.2.1, measures the sum of the elevation and pressure

heads, known as piezometric head, i.e., $z + \dfrac{p_0}{\rho g}$. Thus, EGL is the locus of elevations provided by a series of total head tubes while HGL is the locus of elevations provided by the static pressure tubes. In a frictionless flow with no work done on or by the flowing fluid, the EGL would remain constant along the streamline or stream-tube or flow, Fig. 5.4.1. However, the elevation, pressure head, and kinetic (or velocity) head may vary along the flow. Obviously, the EGL lies above the HGL and the distance between the two at any location along the flow equals the velocity head at that location. In real fluid flow, however, the EGL would fall due to frictional losses along the flow and would rise at the location of pump supplying energy to the flow. On the other hand, HGL may fall or rise depending upon the flow cross-section. In case of negative gauge pressures, HGL would fall below the axis (or centre line) of the flow section.

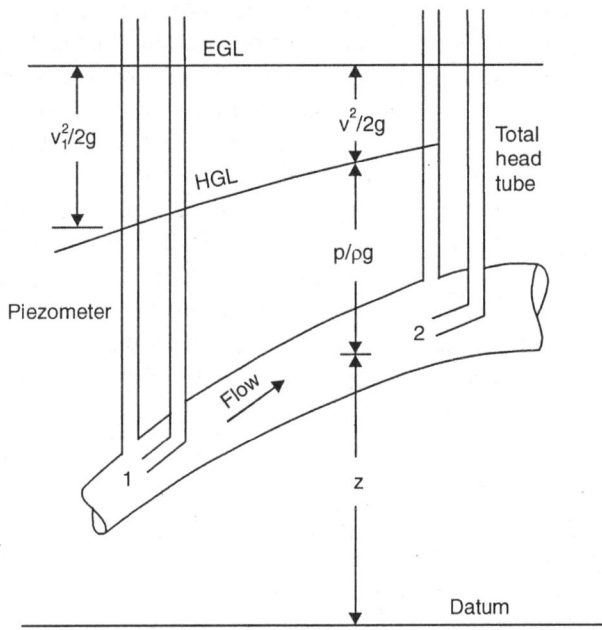

Fig. 5.4.1. HGL and EGL for flow in a pipe

5.5 FLOW THROUGH A SHARP-EDGED ORIFICE

An *orifice* is an aperture or opening through which a fluid passes. Generally, it is made in the wall or base of a container. An orifice, to be used for measuring discharge through it, would have a sharp edge, Fig. 5.5.1, so as to have minimum contact with the fluid and, therefore, minimum frictional effects. In the absence of the sharp edge, flow through orifice depends also on the thickness of the orifice and the roughness of the orifice surface.

Figure 5.5.1 shows a tank that has an orifice in one of its walls. The liquid is discharged in the atmosphere through the orifice in the form of a free jet as the jet is not affected by any liquid on the other side in the presence of which the jet would have been a submerged jet. Liquid approaching the orifice converges towards it. The streamlines continue to converge beyond the orifice edge until they become parallel at the section cc which is at a short distance (about half the orifice diameter) downstream from the orifice. Downstream of the section cc the jet may diverge and, therefore, cc is the section of minimum area of the jet. The section cc is termed *vena contracta*. Some additional, but negligible, curvature

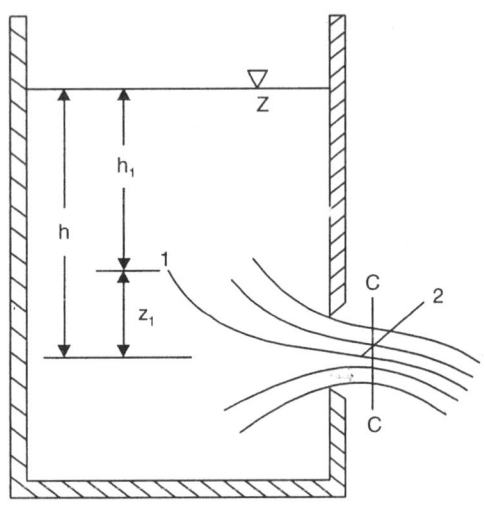

Fig. 5.5.1. Flow through an orifice

of the streamlines due to the downward deflection of the jet by gravity, may result in poorly defined vena contracta. This effect is usually ignored.

Since the streamlines are parallel at the vena contracta, the pressure in the jet there is uniform as, otherwise, there would be accelerations perpendicular to the axis making streamlines curved and non-parallel. Ignoring the effects of surface tension (due to which pressure in the jet is slightly higher than the surrounding pressure, Eq.(1.4.8)), the pressure in the jet at the vena contracta equals the atmospheric pressure surrounding the jet. Incidentally, vena contracta is the only section of the jet at which the pressure is known completely.

If the flow is considered steady (requiring the level in the tank to be invariant with time) and frictional effects are assumed to be negligible, one may apply Bernoulli's equation between any two points on a particular streamline. Since the energy content everywhere in the tank is the same, one can choose point 1 anywhere in the reservoir and point 2 at the vena contracta on any of the streamlines. Selecting a horizontal plane through the centre of the orifice as datum, one can write Bernoulli's equation for points 1 and 2 in Fig. 5.5.1 as

$$\frac{p_1}{\rho g} + \frac{V_1^2}{2g} + z_1 = \frac{p_{cc}}{\rho g} + \frac{V_2^2}{2g} + 0 \qquad \qquad ...(5.5.1)$$

It is assumed that the point 1 in the tank is sufficiently away from the orifice so that the velocity V_1 is negligible. Then, since hydrostatic conditions exist in the tank, the pressure p_1 corresponds to the depth of the point 1 below the free surface and, therefore, equals $\rho g h_1$. With atmospheric pressure taken as zero and its variation from the free surface in the tank to the orifice being negligible, Eq. (5.5.1) reduces to

$$h_1 + z_1 = h = \frac{V_2^2}{2g} \qquad \qquad ...(5.5.2)$$

as

$$p_{cc} = p_{atm} = 0$$

Hence,

$$V_2 = \sqrt{2gh} \qquad \qquad ...(5.5.3)$$

If the orifice is relatively small compared to h, the velocity of the jet is uniform across the vena contracta. Equation (5.5.3) gives the velocity at the vena contracta. Based on experiments, Evangelista Torricelli (1608-47), a student of Galileo, showed that the velocity of a liquid jet from a small orifice is proportional to the square root of the head above the orifice (2). Therefore, Eq. (5.5.3) is known as Torricelli's formula. It may be noted that in the plane of the orifice, neither the velocity nor the pressure is uniform and the average velocity is less than that at the vena contracta.

Since friction (or viscous) and surface tension effects have been excluded from the above analysis, the velocity V_2, obtained from Eq. (5.5.3), is an ideal velocity and is greater than the actual velocity attained at the vena contracta. The ratio of actual (average) velocity at the vena contracta to the ideal velocity V_2 is termed *coefficient of velocity* C_v. Similarly, *coefficient of contraction* C_c is defined as the ratio of the area of cross-section of the jet at the vena contracta to the area of the orifice. Again, due to the effects of friction and contraction, the actual discharge through the orifice is less than the ideal value of the discharge, i.e., $a\sqrt{2gh}$ in which, a is the area of the orifice. The *coefficient of discharge* C_d is the ratio of the actual discharge Q and the ideal value of the discharge, i.e., $a\sqrt{2gh}$. Hence,

$$Q=(C_c a)(C_v\sqrt{2gh}) = C_d\, a\sqrt{2gh} \qquad ...(5.5.4)$$

Thus,
$$C_d = C_c\, C_v \qquad ...(5.5.5)$$

For a relatively large vertical orifice, the velocity at the vena contracta section would vary with the depth below the free surface in accordance with Eq. (5.5.3). Therefore, the discharge Q through a large vertical orifice can only be obtained by integrating the discharge through the small elements of the orifice cross-section. Thus, from Fig. 5.5.2 and Eq. (5.5.4)

$$\delta Q=\left(C_c b\delta h\right)\left(C_v\sqrt{2gh}\right)$$

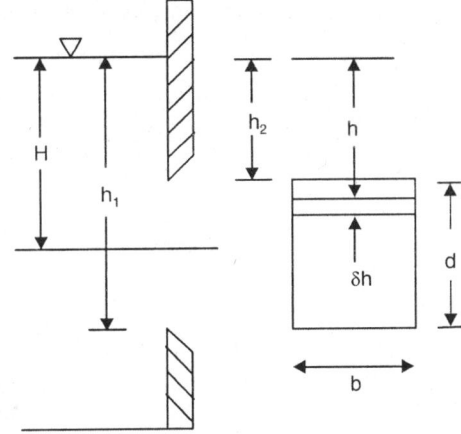

Hence,
$$Q= \int dQ = C_c C_v b\sqrt{2g}\int_{h_2}^{h_1} h^{1/2}.dh \quad ...(5.5.6)$$

where,
$$h_1 =H+\frac{d}{2} \text{ and } h_2 =H-\frac{d}{2}$$

Fig. 5.5.2. A large vertical orifice

On simplifying, Eq. (5.5.6) reduces to

$$Q=\frac{2}{3}\, C_d\, b\sqrt{2g}\left\{\left(H+\frac{d}{2}\right)^{\frac{3}{2}}-\left(H-\frac{d}{2}\right)^{\frac{3}{2}}\right\} \qquad ...(5.5.7)$$

If H is much larger than d, the orifice may be treated as small (even if d is rather large) so that the ideal velocity at the vena contracta has the uniform value of $\sqrt{2gH}$. The discharge computed from $Q=C_d bd\sqrt{2gH}$ may have some negligible error depending upon the relative magnitudes of H and d. In order to be able to measure the discharge passing through an orifice, one needs to determine the coefficient of discharge, C_d which is the ratio of actual discharge $\dot{Q}$ to the ideal discharge $a\sqrt{2gH}$. Thus, C_d can be easily determined by measuring the volume of liquid emerging from the orifice in a known duration and thus measuring the actual discharge Q (i.e., volumetric rate of flow) when the head above the centre of the orifice is H and is maintained at the same level during the experimentation.

If one measures the cross-section of the jet at vena contracta with calipers, the coefficient of contraction C_c can be estimated. The accuracy of this method is not so good. The coefficient of contraction can, however, be estimated by Eq. (5.5.5) if one knows C_d and C_v. The coefficient of velocity

C_v can be obtained by determining the actual velocity of the jet. Figure 5.5.3 shows the horizontal distance x and vertical distance y (both from the vena contracta) that a particle has travelled in time t. Neglecting air resistance, one can assume that the horizontal component of velocity u remains unchanged and, therefore,

$$x = ut$$

The vertical component of velocity is initially zero for the horizontal jet at the vena contracta and the jet falls with a uniform downward acceleration g. Hence,

$$y = \frac{1}{2}gt^2 = \frac{1}{2}g\left(\frac{x}{u}\right)^2$$

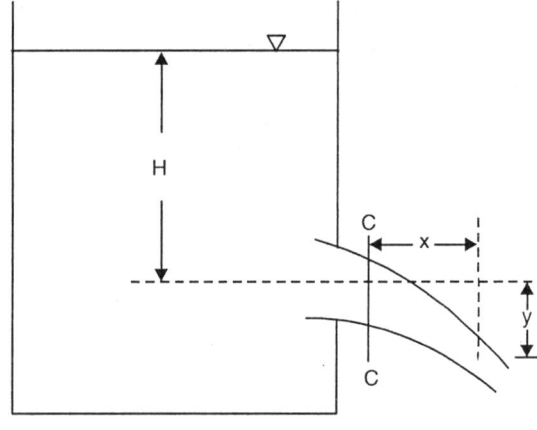

Fig. 5.5.3. Liquid jet with falling reservoir level

Thus,

$$u = x\sqrt{\frac{g}{2y}}$$

Hence,

$$C_v = \frac{x\sqrt{\dfrac{g}{2y}}}{\sqrt{2gH}} = \frac{x}{\sqrt{4yH}}$$

or

$$C_v = \sqrt{\frac{x^2}{4yH}} \qquad\qquad ...(5.5.8)$$

Thus, by simply measuring the coordinates of any point of a liquid jet issuing from an orifice into the atmosphere under a constant head H, one can determine C_v using Eq. (5.5.8). Typical values of the coefficient of velocity are in the range 0.97 to 0.99 for well-made, sharp-edged, circular orifices producing free liquid jets. The coefficient of contraction for a circular sharp-edged orifice is generally in the range 0.61 to 0.66 and the coefficient of discharge is in the range 0.6 to 0.65.

An orifice is said to be submerged when it discharges a jet into a fluid of the same kind, Fig. 5.5.4. A vena contracta is formed in this case too at point 2. Applying Bernoulli's equation between points 1 and 2.

$$\frac{p_1}{\rho g} + z_1 + \frac{V_1^2}{2g} = \frac{p_2}{\rho g} + z_2 + \frac{V_2^2}{2g}$$

Using the explanation for the derivation of Eq. (5.5.3),

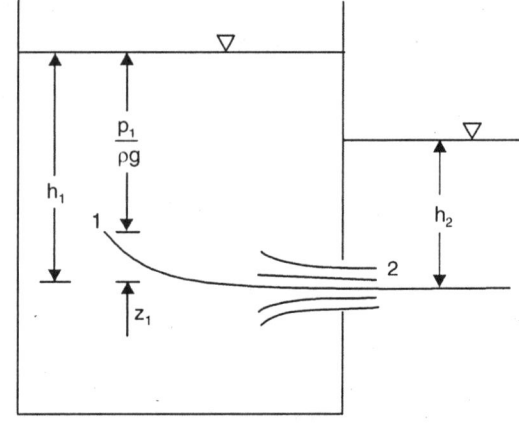

Fig. 5.5.4. A submerged orifice

$$h_1 = h_2 + \frac{V_2^2}{2g}$$

Hence, $$V_2 = \sqrt{2g(h_1 - h_2)}$$ ...(5.5.9)

This means that the Torricell's formula, Eq. (5.5.3), is still applicable provided that the head h is replaced with the effective head, i.e., $(h_1 - h_2)$. The coefficients of submerged orifices remain almost the same as in case of orifices producing a free jet.

Unless the tank of Fig. 5.5.1 is continuously replenished, the level of the free surface falls as the liquid escapes through the orifice. If the free surface of the tank is very large compared to the size of the orifice, the rate of fall of the free surface (or the liquid level) is small and the error involved in using the Bernoulli's equation (applicable for steady flow) is rather negligible. Such condition may be termed as quasi-steady. Figure 5.5.5 shows a tank which is being emptied through an orifice. Let the free surface level fall by a distance δh in time δt when the head over the orifice is h. If A is the surface area of the free surface and a is the area of orifice, one can write

$$(-) A \delta h = \left(C_d a \sqrt{2gh} \right) \delta t$$

$$\therefore \quad \delta t = (-) \frac{A}{C_d a \sqrt{2g}} \frac{\delta h}{\sqrt{h}}$$

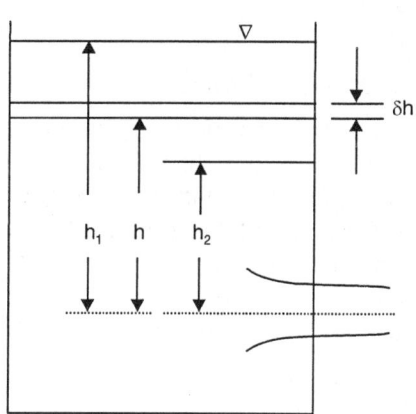

Hence, if time required for lowering the free surface level from h_1 to h_2 is T then,

$$T = \int_0^T dt = (-) \frac{A}{C_d a \sqrt{2g}} \int_{h_2}^{h_1} \frac{dh}{\sqrt{h}}$$

$$\therefore \quad T = \frac{2A}{C_d a \sqrt{2g}} \left(h_1^{\frac{1}{2}} - h_2^{\frac{1}{2}} \right)$$ ...(5.5.10)

Fig. 5.5.5. Unsteady flow through an orifice

Thus, one finds that the Bernoulli's equation can be used for some class of unsteady flows too provided that unsteadiness is not large. Equation (5.5.10) enables determination of C_d of an orifice by noting down the time taken in lowering the free surface level from h_1 to h_2.

5.6 FLOW MEASUREMENT IN PIPES

The flow quantities to be usually measured in a flow system are mainly pressure, velocity, and flow rate besides other quantities such as turbulence, shear etc. Pressure (or pressure difference) can be measured with manometers (Chapter 2). Velocity can be measured with Pitot-static tube (Art.5.3). There are other instruments too for measurement of pressure (such as pressure transducer) and velocity (such as current meter and hot-wire anemometer) that have been described in Art. 6.8.

The choice of a flow meter to measure the flow rate (i.e., discharge) is governed by the range of discharge to be measured, accuracy required, cost of the meter, convenience of reading and estimation, and maintenance aspects. Obviously, one should choose a device that gives the desired accuracy besides being the simplest and cheapest.

5.6.1 Direct methods

Flow rate or discharge of a steady liquid flow can be easily measured by collecting a volume or mass of liquid in a suitable duration. If the duration of collection is long enough, one can determine the flow rate precisely.

Volume measurements for gas flows must take into account compressibility of the gas. A volume of gas sample can be collected in an inverted jar over water. Pressure in the jar must be kept constant and equal to the static pressure of the flow or, alternatively, using Boyle's law, the volume of gas collected must be changed to the volume corresponding to the static pressure in the flow.

By measuring the velocity distribution in a flow cross-section, one can compute the discharge Q by integrating the discharges through elemental area δA through which the flow velocity v has been measured using a Pitot-static tube. Thus,

$$Q = \sum_A v\delta A = \int_A v\, dA$$

...(5.6.1)

Other discharge measuring devices or structures, requiring calibration, have been briefly described here.

5.6.2 Venturi Meter

A venturi meter measures the volumetric rate of flow, i.e., discharge in a pipe. The principle of the venturi meter was first demonstrated in 1797 by Giovanni Battista Venturi (1746-1822) (2). Venturi meter is a casting, Fig. 5.6.1, comprising (i) an upstream cylindrical section having the same size as that of the pipe and has a bronze liner containing piezometer ring to measure the static pressure, (ii) a converging conduit, (iii) a cylindrical throat with a bronze liner containing a piezometer ring to measure the static pressure, and (iv) and a gradually diverging conduit leading to a downstream cylindrical section of the size of the pipe. A differential U-tube manometer is connected to the two piezometer rings

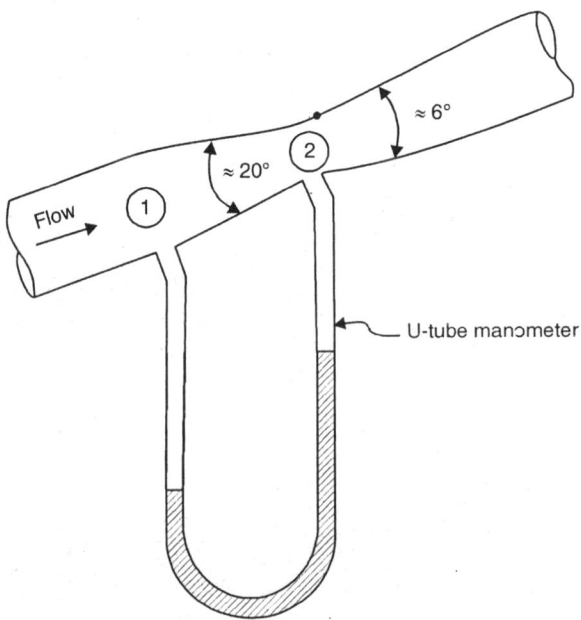

Fig. 5.6.1. Venturi meter

to measure the pressure head difference, $\dfrac{(p_1 - p_2)}{\rho g}$, which is a consequence of greater velocity at the throat than the velocity at the inlet. Assuming steady, incompressible, and frictionless flow, one can combine Bernoulli's equation with continuity equation and write, for flow in a venturi meter, Fig. 5.6.1,

$$\frac{p_1}{\rho g} + \frac{Q^2}{2gA_1^2} + z_1 = \frac{p_2}{\rho g} + \frac{Q^2}{2gA_2^2} + z_2 \qquad \text{...(5.6.2)}$$

so that

$$Q = \frac{A_2}{\sqrt{1 - \left(\dfrac{A_2}{A_1}\right)^2}} \sqrt{2g\left\{\left(\frac{p_1}{\rho g} + z_1\right) - \left(\frac{p_2}{\rho g} + z_2\right)\right\}} \qquad \text{...(5.6.3)}$$

or

$$Q = \frac{A_2}{\sqrt{1 - \left(\dfrac{A_2}{A_1}\right)^2}} \sqrt{2g\Delta h} \qquad \text{...(5.6.4)}$$

in which Δh is the piezometeric head difference $\left\{\left(\dfrac{p_1}{\rho g} + z_1\right) - \left(\dfrac{p_2}{\rho g} + z_2\right)\right\}$ that can be measured directly by a differential U-tube manometer. In Eqs. (5.6.2 – 5.6.4), Q is the ideal value of discharge for frictionless flow. In reality, however, there would be frictional loss between sections 1 and 2 resulting in greater Δh and, thus, giving Q higher than the true one. Hence, Eq. (5.6.4) is modified to

$$Q = C_d \, \frac{A_2}{\sqrt{1 - \left(\dfrac{A_2}{A_1}\right)^2}} \sqrt{2g\Delta h} \qquad \text{...(5.6.5)}$$

in which C_d is the coefficient of discharge of the venturi meter and depends on the area ratio A_2/A_1, the flow rate Q, viscosity of the flowing fluid, and the surface roughness. It should be noted that Δh is the difference of the piezometric head of the fluid in the venturi meter and not the difference of levels of the manometer liquid.

For good measurement accuracy, the pipe upstream of the venturi meter should be straight and of sufficient length so that the fluid entering the meter is free from disturbances. The diverging part of the venturi meter reduces the velocity gradually and restores the pressure as nearly as possible to its original value. A larger angle of divergence increases the energy dissipation. A small angle of divergence results in longer length (and cost) of the meter and, hence, larger friction loss. The diameter at the throat should not be so low as to cause liberation of dissolved gases or vaporization of the flowing liquid due to the reduced pressure.

5.6.3 Flow Nozzle

The flow nozzle or nozzle meter, Fig. 5.6.2, is similar to a venturi meter with its divergent part omitted. Therefore, the discharge equation, Eq. (5.6.5), derived for venturi meter holds good for flow nozzle too.

In the absence of the divergent part of venturi meter, the dissipation of energy downstream of the nozzle throat is greater than that of venturi meter. However, the cost of flow nozzle is much lesser than the cost of venturi meter. The value of the coefficient of discharge of a nozzle meter too depends on the flow rate, area ratio, viscosity of the flowing fluid, and surface roughness and is only a little different from that for venturi meter.

5.6.4 Orifice Meter

Compared to venturi meter and flow nozzle, an orifice meter is simpler and cheaper flow meter for measurement of flow rate in pipe flow. An orifice meter is a sharp-edged orifice plate that is clamped between pipe flanges, Fig. 5.6.3. The orifice is usually kept concentric in pipe. While it is low in cost and easy to install or replace, the orifice meter causes high head loss due to uncontrolled expansion downstream of the orifice plate. Considering flow through orifice meter as steady, incompressible and inviscid, one can apply Bernoulli's equation between sections 1 sufficiently upstream of the orifice plate and 2 at the vena contracta. Thus,

$$\frac{p_1}{\rho g}+\frac{V_1^2}{2g}+z_1=\frac{p_2}{\rho g}+\frac{V_2^2}{2g}+z_2 \qquad ...(5.6.6)$$

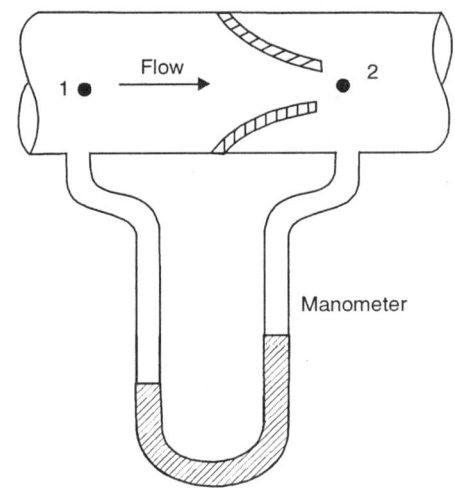

Fig. 5.6.2. Nozzle meter

The continuity equation, Eq. (5.1.2) gives

$$A_1 V_1 = C_c A_2 V_2 \qquad ...(5.6.7)$$

in which C_c is the coefficient of contraction for the orifice and A_2 is the area of cross-section of the orifice. Eliminating V_1 from Eqs. (5.6.6) and (5.6.7), one gets

$$\frac{V_2^2}{2g}\left\{1-C_c^2\left(\frac{A_2}{A_1}\right)^2\right\}=\frac{p_1^*-p_2^*}{\rho g}$$

with $p_1^*=p_1+\rho g z_1$ and $p_2^*=p_2+\rho g z_2$

Thus, $$V_2=\sqrt{\frac{\dfrac{2\left(p_1^*-p_2^*\right)}{\rho}}{1-C_c^2\left(\dfrac{A_2}{A_1}\right)^2}}$$

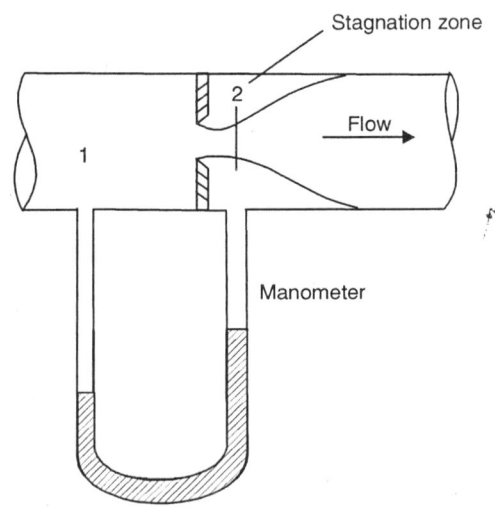

Fig. 5.6.3. Orifice meter

The velocity V_2 is an ideal velocity that neglects friction loss. Thus, the actual velocity would be $C_v V_2$ where, C_v is the coefficient of velocity. At section 2 (vena contracta), the cross-sectional area would be $C_c A_2$. Hence, the discharge or flow rate

$$Q = (C_c A_2) C_v \sqrt{\dfrac{\dfrac{2\left(p_1^* - p_2^*\right)}{\rho}}{1 - C_c^2 \left(\dfrac{A_2}{A_1}\right)^2}}$$

or

$$Q = \dfrac{C_c C_v A_2}{\sqrt{1 - C_c^2 \left(\dfrac{A_2}{A_1}\right)^2}} \sqrt{\dfrac{2\left(p_1^* - p_2^*\right)}{\rho}}$$

or

$$Q = C_d A_2 \sqrt{\dfrac{2\left(p_1^* - p_2^*\right)}{\rho}}$$

or

$$Q = C_d \dfrac{\pi}{4} d^2 \sqrt{2g\Delta h} \qquad\qquad ...(5.6.8)$$

Here, Δh is the piezometric head difference, $\left\{\left(\dfrac{p_1}{\rho g} + z_1\right) - \left(\dfrac{p_2}{\rho g} + z_2\right)\right\}$, that can be measured with a U-tube manometer, d is the diameter of orifice, and C_d is the coefficient of discharge given as

$$C_d = \dfrac{C_c C_v}{\sqrt{1 - C_c^2 \left(\dfrac{A_2}{A_1}\right)^2}} \qquad\qquad ...(5.6.9)$$

It should be noted that Δh is the difference of the piezometric head of the fluid in the orifice meter and not the difference of levels of the manometer liquid.

5.7 WEIRS

A *weir* is an obstruction placed in a channel such that the flowing liquid behind it deflects so as to flow over it. For simple geometries of weir, the discharge Q in the channel is related to the height of the upstream liquid surface above the weir. Thus, weir is a simple but effective flow meter for open channel flows.

Figure 5.7.1 shows two principal types of weirs, sharp-crested weir, Fig. 5.7.1(a) and (e), and broad-crested weir, Fig. 5.7.1(b), that are commonly used for flow measurement in open channels. Weirs are constructed from a sheet of metal or other material. Sharp-crested weir (also known as *notch*, if the weir is small) is usually made in a smooth, plane, vertical plate with its edge bevelled on the downstream side so as to minimize friction loss. When the weir does not cover the full width of channel (i.e, length of weir crest is less than the width of channel), it has *end contractions* and the weir is called *contracted weir*, Fig. 5.7.1(d). A weir, having its length of crest equal to the width of channel, suppresses its end contractions and is called *suppressed weir*, Fig. 5.7.1(c).

The sheet of liquid escaping over the weir crest is known as nappe. For a suppressed weir, atmospheric air may not reach underside of the nappe and pressure there may be less than the atmospheric pressure following the underside air getting dissolved in or escaped with the flowing water. In such condition, the discharge becomes unpredictable. Hence, the space under the nappe must be ventilated by providing an air vent, if required, in the side walls of the channel immediately downstream of the suppressed weir.

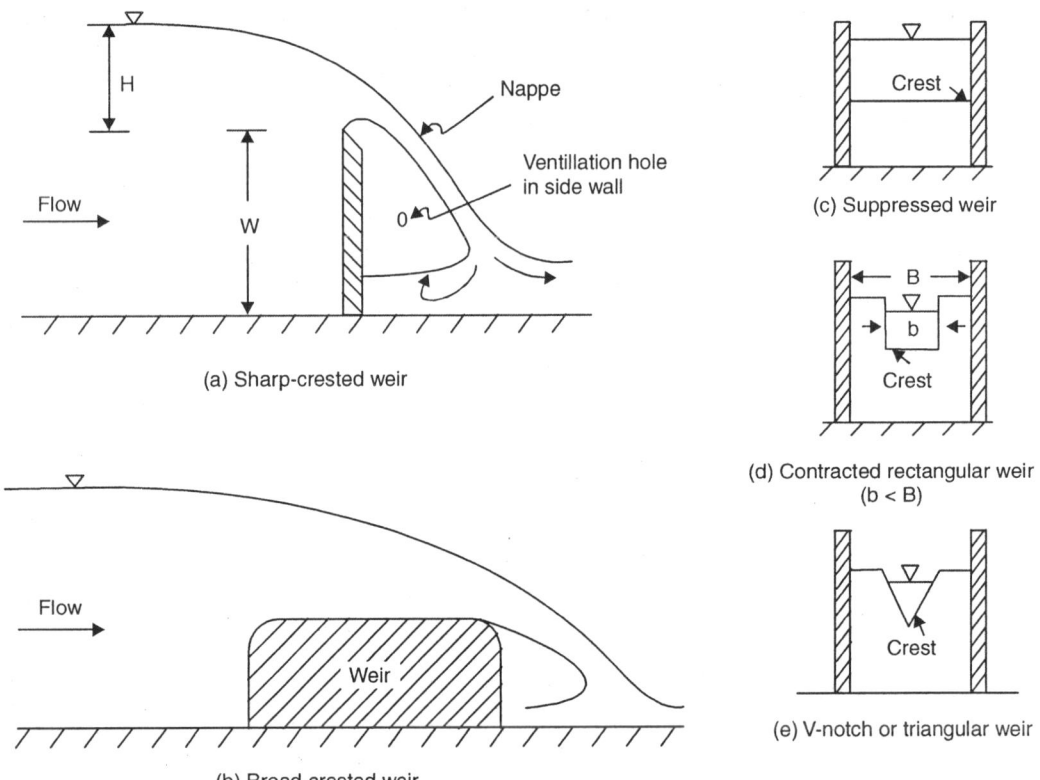

Fig. 5.7.1. Main types of weirs

Flow over a weir has curved streamlines and, hence, there is no cross-section of the flow over which the pressure can be considered as uniform (as at vena contracta section associated with orifices). Besides, viscosity and surface tension also affect the discharge particularly at low flows. Consequently, flow rate over a weir, determined analytically after simplifying assumptions, requires correction through empirically determined discharge coefficient.

The nappe for a sharp-crested rectangular weir is always contracted at top and bottom. For hypothetical ideal condition of no contraction, the flow over sharp-crested weir would appear as is shown in Fig. 5.7.2. The nappe, therefore, has parallel streamlines with uniform atmospheric pressure throughout (neglecting surface tension effects due to which pressure inside the nappe should be higher). At point 1, datum head with respect to the weir crest is H and the velocity head is negligible due to large flow

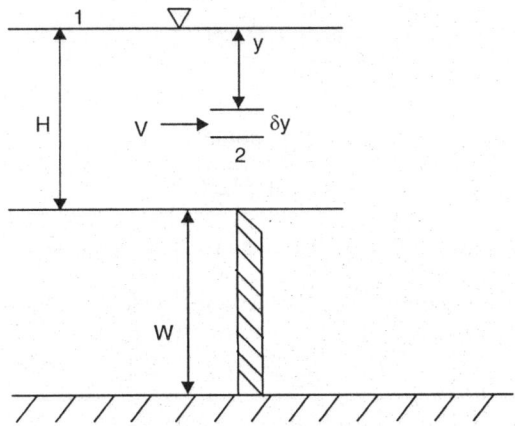

Fig. 5.7.2. Weir nappe without contraction

cross-section. At point 2, the datum head is $H-y$ and the velocity head is $\dfrac{v^2}{2g}$. Bernoulli's equation, applied between sections 1 and 2, yields

$$H + 0 + 0 = (H - y) + \frac{v^2}{2g} + 0$$

or
$$v = \sqrt{2gy}$$

Therefore, the ideal discharge Q_i is given as

$$Q_i = \int_0^H v\, b\ dy = \sqrt{2g}\ b \int_0^H y^{1/2}\ dy = \frac{2}{3} b \sqrt{2g}\ H^{3/2}$$

Here, b is the length of the weir crest. In order to account for the fact that in real situation, the nappe would be contracted and friction loss would also occur, the ideal discharge Q_i is multiplied by a coefficient of discharge. Thus, the discharge over a sharp-crested weir is

$$Q = \frac{2}{3} C_d\, b \sqrt{2g}\ H^{3/2} \qquad\qquad \text{...(5.7.1)}$$

in which C_d, primarily, depends on approach velocity (see Art. 16.3) and, based on experiments of Rehbock (3), is expressed empirically as

$$C_d = 0.611 + 0.075 \frac{H}{W} \qquad\qquad \text{...(5.7.2)}$$

The head H, over the weir is measured upstream from the weir at a location where there is negligible surface contraction. Equation (5.7.1) can, alternatively, be derived as follows:

The area of flow section over a weir is proportional to bH and the velocity of flow across the weir may be determined approximately from the Bernoulli's equation as equal to $\sqrt{2gH}$. Therefore, the discharge

$$Q \propto b \sqrt{2g}\, H^{3/2} \qquad \qquad ...(5.7.3)$$

Introducing the coefficient of discharge C_d' to account for the non-uniformity of velocity, contraction of nappe and frictional losses, Equation (5.7.3) becomes

$$Q = C_d'\, b \sqrt{2g}\, H^{3/2} \qquad \qquad ...(5.7.4)$$

Obviously, Eq. (5.7.4) is the same as Eq. (5.7.1) when C_d' is replaced with $(2/3)C_d$.

Similarly, for triangular weir, Fig. 5.7.3, the velocity at depth y approximately equals to $\sqrt{2gy}$. Hence,

$$Q_i = \int_0^H v\, dA = \int_0^H \sqrt{2gy} \left(\frac{L}{H}\right)(H-y)\, dy = \sqrt{2g}\, \frac{L}{H} \int_0^H y^{1/2} (H-y)\, dy$$

$$\therefore \qquad \qquad Q_i = \frac{4}{15}\sqrt{2g}\left(\frac{L}{H}\right) H^{5/2} \qquad \qquad ...(5.7.5)$$

Introducing the coefficient of discharge C_d to account for non-uniformity of velocity, nappe contraction and frictional effects, Eq. (5.7.5) becomes

$$Q = \frac{8}{15}C_d \sqrt{2g}\, H^{5/2} \tan\frac{\theta}{2} \qquad ...(5.7.6)$$

since $\dfrac{L}{2H} = \tan\dfrac{\theta}{2}$. The discharge coefficient C_d in Eq. (5.7.6) can be obtained experimentally. The discharge obtained from Eqs. (5.7.1 or 5.7.4) and (5.7.6) does not account for viscosity and surface tension effects. These and other aspects related to weirs have been included in Art. 16.3.

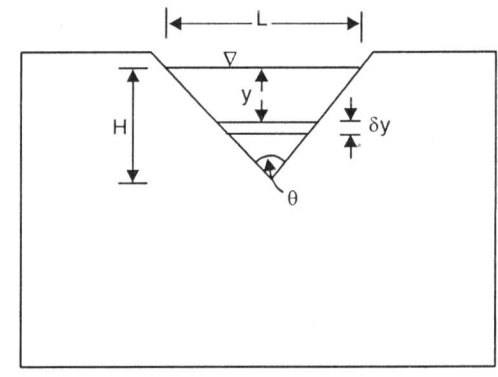

Fig. 5.7.3. V-notch or triangular weir

5.8 SLUICE GATE

A *sluice gate*, Fig. 5.8.1, is a control structure with an opening in a wall, performing as a gate, at its lower end. The sluice gate is commonly used to regulate and also measure the discharge at the crest of an overflow spillway, or at the entrance of an open channel from a river or at the entrance of a river or a canal from a lake. On applying the Bernoulli's equation between point 1, sufficiently upstream of the gate where the velocity is uniform, and point 2 that is located at the vena contracta, one gets

$$p_{atm} + \frac{1}{2}\rho V_1^2 + \rho g h_1 = p_{atm} + \frac{1}{2}\rho V_2^2 + \rho g h_2 \qquad \qquad ...(5.8.1)$$

One could, alternatively, choose points 3 and 4 as the streamline passing through these points is straight. In this case, instead of potential energy ρgh at points 1 and 2, one would use pressure contributions ρgh at 3 and 4. The result remains the same. The continuity equation between the sections through any pair of these points gives

$$Q = bh_1 V_1 = b h_2 V_2 \qquad \qquad ...(5.8.2)$$

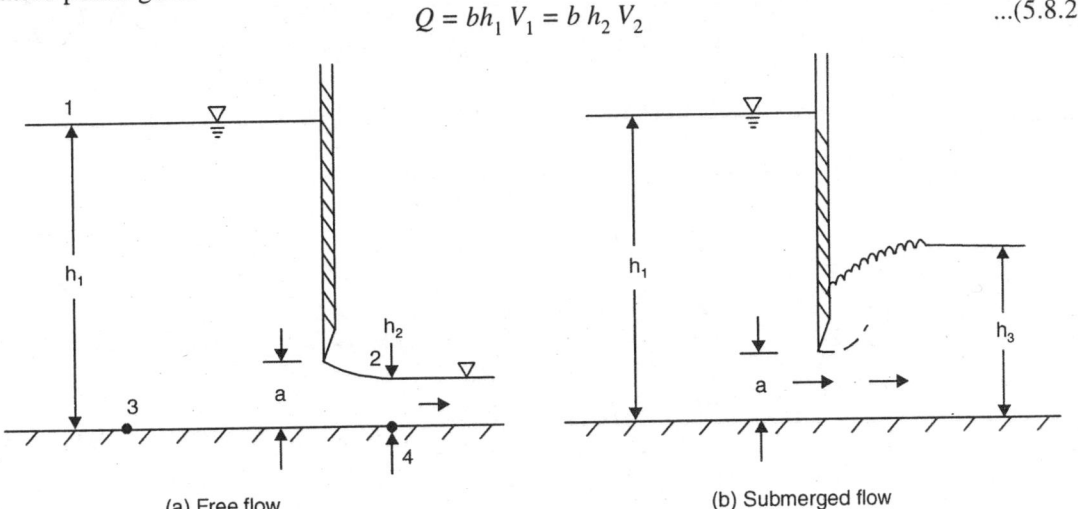

(a) Free flow   (b) Submerged flow

Fig. 5.8.1. Flow through a sluice gate

Substituting the values of V_1 and V_2 from Eq. (5.8.2) in Eq. (5.8.1), one obtains

$$Q = bh_2 \sqrt{\dfrac{2gh_1}{1+\left(\dfrac{h_2}{h_1}\right)}} = bh_2 \sqrt{\dfrac{2g(h_1 - h_2)}{1-\left(\dfrac{h_2}{h_1}\right)^2}} \qquad ...(5.8.3)$$

When $h_1 \gg h_2$ (i.e., the fluid velocity and, hence, the kinetic energy at point 1 upstream of the gate is negligible and, the fluid velocity, after the fluid level has fallen a distance $(h_1 - h_2) \approx h_1$, is approximately $V_2 \approx \sqrt{2gh_1}$), Eq. (5.8.3) becomes

$$Q = bh_2 \sqrt{2gh_1} \qquad \qquad ...(5.8.4)$$

At the vena contracta, the depth h_2 would be less than the gate opening a. If coefficient of contraction for the gate is C_c (typical value $\cong 0.61$) then $h_2 = C_c a$. Also, while applying the Bernoulli's equation, frictional effects were completely neglected and $h_1 - h_2$ was approximated as equal to h_1. Therefore, to account for these deviations, the ideal discharge equation, Eq. (5.8.4), is modified to

$$Q = C_d \, b \, a \sqrt{2gh_1} \qquad \qquad ...(5.8.5)$$

or

$$q = \dfrac{Q}{b} = C_d a \sqrt{2gh_1}. \qquad \qquad ...(5.8.6)$$

in which q is the discharge per unit width of gate or channel and C_d is the coefficient of discharge of the sluice gate that depends on C_c and ratio $\dfrac{h_1}{a}$ for free flow condition, Fig. 5.8.1(a), which means that the fluid jet issuing from the gate remains unaffected by the depth of flow in the channel downstream of the gate. The values of C_d can be obtained from Fig. 5.8.2 (1). When the depth of flow in the channel downstream of the gate is larger than the gate opening a, the jet is drowned or submerged, Fig. 5.8.1(b), and the value of C_d depends on $\dfrac{h_1}{a}$ and $\dfrac{h_3}{a}$ as shown in Fig. 5.8.2. Obviously, when $\dfrac{h_1}{a}$ and $\dfrac{h_3}{a}$ are equal, i.e., $h_1 = h_3$, there is no head to cause the flow and, hence, $C_d = 0$.

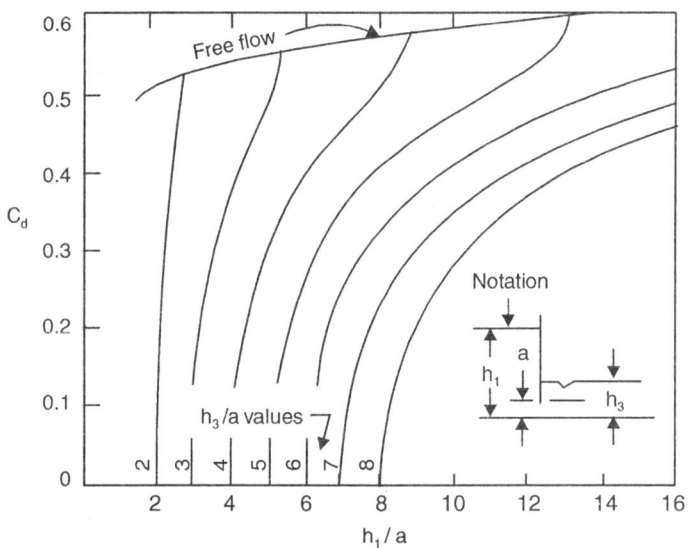

Fig. 5.8.2. Variation of C_d with h_1/a for sluice gates (1)

5.9 LOSSES DUE TO SUDDEN EXPANSION IN A PIPE

The energy (or head) loss due to sudden enlargement in a pipeline, Fig. 5.9.1, can be computed by theoretical analysis using the three basic equations of fluid motion, viz., the continuity, energy (i.e., Bernoulli's) and momentum equations. For incompressible, steady turbulent flow through a sudden expansion in a pipeline, the shear stress in the corner region (of separated flow) is negligible and the flow velocity is considered uniform at sections 1, 2 and 3, Fig. 5.9.1. The pressure at section 2 of the control volume between sections 2 and 3 is assumed to be the same as at section 1, i.e., $p_a = p_b = p_c = p_1$. Applying continuity, energy, and momentum equations, one obtains

$$A_1 V_1 = A_3 V_3 \qquad \qquad ...(5.9.1)$$

$$\frac{p_1}{\rho g} + \frac{V_1^2}{2g} = \frac{p_3}{\rho g} + \frac{V_3^2}{2g} + h_L \qquad \qquad ...(5.9.2)$$

and

$$p_1 A_2 - p_3 A_3 = p_1 A_3 - p_3 A_3 = \rho A_1 V_1 (V_3 - V_1) \qquad \qquad ...(5.9.3)$$

in which h_L is the head loss due to sudden expansion. Combining Eqs. (5.9.1) and (5.9.3), one gets

$$\frac{p_3 - p_1}{\rho} = V_3 (V_1 - V_3)$$

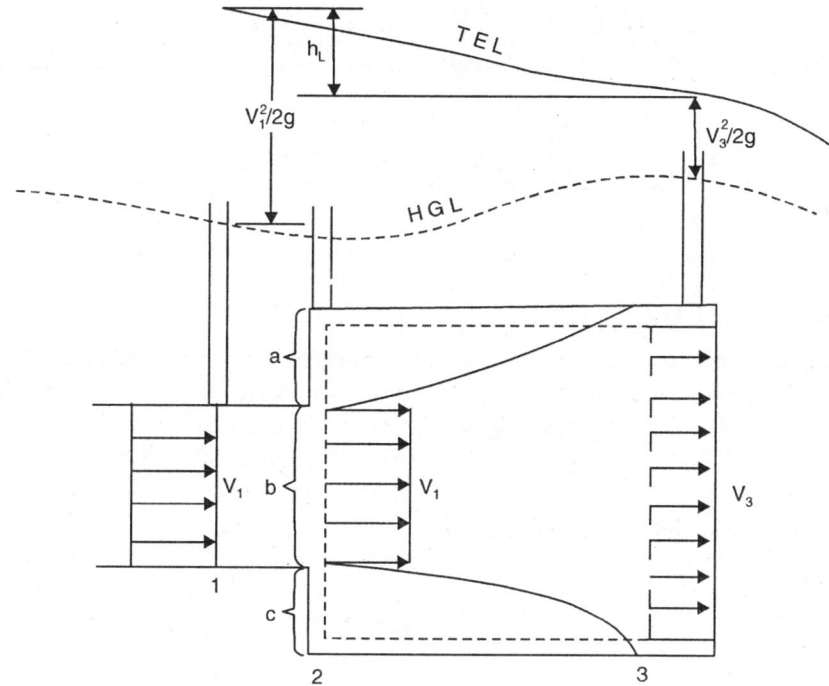

Fig. 5.9.1

From Eq. (5.9.2), one obtains

$$\frac{p_3 - p_1}{\rho} = \frac{V_1^2 - V_3^2}{2} - gh_L$$

Hence,

$$V_3 V_1 - V_3^2 = \frac{V_1^2 - V_3^2}{2} - gh_L$$

or

$$2gh_L = V_1^2 - V_3^2 + 2V_3^2 - 2V_3V_1 = (V_1 - V_3)^2$$

$\therefore$

$$h_L = \frac{(V_1 - V_3)^2}{2g} \qquad \qquad ...(5.9.4)$$

Equation (5.9.4) can, alternatively, be rearranged as

$$h_L = \frac{V_1^2}{2g}\left(1 - \frac{V_3}{V_1}\right)^2$$

Using Eq. (5.9.1), one gets

$$h_L = \left(1 - \frac{A_1}{A_3}\right)^2 \frac{V_1^2}{2g} \qquad \qquad ...(5.9.5)$$

or
$$h_L = K_L \frac{V_1^2}{2g} \qquad\qquad ...(5.9.6)$$

in which K_L is termed the loss coefficient that equals $\left(1 - \dfrac{A_1}{A_3}\right)^2$ or $\left(1 - \dfrac{A_1}{A_2}\right)^2$ as $A_2 = A_3$ = area of cross-section of the enlarged pipe.

5.10 FORCE EXERTED BY JET ON A FLAT PLATE

If a fluid jet strikes a plate, plane or curved, a stream of fluid forms over the surface of the plate until the stream reaches the boundaries of the plate. The fluid, then, leaves the plate tangentially. Consider a fluid jet that strikes a flat plate as shown in Fig. 5.10.1. Chosen control volume is shown by dotted lines. Let x-direction be normal to the plate and y – direction along the plate. Hence, the fluid of the jet, flowing over the surface of the plate, would have no component of velocity and, hence, momentum in the x-direction. If v_x is the velocity of jet in x-direction then $v_x = v \cos \theta$ in which v is the average velocity of the fluid in the jet. Thus, the rate of x-direction momentum entering the control volume is $\rho v A v \cos \theta$ in which A is the area of cross-section of the jet at the inlet end of the control volume. Hence, the rate of change of x-direction momentum for the control volume is $-\rho Q v \cos \theta$ that should be equal to the force on the fluid in the chosen x-direction. Since the pressure around the control volume including the flow region, is atmospheric and gravity effect being neglected (i.e., the problem is considered to be in horizontal plane), the force on the fluid in the control volume can be exerted only by the plate. Therefore, the force exerted by the jet fluid on the plate is $\rho Q v \cos \theta$ in the chosen x-direction.

Likewise, the jet carries, in y-direction, a momentum of $\rho Q v \sin \theta$ per unit time. For this y-direction momentum rate to undergo a change, a force in the y-direction (i.e., parallel to the plate) would have to be applied on the fluid. Such a force can be only the shear force exerted by the plate on the fluid. If the fluid is non-viscous and the surface of the plate is smooth, no shear force would exist. Therefore, the y-direction momentum rate $\rho Q v \sin \theta$ remains unchanged and equals the rate of y-direction momentum leaving the control volume. The resultant force can now be estimated. For real fluids, however, the rate at which y-direction momentum leaves the control

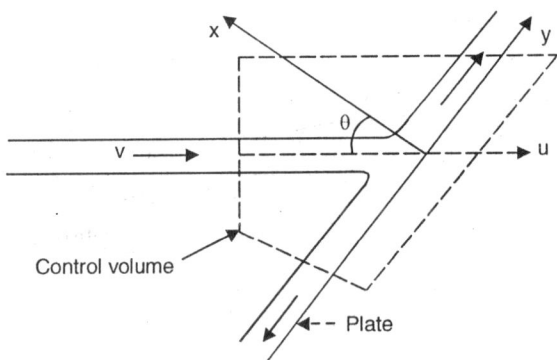

Fig. 5.10.1. Force exerted on a flat plate

volume would differ from the rate at which it enters the control volume. This would, obviously, require information about the velocity at the outlet end for further computations. The jet spreads symmetrically only when $\theta = 0$.

If the plate itself were moving with velocity u in the direction of the original jet velocity, the simplest way to solve the problem is to bring the plate and, therefore, the associated control volume to rest by superimposing on the system a velocity equal and opposite to that of the plate. This way, the problem is similar to that of a jet striking a stationary plate and the mass rate entering the control volume is $\rho A (v - u)$. The rate of change of x-direction momentum is $\rho A (v - u) (v - u) \cos \theta$. Hence, force exerted by the fluid on the plate is $\rho A \ (v - u) \ (v - u) \cos \theta$ acting in the x-direction (i.e., normal to the plate). The force in the y-direction (i.e., parallel to the plate) would depend upon the shear stress between the fluid and the surface of the plate.

If the fluid flows over a curved surface (as in vanes or in a pipe bend), similar technique of analysis can be employed for the computations of force exerted on the curved surface.

EXAMPLES

E5.1. In a pipeline carrying water, section B is 2m higher than section A. The diameter and pressure at sections A and B are as follows:

Section	Diameter (cm)	Pressure (kPa)
A	80	90
B	40	20

If the velocity of flow at section A is 1 m/s, determine the direction of flow. Neglect losses.

Solution: From the continuity principle, Eq. (5.1.2)

$$Q = A_A \, V_A = A_B \, V_B$$

or

$$\frac{\pi}{4}(0.8)^2 \, (1) = \frac{\pi}{4}(0.4)^2 \, (V_B)$$

∴

$$V_B = 4 \text{ m/s}$$

Total head, H_A at section A equals (with $Z_A = 0$; A being treated as reference datum)

$$\frac{p_A}{\rho g} + \frac{V_A^2}{2g} + Z_A$$

$$= \frac{90 \times 1000}{1000 \times 9.8} + \frac{(1)^2}{2 \times 9.8} + 0$$

$$= 9.235 \text{ m}$$

Total head, H_B at section B equals (with Z_B as datum head)

$$= \frac{p_B}{\rho g} + \frac{V_B^2}{2g} + Z_B$$

$$= \frac{20 \times 1000}{1000 \times 9.8} + \frac{(4)^2}{2 \times 9.8} + 2$$

$$= 4.857 \text{ m}$$

Since H_A is greater than H_B, the flow is from A to B.

E5.2. A manometer connected to a Pitot-static tube, placed in an air flow, indicates a difference in pressure head, between the tapings, of 5 mm of water. Compute the air velocity.

Solution: Pressure head difference in terms of the flowing fluid air, Δh_a is given by

$$\rho_a \, g \, \Delta h_a = \rho_w \, g \, \Delta h_w$$

$$\therefore \qquad \Delta h_a = \frac{\rho_w}{\rho_a} \Delta h_w$$

$$= \frac{1000}{1.2}(5)\,\text{mm}$$

$$= 4.17 \text{ m}$$

$\therefore$ The required velocity

$$= \sqrt{2g\,\Delta h_a}$$

$$= \sqrt{2 \times 9.8 \times 4.17}$$

$$= 9.04 \text{ m/s}$$

E5.3. A venturimeter is installed in a pipeline of diameter 15 cm that is inclined upwards at an angle of 30° with the horizontal. The distance between the pipe and throat tapings is kept equal to 50 cm. The maximum expected discharge in the pipeline is 0.2 m³/s. The pressure at the upstream end of the venturimeter is known to be 1 m of water at the maximum discharge. Compute the minimum permissible diameter of the venturimeter at the throat section. Assume vapour pressure of water as 2m of water (absolute) and atmospheric pressure as 10.13 m of water.

Solution: Pressure at the throat section should not be allowed to be lower than the vapour pressure of the flowing liquid. Applying Bernoulli's equation, Eq. (5.1.4), between sections 1 and 2, (Fig. 5.6.1) for the given data, one gets

$$(1.0+10.13)+\frac{(0.2)^2}{\left[\frac{\pi}{4}(0.15)^2\right]^2 (2\times9.8)}+0=2.0+\frac{(0.2)^2}{\left[\frac{\pi}{4}d^2\right]^2 (2\times9.8)}+0.5\sin30^o$$

or $\qquad 11.13 + 6.535 = 2 + \dfrac{3.308\times10^{-4}}{d^4} + 0.25$

$\therefore \qquad d = 0.121\ m = 12.1 \text{ cm}$

E5.4. Water is to be pumped at the rate of 0.1 m³/s from a sump in which the water level is 5 m below the ground. The diameter of the suction pipe is 15 cm. Assuming the head loss on the suction side of the pump to be 3m, estimate the highest permissible position of the pump with respect to the ground surface.

Solution: Applying Bernoulli's equation, Eq. (5.1.5), between sections 1 and 2, Fig. E5.4,

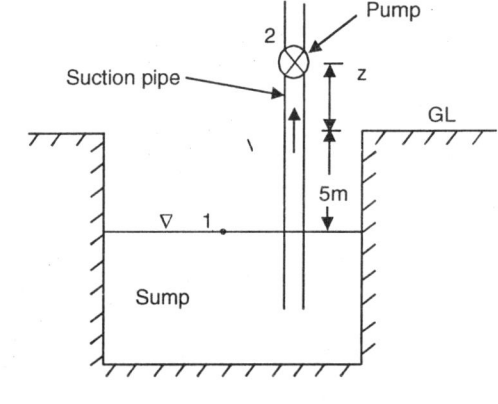

Fig. E 5.4.

$$(0+0+0)=(2-10.13)+\frac{(0.1)^2}{\left[\frac{\pi}{4}(0.15)^2\right]^2(2\times9.8)}$$

$$+(Z+5)+3$$

$\therefore$ $\qquad\qquad\qquad$ $Z=-1.504\ m$

i.e., the pump should be located 1.504 m below the ground level.

E5.5. $\quad$ A 15 cm diameter pipe is reduced to a 5 cm diameter nozzle discharging freely into the atmosphere. The pipe carries a water discharge of 15 l/s. Estimate the force required to hold the nozzle in place. Neglect friction.

Solution: $\quad$ Let F_x be the force required to hold the nozzle in place and is as shown in the Fig. E5.5. Applying momentum principle for the control volume shown dotted in Fig. E5.5,

$\qquad\qquad$ $p_1A_1-p_2A_2-F_x=\rho Q(v_2-v_1)$

$\therefore$ $\qquad$ $F_x=p_1A_1-\rho Q\ (v_2-v_1)\ as\ p_2=0$ (being atmospheric pressure)

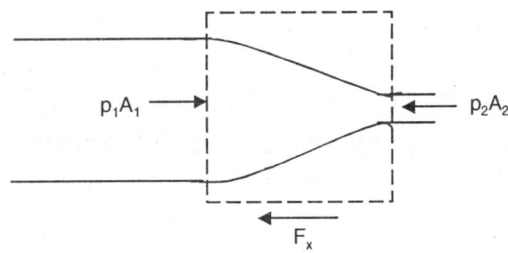

Fig. E 5.5.

From given data,

$$A_1=\frac{\pi}{4}(0.15)^2=0.018\,m^2\ and\ A_2=1.96\times10^{-3}\ m^2$$

$\therefore$ $\quad$ $v_1=15\times10^{-3}/0.018=0.83$ m/s and $v_2=7.65$ m/s

Hence, using Bernoulli's equation

$$p_1=\frac{1}{2}\rho\left(v_2^2-v_1^2\right)=\frac{1}{2}(1000)\left[(7.65)^2-(0.83)^2\right]$$

$\qquad\qquad$ $=28916.8\ N/m^2$

$\therefore$ $\qquad\qquad$ $F_x=28916.8\times0.018-1000\times(0.015)(7.65-0.83)$

$\qquad\qquad\qquad$ $=520-102.3$

$\qquad\qquad\qquad$ $=417.7\ N$

E5.6. $\quad$ An open water tank has a hole of 5 cm diameter in one of its vertical walls and contains water to a depth of 1.6 m above the centre of the hole, compute the reaction of the jet on the tank and its water when (i) it is stationary, and (ii) it is moving with a velocity equal to 1.5 m/s in the direction opposite to that of the jet while the velocity of the jet relative to that of the tank remains unchanged. What would be the work done per second in the case (ii)? Assume coefficients of contraction and velocity as 0.64 and 0.9.

Solution: Velocity of the jet $= C_v \sqrt{2gh}$

$$= 0.9 \sqrt{19.6 \times 1.6}$$

$$= 5.04 \text{ m/s}$$

Area of cross-section of the jet $= 0.64 \dfrac{\pi}{4} \left(5 \times 10^{-2}\right)^2$

$$= 1.26 \times 10^{-3} \text{ m}^2$$

Therefore, discharge $Q = 5.04 \times 1.26 \times 10^{-3} \text{ m}^3/\text{s}$

$$= 6.35 \times 10^{-3} \text{ m}^3/\text{s}$$

Consider the control volume as shown. Force exerted by the fluid in the direction of jet flow (when the tank is stationary),

$$R = -F = -\rho Q \, (v_2 - v_1)$$

$$= -1000 \times 6.35 \times 10^{-3} \, (5.04 - 0)$$

$$= -32 \text{ N}$$

$\therefore$ Force exerted by the jet on the tank is 32 N in the direction opposite to that of the jet.

When the tank is moving with a velocity equal to 1.5 m/s in the direction opposite to that of the jet, the effect is to superimpose a negative (relative to the jet) velocity of 1.5 m/s on the whole system. This means,

$$v_2 = 5.04 - 1.5 = 3.54 \text{ m/s}$$

and $\qquad\qquad v_1 = -1.5 \text{ m/s}$

Hence, $\qquad v_2 - v_1 = 3.54 - (-1.5) = 5.04 \text{ m/s}$

Hence, the reaction of the jet remains the same at 32N. Therefore, the work done per second equals force × velocity of the tank

$$= 32 \times 1.5$$

$$= 48 \text{ Nm/s or } 48 \text{ W}$$

E5.7.   A horizontal tapering 45° pipe bend, Fig. E5.7, has an inlet diameter of 600 mm and outlet diameter of 300 mm. The pressure at its inlet end is 150 kPa and the discharge through the bend is 0.5 m³/s. Compute the net horizontal force exerted by water on the bend. Neglect friction.

Solution: Assuming uniform velocity distribution (i.e. $\beta = 1$) at inlet and outlet, one can write the momentum equations in x-, and y-directions for the fluid in the pipe bend that constitutes the control volume.

$$p_1 A_1 - p_2 A_2 \cos 45° + F_x = \rho Q \, (v_2 \cos 45° - v_1)$$

and $\qquad\qquad (-) p_2 A_2 \sin 45° + F_y = \rho Q \, (v_2 \sin 45°)$

Now $\qquad\qquad v_1 = \dfrac{0.5}{\dfrac{\pi}{4}(0.6)^2} = 1.768 \; m/s$

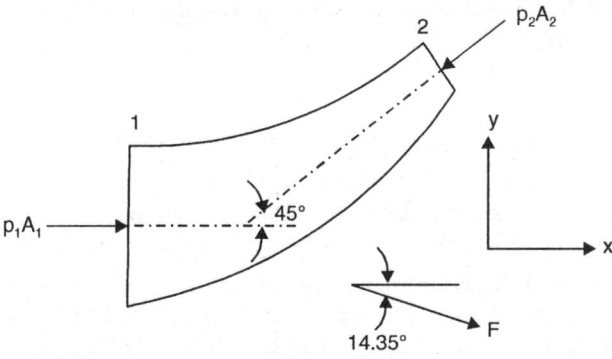

Fig. E 5.7.

$$v_2 = \frac{0.5}{\frac{\pi}{4}(0.3)^2} = 7.074 \ m/s$$

and by the Bernoulli's equation,

$$p_2 = p_1 + \frac{1}{2}\rho\left(u_1^2 - u_2^2\right)$$

$$= 150 \times 10^3 + \frac{1}{2}(1000)\left[(1.768)^2 - (7.074)^2\right]$$

$$= 126.5 \times 10^3 \ pa$$

Hence, from the x-direction momentum equation

$$F_x = 1000 \times 0.5(7.074 \times 0.707 - 1.768) - 150 \times 10^3 \left[\frac{\pi}{4}(0.6)^2\right] + 126.3 \times 10^3 \left[\frac{\pi}{4}(0.3)^2\right](0.707)$$

$$= 1616.66 - 42411.50 + 6311.83$$
$$= -34483 \ N$$

Likewise, from the y-direction momentum equation,

$$F_y = 1000 \ (0.5)(7.074 \times 0.707) + 126.5 \times 10^3 \left[\frac{\pi}{3}(0.3)^2\right](0.707)$$

$$= 2500.66 + 6321.82$$
$$= 8822.48 \ N$$

Therefore, the net force exerted on water,

$$= \sqrt{(34483)^2 + (8822.48)^2}$$

$$= 35593.73 \ N$$

acting in the direction, $\tan^{-1}\left\{\dfrac{8822.48}{(-34483)}\right\}$

$$180 - 14.35°$$

Hence, force F exerted on the bend is equal to 35593.73 N acting in the opposite direction as shown in Fig. E5.7.

PROBLEMS

P5.1   A pipe carrying water tapers from a cross-section of 0.3 m^2 at A to 0.15 m^2 at B. At A, the velocity, assumed uniform, is 2.0 m/s and the pressure 1.15 kPa gauge. If frictional effects are negligible, determine the pressure at B, which is 5 m above the level of A.

P5.2   A submarine, submerged in seawater, travels at 15 km/h. Calculate the pressure at the front stagnation point situated 20 m below the sea surface. (Density of sea water = 1026 kg/m^3).

P5.3   The suction pipe of a pump rises from the sump at a slope of 1 vertical in 4 along the pipe and water flows through it at 2 m s^{-1}. If air dissolved in the flowing water is released when the pressure falls to more than 70 kNm^{-2} below atmospheric pressure, find the maximum length of the pipe neglecting friction.

P5.4   A jet of water is initially 10 cm in diameter and, when directed vertically upwards, reaches a maximum height of 20 m. Assuming that the jet remains circular, determine the discharge of water flowing and the diameter of the jet at a height of 10 m.

P5.5   A pipe AB carries water and tapers uniformly from a diameter of 0.1 m at A to 0.2 m at B over a length of 3 m. Pressure gauges are installed at A, B and also at C, the mid-point of AB. If the pipe centerline slopes upwards from A to B at an angle of 30° and the pressure recorded at A and B are, respectively, 2.0 and 2.3 bar, determine the flow rate through the pipe and pressure recorded at C. Neglect all losses.

P5.6   An orifice plate is to be used to measure the volumetric rate of airflow through a 2m diameter duct. The mean velocity in the duct will not exceed 12 m s^{-1} and a water tube manometer, having a maximum difference between water levels of 160 mm, is to be used. Assuming the coefficient of discharge of the orifice to be 0.64, determine a suitable orifice diameter to make full use of the manometer range. The density of air is 1.2 kg m^{-3}.

P5.7   In an open rectangular channel the velocity, although uniform across the width varies linearly with depth, the value at the free surface being twice that at the base which is moving. Show that the value of the kinetic energy correction factor is 10/9.

P5.8   The diameter of a pipe-bend is 300 mm at inlet and 150 mm at outlet and the flow is turned through 120° in a vertical plane. The axis at inlet is horizontal and the centre of the outlet section is 1.5 m below the centre of the inlet section. The total volume of fluid contained in the bend is 0.085 m^3. Neglecting friction, calculate the magnitude and direction of the net force exerted on the bend by water flowing through it at 0.25 m^3/s when the inlet pressure is 150 kPa.

REFERENCES

(1)   Henry HR: Discussion on *"Diffusion of submerged Jets"* by Albertson, ML et al., Trans. ASCE, Vol. 115, 1950 [Henderson, FM: Open Channel Flow, The Macmillan Company, New York, 1966].

(2)   Massey, BS (revised by John Ward-Smith): *Mechanics of Fluids*, 7th Edition, Chennai Micro Print Pvt. Ltd., Chennai, India, 1998.

(3)   Rehbock, T: Discussion on *"Precise Weir Measurements"* by Schoder, EW and Turner, KB, Trans. ASCE, Vol. 93, 1929 [Henderson, FM: Open Channel Flow, The Macmillan Company, New York, 1966].

CHAPTER 6

Physical Modelling and
Flow Measurement

6.1 DIMENSIONAL ANALYSIS

Only a very few real fluid flows can be solved exactly by analytical methods alone. Combining analytical and experimental techniques, one can, however, solve most of the real fluid flows. This requires approximate mathematical (or analytical) solution, based on simplifying assumptions, of a real flow problem. Thereafter, the experimental verification of the mathematical solution is carried out and refinements in the analytical solution are made. Since experimental work is both time-consuming as well as costly affair, one would, obviously, aim to obtain the maximum information from the minimum experimental work. The technique of dimensional analysis helps one in achieving this aim (2).

Primarily, dimensional analysis reduces the number and complexity of the physical variables (or quantities) that affect a specified physical phenomenon (8). For example, one wants to obtain information about the drag force (i.e., the force exerted by a flowing fluid in the direction of motion) on a smooth sphere. Obviously, the drag force, F should depend on the size of the sphere (that may be characterized by the diameter, D), the fluid properties such as viscosity μ, and mass density ρ, and the fluid velocity V. For the experimental analysis, one would require to vary each of the four independent variables V, D, ρ, and μ, while keeping other three constants and, thus, obtain the information on the variation of the dependent variable F with the variation of the variable that was varied. Supposing that one requires a minimum of ten observations to define or plot a curve to obtain the variation between two variables, one would need to conduct 10^4 i.e., 10,000 experiments. Not only this, to be able to predict F for the specified values of D, ρ, μ and V, one would end up in having 1000 curves or 100 figures each containing 10 curves representing the variation of F with V for 10 different values of D when ρ and μ are held constant. Since ρ and μ, together, are to be varied 100 times, there will be 100 such figures. One can, therefore, easily appreciate that the experimental programme, carried out in this manner, is both time-consuming as well as expensive. Besides, the obtained information has not been presented in a comprehensible manner. Further, one would require variety of fluids to enable one to vary ρ and μ. By adopting the dimensional analysis technique for this problem, one would have required dealing with only two dimensionless variables, $\dfrac{F}{\rho D^2 V^2}$ and $\dfrac{VD\rho}{\mu}$, instead of five dimensional

variables, viz., F, D, ρ, μ and V (see Art.6.3). This means that the dimensionless force coefficient $\dfrac{F}{\rho D^2 V^2}$

is a function only of the dimensionless variable $\dfrac{VD\rho}{\mu}$ that is termed the Reynolds number. The nature

of the functional relationship can be obtained by experimentation. One needs to vary only $\dfrac{VD\rho}{\mu}$ (and not

the individual dimensional variables) and one can achieve this variation by simply varying the velocity of flow in, say, a wind tunnel and working with one sphere (i.e., one value of diameter D) only. There is absolutely no need to have 10 spheres of different diameters or to have 100 different fluids with 10 different densities and 10 different viscosities,

Another important benefit of working with dimensional analysis is that it provides scaling laws that are so useful in predicting the behaviour or response of a prototype based on the information collected on its model in a laboratory environment.

Dimensional analysis has a strong physical and mathematical basis. Nevertheless, this technique has many subtleties and nuances and, therefore, can be perfected only by practice.

6.2 BUCKINGHAM'S Pi THEOREM

The dimensional analysis is based on the principle of dimensional homogeneity that can be stated as follows (8):

An equation, describing true relationship amongst variables governing a physical phenomenon, is always dimensionally homogeneous. This means that each of the additive terms of the equation will have the same dimensions.

Buckingham's Pi theorem (1) enables one to obtain important dimensionless parameters from the dimensional variables governing a physical phenomenon. Consider a physical phenomenon for which the dependent parameter p_1 is dependent on n-1 independent parameters (p_2, p_3, ...p_n) governing the phenomenon. This relationship can be expressed in functional form as either

$$p_1 = \phi\ (p_2, p_3 \ldots\ldots\ldots p_n) \qquad\qquad ...(6.2.1)$$

or

$$\psi\ (p_1, p_2, p_3, \ldots\ldots\ldots\ p_n) = 0 \qquad\qquad ...(6.2.2)$$

where, ϕ and ψ are different unspecified functions. Buckingham's Pi theorem states that the n parameters of Eq. (6.2.2) may be grouped into n-m independent dimensionless ratios that are usually termed Π parameters and are expressible in functional form as

$$f(\Pi_1, \Pi_2, \ldots\ldots\ldots\ldots\Pi_{n-m}) = 0 \qquad\qquad ...(6.2.3)$$

or, alternatively,

$$\Pi_1 = f_1\ (\Pi_1, \Pi_2, \ldots\ldots\ldots\ldots\Pi_{n-m}) \qquad\qquad ...(6.2.4)$$

Here, m is the number of the basic or primary or fundamental dimensions (i.e., mass M, length L, time T, and temperature θ) that appear in the variables (p_1 ....p_n) governing the phenomenon. For example, in the problem of fluid flow past a smooth sphere (Art. 6.1) n is 5 and m is 3, since the primary dimension θ does not appear in any of the five variables F, D, ρ, μ, and V governing the phenomenon.

The *n-m* $\prod$ parameters of Eq. (6.2.3 or 6.2.4) are not unique. A $\prod$ parameter is not independent if it can be formed from a product or quotient of any combination of the other $\prod$ parameters of the problem. For example, if

$$\Pi_3 = \frac{\Pi_1}{2\Pi_2\Pi_3} \text{ or } \Pi_4 = \frac{\Pi_1^{3/4}}{\Pi_2^3}$$

then, neither $\prod_3$ nor $\prod_4$ are independent $\prod$ parameters. Further, the theorem does not predict the nature of functions f or f_1 and, hence, the relationship among the independent $\prod$ parameters has to be determined experimentally.

6.3 DETERMINATION OF $\prod$ PARAMETERS

The following steps describe one of the recommended procedures for determining $\prod$ parameters:

1. Select all the pertinent variables that govern the physical phenomenon. If all the relevant variables were not included, the relation obtained would not be the correct one. If an additional variable, not affecting the phenomenon, is included in the list, the result may have an additional $\prod$ parameter that will be dropped when the experimental results indicate this parameter to be redundant or extraneous. For the problem of fluid flow past a sphere, the variables are F, D, ρ, μ, and V, and the functional relation for these five dimensional parameters is

$$F = \phi (D, \rho, \mu, V) \qquad \qquad ...(6.3.1)$$

or $\qquad \psi (F, D, \rho, \mu, V) = 0 \qquad \qquad ...(6.3.2)$

Thus, $n = 5$

2. The primary dimensions that appear in these variables (F, D, ρ, μ, and V) are M, L, and T. Hence, $m = 3$.

3. Select from the list of pertinent variables a number of repeating variables equal to the number of primary dimensions. The repeating variables should be chosen using the following guidelines:

   (a) The dependent variable is not included in the list of the repeating variables.
   (b) No two repeating variables may have the same net dimensions differing by only an exponent. For example, length [L] and second moment of area [L^4] together should not be included as the repeating variables.
   (c) The repeating variables do not combine among themselves to form a dimensionless group.
   (d) Together they contain all the primary dimensions that appear in the pertinent variables governing the phenomenon.
   (e) Since repeating variables would appear in all the dimensionless $\prod$ parameters, one should prefer to select more common or popular variables (such as flow velocity, body or flow dimension, mass density) and not obscure variables (such as velocity of sound, surface roughness, surface tension etc.).

   Normally, for fluid flow problems, one flow (or kinematic) property, one geometric (or scale) characteristic, and one fluid property relating to forces or mass of the system constitute a good set of repeating variables (3). For the problem of sphere chosen for illustration, and also many other problems, one can choose V, D, and ρ as repeating variables.

4. Combine the set of repeating variables with each of the non-repeating variables of the phenomenon, in turn, to form *n-m* dimensional equations. For the problem chosen for illustration, the two

dimensionless parameters would be formed by combining V, D, and ρ once with F and next with μ such that

$$\Pi_1 = V^a \, D^b \, \rho^c \, F \Rightarrow \left[\left(\frac{L}{T} \right)^a (L)^b \left(\frac{M}{L^3} \right)^c \left(\frac{ML}{T^2} \right) \right]$$

i.e.,

$$\left[M^{c+1} \, L^{a+b-3c+1} \, T^{-a-2} \right]$$

and

$$\Pi_2 = V^d \, D^e \, \rho^f \, \mu \Rightarrow \left[\left(\frac{L}{T} \right)^d (L)^e \left(\frac{M}{L^3} \right)^f \left(\frac{M}{LT} \right) \right]$$

i.e.,

$$\left[M^{f+1} \, L^{d+e-3f-1} \, T^{-d-1} \right]$$

in which, $\Rightarrow [\ ]$ means "has dimensions of".

For Π_1 and Π_2 to be dimensionless, the resulting exponents of M, L, and T should be zero. This means, for Π_1

$$c + 1 = 0$$
$$a + b - 3c + 1 = 0$$

and
$$-a - 2 = 0$$

Hence,
$$c = -1, \, a = -2, \text{ and } b = -2$$

Therefore,
$$\Pi_1 = V^{-2} \, D^{-2} \rho^{-1} \, F = \frac{F}{\rho V^2 D^2}$$

Similarly, for Π_2

$$f + 1 = 0$$
$$d + e - 3f - 1 = 0$$

and
$$-d - 1 = 0$$

Hence, $f = -1$, $d = -1$, and $e = -1$

Therefore,
$$\Pi_2 = V^{-1} \, D^{-1} \rho^{-1} \, \mu = \frac{\mu}{VD\rho}$$

One may check the dimensions of Π_1 and Π_2 to see if they are dimensionless.

5. The functional relationship can now be expressed. For the problem chosen for illustration, the functional relation is:

$$f(\Pi_1, \Pi_2) = 0 \qquad \qquad ...(6.3.3)$$

or
$$\Pi_1 = f_1(\Pi_2) \qquad \qquad ...(6.3.4)$$

or
$$\frac{F}{\rho V^2 D^2} = f_1 \left(\frac{\mu}{VD\rho} \right) \qquad \qquad ...(6.3.5)$$

Since inverse of a dimensionless parameter is also dimensionless, Eq.(6.3.5) is, preferably, written as,

$$\frac{F}{\rho V^2 D^2} = f_2 \left(\frac{VD\rho}{\mu} \right) \qquad \qquad ...(6.3.6)$$

6.3.1 Alternative Method for Determination of Π Parameters

Another alternative method (4) for obtaining Π parameters expresses primary dimensions M, L, and T in terms of the chosen repeating variables. Thus, for the repeating variables V, D, and ρ, one gets,

$$V = LT^{-1}, D = L, \text{ and } \rho = ML^{-3}$$

Therefore, $L = D$, $T = DV^{-1}$ and $M = \rho D^3$

The next step is to write the dimensions of non-repeating variables (F and μ for the chosen problem for illustration) in terms of the repeating variables. Thus,

$$F = MLT^{-2} = \left(\rho D^3\right)(D)\left(D^{-2}V^2\right) = \rho V^2 D^2$$

Hence, the corresponding Π parameter is

$$\Pi_1 = \frac{F}{\rho V^2 D^2}$$

Similarly, $\mu = ML^{-1}T^{-1} = \left(\rho D^3\right)\left(D^{-1}\right)\left(D^{-1}V\right) = VD\rho$

$$\Pi_2 = \frac{\mu}{VD\rho}$$

The advantage of this method is that it does not require the repeated solution of three equations in three unknowns for the determination of each Π parameter.

6.4 SIGNIFICANCE OF MAJOR DIMENSIONLESS PARAMETERS IN FLUID MECHANICS

Consider a fluid flow phenomenon that depends on the velocity V, mass density ρ, viscosity μ, pressure drop Δp, gravitational acceleration g, surface tension σ, bulk modulus of elasticity K, and linear dimensions l, l_1 and l_2. The functional relation for these pertinent variables can, therefore, be written as

$$\Psi\left(V, \rho, \mu, \Delta p, g, \sigma, K, l, l_1, l_2\right) = 0 \qquad \text{...(6.4.1)}$$

Thus, $n = 10$ and $m = 3$. Therefore, according to Buckingham's PI theorem, there would be (n-m), i.e., seven dimensionless Π parameters. Selecting velocity V, characteristic length l, and mass density ρ as three repeating variables, one can work out seven independent dimensionless Π parameters. These would be as follows:

$$\Pi_1 = \frac{\Delta p}{\rho V^2}, \qquad \Pi_2 = \frac{Vl\rho}{\mu}, \qquad \Pi_3 = \frac{V}{\sqrt{gl}}, \qquad \Pi_4 = \frac{\rho l V^2}{\sigma},$$

$$\Pi_5 = \frac{V}{\sqrt{\dfrac{K}{\rho}}}, \qquad \Pi_6 = \frac{l_1}{l}, \qquad \Pi_7 = \frac{l_2}{l}$$

Therefore, functional relation for the fluid flow phenomenon in terms of the dimensionless Π parameter is

$$f\left(\frac{\Delta p}{\rho V^2}, \frac{Vl\rho}{\mu}, \frac{V}{\sqrt{gl}}, \frac{\rho l V^2}{\sigma}, \frac{V}{\sqrt{K/\rho}}, \frac{l_1}{l}, \frac{l_2}{l}\right) = 0 \qquad \qquad ...(6.4.2)$$

Each of the first five Π parameters of Eq. (6.4.2) is a dimensionless number that has been given the name of a scientist or engineer who pioneered the use of that dimensionless number. It is worth understanding the physical significance of these dimensionless numbers. This would help one understand the flow phenomenon under one's consideration.

Forces that may affect a fluid phenomenon, generally, include those due to inertia, pressure, viscosity, gravity, surface tension, and compressibility. From the Newton's second law of motion, force is the product of mass m and acceleration a that can be, respectively, written as

$$m = \rho \, \forall$$

and

$$a = \frac{dv}{dt} \cong v \frac{dv}{ds}$$

Since $\forall$ is the volume and has the dimension of L^3, one can say that $m \propto \rho l^3$.

Similarly, $a \propto \dfrac{V^2}{l}$. Hence,

Inertial force $\quad = m \, a \propto \rho V^2 l^2$

Similarly, $\qquad$ Pressure force $\ = (\Delta p)A \propto \Delta p l^2$

$$\text{Viscous force} \quad = \tau A = \mu \frac{du}{dy} A \propto \mu \frac{V}{l} l^2 = \mu V l$$

Gravity force $\ = m \, g \propto g \, \rho \, l^3$

Surface tension force $\propto \sigma \, l$

Compressibility force $= KA \propto K l^2$

Inertial force is always present in all fluid flow problems. The ratio of the inertial force with each of the five remaining forces listed above leads to five basic dimensionless Π parameters (or numbers) that are usually encountered in fluid flow problems.

6.4.1 The Euler Number

The first Π parameter in Eq. (6.4.2), $\dfrac{\Delta p}{\rho V^2}$ is the ratio of the pressure force and inertial force. This Π

parameter is often written as $\dfrac{\Delta p}{\dfrac{1}{2}\rho V^2}$ so as to make the denominator equal to the dynamic pressure. This

ratio $\dfrac{\Delta p}{\dfrac{1}{2}\rho V^2}$ is called the *Euler Number* after the Swiss Mathematician, Leonhard Euler (1707 – 1783),

who did much of the initial analytical work in fluid mechanics. Euler number is often called the *pressure coefficient*. The term pressure difference, Δp is with respect to a reference pressure that is usually the static pressure in the flow region.

In some situations of liquid flow, the pressure in some flow region may reduce to vapour pressure of the liquid giving rise to the cavitation phenomenon. In such cases, the pressure difference is taken relative to the vapour pressure and the resulting dimensionless number is the *cavitation number.*

6.4.2 The Reynolds Number

The second Π parameter in Eq. (6.4.2), $\dfrac{Vl\rho}{\mu}$ is the ratio of the inertial force and viscous force and is called the *Reynolds Number, R_e* after the British engineer Osborne Reynolds (1842 - 1912) who, in 1880's, studied the transition between laminar and turbulent flow regimes in a small diameter pipe. A critical value of the Reynolds number distinguishes between laminar and turbulent flow in pipes, in the boundary layer, or around immersed bodies. Relatively large Reynolds number flows have larger inertial force compared with the viscous force and are usually turbulent. On the other hand, flows, having relatively smaller inertial force compared with the viscous force, are laminar.

6.4.3 The Froude Number

The square root of the ratio of the inertial force and the gravity force, i.e., $\dfrac{V}{\sqrt{gl}}$ is called the *Froude number F_r* after a British naval architect, William Foude (1810 – 1879). This number is important in flows that have a free surface and are, therefore, affected by gravity. The characteristic length for open channel or free surface flow is taken as the depth of flow. Flow in open channel is termed either subcritical, or critical, or supercritical depending upon whether the Froude number is less than, or equal to, or greater than unity.

6.4.4 The Weber Number

The fourth Π parameter of Eq. (6.4.2) represents the ratio of the inertial force and the surface tension force and is called the *Weber number W* named after Moritz Weber (1871-1951) of the Polytechnic Institute of Berlin who developed the laws of similitude. The Weber number is important when it is of the order of unity or less. This would typically be so when the surface curvature is comparable in size to the liquid depth such as in droplets, capillary flows, small depths in hydraulic models etc.

6.4.5 The Mach Number

The fifth Π parameter in Eq. (6.4.2), $\dfrac{V}{\sqrt{K/\rho}}$ can be obtained by taking the square root of the ratio of the inertial force and the compressibility force. This ratio is named as the *Mach number M* after Ernst Mach (1838-1916), an Austrian Physicist. The speed of sound in a liquid is equal to $\sqrt{K/\rho}$. Thus, the Mach number compares the velocity of fluid flow with the velocity of sound in the fluid. Flow is, therefore, termed sonic when $M=1$, subsonic when $M<1$, and supersonic when $M>1$.

6.4.6 Other Dimensionless Parameters

In addition to the dimensionless Π parameters, described above, there are many more dimensionless parameters one comes across while dealing with fluid flows. Some of these and the ones, described above, have been listed in Table 6.4.1.

Table 6.4.1: Dimensionless Groups in Fluid Mechanics (8)

Parameter	Definition	Qualitative Ratio of Effects	Importance
Euler Number	$E_u = \dfrac{\Delta p}{(1/2)\rho V^2}$	$\dfrac{\text{Pressure Force}}{\text{Inertial Force}}$	Fluid dynamics
Reynolds Number	$R_e = \dfrac{VL\rho}{\mu}$	$\dfrac{\text{Inertial Force}}{\text{Viscous Force}}$	Pipe flow and viscous flow
Froude Number	$F_r = \dfrac{V}{\sqrt{gL}}$	$\dfrac{\text{Inertial Force}}{\text{Gravity Force}}$	Free-surface flow
Weber Number	$W = \dfrac{\rho V^2 L}{\sigma}$	$\dfrac{\text{Inertial Force}}{\text{Surface Tension Force}}$	Free-surface flow
Mach Number	$M = \dfrac{V}{c}$	$\dfrac{\text{Flow Speed}}{\text{Sound Speed}}$	Compressible flow
Roughness Ratio	$\dfrac{\in}{L}$	$\dfrac{\text{Wall Roughness}}{\text{Body Length}}$	Turbulent, rough walls
Temperature Ratio	$\dfrac{T_w}{T_0}$	$\dfrac{\text{Wall Temperature}}{\text{Stream Temperature}}$	Heat transfer
Pressure Coefficient	$C_p = \dfrac{p - p_0}{(1/2)\rho V^2}$	$\dfrac{\text{Static Pressure Force}}{\text{Dynamic Force}}$	Aerodynamics, hydrodynamics
Lift Coefficient	$C_L = \dfrac{F_L}{(1/2)\rho V^2 A}$	$\dfrac{\text{Lift Force}}{\text{Dynamic Force}}$	Aerodynamics, hydrodynamics
Drag Coefficient	$C_D = \dfrac{F_D}{(1/2)\rho V^2 A}$	$\dfrac{\text{Drag Force}}{\text{Dynamic Force}}$	Aerodynamics, hydrodynamics
Friction Factor	$f = \dfrac{h_f}{\left(\dfrac{V^2}{2g}\right)\left(\dfrac{L}{D}\right)}$	$\dfrac{\text{Friction Head Loss}}{\text{Velocity Head}}$	Pipe flow
Skin Friction Coefficient	$C_f = \dfrac{\tau_{wall}}{\rho V^2/2}$	$\dfrac{\text{Wall Shear Stress}}{\text{Dynamic Pressure}}$	Boundary layer flow
Cavitation Number (Euler Number)	$Ca = \dfrac{p - p_v}{1/2\,\rho V^2}$	$\dfrac{\text{Pressure}}{\text{Dynamic Pressure}}$	Cavitation

(Contd.)

Parameter	Definition	Qualitative Ratio of Effects	Importance
Prandtl Number	$\Pr = \dfrac{\mu C_p}{k}$	$\dfrac{\text{Dissipation}}{\text{Conduction}}$	Heat convection
Eckert Number	$Ec = \dfrac{V^2}{c_p T_0}$	$\dfrac{\text{Kinetic Energy}}{\text{Enthalpy}}$	Dissipation
Specific-heat Ratio	$k = \dfrac{c_p}{c_v}$	$\dfrac{\text{Enthalpy}}{\text{Internal Energy}}$	Compressible flow
Strouhal Number	$St = \dfrac{\omega L}{V}$	$\dfrac{\text{Oscillation}}{\text{Mean Speed}}$	Oscillating flow

6.5 EMPIRICAL RELATION BASED ON DIMENSIONAL ANALYSIS

Consider a flow of fluid in a pipe. The functional relation based on dimensional analysis, Eq. (6.4.2), includes all the relevant parameters affecting a typical flow besides three characteristic length parameters l, l_1 and l_2, which may, respectively, be considered as diameter of the pipe D, length of pipe L and the effective height of the surface roughness ϵ of the pipe. Thus, Eq. (6.4.2) can be rewritten as

$$f_1\left(\frac{\Delta p}{(1/2)\rho V^2}, R_e, F_r, W, M, \frac{L}{D}, \frac{\epsilon}{D}\right) = 0 \qquad \qquad ...(6.5.1)$$

For flow in a pipe, there is no free surface and, hence, F_r is not important and can be dropped for further analysis. Similarly, surface tension effects are not relevant for pipe flow. Therefore, W too can be dropped. If the velocity of flow in pipe is much less than that of sound in the flowing fluid, then compressibility effects too can be neglected. This means, M too can be neglected for further analysis. Equation (6.5.1), thus, reduces to

$$f_2\left(\frac{\Delta p}{(1/2)\rho U^2}, R_e, \frac{L}{D}, \frac{\epsilon}{D}\right) = 0 \qquad \qquad ...(6.5.2)$$

or

$$\frac{\Delta p}{(1/2)\rho V^2} = f_3\left(R_e, \frac{L}{D}, \frac{\epsilon}{D}\right) \qquad \qquad ...(6.5.3)$$

To obtain the nature of the functional relation f_3 one has to resort to experimentation. Based on experimentation, it has been observed that the pressure drop Δp varies linearly with the length L.

Hence,

$$\frac{\Delta p}{(1/2)\rho V^2} = \frac{L}{D} f_4\left(R_e, \frac{\epsilon}{D}\right)$$

or

$$\frac{\Delta p}{\dfrac{1}{2}\rho V^2 \left(\dfrac{L}{D}\right)} = f_4\left(R_e, \frac{\epsilon}{D}\right) \qquad \qquad ...(6.5.4)$$

The left-hand-side term of Eq. (6.5.4) is defined as the *friction factor f* and, hence,

$$f = f_4\left(R_e, \frac{\epsilon}{D}\right) \qquad \qquad ...(6.5.5)$$

One can now carry out experiments and, based on the experimental data, prepare a plot with f as ordinate, R_e as abscissa, and ϵ/D (termed as *relative roughness*) as third variable. One should, however, plan experimentation in such a manner that the effect of the independent dimensionless parameters (i.e., R_e and ϵ/D) on the dependent parameter (i.e., f) is obtained in isolation of each other. This means that while studying the variation of f with R_e, the third dimensionless parameter ϵ/D should be kept constant so as to obtain the variation of f with R_e for the chosen value of ϵ/D. Thereafter, set another ϵ/D and, again, vary R_e to obtain the variation of f with R_e for this value of ϵ/D. If these data too plot on the previous variation (f v/s R_e), in spite of much different value of ϵ/D for the second set compared to that of the first set, one may conclude that f is not dependent on ϵ/D for the range of values of R_e and ϵ/D employed during the experimentation. Otherwise, one obtains the relation f v/s R_e for different values of ϵ/D. If the experimentation were carried out by changing R_e and ϵ/D arbitrarily and simultaneously, one might have obtained spurious variation of f v/s R_e indicating no effect of ϵ/D on f.

Experimentation, carried out in the aforesaid proper manner, has, indeed, revealed that when the Reynolds number R_e, for pipe flows, is less than about 2000, all the data corresponding to different values of ϵ/D yield one straight line and, therefore, f is independent of ϵ/D for R_e less than 2000. For this condition, the end result, therefore, is

$$f = f_5(R_e) \qquad \qquad ...(6.5.6)$$

It would be observed (Chapter 9) that this relation can be obtained based on theoretical considerations alone. However, the experimental verification would still be required for the verification and modification, if needed, of the theoretical prediction.

For higher Reynolds number (i.e., $R_e > 2000$), however, the friction factor would depend on the Reynolds number and/or relative roughness. This would be further dealt with later in Chapter 9.

6.6 PHYSICAL SIMILARITY

Model studies of proposed hydraulic structures are often undertaken so as to have visual observations of the flow and also collect numerical data that would yield the desired information for the prototype structure. Typical examples are calibration of weirs and gates, observing flow patterns, estimation of forces on industrial structures etc.

Once the dimensional analysis for a physical phenomenon has been carried out, the next step is, obviously, to establish suitable relationship amongst the relevant dimensionless Π parameters. This would require working with experimental set-ups that would simulate the physical phenomenon on a scale different, usually smaller, than that of the phenomenon. So that this exercise is useful, the model tests in the laboratory must yield data or information that can be scaled, or interpreted, to obtain the desired quantities such as forces, moments, dynamic loads etc. that would exist on a full-scale prototype.

Therefore, the investigator must ensure *physical similarity* between the model being tested and prototype to be designed and constructed. Two systems are considered physically similar in respect of specified physical quantities when the ratio of the corresponding magnitudes of these quantities between the two systems is the same (and the same direction, if the quantity is a vector quantity) everywhere in the system. Complete similarity between flow condition for a model and its prototype requires that all relevant dimensionless Π parameters have the same corresponding values for the model and the prototype. This means that a physical phenomenon governed by a functional relation

$$\Pi_1 = f\left(\Pi_2, \Pi_3, \ldots\ldots\ldots\ldots\Pi_n\right)$$

would be similar in model and prototype, if

$$\Pi_{2m} = \Pi_{2p}, \Pi_{3m} = \Pi_{3p}, \ldots\ldots\ldots\ldots, \Pi_{nm} = \Pi_{np}$$

and, therefore, the model value of Π_1, i.e., Π_{1m} would be equal to the desired value of Π_1 for the prototype, i.e., Π_{1p}. Here, the subscripts m and p refer, respectively, to the model and prototype. Attaining complete similarity (i.e., equality of all Π parameters, governing the specified physical phenomenon, between model and prototype), however, is impossible for most of the engineering problems (6). In engineering, therefore, one attempts to obtain particular types of similarity, the most common being geometric, kinematic, and dynamic similarities. While dealing with thermodynamic phenomena, one considers thermal similarity too in addition to the geometric, kinematic, and dynamic similarities.

6.6.1 Geometric Similarity

A model would be geometrically similar to its prototype if and only if all linear dimensions in all three coordinates have the same linear scale or scale factor. Hence, if the size of a model is to be one-tenth the size of the prototype, the model's length, width, and height must each be one-tenth the prototype's length, width and height, respectively. This applies to the fluid flow geometry as well as model geometry. All angles are to be preserved in geometric similarity. The orientations of flow directions with respect to the surroundings must be identical in model and prototype. It is like taking a photograph of the prototype and reducing or enlarging it until it fits the chosen size for the model. Geometric similarity is, therefore, similarity of shape and extends to the actual surface roughness of the model and prototype.

Geometric similarity is, obviously, the prime requirement for attaining similarity of other physical quantities in the model and prototype. In many situations, however, it is not easy to attain perfect geometric similarity. For a small model, for example, the surface roughness may not be as per the chosen scale factor, unless the model surfaces can be made much smoother than those of the prototype. A small model of alluvial river may require, according to the chosen scale factor, the use of a powder of impossible fineness to represent sand in the prototype alluvial river. While modelling long reach of an alluvial river, if one uses the same scale factor for horizontal lengths and vertical lengths, the depth of flow in the model may be so shallow that flow may be laminar instead of turbulent of the prototype river and surface tension effects in the model become considerable. In such situations, distorted model may be necessary.

6.6.2 Kinematic Similarity

Kinematic similarity implies similarity of lengths (i.e., geometric similarity) besides the similarity of time intervals. Kinematic similarity is the similarity of motion and, therefore, requires that just as the length scale factor is a fixed value, the corresponding time scale factor (or ratio) must also be a fixed value.

According to Langhaar (5), the motion of two systems are kinematically similar if homologous particles lie at homologous points at homologous times. If the ratio of a length in the model l_m to its corresponding length in the prototype, l_p is $l_r\ (=l_m/l_p)$ and ratio of the corresponding time intervals is $t_r\ (=t_m/t_p)$, then the corresponding velocities and acceleration would, respectively, be in the ratios l_r/t_r

and $\dfrac{l_r}{t_r^2}$. A familiar example of kinematic similarity can be seen in a planetarium in which heavenly bodies (planets) are produced in a certain length scale factor. For reproducing the motion of the planets, a fixed ratio of time intervals (and, hence, velocities and accelerations) must be used.

The streamline patterns of two kinematically similar motions would, obviously, be geometrically similar at the corresponding times. Since the boundaries constitute the bounding streamlines, kinematically similar motions are possible only when geometric similarity exists.

While attempting to obtain kinematic similarity, it must also be ensured that flow regimes remain the same in model and prototype. Thus, the compressibility, surface tension, and cavitation effects, if not present in the prototype flow, must be avoided in the model flow too.

6.6.3 Dynamic Similarity

Dynamic similarity means similarity of forces in magnitude as well as direction. Similarity of forces (i.e., dynamic similarity) is necessary because the direction taken by any fluid particle is decided by the resultant of all the forces acting on the fluid particle. Consequently, the magnitude ratio of any two forces (such as due to viscosity, gravitational acceleration etc.) in model must be the same as the magnitude ratio of the corresponding forces in the prototype. Two flow systems must have geometric as well as kinematic similarity in order to have dynamic similarity. However, a geometrically similar model may not be necessarily dynamically similar. In a fluid flow system, there may be several kinds of forces, viz., pressure, viscous, gravitational, surface tension, elasticity, etc.

Any of these forces acting, in combination, on a fluid particle in a flow system would have a resultant that, in accordance with Newton's law $F_i = ma$, would cause the particle to accelerate in the same direction as the resultant force F_i. If a force polygon is drawn for the forces acting on a fluid particle, then, this inertial force F_i corresponds to the line required to close the force polygon. Having the force polygons for the model and prototype similar ensures dynamic similarity. The force polygons for the prototype and model of a sluice gate or any other structure would have exactly the same shape (i.e., similar) if

$$\frac{F_{pp}}{F_{ip}} = \frac{F_{pm}}{F_{im}} \qquad ...(6.6.1)$$

$$\frac{F_{gp}}{F_{ip}} = \frac{F_{gm}}{F_{im}} \qquad ...(6.6.2)$$

and
$$\frac{F_{vp}}{F_{ip}} = \frac{F_{vm}}{F_{im}} \qquad ...(6.6.3)$$

excluding surface tension and cavitation. Here, F represents the forces and the first subscripts p, g, v and i represent, respectively, the cause of the force, i.e., pressure, gravity, viscosity, and inertia. The second subscripts p and m stand, respectively, for prototype and model terms.

Since it is the force that governs the motion of a fluid particle, the kinematic similarity and, hence, the geometric similarities are also achieved by satisfying the above model laws. Equations (6.6.1 – 6.6.3) can, alternatively, be written as

$$E_{um} = E_{up} \qquad ...(6.6.4)$$

$$F_{rm} = F_{rp} \qquad ...(6.6.5)$$

$$R_{em} = R_{ep} \qquad ...(6.6.6)$$

This means that ensuring equality of the corresponding non-dimensional Π parameters of model and prototype attains similarity between model and prototype.

If it were a problem of drag around a cylinder or sphere, the Π parameter involving the drag force would have been the drag coefficient $C_D \left(= \dfrac{F_D}{(1/2)\,\rho V^2 A} \right)$ and for the dynamic similarity,

$$C_{Dm} = C_{Dp} \qquad \qquad ...(6.6.7)$$

Some of the forces are not relevant at all or have negligible effect. Therefore, one needs to satisfy dynamic similarity requirement only in respect of the most important forces affecting the flow being modelled.

6.7 MODELLING

Wind tunnels and water tunnels or water channels are often used to examine the flow pattern and predict the forces that would be exerted on a fully submerged body by the fluid flowing past it. Since kinematic viscosity of water is about one-tenth the kinematic viscosity of air, water tunnels can be used for relatively higher Reynolds number models. High-speed wind tunnels are used for testing the models of an aircraft. Boundary layer wind tunnels, having relatively longer test section, are used for prediction of wind forces on industrial structures such as tall buildings, cooling towers, chimneys, bridges, antennas etc.

In pipe flows, viscous and inertial forces are the significant ones to be simulated. That is, one should have the same Reynolds number in model and prototype for achieving dynamic similarity. The pressure coefficient would, then, be the same in model and prototype. If the prototype and model fluids are the same, the product of velocity and diameter has to be the same for model and prototype. This means, one would require very high velocity in a small model.

Hydraulic structures such as spillways, channel transitions, weirs etc. would have gravity and inertial forces much greater than other forces. Froude number equality in model and prototype would, therefore, attain the dynamic similarity.

6.7.1 Model Scales

Depending upon the modelling criterion chosen, ratio of a physical quantity in the model and the corresponding quantity in the prototype can be determined. This ratio can be termed as model scale or simply scale (or ratio) for the quantity, e.g., velocity ratio etc. For models based on Reynolds number equality, i.e., $V_m l_m \rho_m / \mu_m = V_p l_p \rho_p / \mu_p$, one can write

length scale,
$$l_r = \frac{l_m}{l_p}$$

velocity scale,
$$V_r = \frac{V_m}{V_p}$$

mass density scale,
$$\rho_r = \frac{\rho_m}{\rho_p}$$

and viscosity scale,

$$\mu_r = \frac{\mu_m}{\mu_p}$$

so that equality of R_e yields

$$\frac{V_r l_r \rho_r}{\mu_r} = 1$$

Therefore, for known values of l_r, ρ_r and μ_r, one obtains

the velocity scale,

$$V_r = \frac{\mu_r}{l_r \rho_r}$$

the time scale,

$$t_r = \frac{l_r}{V_r} = \frac{l_r^2 \rho_r}{\mu_r}$$

the acceleration scale,

$$a_r = \frac{V_r}{t_r} = \frac{\mu_r^2}{l_r^3 \rho_r^2}$$

the force scale,

$$F_r = m_r a_r = \rho_r l_r^3 \frac{\mu_r^2}{l_r^3 \rho_r^2} = \frac{\mu_r^2}{\rho_r}$$

and the discharge scale,

$$Q_r = \frac{l_r^3}{t_r} = \frac{l_r \mu_r}{\rho_r}$$

Likewise, for models based on Froude number equality,

$$\frac{V_m}{\sqrt{g_m l_m}} = \frac{V_p}{\sqrt{g_p l_p}}$$

$$\therefore \qquad V_r = l_r^{0.5} g_r^{0.5} = l_r^{0.5} \qquad \text{if} \qquad g_m = g_p$$

$$t_r = \frac{l_r}{V_r} = \frac{l_r^{0.5}}{g_r^{0.5}} = l_r^{0.5} \qquad \text{if} \qquad g_m = g_p$$

$$a_r = \frac{V_r}{t_r} = g_r = 1 \qquad \text{if} \qquad g_m = g_p$$

$$F_r = \rho_r l_r^3 g_r = \rho_r l_r^3 \qquad \text{if} \qquad g_m = g_p$$

$$Q_r = \frac{l_r^3}{t_r} = l_r^{2.5} g_r^{0.5} = l_r^{2.5} \qquad \text{if} \qquad g_m = g_p$$

Thus, knowing l_r, μ_r and ρ_r for any model test, the desired quantities for the prototype can be predicted on the basis of measurement of the same quantity in the model.

6.7.2 Limitations of Modelling

For perfect dynamic similarity, one has to satisfy modelling laws similar to Eqs. (6.6.1 – 6.6.3) not only for pressure, gravity, viscous, and inertial forces but also for other forces such as elastic forces, surface tension forces etc. For example, consider a hydraulic model for testing a flow system with free surface. Dynamic similarity for this demands equality of Froude number and Reynolds number between the model and prototype. This means

$$\frac{V_m}{\sqrt{g_m l_m}} = \frac{V_p}{\sqrt{g_p l_p}} \qquad \qquad ...(6.7.1)$$

and
$$\frac{V_m l_m}{\nu_m} = \frac{V_p l_p}{\nu_p} \qquad \qquad ...(6.7.2)$$

It should be noted that Froude number contains only length and time dimensions and is, therefore, a purely kinematic parameter as far as modelling is concerned. If the length scale ratio is $l_r \ (= l_m/l_p)$ and since $g_m = g_p$, Eq. (6.7.1) yields

$$V_r = \frac{V_m}{V_p} = \left(\frac{l_m}{l_p}\right)^{1/2} = l_r^{1/2} \qquad \qquad ...(6.7.3)$$

and, hence, Eq. (6.7.2) yields

$$\frac{\nu_m}{\nu_p} = \frac{V_m l_m}{V_p l_p} = l_r^{3/2} \qquad \qquad ...(6.7.4)$$

For a one-tenth-scale model, i.e., $l_r = 0.1$

$$\nu_m = (0.1)^{3/2} \nu_p = 0.032 \nu_p$$

If the prototype fluid is water with $\nu_p = 1 \times 10^{-6} \ m^2/s$, one would require, for the model, a liquid whose kinematic viscosity is only 0.032 times the kinematic viscosity of water. Looking at the properties of the fluids, Table 1.5, one notes that only mercury has kinematic viscosity that is smaller than that of water only by a factor of about 10. And mercury model would be expensive besides being a source of health hazard. Therefore, in practice, water is used for the model too and the Reynolds number similarity is unavoidably violated (8). In free surface flows, the Froude number is the dominant parameter and is, therefore, made equal in both model and prototype. The low-Reynolds number model data are extrapolated to predict high-Reynolds number prototype data. Obviously, there is considerable uncertainty in such extrapolation, but there is no other viable practical alternative method for model testing.

Similarly, for aerodynamic model testing of an air flow system with no free surface, the dynamic similarity of the model and prototype requires the equality of Reynolds number and Mach number (for compressibility effects) between model and prototype. The compressibility criterion is

$$\frac{V_m}{c_m} = \frac{V_p}{c_p} \qquad \qquad ...(6.7.5)$$

Here, c represents the speed of sound in the flowing fluids. Eliminating V_m and V_p from Eqs. (6.7.2) and (6.7.5), one gets

$$\frac{v_m}{v_p} = \frac{l_m}{l_p} \frac{c_m}{c_p}$$,

Since, the prototype fluid is air, one would need a wind tunnel fluid of low viscosity and high velocity of sound. Hydrogen, too costly and dangerous to work with, is the only practical fluid that meets the requirements. Therefore, wind tunnels are operated with air as the working fluid and, once again, violating the Reynolds number equality in aerodynamic testing too. Even for wind tunnel testing of industrial structures (such as chimneys, cooling towers etc.) the Reynolds number equality demands

$$\frac{V_m l_m}{v_m} = \frac{V_p l_p}{v_p}$$

Since

$$v_m = v_p$$

$$V_m = V_p \frac{l_p}{l_m}$$

Even for a model scale of 1/50, one would require velocity of air in the model to be equal to 50 times the design velocity for the prototype. This may not be attainable and even if attainable, the flow regime may change from subsonic to supersonic. Therefore, only recourse is to have similarity of flow pattern, i.e., velocity distribution and turbulence structure for extrapolating the model results to prototype conditions (6).

In many model studies, only having equality of more than one dimensionless number can attain dynamic similarity and this may not be attainable. For example, the drag force (or resistance) of a ship is due to skin friction (i.e., viscous force) on its hull, and surface wave resistance (gravity force). Complete dynamic similarity requires equality of Reynolds number (for viscous effects) and Froude number (for surface wave effects) for both model and prototype. Equation (6.7.4), for this condition, yields that if model fluid is also water (which is the only practical fluid for most of the free surface flow modelling), the model must be of the size of the prototype. Using a small model, based on Froude number equality, and measuring the total drag on it overcome this difficulty (7). From the measured total drag is subtracted the computed skin-friction drag on the model to obtain the surface wave resistance for the model. The model's surface wave resistance is used for the prediction of the surface wave resistance for the prototype that could not have been estimated otherwise. To this is added the computed skin-friction resistance for the prototype and thus total resistance on the ship is obtained.

If one wants to obtain data for drag around a body moving in atmosphere (i.e., infinite flow field), one would require a wind tunnel of infinite cross-section. However, in practice, this restriction may be relaxed considerably to be able to use wind tunnels of a reasonable cross-section.

Fortunately, in many flow conditions, only two or three forces are of considerable significance and other forces are either not relevant at all or have negligible effects on the flow system. Therefore, one needs to satisfy dynamic similarity requirements only in respect of the most important forces influencing the flow being modelled. For example, if the flow is incompressible and there exists a free surface, equality of Froude number between model and prototype is considered adequate. But, incompressible flow with no free surface would require equality of Reynolds number between model and prototype and this too may not be attainable at times.

6.7.3 General Comments on Model Testing

Model testing, although useful and powerful method for analysis of flow, does not yield automatically accurate, complete and easily interpreted results. Careful planning, accurate model construction, correct measurements with precise instruments, and correct interpretation of the results are required for successful model testing.

6.8 FLOW MEASUREMENT

There will always be a need of accurate measurement of quantities relating to fluid flow for any practical fluid engineering problem. Whether one deals with laboratory and/or field studies relating to fluid flow, one needs to measure local properties (such as pressure, velocity, temperature, viscosity, turbulence intensity, density), integrated properties (such as mass flow rate or volume flow rate), and global properties (such as visualization of the entire flow field). Measurements of pressure, velocity, and volume flow rate are commonly made in almost all the fluid flow problems and, accordingly, measurement techniques for pressure, velocity, and volume flow rate have been briefly dealt with in this section.

6.8.1 Pressure Measurement

There is no pressure variation normal to straight streamlines. It is, therefore, possible to measure the static pressure in a flowing fluid using a wall pressure "tap" (a small hole, drilled in the wall, with its axis perpendicular to the wall surface, i.e., streamlines) placed in a region where the streamlines are straight, Fig. 6.8.1. The hole should be free from burrs. The tap is connected to a suitable measuring instrument such as piezometer or manometer (Chapter 2), or suitable pressure transducer. In a fluid flow far from its boundary, or where streamlines are curved, one may use a static pressure probe.

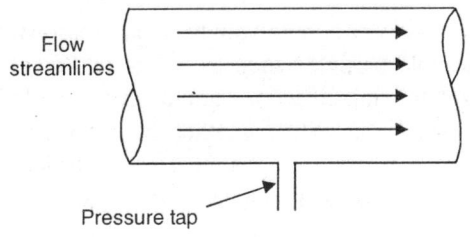

Fig. 6.8.1. Pressure tap for measurement of static pressure

The stagnation pressure is obtained when the velocity of a flowing fluid is decelerated to zero by a frictionless process. Using Bernoulli's equation, Eq. (4.6.8), the stagnation pressure p_s equals the sum of the static pressure p and the dynamic pressure $(\frac{1}{2}) \rho V^2$. That is,

$$p_s = p + \frac{1}{2}\rho V^2 \qquad \qquad ...(6.8.1)$$

Stagnation pressure can be measured by a total head tube, Fig. 5.2.1. Pressures can also be measured by Bourdon gauge (Chapter 2).

Electrical pressure transducers, based on the principle of variable capacitance or resistance or inductance, have been developed for effective pressure measurements even when pressure fluctuates. Transducer is a device used for converting a physical quantity such as pressure, force, temperature etc., into a measurable electrical parameter like voltage, current, capacitance, resistance etc. Variable capacitance pressure transducer, very suitable device for measuring very low pressures, has a stretched metal diaphragm positioned symmetrically between two stationary plates. The capacitance between these plates and the electrically-grounded diaphragm varies with the deflection of the diaphragm that is proportional to the differential pressure applied on the diaphragm. The changes in the capacitance can be suitably measured and, thus, pressure difference estimated. Capacitance transducers have good linear and high frequency response.

Certain materials generate an electrostatic charge or voltage when mechanical stresses are applied across them. This property of piezoelectricity has been utilized in the design of pressure transducers wherein the mechanical stress is generated by the diaphragm subjected to pressure. Most commonly used materials for this purpose are natural quartz and variety of synthetic ceramic materials. Piezoelectric transducers are widely used for measuring rapidly-fluctuating pressures.

Another popular pressure transducer uses LVDT (linear variable differential transformer) as the sensing element. An elastic diaphragm is suitably coupled to the core of LVDT. On application of pressure, the diaphragm deflects and the movement of the core is sensed precisely and the electrical output is interpreted in terms of pressure values.

6.8.2 Velocity Measurement

Velocity averaged over a small region, or point can be measured by several different methods such as (8)

1. Trajectory of floats
2. Rotating electro-mechanical devices
3. Pitot-static tube
4. Electromagnetic current meter
5. Thermal anemometer
6. Laser-doppler anemometer

A very simple way to obtain reasonable estimate of flow velocity is by monitoring the motion of visible particles such as flakes or floats on the surface of a channel flow, small neutrally buoyant spheres mixed with a liquid, hydrogen bubbles, entrained dust particles in gas flows etc. One must, however, establish whether the particle motion truly simulates the fluid motion.

A rotating electro-mechanical device consists of a series of cups or vanes (similar to a windmill), mounted on a shaft held in bearings. The angular speed of rotation of the rotating element is correlated with the linear velocity of flow through calibration. These devices are known as anemometers, when used for air flow, and current meters (Art. 16.5) if used for water flow. The shaft is connected to a revolution counter that helps in making observations of the number of revolutions in a suitable time interval.

Pitot-static tube has already been described in Art. 5.3. It should be noted that the Pitot tube is not suitable for measuring low velocity. Further, because of the slow response of the fluid-filled tubes that connect the Pitot tube with pressure sensors, like manometer, the Pitot tube is not suitable for unsteady flow measurements.

Electromagnetic flow meter uses the principle of magnetic induction. When a magnetic field is created across a conducting fluid that is flowing, the flow will induce a voltage at right angles to the field or velocity vectors. Electrodes placed (in or near the flow) on a pipe diameter are used to detect the resulting voltage that is proportional to the average axial flow velocity. The electromagnetic flow meter requires calibration.

Thermal anemometers use tiny elements (a wire or a film), Fig. 6.8.2, that are heated electrically and, hence, are commonly known as hot-wire anemometer (suitable for gas flows) or hot-film anemometer (suitable for liquid flows). These tiny elements (about 0.1mm long and 0.002 mm diameter and made of tungsten or platinum) constitute one arm of a balanced Wheatstone bridge. When the heated element is placed in a flowing fluid, it gets cooled and, therefore, its resistance is reduced. The bridge is no longer balanced. Sophisticated electronic feedback circuits maintain the temperature and, hence, resistance of the element constant. The feedback current, therefore, is the measure of the flow velocity and the relation between them can be established by calibration. The feedback current, *I*, and the flow velocity, *V* are related as

$$I^2 \propto \sqrt{V}$$

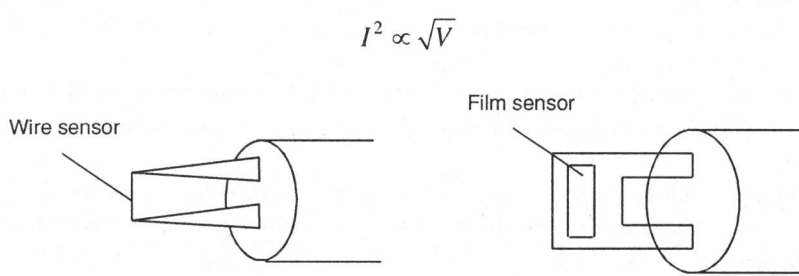

Fig. 6.8.2. Hot-wire and Hot-film sensors

Since the thermal mass (and, hence, inertia) of sensors used for hot-wire anemometry is extremely small, their response to fluctuations in flow velocity is fast. This makes thermal anemometers ideal for measurement of turbulence quantities. These are extensively used for research purposes. Data acquisition and on-line data processing helps in obtaining complete information about turbulent flows. Hot wire can easily be arranged in groups to measure two and three-dimensional velocity components.

In the laser Doppler anemometer (LDA), a laser beam provides highly focussed, coherent monochromatic light that passes through a small volume in the flow field at the location where the velocity is to be measured. Laser light is scattered from moving particles that are present in the flow or introduced for this purpose. A change in frequency of the scattered light is caused and this is called Doppler effect. The frequency shift (or change) Δf is proportional to the velocity of the particle and, hence, the flow.

As shown in Fig. 6.8.3, a focussing device splits the laser into two beams that cross the flow at an angle θ. Their intersection forms the measuring volume or resolution of the instrument. Particles passing through this volume scatter the beams. The scattered beams, then, pass through receiving optics to a photodetector and signal processor that convert the light to an electric signal that can be measured or displayed or even stored for further analysis. The velocity of flow *V* is given as

$$V = \frac{\lambda \Delta f}{2 \sin(\theta/2)}$$

in which λ is the wavelength of the laser light.

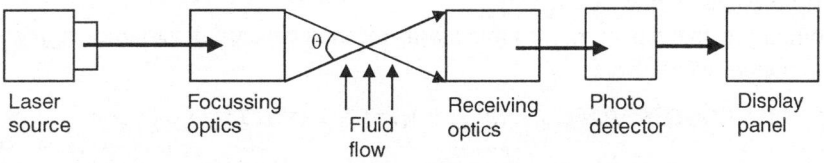

Fig. 6.8.3. Units of Laser-Doppler anemometer

By using more than one photodetector and other operating procedures, one can measure two-or three-dimensional velocity components. Both liquid as well as gas flows can be measured. Only requirement is that scattering particles must exist in the flow or one has to seed the flow. LDA does not disturb the flow. It does not require calibration and the output voltage is linear with velocity. Besides, the output is independent of the thermodynamic properties of the fluid. But, LDA is a costly system and requires the fluid and the apparatus to be transparent.

6.8.3 Volume Flow Rate Measurements

While working with flowing fluids, one is often required to measure accurately the volume flow rate, i.e., discharge. The methods used for measuring discharge can be either gravimetric or volumetric or methods using head loss fittings.

In gravimetric and volumetric methods, mass (or weight) or volume of the flowing liquid is collected in a container for a known duration of time. And, thus, the volumetric rate of flow i.e., discharge can be measured.

Head loss fittings (such as orifice meter, venturi meter, nozzle meter) obstruct the flow and cause a pressure drop that is a measure of the discharge. Principles involved for such meters have already been dealt with in Chapter 5.

One can also estimate the discharge by integrating the velocity profile that may be obtained by a suitable device that measures flow velocity.

For open channel flows, one uses weirs and flumes that would be discussed in Chapter 16.

EXAMPLES

E6.1:   The power input P required by a centrifugal pump depends on the discharge Q, impeller diameter D, rotational rate (i.e, revolutions per unit time) Ω, and the density ρ and viscosity μ of the fluid. Obtain the functional relationship for P in terms of the dimensionless quantities.

Solution: Given $P = f_1 (Q, D, \Omega, \rho, \mu)$

Dimensions of each of these variables are

P	Q	D	Ω	ρ	μ
ML^2T^{-3}	L^3T^{-1}	L	T^{-1}	ML^{-3}	$ML^{-1}T^{-1}$

Therefore, number of variables, $n = 6$, and

number of primary dimensions, $m = 3$

Hence, number of repeating variables = 3, and

Number of Π parameters = 3

Variables D, Ω and ρ represent, respectively, the geometric, flow, and fluid characteristics and they cannot form a dimensionless group because only Ω contains time dimension and only ρ contains mass dimension.

Combining the repeating variables with the non-repeating variables, three dimensionless parameters would be

$$\Pi_1 = D^a \Omega^b \rho^c P = L^a T^{-b} M^c L^{-3c} ML^2T^{-3} = M^0 L^0 T^0$$

$$\Pi_2 = D^a \Omega^b \rho^c Q = L^a T^{-b} M^c L^{-3c} L^3 T^{-1} = M^0 L^0 T^0$$

$$\Pi_3 = D^a \Omega^b \rho^c \mu = L^a T^{-b} M^c L^{-3c} ML^{-1}T^{-1} = M^0 L^0 T^0$$

Equating exponents for Π_1

$$c + 1 = 0$$
$$a - 3c + 2 = 0$$
$$-b - 3 = 0$$

On solving these, one gets $a = -5$, $b = -3$, and $c = -1$

Therefore, $$\Pi_1 = D^{-5}\Omega^{-3}\rho^{-1} P = \frac{P}{\rho\Omega D^5}$$

This Π parameter is known as the power coefficient of a pump.
Similarly, on equating exponents of Π_2
$$c = 0$$
$$a - 3c + 3 = 0$$
$$-b - 1 = 0$$
On solving these equations, one gets $a = -3$, $b = -1$, and $c = 0$

Therefore, $$\Pi_2 = D^{-3}\Omega^{-1}\rho^0 Q = \frac{Q}{\Omega D^3}$$

This Π parameter is known as flow coefficient of a pump.
Likewise, on equating exponents of Π_3
$$c + 1 = 0$$
$$a - 3c - 1 = 0$$
$$-b - 1 = 0$$
On solving these equations, one gets $a = -2$, $b = -1$, and $c = -1$

Therefore, $$\Pi_3 = D^{-2}\Omega^{-1}\rho^{-1}\mu = \frac{\mu}{\rho\Omega D^2}$$

This Π parameter is a sort of Reynolds number.
Thus, the desired functional relationship is

$$\frac{P}{\rho\Omega D^5} = f\left(\frac{Q}{\Omega D^3}, \frac{\mu}{\rho\Omega D^2}\right)$$

E6.2: A ship whose hull length is 140 m travels at 7.6 m/s. At what velocity should its model (scale =1:30) be towed through water for achieving dynamic similarity?

Solution: For the desired dynamic similarity, one has to ensure Froude number equality. Therefore,

$$\frac{V_p}{\sqrt{gL_p}} = \frac{V_m}{\sqrt{gL_m}}$$

$\therefore$ $$V_m = V_p \sqrt{\frac{L_m}{L_p}}$$

$$= \frac{7.6}{\sqrt{30}}$$

$$= 1.388 \text{ m/s}$$

E6.3: A torpedo is expected to attain a velocity of 6 m/s in water. Its model is to be tested in a towing tank at a velocity of 24 m/s. What should be the model scale? If this model is to be tested in a wind tunnel under a pressure of 20 atm and temperature of 20°C, what should be the wind speed in the tunnel? The absolute viscosity of air can be taken as 1.85×10^{-5} kg/ms. Gas constant for air = 29.3.

Solution: For the desired dynamic similarity, one has to ensure Reynolds number equality. If L is the length of the prototype torpedo, then

$$\frac{6 \times L_p}{v_p} = \frac{24 \times L_m}{v_m}$$

Since

$$v_m = v_p, \quad \frac{L_m}{L_p} = \frac{6}{24} = \frac{1}{4}$$

i.e., model scale is 1:4.

The density of air at pressure equal to 20 atm is obtained from

$$pV = RT$$

or

$$p/\rho g = RT$$

or

$$\rho = \frac{p}{gRT} = \frac{20 \times 101300}{(9.80)(29.3)(273+20)} = 24.08 \text{ kg/m}^3$$

$\therefore$

$$\frac{6 \times L_p}{1.0 \times 10^{-6}} = \frac{V \times L_p / 4}{1.85 \times 10^{-5} / 24.08}$$

$\therefore$

$$V = 18.44 \text{ m/s}$$

Hence, the speed in the wind tunnel should be 18.44 m/s

E6.4:   A sonar transducer (spherical in shape, and diameter 400 mm) will be towed at 5 knots (nautical miles per hour; one nautical mile is 1852 meters) in sea water whose mass density is 1025 kg/m³ and kinematic viscosity is 1.4×10^{-6} m²/s. The model of the prototype sonar transducer is 200 mm in diameter and is to be tested in a wind tunnel for the prediction of drag on the prototype sonar transducer. Density and kinematic viscosity of air at the test condition are, respectively, 1.2 kg/m³ and 1.4×10^{-5} m²/s. Determine the required wind speed in the tunnel. If the drag on the model is measured as 30N, estimate the drag on the prototype sonar transducer.

Solution: Assuming that cavitation effects are absent in the prototype and compressibility effects are absent in the model test, the functional relation for F would be Eq. (6.3.6), i.e.,

$$\frac{F}{(1/2)\rho V^2 D^2} = f\left(\frac{VD\rho}{\mu}\right)$$

and similarity laws applicable would be Eqs. (6.6.6) and (6.6.7) to ensure dynamic similarity.

Hence, $R_{em} = R_{ep}$ and $\left(\dfrac{F}{(1/2)\rho V^2 D^2}\right)_m = \left(\dfrac{F}{(1/2)\rho V^2 D^2}\right)_p$

From $R_{em} = R_{ep}$

$$\left(\frac{VD}{\nu}\right)_m = \left(\frac{VD}{\nu}\right)_p$$

$$V_p = 5 \times 1852/3600 = 2.57 \text{ m/s}$$

$$\therefore \quad V_m = \frac{\nu_m}{\nu_p} \frac{D_p}{D_m} V_p$$

$$= \frac{1.4 \times 10^{-5}}{1.4 \times 10^{-6}} \frac{400}{200} (2.57) \text{ m/s}$$

$$= 51.4 \text{ m/s}$$

Thus, the wind speed in the tunnel is 51.4 m/s. Since the speed is low, the compressibility effects are negligible.

Model and prototype flows are dynamically similar. Therefore,

$$\left(\frac{F}{(1/2)\rho V^2 D^2}\right)_m = \left(\frac{F}{(1/2)\rho V^2 D^2}\right)_p$$

$$\therefore \quad \frac{30}{1.2(51.4)^2 (200)^2} = \frac{F_p}{1025(2.57)^2 (400)^2}$$

$$\therefore \quad F_p = \frac{30 \times 1025 \times (2.57)^2 (400)^2}{1.2 \times (51.4)^2 (200)^2} = 256.25 \text{ N}$$

PROBLEMS

P6.1. The pressure drop for flow through a sudden contraction in a circular pipe may be expressed as $\Delta p = p_1 - p_2 = f(\rho, \mu, V, d, D)$. Obtain the resulting dimensionless parameters to organize experimentation for the problem.

P6.2. A smooth flat plate is placed in an incompressible flow with zero pressure gradient. The boundary-layer thickness, δ, on the plate depends on the free stream velocity V, the fluid viscosity μ, the fluid density ρ, and the distance from the leading edge of the plate x. Obtain dimensionless Π parameters.

P6.3. The wall shear stress τ_w, in a boundary layer depends on distance from the leading edge of the body x, the fluid density ρ, the fluid viscosity μ, and the free stream speed of the flow V. Express these variables in the form of dimensionless Π parameters and thus, obtain the functional relationship among these.

P6.4. The mean velocity $\bar{u}$, for turbulent flow in a pipe or a boundary layer may be correlated using the wall shear stress τ_w, distance from the wall y, and the fluid properties ρ and μ. Use dimensional analysis to show that the functional relationship between these parameters may be expressed as

$$\frac{\overline{u}}{u_*} = f\left(\frac{yu_*}{\nu}\right)$$

where $u_* = \left(\dfrac{\tau_w}{\rho}\right)^{1/2}$ and is the *friction velocity*.

P6.5 The speed of a capillary wave, forming on the free surface of a liquid due to surface tension, depends on surface tension σ, wavelength λ, and liquid density ρ. Obtain the wave speed as a function of these variables in dimensionless form.

P6.6 An airplane is to operate at 25 m/s in air at standard condition. A model is constructed to 1/30 scale and tested in a wind tunnel at the same air temperature to determine drag on the airplane. What criterion should be considered to obtain dynamic similarity? If the model is tested at 70 m/s, what pressure should be used in the wind tunnel? If the model drag force is 250 N, what will be the drag on the prototype?

P6.7 A 1:30 scale model of a submarine is tested in a water tunnel. The drag force F_D, depends on water speed V, density ρ, and viscosity μ, and on model volume, $\forall$. Determine the drag of the full-scale submarine at 27 Km/s if the model test at 10 Km/s yielded a measured drag force of 15 N.

P6.8 The pressure rise Δp, of a liquid flowing steadily through a centrifugal pump is expected to depend on pump diameter D, angular speed of the rotor ω, volume flow rate Q and density ρ. The table gives data for the prototype and for a geometrically similar model pump. For conditions corresponding to dynamic similarity between the model and prototype pumps, calculate the missing values in the table.

Variable	Prototype	Model
Δp		74.9 kPa
Q	1.20 m^3 / min	
ρ	800 kg/m^3	1000 kg/m^3
ω	10 rad/s	100 rad/s
D	60 mm	120 mm

P6.9 The period of oscillation T of a water surface wave is expected to be dependent on density ρ, wavelength λ, depth h, gravity g, and surface tension Y. Taking λ, ρ, and g as repeating variables rewrite the functional relationship for T in dimensionless form.

P6.10 The velocity at a point in a model of a spillway for a dam is 1.5 m/s. For a geometric scale (between prototype length and model length) of 8:1 what would be the velocity at the corresponding point in the prototype under similar conditions?

P6.11 The wave drag on a model of a ship is 25 N at a speed of 4 m/s. For a prototype ten times longer than the model, determine the corresponding speed and wave drag if the liquid is the same in both cases?

REFERENCES

(1) Buckingham, E: "*On physically similar systems: Illustrations of the use of dimensional equations*", Physical Review, vol. 4, 1914, pp. 345-376.

(2) Fox, RW and McDonald, AT: *"Introduction to Fluid Mechanics"*, John Wiley & Sons, New York, 1994.

(3) Garde, RJ, and Asawa GL: *"Atmospheric boundary layer and its simulation in wind tunnels"*, Proc. Int. Symposium on Wind Loads on Structures, Oxford & IBH Publishing Co. Pvt. Ltd., 1990.

(4) Humsaker, J.C. and Rightmire, BG: *"Engineering Applications of Fluid Mechanics"*, McGraw-Hill, New York, 1961 [Streeter et al. (7)]

(5) Langhaar, HL: *"Dimensional Analysis and the Theory of Models"*, John Wiley, New York, 1951 [Streeter et al. (7)].

(6) Plate, EJ: *"Wind tunnel modelling of wind effects in Engineering"*, Chapter 13 in Engineering Meteorology edited by Plate, EJ, Elsevier Scientific Publishing Company, Amsterdam, 1982.

(7) Streeter, VL Wylie, EB and Bedford KW: *"Fluid Mechanics"*, McGraw-Hill Book Co., Singapore, 1998.

(8) White, F.M.: *"Fluid Mechanics"*, McGraw-Hill Book Co., New York, 2003.

CHAPTER 7 _____

Viscous Flows

7.1 LAMINAR AND TURBULENT FLOWS

Basic equations governing the fluid motion have been dealt with in Chapters 3 and 4. Based on viscous effects, a flow can be treated as either an inviscid (or non-viscous or ideal) flow or viscous flow. Viscosity of the real fluid causes shear stress in a moving fluid besides resulting into irreversible losses. Viscous flows (i.e., flows of real fluid) are classified into two categories: 1) laminar flows, and 2) turbulent flows.

In laminar flow, the fluid particles move in parallel layers (or laminas) with only molecular interchange of momentum between the adjacent layers. The viscous shear that opposes the relative motion of adjacent fluid layers also dampens any disturbance that tends to make the flow turbulent.

In turbulent flow, however, the fluid particles have very erratic motion causing transverse interchange of fluid particles and the associated momentum.

Since it is the viscous shear force that dampens the tendencies of the flow becoming turbulent, the ratio of the inertial force and viscous shear force (i.e., the Reynolds number) becomes a useful criterian to indicate whether the flow is laminar (i.e., when viscous shear force is relatively large) or turbulent (i.e., when the viscous shear force is rather small). As mentioned in Chapter-6, the Reynolds number equals $\dfrac{VL\rho}{\mu}$ in which V and L are, respectively, the characteristic velocity and length of flow, and ρ and μ are, respectively, the mass density and viscosity of the flowing fluid.

The distinction between laminar flow and turbulent flow can be understood through the Reynolds' experiment. The set-up, Fig. 7.1.1 (a), used by Osborne Reynolds in 1883 (3), consisted of a water tank to which was fitted a small diameter glass tube with a bell-mouth inlet end and a valve at its outlet end. An arrangement was made for injecting a coloured dye into the glass tube from its inlet end. Reynolds considered the average velocity of flow, V in the tube and diameter of the tube, D as the characteristic velocity and length, respectively. Since, the mass density ρ and viscosity μ of the flowing fluid, and the diameter of the tube D are fixed for a set-up, the Reynolds number can be varied by varying the discharge (and, hence, the velocity V) in the tube with the help of the valve.

For small discharges (and, hence, the Reynolds number), the coloured dye would remain straight and pass through the tube without getting mixed with water, Fig. 7.1.1(b) indicating that the flow is laminar. If one were to open the valve further gradually, this pattern of flow persists until the velocity (and, hence, the Reynolds number) attains a value at which the coloured dye begins to waver, Fig. 7.1.1(c) and the flow is tending to be turbulent. With further opening of the valve, the velocity increases further and the coloured dye mixes with water, Fig. 7.1.1(d), indicating violent transverse interchange of fluid particles and, therefore, the turbulent flow. With very careful manipulation (such as allowing the water to stand in the tank for several days before experimentation and opening the valve very gradually and taking all precautions to avoid vibration of the set-up), some researchers have maintained laminar flow even upto the Reynolds number of about 40,000 (5). However, in normal installation, the flow turns to turbulent at much smaller Reynolds number.

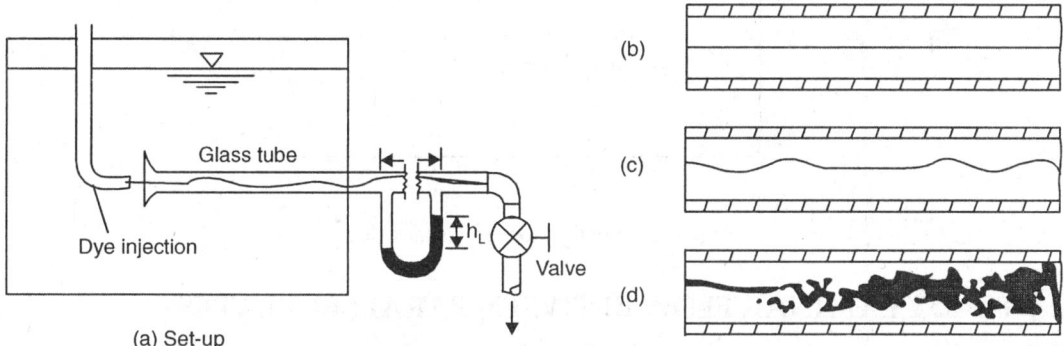

Fig. 7.1.1. Reynolds Experiment

Starting with turbulent flow in the glass tube, Reynolds discovered that the flow in the tube is always laminar if the discharge is reduced so that the Reynolds number R_e is less than 2000. Thus, the lower critical Reynolds number[1] for pipe flow, under normal conditions, is 2000. The pipe flow turns turbulent in the range of Reynolds number from 2000 to 4000. The Reynolds number at which flow becomes fully turbulent, is called the upper critical Reynolds number. The flow is said to be in transition when the flow sometimes is laminar and at other times it is turbulent. The transition range of Reynolds number is, obviously, from 2000 to 4000.

It should be noted that the characteristic length (and velocity too) would be different for different flows. For example, flow past a sphere or cylinder would be described by Reynolds numbers with characteristic length as the diameter of the sphere or cylinder. Therefore, the value of the lower and upper critical Reynolds numbers would also be different for different flows.

If a manometer is connected to the pipe as shown in Fig. 7.1.1(a), one can determine the head loss h_L in a length l of the pipe for the prevailing velocity of flow, U in the tube (or pipe) and prepare a plot of $\log \dfrac{h_L}{l}$ versus $\log U$, Fig. 7.1.2. The data corresponding to laminar flow would fall along AB, while those for turbulent flow would fall along CD, thus, indicating that in laminar flow, the head loss is proportional to U and in turbulent flow, the head loss is proportional to U^n such that n may vary from 1.8 to 2.0. For the transition condition, data would not follow a definite pattern.

---

[1] In the literature, one finds the lower critical Reynolds number ranging between 2000 to 2300.

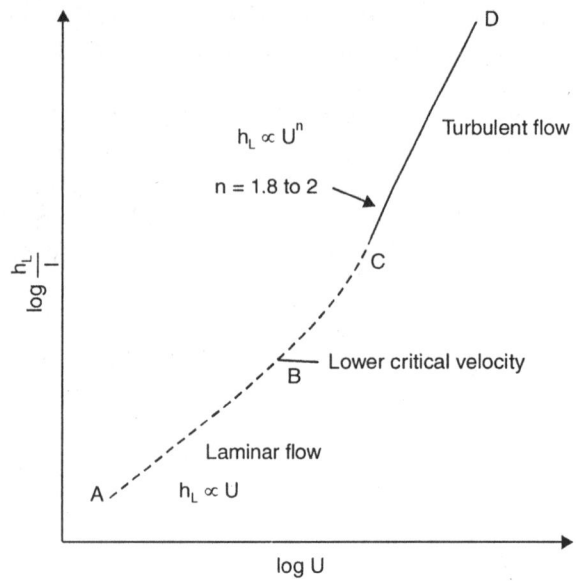

Fig. 7.1.2. Friction loss in viscous flows

7.2 STEADY LAMINAR FLOW BETWEEN PARALLEL PLATES

Consider steady laminar flow between inclined parallel plates such that the lower plate is stationary and the upper plate is moving with a velocity U, Fig. 7.2.1, parallel to the flow direction. The flow is

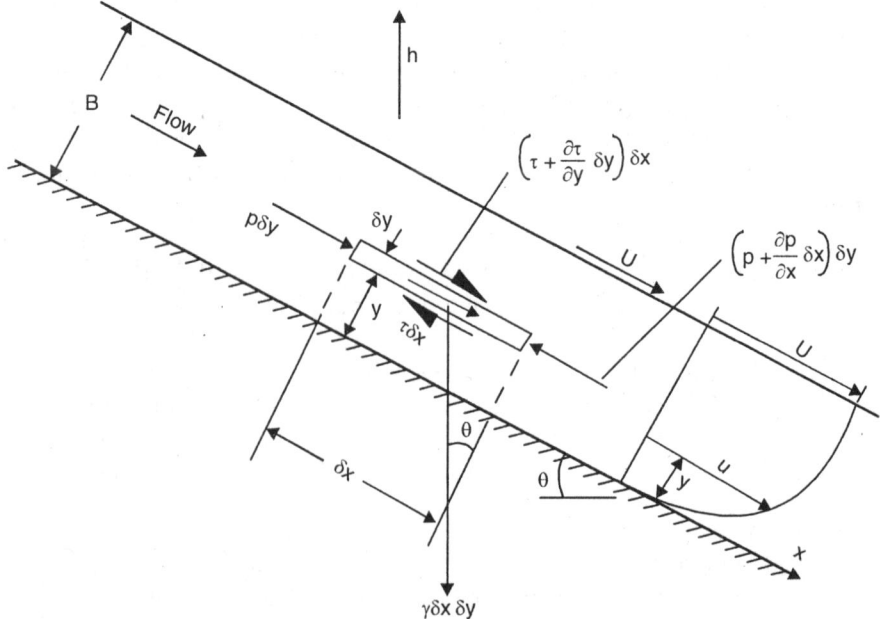

Fig. 7.2.1. Flow between inclined parallel plates with the upper plate in motion

considered two-dimensional. Therefore, the plates are sufficiently long and wide so that the 'end effects' may be ignored. There is no movement of fluid in the direction perpendicular to the flow direction. Hence, the pressure varies only along the flow direction.

For the fluid element (of unit width perpendicular to the plane of paper) moving with a constant velocity $u(y)$, one can write the following equation:

$$p\delta y - \left(p\delta y + \frac{\partial p}{\partial x}\delta x\delta y \right) - \tau\delta x + \left(\tau\delta x + \frac{\partial \tau}{\partial y}\delta y\delta x \right) + \rho g\,\delta x\delta y\sin\theta = 0$$

or

$$-\frac{\partial p}{\partial x} + \frac{\partial \tau}{\partial y} - \rho g\frac{\partial h}{\partial x} = 0$$

since

$$\sin\theta = -\frac{\partial h}{\partial x}$$

i.e.,

$$\frac{\partial \tau}{\partial y} = \frac{\partial}{\partial x}(p+\rho gh) = \frac{\partial p^*}{\partial x} \qquad \qquad ...(7.2.1)$$

in which p^* is the piezometric pressure $(p + \rho gh)$. Since the velocity u is dependent only on y and the shear stress τ depends only on the viscosity, μ and velocity gradient $\dfrac{\partial u}{\partial y}$

$$\frac{\partial \tau}{\partial y} = \frac{d\tau}{dy} = \mu\frac{d^2u}{dy^2}$$

Likewise, p^* or $(p + \rho gh)$ varies only in the x-direction. Therefore,

$$\frac{\partial p^*}{\partial x} = \frac{dp^*}{dx}$$

Hence,

$$\mu\frac{d^2u}{dy^2} = \frac{dp^*}{dx} \qquad \qquad ...(7.2.2)$$

On integrating Eq. (7.2.2) twice, one obtains,

$$u = \frac{y^2}{2\mu}\left(\frac{dp^*}{dx} \right) + Ay + C$$

in which A and C are constants of integration which can be evaluated from the conditions:
$$u = 0 \text{ at } y = 0 \text{ and } y = U \text{ at } y = B$$

Hence,

$$C = 0$$

and

$$A = \frac{U}{B} - \frac{B}{2\mu}\left(\frac{dp^*}{dx} \right)$$

$\therefore$

$$u = \frac{U}{B}y - \frac{1}{2\mu}\left(\frac{dp^*}{dx} \right)\left(By - y^2 \right) \qquad \qquad ...(7.2.3)$$

or

$$u = \frac{U}{B}y - \frac{1}{2\mu}\left[\frac{d}{dx}(p+\rho gh)\right]\left(By-y^2\right)$$
...(7.2.4)

For horizontal plates, h will be the same at any value of y. Hence, $p*$ can be replaced with pressure p only. If there is no pressure gradient either, Eq. (7.2.4) reduces to

$$u = \frac{U}{B}y$$
...(7.2.5)

Equation (7.2.5) is the linear distribution of velocity of *Couette* flow. If the upper plate is stationary, $U = 0$ and, hence, Eq. (7.2.3) reduces to

$$u = -\frac{1}{2\mu}\left(\frac{dp*}{dx}\right)\left(By-y^2\right)$$
...(7.2.6)

If the plates are horizontal, Eq. (7.2.6) reduces to

$$u = -\frac{1}{2\mu}\left(\frac{dp}{dx}\right)\left(By-y^2\right)$$
...(7.2.7)

The velocity variations for Eqs. (7.2.5) and (7.2.7) are shown in Fig. 7.2.2.

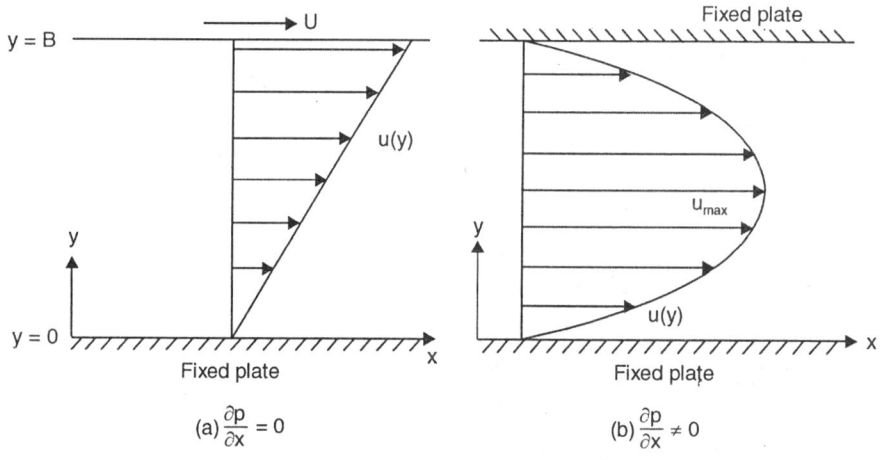

Fig. 7.2.2. Incompressible viscous flow between parallel plates

The average velocity V between the stationary parallel plates is obtained from Eq. (7.2.6) as follows:

$$V = \frac{1}{B}\int_0^B u\,dy$$

$$= \frac{1}{B}\left(-\frac{1}{2\mu}\right)\left(\frac{dp*}{dx}\right)\int_0^B\left(By-y^2\right)dy$$

$$V = -\frac{B^2}{12\mu}\frac{dp*}{dx}$$
...(7.2.8)

If one considers two sections 1 and 2 which are l distance apart along the flow and at which the piezometric pressures are p_1^* and p_2^*, then Eq. (7.2.8) yields

$$V = -\frac{B^2}{12\mu}\frac{p_2^*-p_1^*}{l}$$

or

$$p_1^* - p_2^* = \frac{12\mu Vl}{B^2} \qquad\qquad ...(7.2.9)$$

Equation (7.2.9) indicates that the loss in piezometric pressure and, hence, energy (as the kinetic energy is the same at the two sections) is proportional to the velocity V. Equation (7.2.9) can, alternatively, be written as

$$H_1 - H_2 = \frac{12\mu Vl}{\rho gB^2} \qquad\qquad ...(7.2.10)$$

where, H_1 and H_2 are the piezometric heads at sections 1 and 2, respectively.

Steady uniform laminar flow in an infinitely wide open channel can be treated as lower half flow between inclined parallel plates. Hence, the depth of flow in open channel, h equals $B/2$. Substituting $B = 2h$ in Eq. (7.2.8), one gets

$$U = -\frac{h^2}{3\mu}\frac{dp^*}{dx} \qquad\qquad ...(7.2.11)$$

Open channel flow has a free surface and no pressure gradient. The flow is maintained by the gravity and there is decrease in the potential energy in the direction of flow. This means,

$$\frac{dp^*}{dx} = \frac{\rho g(z_1 - z_2)}{l}$$

where l indicates the distance between two sections with datum head as z_1 and z_2. Thus, $\dfrac{(z_1-z_2)}{l}$ is the slope S of the channel bed (or water surface as the flow is uniform). The slope S is assigned negative sign as z decreases in the direction of flow (i.e., x-axis). Therefore, Eq. (7.2.11) can be written as

$$U = \frac{h^2(\rho gS)}{3\mu} \qquad\qquad ...(7.2.12)$$

7.3 DISSIPATION OF ENERGY THROUGH VISCOUS SHEAR

In Art. 4.6, it was shown that the term p/ρ in the Bernoulli's equation, Eq. (4.6.5), represents the work done on unit mass of the flowing fluid by the unbalanced pressure forces to produce a change in either the kinetic energy or the potential energy of the fluid. In the case of steady, uniform, and horizontal flow, there would be no change in either the kinetic energy (as the flow is steady and uniform) or the potential energy (as the flow is in horizontal direction). Therefore, the work done by the pressure forces in maintaining viscous flow must result in continuous generation of heat just as in the case of solid friction. In other words, the deformation of a fluid element during fluid motion causes viscous stresses within the element that tend to resist such deformation, and the work done by the external forces in deforming the element must, therefore, be equal to the heat generated by the viscous shear (4).

For steady uniform flow, the forces that do work upon fluid elements are caused by some external source such as hydraulic machinery or weight of the fluid itself either of which can be treated as a source of mechanical energy. Also, if a fluid were in motion, work would be done within the fluid regardless of whether or not external source acts upon the fluid. This means that generation of heat may also take place at the expense of the kinetic energy of the flow and, in that case, the velocity of flow will gradually approach zero as the kinetic energy is being converted into heat. Therefore, the process of viscous resistance to fluid deformation can be regarded as a general means by which mechanical energy is converted into thermal energy (4).

Consider a fluid element of steady uniform flow, Fig. 7.3.1. After equating the algebraic sum of the longitudinal forces to zero, there remains the couple due to shear stresses in opposite directions along the horizontal faces of the fluid element. The product of the moment of the couple, $(\tau\delta x\delta z)\delta y$, and angular velocity $\delta u/\delta y$ must, therefore, be equal to the rate at which work is done by the couple that equals the rate of dissipation of energy. Dividing the product by the volume of the element and letting this volume, $\delta x\,\delta y\,\delta z$ approach zero, the rate of dissipation of energy per unit volume, equals $\tau\dfrac{du}{dy}$, i.e.,

the product of the shear intensity and the velocity gradient. For laminar flow, $\tau = \mu\dfrac{du}{dy}$ and, hence, rate of dissipation of energy per unit volume equals $\mu\left(\dfrac{du}{dy}\right)^2$ and its representative variation for parallel flow between stationary boundaries is shown in Fig. 7.3.2. It is, therefore, obvious that the zone of maximum dissipation of energy (and, hence, generation of heat too) is in the vicinity of the boundary where the shear and the velocity gradient too are highest (4).

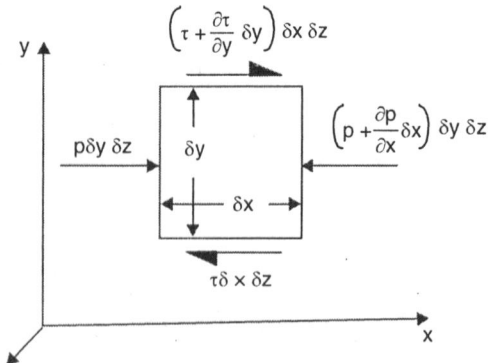

Fig. 7.3.1. Pressure and shear on a fluid element

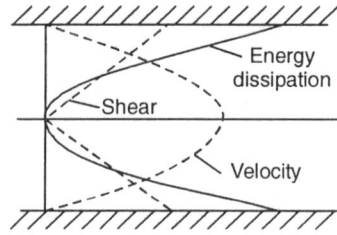

Fig. 7.3.2. Typical variation of velocity, shear, and energy dissipation for laminar flow in a circular pipe

7.4 INSTABILITY OF LAMINAR FLOW

In any given flow, there exist some disturbances that are caused due to inlet conditions, flow geometry, etc. The disturbances are in the form of longitudinal and transverse velocity components that are superimposed on the main flow velocity. The inertial forces present in the flow tend to magnify these disturbances while the viscous forces dampen these disturbances. Consider two adjacent fluid layers of Fig. 7.4.1(a) that are moving with different velocities and, hence, there is a surface of discontinuity. If a disturbance is caused at this surface, the spacing between the adjacent streamlines is varied, Fig. 7.4.1(b). This, in turn, would cause variation in velocity and pressure. The resulting pressure difference across the surface of discontinuity would further increase the displacement of streamlines, Fig. 7.4.1(c), thereby, ultimately, producing a series of vortices or eddies at the surface of discontinuity, Fig. 7.4.1(d). If the energy of the disturbance is absorbed by the main flow, the disturbance is dampened. If the disturbance extracts energy from the main flow, it gets amplified. The amplification or damp-ening of the disturbance depends on the fluid and flow characteristics (such as viscosity, mass density, velocity gradient, and proxim-ity of the boundary). Flow remains laminar when the disturbances are dampened and be-comes unstable and, then, turbulent if the dis-turbances are amplified. Rouse (4) expressed the instability parameter χ (Chi) as

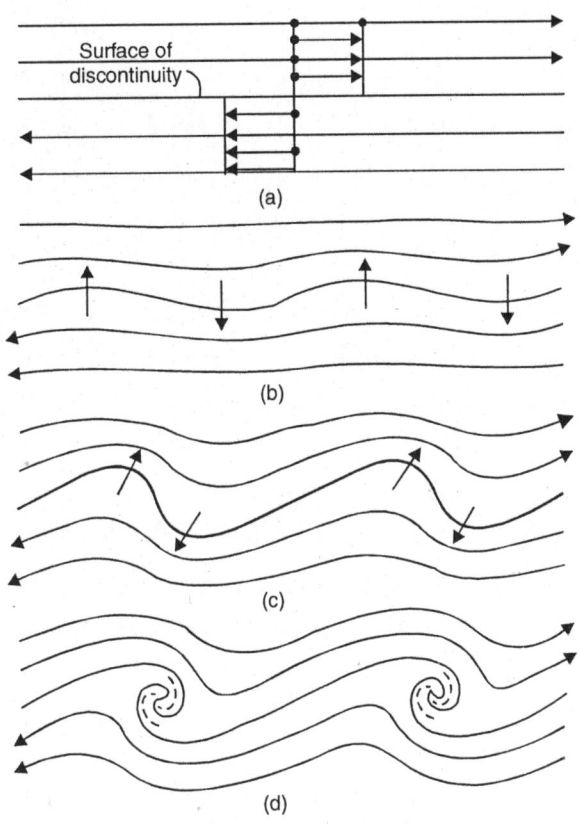

Fig. 7.4.1. Instability of flow at a surface of discontinuity (4)

$$\chi = \frac{y^2 \rho \left(\dfrac{du}{dy} \right)}{\mu} \qquad ...(7.4.1)$$

in which y is the distance from the boundary and $\dfrac{du}{dy}$ is the velocity gradient. A flow with χ < 500 in the entire flow region is inherently stable and flow remains laminar. If the instabil-ity parameter χ exceeds 500, the flow becomes unstable and may change to turbulent.

7.5 TURBULENT FLOW

Turbulent flow results from instability of laminar flow. Eddies that form on account of amplification of disturbances spread over the entire flow region and the mean motion is superimposed with the fluctu-ating components of velocity and pressure. As a result, there is transverse movement of fluid particles between adjacent layers in addition to the molecular intermixing. Such large scale intermixing or diffusion is an important feature of a turbulent flow.

Turbulence is defined as an irregular condition of motion in which various flow quantities (veloci-ties and pressure) vary randomly with time and space coordinates so that statistically average quantities can be obtained (1). The random fluctuations of velocity components and pressure make analysis of turbulent motion very difficult even with numerical and experimental methods. While dealing with turbulent motion, it is more convenient to separate the instantaneous quantities into mean or time-averaged values and fluctuating values. Figure 7.5.1 shows the variation of x-direction velocity u with time and this can be expressed as

$$u = \bar{u} + u'$$ ...(7.5.1a)

and likewise, $$v = \bar{v} + v'$$ ...(7.5.1b)

$$w = \bar{w} + w'$$ ...(7.5.1c)

and $$p = \bar{p} + p'$$ ...(7.5.1d)

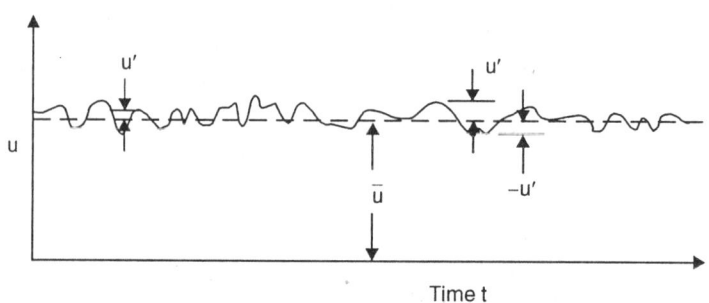

Fig. 7.5.1. Variation of u with time in turbulent flow

Here, $\bar{u}, \bar{v}, \bar{w}$ and $\bar{p}$ are the time-averaged values of $u, v, w,$ and p, respectively, and $u', v', w',$ and p' are the corresponding fluctuating parts. The mean or time-averaged value of the instantaneous value u is expressed as

$$\bar{u} = \frac{1}{T_0} \int_t^{t+T_0} u \, dt$$ ...(7.5.2)

in which T_0 is an averaging time period that should be sufficiently large and greater than any period of fine-scale turbulent variations. It is obvious that time-average of the fluctuating component would be zero. This means that for the fluctuations in x-direction velocity,

$$\bar{u'} = \frac{1}{T_0} \int_t^{t+T_0} (u - \bar{u}) \, dt = \bar{u} - \bar{u} = 0$$

Therefore, $$\bar{u'} = \bar{v'} = \bar{w'} = 0$$

However, the mean or time-average of the square of the fluctuating part is not zero. That is,

$$\overline{u'^2} = \frac{1}{T_0} \int_t^{t+T_0} (u - \bar{u})^2 \neq 0$$

The square root of this quantity, i.e., the root mean square of the measured values of the fluctuating component is a quantitative measure of the level or intensity of turbulence. The turbulence intensity is expressed as either $\dfrac{\sqrt{\overline{u'^2}}}{\bar{u}}$ or $\dfrac{\sqrt{\frac{1}{3}\left(\overline{u'^2}+\overline{v'^2}+\overline{w'^2}\right)}}{\bar{u}}$ for the main flow direction. Similarly, for y-direction,

turbulence intensity is expressed as $\dfrac{\sqrt{\overline{v'^2}}}{\bar{u}}$ and $\dfrac{\sqrt{\overline{w'^2}}}{\bar{u}}$, respectively.

It may be noted that the mean values or time-averaged values of the products of the fluctuation components such as $\overline{u'v'}$ or $\overline{u'w'}$ etc. are not zero.

Consider two adjacent fluid layers moving with velocities $\bar{u}_1$ and $\bar{u}_2$ such that $\bar{u}_1 > \bar{u}_2$ and $u' = \bar{u}_1 - \bar{u}_2$. Because of turbulence there would be transverse movement of fluid particles between the fluid layers. Assume that a small mass of the fluid from the lower layer (moving with velocity $\bar{u}_2$ and whose surface area is dA) moves upward due to the fluctuating velocity v' and gets embedded in the upper layer. Thus, rate of change of momentum of this mass is $\rho Av'\left(\bar{u}_1 - \bar{u}_2\right) = \rho Av'u'$. Since the fluid of mass $\rho Av'$ moved from smaller velocity layer to a larger velocity layer, the upper layer gets retarded. Similarly, a mass $\rho Av'$ coming from upper layer to the lower layer (i.e., v' is negative) would accelerate the lower layer (i.e., u' is positive). This means, positive v' is, generally, associated with negative u' and vice versa. In other words, the product $\rho A\overline{u'v'}$ would, generally, be negative. As a result of the change in momentum rate there would exist a shear force in the x-direction on the surface of the layer. Thus, shear stress on account of the turbulent fluctuations can be written as

$$\tau = \frac{-\rho A\overline{u'v'}}{A} = -\rho\overline{u'v'} \qquad \qquad ...(7.5.3)$$

Since $\overline{u'v'}$ is −ve, negative sign is introduced to make τ positive.

7.6 REYNOLDS EQUATIONS FOR TURBULENT MOTION

Equations governing the incompressible fluid motion, i.e., continuity equation, Eq. (3.8.5) and momentum equations, Eqs. (4.4.4 – 4.4.6) can be rewritten as follows:

$$\frac{\partial u}{\partial x} + \frac{\partial v}{\partial y} + \frac{\partial w}{\partial z} = 0 \qquad \qquad ...(7.6.1)$$

$$\rho\left[\frac{\partial u}{\partial t} + \frac{\partial u^2}{\partial x} + \frac{\partial(uv)}{\partial y} + \frac{\partial(uw)}{\partial z}\right] = -\frac{\partial p}{\partial x} + \mu\left(\frac{\partial^2 u}{\partial x^2} + \frac{\partial^2 u}{\partial y^2} + \frac{\partial^2 u}{\partial z^2}\right) \qquad ...(7.6.2)$$

$$\rho\left[\frac{\partial v}{\partial t} + \frac{\partial(uv)}{\partial x} + \frac{\partial v^2}{\partial y} + +\frac{\partial(vw)}{\partial z}\right] = -\frac{\partial p}{\partial y} + \mu\left(\frac{\partial^2 v}{\partial x^2} + \frac{\partial^2 v}{\partial y^2} + \frac{\partial^2 v}{\partial z^2}\right) \qquad ...(7.6.3)$$

$$\rho\left[\frac{\partial w}{\partial t} + \frac{\partial(uw)}{\partial x} + \frac{\partial(vw)}{\partial y} + \frac{\partial(w^2)}{\partial z}\right] = \rho g_z - \frac{\partial p}{\partial z} + \mu\left(\frac{\partial^2 w}{\partial x^2} + \frac{\partial^2 w}{\partial y^2} + \frac{\partial^2 w}{\partial z^2}\right) \qquad ...(7.6.4)$$

It should be noted that

$$\frac{\partial u^2}{\partial x} + \frac{\partial (uv)}{\partial y} + \frac{\partial (uw)}{\partial z}$$

$$\equiv 2u\frac{\partial u}{\partial x} + u\frac{\partial v}{\partial y} + v\frac{\partial u}{\partial y} + u\frac{\partial w}{\partial z} + w\frac{\partial u}{\partial z}$$

$$\equiv u\left[\frac{\partial u}{\partial x} + \frac{\partial v}{\partial y} + \frac{\partial w}{\partial z}\right] + u\frac{\partial u}{\partial x} + v\frac{\partial u}{\partial y} + w\frac{\partial u}{\partial z}$$

$$\equiv u\frac{\partial u}{\partial x} + v\frac{\partial u}{\partial y} + w\frac{\partial u}{\partial z}$$

and, similarly, for other convective acceleration terms in y- and z-directions.

On substituting the values of the instantaneous velocity components in terms of the time-averaged mean and fluctuating values in the continuity equation, Eq. (7.6.1), for incompressible fluid flow, and, then, taking the time-average, one obtains,

$$\frac{\partial \bar{u}}{\partial x} + \frac{\partial \bar{v}}{\partial y} + \frac{\partial \bar{w}}{\partial z} = 0 \qquad \qquad ...(7.6.5)$$

Similarly, the time-averaged momentum equation for x-direction can be obtained by substituting the values of the instantaneous velocity components in terms of the time-averaged mean and fluctuating values in the momentum equation for incompressible fluid flow, Eq. (7.6.2). This yields,

$$\rho\left[\frac{\partial(\bar{u}+u')}{\partial t} + \frac{\partial(\bar{u}+u')(\bar{u}+u')}{\partial x} + \frac{\partial(\bar{u}+u')(\bar{v}+v')}{\partial y} + \frac{\partial(\bar{u}+u')(\bar{w}+w')}{\partial z}\right]$$

$$= -\frac{\partial}{\partial x}(\bar{p}+p') + \mu\left[\frac{\partial^2(\bar{u}+u')}{\partial x^2} + \frac{\partial^2(\bar{u}+u')}{\partial y^2} + \frac{\partial^2(\bar{u}+u')}{\partial z^2}\right] \qquad ...(7.6.6)$$

Now the quantities such as

$$\overline{(\bar{u}+u')(\bar{u}+u')} = \overline{\bar{u}\bar{u}} + \overline{u'u'} = \bar{u}\bar{u} + \overline{u'u'}$$

Similarly,

$$\overline{(\bar{u}+u')(\bar{v}+v')} = \bar{u}\bar{v} + \overline{u'v'}$$

and

$$\overline{(\bar{u}+u')(\bar{w}+w')} = \bar{u}\bar{w} + \overline{u'w'}$$

Therefore, the L.H.S. of the x-direction momentum equation, Eq. (7.6.6) yields

$$\rho\left[\frac{\partial \bar{u}}{\partial t} + \frac{\partial}{\partial x}\left(\bar{u}\bar{u} + \overline{u'u'}\right) + \frac{\partial}{\partial y}\left(\bar{u}\bar{v} + \overline{u'v'}\right) + \frac{\partial}{\partial z}\left(\bar{u}\bar{w} + \overline{u'w'}\right)\right]$$

$$\equiv \rho \left[\frac{\partial \overline{u}}{\partial t} + \overline{u}\frac{\partial \overline{u}}{\partial x} + \overline{v}\frac{\partial \overline{u}}{\partial y} + \overline{w}\frac{\partial \overline{u}}{\partial z} + \overline{u}\left(\frac{\partial \overline{u}}{\partial x} + \frac{\partial \overline{v}}{\partial y} + \frac{\partial \overline{w}}{\partial z} \right) + \frac{\partial \overline{u'^2}}{\partial x} + \frac{\partial \overline{u'v'}}{\partial y} + \frac{\partial \overline{u'w'}}{\partial z} \right]$$

$$\equiv \rho \left[\frac{\partial \overline{u}}{\partial t} + \overline{u}\frac{\partial \overline{u}}{\partial x} + \overline{v}\frac{\partial \overline{u}}{\partial y} + \overline{w}\frac{\partial \overline{u}}{\partial z} + \frac{\partial \overline{u'^2}}{\cdot \partial x} + \frac{\partial \overline{u'v'}}{\partial y} + \frac{\partial \overline{u'w'}}{\partial z} \right]$$

Thus, for the *x*-direction, the momentum equation is

$$\rho \frac{D\overline{u}}{Dt} = \rho \left[\frac{\partial \overline{u}}{\partial t} + \overline{u}\frac{\partial \overline{u}}{\partial x} + \overline{v}\frac{\partial \overline{u}}{\partial y} + \overline{w}\frac{\partial \overline{u}}{\partial z} \right] = -\frac{\partial p}{\partial x} + \mu \left[\frac{\partial^2 \overline{u}}{\partial x^2} + \frac{\partial^2 \overline{u}}{\partial y^2} + \frac{\partial^2 \overline{u}}{\partial z^2} \right]$$

$$+ \left[\frac{\partial \left(-\rho \overline{u'^2} \right)}{\partial x} + \frac{\partial \left(-\rho \overline{u'v'} \right)}{\partial y} + \frac{\partial \left(-\rho \overline{u'w'} \right)}{\partial z} \right] \qquad ...(7.6.7)$$

Similarly,

$$\rho \frac{D\overline{v}}{Dt} = \rho \left[\frac{\partial \overline{v}}{\partial t} + \overline{u}\frac{\partial \overline{v}}{\partial x} + \overline{v}\frac{\partial \overline{v}}{\partial y} + \overline{w}\frac{\partial \overline{v}}{\partial z} \right] = -\frac{\partial p}{\partial y} + \mu \left[\frac{\partial^2 \overline{v}}{\partial x^2} + \frac{\partial^2 \overline{v}}{\partial y^2} + \frac{\partial^2 \overline{v}}{\partial z^2} \right]$$

$$+ \left[\frac{\partial \left(-\rho \overline{u'v'} \right)}{\partial x} + \frac{\partial \left(-\rho \overline{v'^2} \right)}{\partial y} + \frac{\partial \left(-\rho \overline{v'w'} \right)}{\partial z} \right] \qquad ...(7.6.8)$$

and

$$\rho \frac{D\overline{w}}{Dt} = \rho \left[\frac{\partial \overline{w}}{\partial t} + \overline{u}\frac{\partial \overline{w}}{\partial x} + \overline{v}\frac{\partial \overline{w}}{\partial y} + \overline{w}\frac{\partial \overline{w}}{\partial z} \right] = \rho g_z - \frac{\partial p}{\partial z} + \mu \left[\frac{\partial^2 \overline{w}}{\partial x^2} + \frac{\partial^2 \overline{w}}{\partial y^2} + \frac{\partial^2 \overline{w}}{\partial z^2} \right]$$

$$+ \left[\frac{\partial \left(-\rho \overline{u'w'} \right)}{\partial x} + \frac{\partial \left(-\rho \overline{v'w'} \right)}{\partial y} + \frac{\partial \left(-\rho \overline{w'^2} \right)}{\partial z} \right] \qquad ...(7.6.9)$$

These equations were first derived by Reynolds and, hence, are known as the Reynolds equations. The terms $-\rho \overline{u'^2}, -\rho \overline{u'v'}$, etc. provide stress-like effect on the flow and are, accordingly, termed as Reynolds stresses (see Art. 7.5) that are, in general, unknown. The estimation of these stresses requires experimentation or some empirical methods. Equations (7.6.7 – 7.6.9) can, alternatively, be written as

$$\rho \frac{D\overline{u}}{Dt} = -\frac{\partial p}{\partial x} + \frac{\partial \tau_{xx}}{\partial x} + \frac{\partial \tau_{yx}}{\partial y} + \frac{\partial \tau_{zx}}{\partial z} \qquad ...(7.6.10)$$

$$\rho \frac{D\overline{v}}{Dt} = -\frac{\partial p}{\partial y} + \frac{\partial \tau_{xy}}{\partial x} + \frac{\partial \tau_{yy}}{\partial y} + \frac{\partial \tau_{zy}}{\partial z} \qquad ...(7.6.11)$$

$$\rho \frac{D\bar{w}}{Dt} = \rho g_z - \frac{\partial p}{\partial z} + \frac{\partial \tau_{xz}}{\partial x} + \frac{\partial \tau_{yz}}{\partial y} + \frac{\partial \tau_{zz}}{\partial z} \qquad \text{...(7.6.12)}$$

For a flow in which $\bar{u} = \bar{u}(y)$, $\bar{p} = \bar{p}(x)$ and $\bar{v} = \bar{w} = 0$, Eqs. (7.6.10 – 7.6.12) yield

$$\rho \frac{\partial \bar{u}}{\partial t} = -\frac{\partial \bar{p}}{\partial x} + \frac{\partial \tau}{\partial y} \qquad \text{...(7.6.13)}$$

in which

$$\tau = \mu \frac{\partial \bar{u}}{\partial y} - \rho \overline{u'v'} = \tau_l + \tau_t \qquad \text{...(7.6.14)}$$

i.e., the total shear is made up of the laminar shear stress, $\tau_l \left(= \mu \dfrac{\partial \bar{u}}{\partial y}\right)$ and the turbulent shear stress,

$\tau_t \left(= -\rho \overline{u'v'}\right)$. One simple way of estimating the shear stress is to express the shear stress τ_t in the form similar to Newton's law of viscosity, Eq. (1.4.4), that is,

$$\tau_t = -\rho \overline{u'v'} = \eta \frac{\partial \bar{u}}{\partial y} \qquad \text{...(7.6.15)}$$

in which η is termed eddy viscosity or eddy dynamic viscosity that depends on fluid as well as flow characteristics. This approach was first suggested by J. Boussinesq in 1877. It, however, did not find much favour among researchers because of the dependence of η on flow conditions.

7.7 PRANDTL'S MIXING LENGTH THEORY

Prandtl(2) presented mixing length theory so as to be able to estimate $-\rho \overline{u'v'}$. Prandtl's mixing length l can be considered as the transverse distance between the two adjacent fluid layers such that the lumps of fluid particles from one layer could move to the other layer and be embedded in it while retaining, during their travel through distance l, their momentum in the direction of the motion of the fluid layers. Prandtl further assumed that

$$u' \approx l \frac{d\bar{u}}{dy}$$

and

$$v' \approx u'$$

Hence,

$$\tau_t = -\rho \overline{u'v'} \approx \rho l^2 \left(\frac{d\bar{u}}{dy}\right)^2 \qquad \text{...(7.7.1)}$$

Transverse movement of fluid always occurs in all parts of turbulent flow except at and in the vicinity of a boundary where this interchange of fluid reduces to zero. Therefore, the fluctuations u' and v', and, hence, the mixing length must approach zero at the boundary.

7.8 VELOCITY DISTRIBUTION IN TURBULENT FLOW

Equation (7.6.14), after substituting for τ_t from Eq. (7.7.1), can also be written as

$$\tau = \mu \left(\frac{d\bar{u}}{dy} \right) + \rho l^2 \left(\frac{d\bar{u}}{dy} \right)^2 \qquad ...(7.8.1)$$

The first term on the right side of Eq. (7.8.1) is the viscous shear which, in turbulent flows, is negligibly small except near the boundary. Therefore, in turbulent flows, shear is assumed to be due to turbulence alone. The overbars denoting time averages will be dropped in further analysis.

Figure 7.8.1 shows a typical variation of velocity and shear stresses τ, τ_l and τ_t in a turbulent flow near a wall. The thin layer near the wall, where laminar shear is dominant, is termed viscous shear layer or, simply, viscous layer. In this layer, the shear stress in the fluid is constant and equals the shear stress at the wall τ_0. The velocity within this layer, u is related to the shear stress $\tau_0(x)$ and absolute viscosity of the fluid μ through Newton's law of viscosity, Eq. (1.4.4), as

$$\tau_0 = \mu \frac{du}{dy} \approx \mu \frac{u}{y}$$

or
$$\frac{\tau_0}{\rho} = \frac{\mu}{\rho} \frac{u}{y}$$

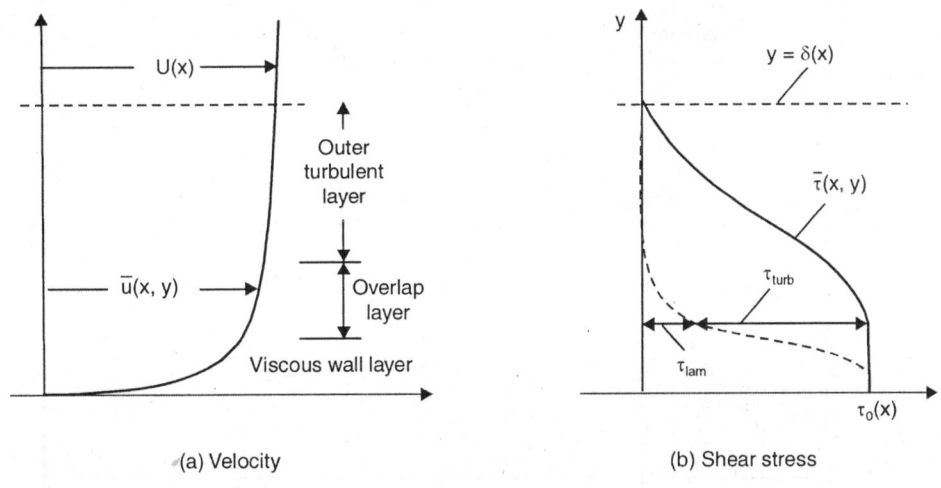

Fig. 7.8.1. Variation of velocity and shear in turbulent flow near a wall

The term $\sqrt{\dfrac{\tau_0}{\rho}}$ has the dimensions of velocity and is, therefore, called shear velocity u_*. Hence,

$$\frac{u}{u_*} = \frac{u_* y}{\nu} \qquad ...(7.8.2)$$

This velocity variation is known as law of the wall and is valid near the wall upto $\dfrac{u_* \, y}{v} \leq 5$. In the overlap layer (between the viscous shear layer and outer turbulent zone, Fig. 7.8.1) too it is assumed that the shear stress is approximately constant and equals the shear stress at the wall, i.e., τ_0. But, in this region, as in the outer turbulent zone, turbulence dominates and the viscous (or laminar) shear stress is negligible. Therefore, Eq. (7.8.1) reduces to

$$\tau_0 = \rho l^2 \left(\frac{du}{dy}\right)^2 \qquad\qquad ...(7.8.3)$$

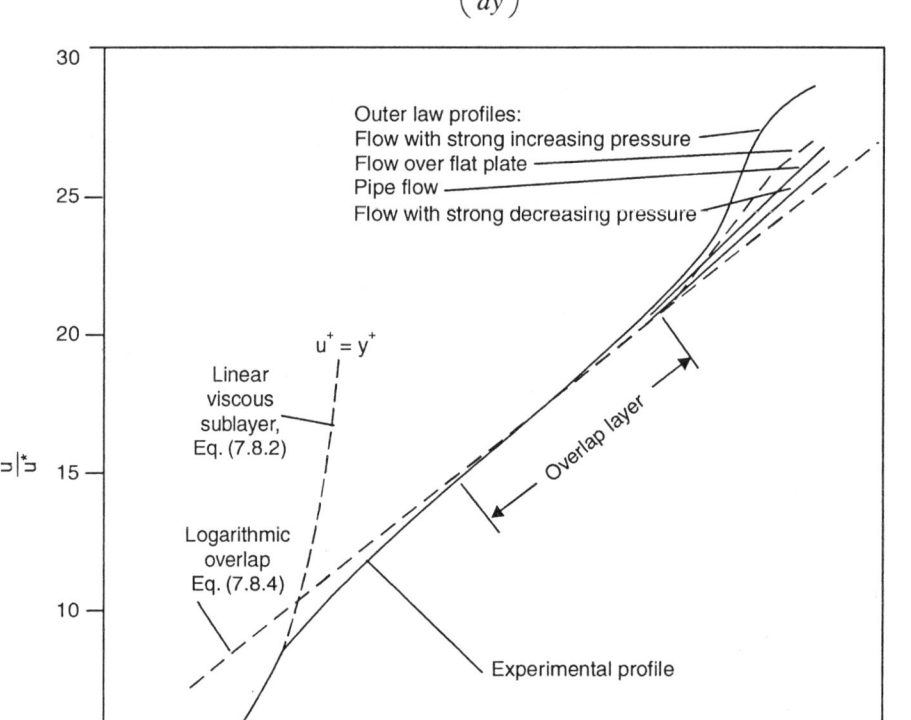

Fig. 7.8.2. Typical velocity variations in turbulent flow (6)

Since l has dimensions of length and it must, like u' and v', also vanish at the wall, l can be assumed as $l = ky$. Therefore, Eq. (7.8.3) can be rearranged to give

$$\frac{du}{u_*} = \frac{1}{k}\frac{dy}{y}$$

On integration, one gets

$$\frac{u}{u_*} = \frac{1}{k} \ln y + \text{constant} \qquad \qquad ...(7.8.4)$$

This equation is called the logarithmic distribution of velocity that actually approximates almost the entire velocity profile, Fig. 7.8.2, since the inner wall law, Eq. (7.8.2), typically extends over less than 2 percent of the profile and can, therefore, be neglected. Therefore, one can use Eq. (7.8.4) as an excellent approximation to solve nearly every turbulent flow problem. Equation (7.8.4), however, does not approximate for the outer law when the pressure increases strongly downstream (i.e., adverse pressure gradient, as in a diffuser) (6).

EXAMPLES

E 7.1.   For laminar flow in a 12.5 mm diameter pipe, determine (a) the maximum discharge if the flowing fluid is water and (b) maximum average velocity if the flowing fluid is air.

Solution: Taking lower critical Reynolds number as 2000,

$$\frac{VD\rho}{\mu} = \frac{4Q}{\pi D \nu} = 2000$$

Therefore, for water, $Q = \dfrac{(2000)(\pi)(0.0125)(10^{-6})}{4}$

$$= 19.64 \times 10^{-6} \text{ m}^3/\text{s}$$

$$= 19.64 \text{ cm}^3/\text{s}$$

and for air, $V = \dfrac{(2000)(1.5 \times 10^{-5})}{0.0125}$

$$= 2.4 \text{ m/s}$$

E 7.2.   A viscous oil ($\mu = 0.8$ Ns/m^2) flows steadily between two stationary parallel plates that are 3 mm apart. Determine the shear stress on the upper plate and also discharge (per metre width) through the gap between the parallel plates. Assume piezometric pressure gradient to be 1000 N/m^2/m.

Solution: For the given data, Eq. (7.2.6) gives

$$u = -\frac{1}{2(0.8)}(-1000)(0.003\, y - y^2)$$

$$= 1.875\, y - 625\, y^2$$

$\therefore$ $\dfrac{du}{dy} = 1.875 - 1250\, y$

Hence, $\tau_0 = \mu \dfrac{du}{dy}\bigg|_{y=0.003} = -1.5 \text{ N/m}^3$

and discharge per meter width q is

$$q = \int_0^{0.003} u\, dy$$

$$= \int_0^{0.003} \left(1.875\, y - 625\, y^2\right) dy$$

$$= \left[1.875 \frac{y^2}{2} - 625 \frac{y^3}{3}\right]_0^{0.003}$$

$$= 8.4375 \times 10^{-6} - 5.625 \times 10^{-6}$$

$$= 2.8125 \times 10^{-6}\ \text{m}^3/\text{s/m}$$

E7.3.　One plate moves relative to another as shown in Fig. E7.3. Determine the velocity distribution, discharge, and the shear stress exerted on the upper plate. Assume $\mu = 0.1$ Ns/m^2 and $\rho = 900$ kg/m^3. Other data are as shown in the figure.

Solution:　At A,　$p + \rho g h = 1500 - (900)\,(9.8)\,(3)$

$$= -26960\ \text{Pa}$$

at B,　$p + \rho g h = 850$ Pa

　　(with respect to the datum at B)

Length $AB = \sqrt{3^2 + 2^2} = \sqrt{13} = 3.606$ m

Therefore,

$$\frac{d(p+\rho g h)}{dx} = \frac{\left[850 - (-24960)\right]}{3.606} = 7157.51\ \text{Pa/m}$$

From Eq. (7.2.4) and Fig. E7.3,

　　　　$B = 0.005$ m and $U = -1.0$ m/s

$\therefore$　　　$u = \dfrac{-1}{0.005} y - \dfrac{1}{2(0.1)} \left[7157.51\right]\left(0.005\, y - y^2\right)$

$$= -200\, y - 178.94\, y + 35787\, y^2$$

$$= 35787\, y^2 - 378.94\, y$$

The discharge per meter width q is

$$q = \int_0^{0.005} u\, dy = \int_0^{0.005} \left(35787\, y^2 - 378.94\, y\right) dy = \left[35787 \frac{y^3}{3} - 378.94 \frac{y^2}{2}\right]_0^{0.005}$$

$$= 1.49 \times 10^{-3} - 4.74 \times 10^{-3} = -3.25 \times 10^{-3}\ \text{m}^3/\text{s/m}$$

Therefore, the discharge is upward, Fig. E7.3

Shear stress on the upper moving plate is

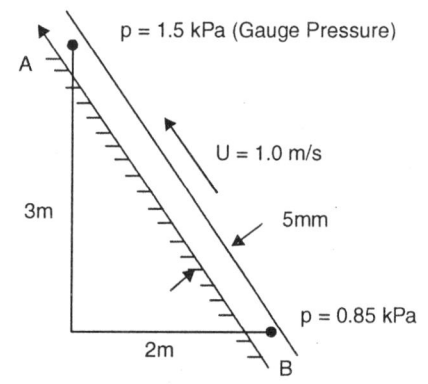

p = 1.5 kPa (Gauge Pressure)

A

U = 1.0 m/s

3m

5mm

2m

p = 0.85 kPa

B

Fig. E 7.3

$$\tau = \mu \frac{du}{dy}\bigg|_{y=0.005} = (0.1)\left[71574\,y - 378.94\right]_{y=0.005} = -2.107\,\text{Pa}$$

E 7.4. The shear stress at the bed of a channel is sometimes determined by observing the velocities at two points along a vertical in the flow section. Obtain the expression for shear stress.

Solution: The velocity variation in turbulent flow is given by Eq. (7.7.4) as

$$\frac{u}{u_*} = \frac{1}{k}\ln y + C$$

At $\qquad y = y_1,\ u = u_1$

and $\qquad y = y_2,\ u = u_2$

Therefore, $\qquad \dfrac{u_1}{u_*} = \dfrac{1}{K}\ln y_1 + C$

and $\qquad \dfrac{u_2}{u_*} = \dfrac{1}{k}\ln y_2 + C$

Hence, $\qquad \dfrac{\left(u_2 - u_1\right)^2}{u_*^2} = \left[\dfrac{1}{k}\ln\dfrac{y_2}{y_1}\right]^2$

Since $\qquad u_* = \sqrt{\dfrac{\tau_0}{\rho}}$

$\therefore \qquad \tau_0 = \rho\,\dfrac{k^2\left(u_2 - u_1\right)^2}{\left(\ln\left(\dfrac{y_2}{y_1}\right)\right)^2}$

or $\qquad \tau_0 = \rho\,\dfrac{\left(u_2 - u_1\right)^2}{33.06\left[\log\dfrac{y_2}{y_1}\right]^2}$

as $\qquad k = 0.4$

PROBLEMS

P7.1 What is the Reynolds number of flow of 0.35 m³/s oil, with specific gravity = 0.85 and $\mu = 0.025$ Ns/m² through a pipe of diameter equal to 400 mm.

P7.2 Determine the time average velocity $\bar{u}$, $\overline{u'^2}$, and $\sqrt{\dfrac{\overline{u'^2}}{\bar{u}}}$ for the following data of instantaneous velocities (in m/s) measured at a given point in a flow field at the intervals of 1 second.

8.18	8.10	8.05	8.13	8.22	8.28	8.12
7.98	7.94	8.15	7.96	8.24	8.03	8.08

P7.3 For flow between stationary parallel plates, determine the ratio of maximum velocity to average velocity.

P7.4 Compute the kinetic energy and momentum correction factors for laminar flow between fixed parallel plates.

P7.5 Laminar flow of a fluid of viscosity 0.8 Ns/m^2 and density 1250 kg/m^3 occurs between two parallel plates that are 10 mm apart and inclined at 45° with the horizontal. The upper plate moves with a velocity of 1.2 m/s relative to the lower plate and in a direction opposite to that of the fluid flow. Pressure gauges mounted at two sections, 1m vertically apart, read pressures of 200 kN/m^2 and 80 kN/m^2 respectively. Compute, the maximum flow velocity, and shear stress on the upper plate.

P7.6 A thin film of water flows over a parking lot of bottom slope 0.003. Find the depth if the flow rate is 0.08 l/s/m.

P7.7 Find the pressure gradient that results in zero shear stress at the lower wall, where $y = 0$, for flow between two parallel plates. The plates are spaced a apart and are horizontal. The upper plate speed is U with respect to the stationary lower plate.

REFERENCES

(1) Hinze, JO: *"Turbulence"*, McGraw-Hill, New York, 1975.

(2) Prandtl, L: *"Essentials of Fluid Dynamics"*, Hafner, New York, 1952 [Streeter, et al. (5)].

(3) Reynolds, O: *"An experimental investigation of the circumstances which determine whether the motion of water shall be direct or sinuous, and of the laws of resistance in parallel channels"*, Trans. Royal Society, London, vol. 174, 1883. [Streeter et al. (5)].

(4) Rouse, H: *"Elementary Mechanics of Fluids"*, John Wiley and Sons, New York, 1946.

(5) Streeter, VL, Wylie, EB, and Bedford, KW: *"Fluid Mechanics"*, 9th Edition, McGraw-Hill Companies, Singapore, 1998].

(6) White, F.M.: *"Fluid Mechanics"*, 5th Edition, McGraw-Hill Co., 2003.

CHAPTER 8

Boundary Layer and External Flows

8.1 BOUNDARY LAYER

Flow of a low-viscosity fluid may be considered irrotational. However, real fluids do not slip at a solid surface. This means that at the solid boundary in a flow domain, there is no relative motion between the fluid and boundary. Therefore, there is a region in the vicinity of a solid boundary where the velocity rises gradually from zero at the boundary (relative to the boundary) to the ambient stream velocity. Hence, there is a large velocity gradient at the boundary resulting into large shear forces in the vicinity of the boundary which reduce the flow speed in that region. The fluid layer that has had its velocity affected by the boundary shear is called the *boundary layer*.

The concept of boundary layer provides link between the idealized irrotational (or potential) flow away from the boundary, and viscous (or rotational) flow of real fluid near the boundary. The concept of boundary layer leads to the hypothesis that even for real fluids, the effect of internal friction (or viscous shear) in a fluid is significant only in a thin region in the vicinity of the solid boundary. Therefore, the flow outside the thin region in the vicinity of the solid boundary can be considered as irrotational (or ideal fluid) flow.

The concept of boundary layer, first introduced by Ludwing Prandtl in 1904(6), is applicable to both *internal* flows (confined by walls) as well as *external* flows such as flow around a body immersed in a fluid stream. In case of internal flows, the boundary layer grows from the confining walls and meet downstream, thus encompassing the entire flow section. External flows, on the other hand, are unconfined such as flow around a body immersed in a fluid or motion of a body immersed in a fluid. Examples of external flow include motion of airplanes, ships, automobiles or flow of air around buildings etc. Thus, concept of boundary layer finds applications in conduit flows, open channel flows, aerodynamics (airplanes, rockets, projectiles), hydrodynamics (ships, submarines, torpedos), transportation (automobiles, trucks, cycles), wind engineering (buildings, bridges, cooling towers, chimneys), and ocean engineering (buoys, breakwaters, cables, moored instruments).

8.2 BOUNDARY LAYER OVER A FLAT PLATE

Figure 8.2.1 shows qualitative growth of a boundary layer that develops over a stationary, smooth, and long flat plate aligned in the ambient flow direction. When the flowing fluid comes in contact with the plate at the leading edge, the fluid particles in contact with the plate come to rest. The particles near the surface of the plate are retarded by the stationary fluid particles at the plate. This retardation effect continues, although diminishingly, in the direction normal to the flow direction and the velocity reaches the ambient stream velocity asymptotically. The thin region in the vicinity of the plate is characterized by considerable retardation effect and grows continuously over the plate along the flow direction. This region is termed *boundary layer*. The flow in the boundary layer is laminar upto a short distance from the leading edge and, accordingly, the boundary layer is termed the *laminar boundary layer* (LBL). The transition of flow within the boundary layer from laminar to turbulent occurs over a region of the plate instead of at a single location across the plate. At the downstream end of the transition region, the flow within the boundary layer becomes fully turbulent. Hence, the boundary layer downstream of the transition region of the plate is termed the *turbulent boundary layer* (TBL). In the turbulent boundary layer region, there exists a very thin layer next to the surface of the plate (or boundary) that has laminar motion. This layer is called *laminar* (or *viscous*) *sublayer*. For rough boundaries, the boundary layer may be turbulent right from the leading edge. Reynolds number for the boundary layer flow is generally expressed as

$$R_{ex} = \frac{Ux\rho}{\mu}$$

...(8.2.1)

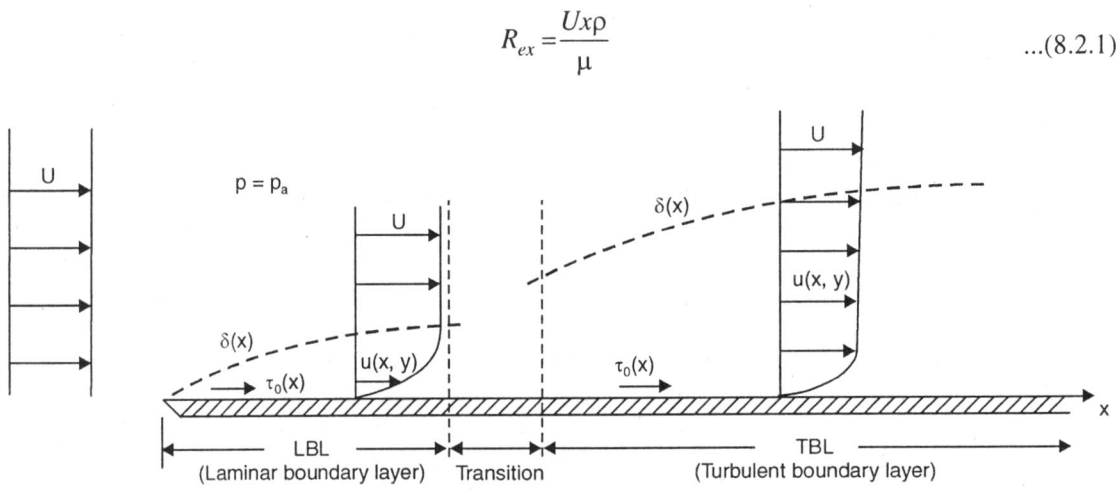

Fig. 8.2.1. Growth of boundary layer on a flat plate (vertical distances exaggerated)

in which U is the free (or ambient) stream velocity, x is the distance, along the plate, from the leading edge. For boundary layer computations, the transition Reynolds number is usually considered to be 500,000. For air, with free stream velocity of 30 m/s, this Reynolds number corresponds to a length of about 0.25 m, along the plate, from the leading edge. Figure 8.2.2 compares the flow past a thin plate at low and high Reynolds numbers.

It is worth considering the behaviour of a small rectangular fluid particle moving in the boundary layer flow. The particle moving in the ambient flow retains its original shape as it moves in the uniform flow. But, the shape of the particle moving inside the boundary layer gets distorted because of the velocity gradient within the boundary layer and, therefore, the upper edge of the particle moves with higher speed compared to the speed with which the lower edge of the particle moves. Therefore, the particles inside the boundary layer rotate while they are moving. Hence, the flow inside the boundary layer is rotational (or viscous) while the flow outside may be treated as irrotational (or non-viscous).

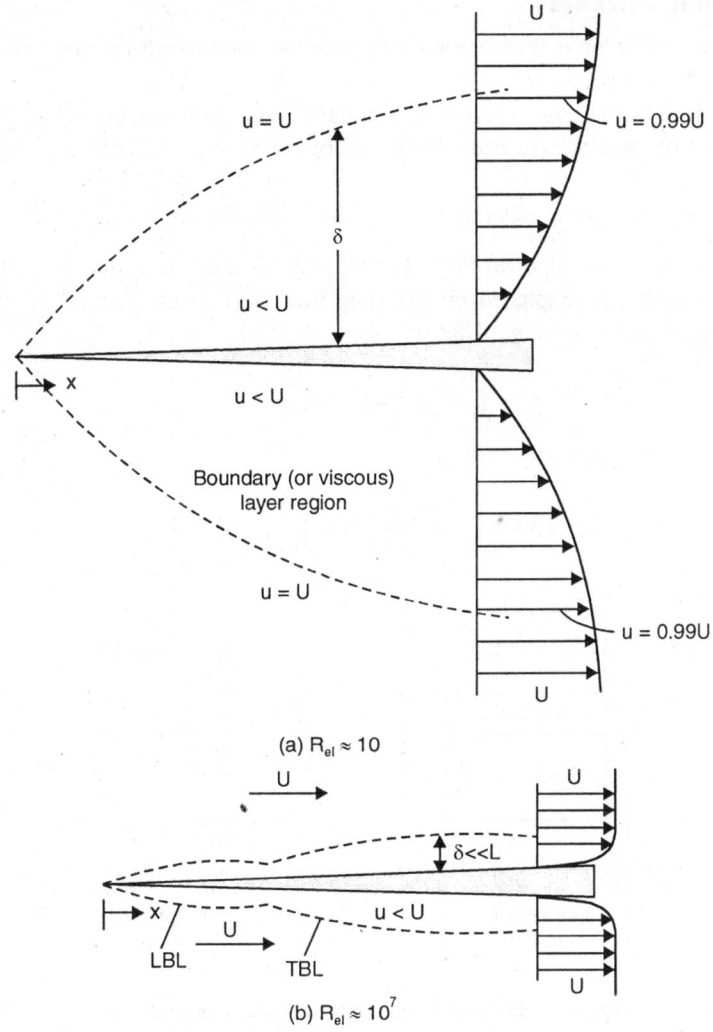

Fig. 8.2.2. Growth of boundary layer on a flat plate at low and high Reynolds number

8.3 BOUNDARY LAYER THICKNESS

8.3.1 Nominal Boundary Layer Thickness

The velocity profile $u = u(x,y)$ within a boundary layer is always asymptotic to the ambient stream velocity $u = U$, i.e., only at $y = \infty$, $u = U$. This means that the boundary layer does not have a sharp edge or well-defined thickness. The thickness of the boundary layer has been defined in different ways. Very often, the boundary layer thickness δ is defined as the distance from the plate (or boundary) at which the fluid velocity is 99 percent of the ambient stream velocity. This means that at $y = \delta$, $u = 0.99U$. This boundary layer thickness δ is also called the *nominal boundary layer thickness* or *boundary layer disturbance thickness*. The velocity profile $u = u(x, y)$ is asymptotic to the ambient stream velocity U. The value "99 percent of the ambient stream velocity" is arbitrary.

8.3.2 Displacement Thickness

Because of the flow retardation due to viscous effects in the boundary layer, the mass flow rate adjacent to a boundary (Fig. 8.3.1a) is less than the mass flow rate that would have existed in the absence of the boundary (or boundary layer), Fig. 8.3.1(b). If viscous forces (or boundary layer effects) were absent, the velocity at a section would have been U instead of $u(x,y)$. The difference in mass flow rates A is,

therefore, $\int_0^\infty \rho(U-u)b\,dy$ where, b is the width of the boundary surface in the direction perpendicular to the flow direction. The *displacement thickness* $\delta*$ is defined as the distance by which the boundary would have to be imaginarily displaced inward (i.e., into the flow) to reduce the mass flow rate of a frictionless fluid by the same amount. That is,

$$\rho U b \delta* = \int_0^\infty \rho(U-u)b\,dy$$

For incompressible fluids, therefore,

$$\delta* = \int_0^\infty \left(1-\frac{u}{U}\right)dy \approx \int_0^\delta \left(1-\frac{u}{U}\right)dy \qquad \text{...(8.3.1)}$$

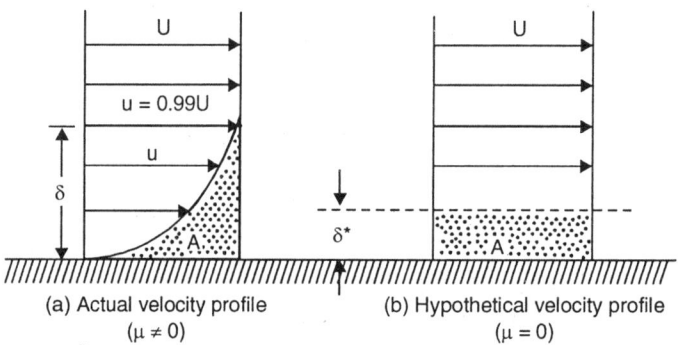

(a) Actual velocity profile
($\mu \neq 0$)

(b) Hypothetical velocity profile
($\mu = 0$)

Fig. 8.3.1. Definition sketch for displacement thickness

The concept of the displacement thickness is useful in analysing a viscous flow as irrotational flow after the boundary is imagined to have been displaced into the flow by a distance $\delta*$.

8.3.3 Momentum Thickness

Because of the retardation of flow within a boundary layer, the momentum flux, like mass flux, at any section is less than the momentum flux that would have existed at the section if the flow adjacent to the boundary were frictionless. The momentum deficiency of the *actual* mass flow rate, $\int_0^\infty \rho ub\,dy$, through

the boundary layer is $\int_0^\infty \rho u(U-u)b\,dy$. If boundary layer effects were to be made absent, the boundary would have to move a distance θ (known as *momentum thickness*) inward into the imaginary frictionless flow to compensate for the momentum deficiency. That is,

$$\rho Ub\theta U = \rho U^2 b\theta = \int_0^\infty \rho u(U-u)bdy$$

For incompressible fluids, therefore,

$$\theta = \int_0^\infty \frac{u}{U}\left(1-\frac{u}{U}\right)dy \approx \int_0^\delta \frac{u}{U}\left(1-\frac{u}{U}\right)dy \qquad ...(8.3.2)$$

Since, the expressions for the displacement thickness and momentum thickness contain integrals across the boundary layer, these thicknesses are integral thicknesses that are appreciably easier to evaluate accurately once the boundary layer thickness has been estimated. Besides, they have physical meaning without any arbitrariness.

8.4 MOMENTUM EQUATION APPLIED TO BOUNDARY LAYER

Flow within the boundary layer region can be analyzed using the relevant equations, Eqs. (4.4.4 – 4.4.6) governing the motion of viscous fluids. However, one can also obtain approximate solutions for boundary layer growth and drag simply by the use of the momentum principle. This approach was first suggested by Hungarian – American engineer Theodore von Karman (1881 – 1963)(10).

Consider a small length AB (=δx) of a plate over which boundary layer has developed, Fig. 8.4.1. It is assumed that the plate is sufficiently wide (perpendicular to the plane of the diagram) so that the flow over the plate can be considered two-dimensional and the 'edge effects' are insignificant. The outer edge of the boundary layer is CD and is not a streamline because the mass flow rate at two different sections of the boundary layer are different and the balance has to flow across the edge of the boundary layer. ACE and BD are the planes passing through A and B and also perpendicular to the plate AB. The point C of ACE is on the edge of the boundary layer and E on the streamline passing through D. No fluid can cross the streamline ED and, therefore, ABDE can be chosen as a control volume for application of the momentum principle.

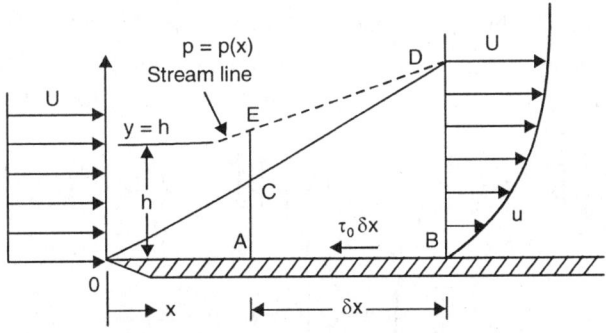

Fig. 8.4.1. Control volume for application of the momentum principle

It is further assumed that the pressures (piezometric) over the faces AE and BD do not vary. Hence, if the mean value of the pressure over the face AE is p, the mean pressure over BD would be $p + \left(\dfrac{\partial p}{\partial x}\right)\delta x$.

Therefore, the net pressure force on the control volume in the x-direction equals

$$p(AE)-\left(p+\frac{\partial p}{\partial x}\delta x\right)(BD)+\left(p+\frac{1}{2}\frac{\partial p}{\partial x}\delta x\right)(BD-AE)$$

Here, the last term is the x-direction component of the force due to the mean pressure $\left(p+\frac{1}{2}\frac{\partial p}{\partial x}\delta x\right)$ on the surface ED. This expression yields

$$-\frac{1}{2}\frac{\partial p}{\partial x}\delta x(BD+AE)$$

$$\equiv -\frac{\partial p}{\partial x}\delta x(BD) \text{ since as } \delta x \to 0, AE \approx BD$$

Therefore, the total *x*-direction force on the control volume is

$$-\tau_0\,\delta x-\frac{\partial p}{\partial x}\delta x(BD) \qquad\qquad ...(8.4.1)$$

in which τ_0 is the shear stress at the plate (or boundary).

Now consider an elementary strip of thickness δy in the plane AC, at a distance y from the plate and let the strip be of unit width perpendicular to the plane of the diagram. The x-direction momentum rate through the strip is $\rho u^2\delta y$ and for the entire boundary layer, the x-direction momentum rate is $\int_0^\delta \rho u^2\,dy$. The corresponding value for the plane (or section) BD is,

$$\int_0^\delta \rho u^2\,dy+\frac{\partial}{\partial x}\left(\int_0^\delta \rho u^2 dy\right)\delta x$$

The *x*-direction momentum rate through CE can be written as $\rho U^2(CE)$ as U represents the velocity (in the *x*-direction) of the ambient stream outside the boundary layer. Hence, the net rate of change of the *x*-direction momentum of the fluid passing through the control volume ABDE is,

$$\left[\int_0^\delta \rho u^2 dy+\frac{\partial}{\partial x}\left(\int_0^\delta \rho u^2 dy\right)\delta x\right]-\left[\int_0^\delta \rho u^2\,dy+\rho U^2(CE)\right]$$

$$\equiv \frac{\partial}{\partial x}\left(\int_0^\delta \rho u^2\,dy\right)\delta x-\rho U^2(CE)$$

Since p is a function of x, U too would depend on x and, hence, U need not be the same as the velocity of flow far upstream in the ambient flow. By continuity principle,

Mass flow rate across AE = mass flow rate across BD

i.e., Mass flow rate across CE = Mass flow rate across BD – Mass flow rate across AC

or

$$\rho U(CE)=\left[\int_0^\delta \rho u\,dy+\frac{\partial}{\partial x}\left(\int_0^\delta \rho u\,dy\right)\delta x\right]-\int_0^\delta \rho u\,dy=\frac{\partial}{\partial x}\left(\int_0^\delta \rho u\,dy\right)\delta x$$

Hence,
$$\rho U^2 (CE) = U \frac{\partial}{\partial x} \left(\int_0^\delta \rho u \, dy \right) \delta x$$

Therefore, the net rate of change of the x-direction momentum of the fluid passing through control volume ABDE is

$$\frac{\partial}{\partial x} \left(\int_0^\delta \rho u^2 \, dy \right) \delta x - U \frac{\partial}{\partial x} \left(\int_0^\delta \rho u \, dy \right) \delta x \qquad \qquad ...(8.4.2)$$

Equating the expressions (8.4.1) and (8.4.2), in accordance with the momentum principle, and, then dividing by $-\delta x$, one obtains

$$\tau_0 + \frac{\partial p}{\partial x} (BD) = U \frac{\partial}{\partial x} \int_0^\delta \rho u \, dy - \frac{\partial}{\partial x} \int_0^\delta \rho u^2 \, dy \qquad \qquad ...(8.4.3)$$

Boundary layer thickness δ is usually very small compared to the development length x. Therefore, except for extremely low Reynolds numbers, the acceleration of the fluid particles in the y-direction is negligible compared to that in the x-direction. Hence, the pressure in the boundary layer may be taken to be the same as that outside it. That is, the pressure in the ambient stream is *impressed* on the boundary layer.

Outside the boundary layer in the ambient stream, the flow is considered irrotational. Therefore, Bernoulli's equation $\left(p + \frac{1}{2} \rho U^2 = \text{constant} \right)$ may be applied to it. Differentiating the Bernoulli's equation for the ambient stream, one gets

$$\frac{\partial p}{\partial x} + \rho U \frac{\partial U}{\partial x} = 0 \qquad \qquad ...(8.4.4)$$

Substituting the value of $\frac{\partial p}{\partial x}$, obtained from Eq. (8.4.4), in Eq. (8.4.3), one obtains

$$\tau_0 - \rho U \frac{\partial U}{\partial x} \left(\int_0^\delta dy \right) = U \frac{\partial}{\partial x} \int_0^\delta \rho u \, dy - \frac{\partial}{\partial x} \int_0^\delta \rho u^2 \, dy$$

Since
$$U \frac{\partial u}{\partial x} = \frac{\partial}{\partial x} (Uu) - u \frac{\partial U}{\partial x}$$

∴
$$\tau_0 = \rho \frac{\partial}{\partial x} \int_0^\delta U u \, dy - \rho \frac{\partial U}{\partial x} \int_0^\delta u \, dy - \rho \frac{\partial}{\partial x} \int_0^\delta u^2 \, dy + \rho \frac{\partial U}{\partial x} \int_0^\delta U \, dy$$

or
$$\tau_0 = \rho \frac{\partial}{\partial x} \int_0^\delta u (U - u) \, dy + \rho \frac{\partial U}{\partial x} \int_0^\delta (U - u) \, dy \qquad \qquad ...(8.4.5)$$

Since $U - u$ becomes zero at the edge of the boundary layer, the upper limit of both integrals in Eq. (8.4.5) may also be changed to ∞. Therefore,

$$\tau_0 = \rho \frac{\partial}{\partial x} \int_0^\infty u (U - u) \, dy + \rho \frac{\partial U}{\partial x} \int_0^\infty (U - u) \, dy \qquad \qquad ...(8.4.6)$$

Combining Eq. (8.4.5) or Eq. (8.4.6) with Eqs. (8.3.1 and 8.3.2) and replacing partial derivatives with full derivatives as the quantities vary with x alone, one obtains

$$\tau_0 = \rho \frac{d\left(U^2 \theta\right)}{dx} + \rho U \delta * \frac{dU}{dx} \qquad \qquad ...(8.4.7)$$

Equation (8.4.7) is known as von Karman's *momentum integral equation* of the boundary layer that forms the basis of many approximate solutions of boundary layer problems. This equation is applicable to laminar, turbulent or transition flow in the boundary layer over a flat plate. However, the use of the equation requires the knowledge of the variation of u with y so that the momentum thickness and displacement thickness can be evaluated. It should be noted that Eq. (8.4.7) has been derived for steady flow in a boundary layer on a flat plate over which there may be variation of pressure in the direction of flow. If pressure remains constant in the x-direction,

$$\frac{\partial p}{\partial x} = 0$$

and, hence, $\dfrac{\partial U}{\partial x} = 0$ from Eq. (8.4.4).

Therefore, Eq. (8.4.7) reduces to

$$\tau_0 = \rho U^2 \frac{d\theta}{dx} \qquad \qquad ...(8.4.8)$$

Equation (8.4.8) is applicable to boundary layer flow with zero pressure gradient.

8.5 APPLICATION OF MOMENTUM INTEGRAL EQUATION

8.5.1 Laminar Boundary Layer over a Flat Plate

For laminar boundary layer flow over a flat plate, velocity profile $u(x,y)$ can be assumed as a polynomial in y:

$$u = a + by + cy^2 \text{ for } 0 \le y \le \delta(x)$$

with boundary conditions as follows:

at $\quad y = 0, u = 0 \qquad \therefore a = 0$

at $\quad y = \delta, u = U \qquad \therefore U = b\delta + c\delta^2$

at $\quad y = \delta, \dfrac{\partial u}{\partial y} = 0 \quad \therefore 0 = b + 2c\delta$

Hence, $b = \dfrac{2U}{\delta}$ and $c = -\dfrac{U}{\delta^2}$

Therefore, the velocity profile is

$$u = 2\frac{U}{\delta}y - \frac{U}{\delta^2}y^2$$

or
$$\frac{u}{U} = 2\left(\frac{y}{\delta}\right) - \left(\frac{y}{\delta}\right)^2 \quad \text{for } 0 \le \frac{y}{\delta} \le 1 \qquad \qquad ...(8.5.1)$$

Substituting η for $\frac{y}{\delta}$ so that $dy = \delta d\eta$ and

$$\frac{u}{U} = 2\eta - \eta^2 = f(\eta) \qquad \qquad ...(8.5.2)$$

Flow being laminar, $\tau_0 = \mu \dfrac{\partial u}{\partial y}\bigg|_{y=0}$

$$= \mu \frac{\partial u}{\partial \eta} \frac{\partial \eta}{\partial y}\bigg|_{\eta=0}$$

$$= \mu U (2 - 2\eta)\left(\frac{1}{\delta}\right)\bigg|_{\eta=0}$$

$\therefore \qquad \qquad \tau_0 = \dfrac{2\mu U}{\delta} \qquad \qquad ...(8.5.3)$

The momentum thickness θ, Eq. (8.3.2), can be obtained as follows:
From Eqs. (8.3.2) and (8.5.2),

$$\theta = \int_0^\delta \frac{u}{U}\left(1 - \frac{u}{U}\right) dy$$

$$= \delta \int_0^1 f(n)\left[1 - f(n)\right] dn$$

$$= \delta \int_0^1 f(1-f) d\eta$$

$\therefore \qquad \qquad \theta = \alpha_1 \delta \qquad \qquad ...(8.5.4)$

where
$$\alpha_1 = \int_0^1 \left(2\eta - \eta^2\right)\left(1 - 2\eta + \eta^2\right) d\eta$$

$$= \int_0^1 \left(2\eta - 5\eta^2 + 4\eta^3 - \eta^4\right) d\eta$$

$$= \left[\eta^2 - \frac{5}{3}\eta^3 + \eta^4 - \frac{\eta^5}{5}\right]_0^1$$

$$= \left[1 - \frac{5}{3} + 1 - \frac{1}{5} \right]$$

$$= \frac{2}{15}$$

Likewise, from Eqs, (8.3.2) and (8.5.2),

$$\delta^* = \int_0^\delta \left(1 - \frac{u}{U} \right) dy$$

$$= \delta \int_0^1 (1 - f) d\eta$$

$$= \delta \int_0^1 \left(1 - 2\eta + \eta^2 \right) d\eta$$

$$= \delta \left[\eta - \eta^2 + \frac{\eta^3}{3} \right]_0^1$$

$$= \frac{\delta}{3}$$

Substituting the values of τ_0 and θ in the von Karman's momentum integral equation, Eq. (8.4.8),

$$\frac{2\mu U}{\delta} = \rho U^2 \left(\frac{2}{15} \right) \frac{d\delta}{dx}$$

or

$$\delta d\delta = \frac{15\mu}{\rho U} dx$$

This is a differential equation for δ which, on integration, yields

$$\frac{\delta^2}{2} = \frac{15\mu x}{\rho U} + C_1$$

The value of C_1 can be obtained by substituting the condition,

$$\delta = 0 \text{ at } x = 0$$

so that

$$C_1 = 0$$

Therefore,

$$\frac{\delta^2}{2} = \frac{15\mu x}{\rho U}$$

or

$$\delta = \sqrt{\frac{30\mu x}{\rho U}}$$

or
$$\frac{\delta}{x} = \sqrt{\frac{30\mu}{\rho U x}}$$

$\therefore$
$$\frac{\delta}{x} = \frac{5.48}{\sqrt{R_{ex}}} \qquad ...(8.5.5)$$

By combining Eqs (8.5.3 and 8.5.5), one can obtain wall shear stress τ_0 or skin-friction coefficient, c_f defined as

$$c_f = \frac{\tau_0}{\frac{1}{2}\rho U^2}$$

$$= \frac{2\mu\left(\frac{U}{\delta}\right)}{\frac{1}{2}\rho U^2}$$

$$= \frac{4\mu}{\rho U \delta} = \frac{4\mu}{\rho U x}\left(\frac{x}{\delta}\right) = \frac{4\mu}{\rho U x}\left(\frac{\sqrt{R_{ex}}}{5.48}\right)$$

$$c_f = \frac{0.73}{\sqrt{R_{ex}}} \qquad ...(8.5.6)$$

Equations (8.5.5) and (8.5.6), respectively, are expressions for the estimation of the boundary layer thickness and local shear stress. Although estimated in an approximate manner, using von Karman's momentum integral equation, the result is almost accurate being about 10 percent higher than the known accepted solutions for laminar boundary layer flow over a flat plate, obtained from the relevant viscous flow equations. The accepted solutions for δ and c_f are,

$$\frac{\delta}{x} = \frac{5.0}{\sqrt{R_{ex}}} \qquad ...(8.5.7)$$

and
$$c_f = \frac{0.664}{\sqrt{R_{ex}}} \qquad ...(8.5.8)$$

Once the variation of τ_0 with x is known, the viscous drag force F_D on the surface of the plate can be estimated by integration over the area of the surface, i.e.,

$$F_D = \int_0^L \tau_0 \, b \, dx = \int_0^L \frac{0.664}{\sqrt{R_{ex}}}\left(\frac{1}{2}\rho U^2\right) b \, dx$$

$\therefore$
$$C_f = \frac{F_D}{bL\frac{1}{2}\rho U^2} = \frac{0.664}{L}\left(\frac{\mu}{U\rho}\right)^{1/2} \int_0^L x^{-1/2} \, dx$$

$$= \frac{0.664}{L}\left(\frac{\mu}{U\rho}\right)^{1/2} \left(2L^{1/2}\right)$$

$$= 1.328 \left(\frac{\mu}{U \rho L} \right)^{1/2}$$

$$\therefore \qquad\qquad C_f = \frac{1.328}{\sqrt{R_{el}}} \qquad\qquad\qquad ...(8.5.9)$$

In Eq. (8.5.9), C_f is the average skin-friction coefficient over the whole length of plate L, and R_{el} is the Reynolds number with characteristic length as L. The skin-friction coefficient c_f, Eq. (8.5.8), is the local value of skin-friction coefficient at distance x from the leading edge. Equations (8.5.7 – 8.5.9) are used for laminar boundary layer over a flat plate. Some typical values of δ/x, c_f and C_f, as obtained from these equations, are listed in Table 8.5.1.

Table 8.5.1: Some typical values of laminar boundary layer parameters

	R_{ex} or R_{el}					
	10	10^2	10^3	10^4	10^5	5×10^5
δ/x	1.58	0.5	0.158	0.05	0.016	0.007
c_f	0.21	0.066	0.021	0.0066	0.0021	0.0009
C_f	0.42	0.133	0.01	0.0133	0.0042	0.0005

Table 8.1 indicates that at higher Reynolds numbers (say, greater than 10^4) the boundary layer thickness is so thin that its displacement effect on the outer inviscid ambient flow is negligible. Therefore, in such situations, the pressure distribution along the plate can be computed from irrotational flow theory as if the boundary layer did not exist there. This external pressure field is said to 'drive' the boundary layer flow (11).

8.5.2 Turbulent Boundary Layer over a Flat Plate

The universal velocity distribution for turbulent flow is logarithmic and too complex mathematically for convenient use with the momentum integral equation. A simpler approach, however, is to use Prandtl's one-seventh power law velocity distribution for pipe flow, i.e.,

$$\frac{u}{u_{\max}} = \left(\frac{y}{R} \right)^{1/7} \qquad\qquad\qquad ...(8.5.10)$$

in which u is the velocity of flow at distance y measured from the pipe wall, R is the radius of pipe, and u_{max} is the centre-line velocity, i.e., at $y = R$. Replacing u_{max} with ambient stream velocity U and R with boundary layer thickness δ, Eq. (8.5.10) becomes

$$\frac{u}{U} = \left(\frac{y}{\delta} \right)^{1/7} = \eta^{1/7} = f \qquad\qquad\qquad ...(8.5.11)$$

which velocity profile is considered suitable for use in von Karman's momentum integral equation for obtaining approximate solution of turbulent boundary layer. Since the flow is turbulent and also Eq.

(8.5.11) gives $\dfrac{du}{dy} = \infty$ at the wall, one cannot use Eq. (8.5.11) for computation of τ_0 for use in the momentum integral equation and, instead, one may adopt the expression for wall shear stress developed, using Eqs. (9.6.5 and 9.6.7), for smooth pipe surfaces (9). This expression, after replacing R with δ and u_{max} with U, is

$$\tau_0 = 0.023 \rho U^2 \left(\frac{v}{U\delta}\right)^{1/4} \qquad \qquad \ldots(8.5.12)$$

From Eq. (8.5.4),

$$\theta = \alpha_1 \delta$$

$$= \delta \int_0^1 \eta^{1/7}\left(1 - \eta^{1/7}\right) d\eta$$

$$= \delta \int_0^1 \left(\eta^{1/7} - \eta^{2/7}\right) d\eta$$

$$= \delta \left[\frac{7}{8}\eta^{8/7} - \frac{7}{9}\eta^{9/7}\right]_0^1$$

$$\therefore \qquad \qquad \theta = \frac{7}{12}\delta$$

Likewise,

$$\delta^* = \int_0^\delta \left(1 - \frac{u}{U}\right) dy$$

$$= \delta \int_0^1 (1 - f) d\eta$$

$$= \delta \int_0^1 \left(1 - \eta^{1/7}\right) d\eta$$

$$= \delta \left[\eta - \frac{7}{8}\eta^{8/7}\right]_0^1$$

$$= \frac{\delta}{8}$$

Substituting the values of τ_0 and θ in Eq. (8.4.8),

$$0.023 \rho U^2 \left(\frac{v}{U\delta}\right)^{1/4} = \rho U^2 \frac{7}{72} \frac{d\delta}{dx}$$

$$\therefore \qquad \delta^{1/4}\, d\delta = 0.24\left(\frac{\nu}{U}\right)^{1/4} dx$$

On integration,

$$\frac{4}{5}\delta^{5/4} = 0.24\left(\frac{\nu}{U}\right)^{1/4} x + C$$

Assuming that the boundary layer is turbulent right from the leading edge, i.e., $\delta = 0$ at $x = 0$.

$$\therefore \qquad\qquad C = 0$$

Hence,

$$\delta^{5/4} = 0.3\left(\frac{\nu}{U}\right)^{1/4} x$$

or

$$\delta = 0.382\left(\frac{\nu}{U}\right)^{1/5} x^{4/5} \qquad\qquad ...(8.5.13)$$

or

$$\frac{\delta}{x} = \frac{0.382}{\left(R_{ex}\right)^{1/5}} \qquad\qquad ...(8.5.14)$$

Equations (8.5.7) and (8.5.13) indicate that the boundary layer thickness increases as $x^{1/2}$ for the laminar boundary layer and as $x^{4/5}$ for the turbulent boundary layer. This means that the turbulent boundary layer grows much faster in comparison with the laminar boundary layer.

Using Eq. (8.5.12), one gets

$$\frac{\tau_0}{\frac{1}{2}\rho U^2} = 0.046\left(\frac{\nu}{U\delta}\right)^{1/4}$$

or

$$c_f = 0.046\left[\frac{\nu}{U}\frac{U^{1/5}}{0.382\,\nu^{1/5}x^{4/5}}\right]^{1/4}$$

$$\therefore \qquad c_f = \frac{0.059}{\left(R_{ex}\right)^{1/5}} \qquad\qquad ...(8.5.15)$$

The viscous drag force on the surface of the plate, F_D can be estimated by integration as follows:

$$F_D = \int_0^L \tau_0\, b\, dx = \int_0^L \frac{0.059}{\left(R_{ex}\right)^{1/5}}\left(\frac{1}{2}\rho U^2\right) b\, dx$$

$$\therefore \qquad C_f = \frac{F_D}{bL\frac{1}{2}\rho U^2} = \frac{0.059}{L}\left(\frac{\nu}{U}\right)^{1/5}\int_0^L x^{-1/5}\, dx$$

or
$$C_f = \frac{0.059}{L}\left(\frac{\nu}{U}\right)^{1/5}\left(\frac{5}{4}\right)L^{4/5}$$

or
$$C_f = \frac{0.074}{\left(\dfrac{UL}{\nu}\right)^{1/5}} = \frac{0.074}{\left(R_{el}\right)^{1/5}} \qquad \text{...(8.5.16)}$$

Experimental investigations have shown that Eqs. (8.5.15) and (8.5.16) predict turbulent skin friction drag on a flat plate very well for $5 \times 10^5 < R_{ex} < 10^7$. This is in spite of the approximate nature of the analysis that assumes the turbulent boundary layer right from the leading edge. Some typical values of δ/x, c_f, and C_f for the turbulent boundary layer, as obtained from Eqs. (8.5.14 – 8.5.16), are given in Table 8.5.2.

Table 8.5.2: Typical values of turbulent boundary layer parameters

	R_{ex} or $R_{el} = 5 \times 10^5$	R_{ex} or $R_{el} = 10^6$	R_{ex} or $R_{el} = 10^7$
δ/x	0.028	0.024	0.015
c_f	0.0043	0.0037	0.0024
C_f	0.0054	0.0047	0.0030

For values of R_{ex} ranging between 10^7 and 10^9, Schlichting (8) considered logarithmic variation of velocity, Eq. (7.8.4), instead of power-law type variation and obtained the following semi-empirical relation:

$$C_f = \frac{0.455}{\left[\log_{10} R_{el}\right]^{2.58}} \qquad \text{...(8.5.17)}$$

Equations (8.5.16) or (8.5.17) and (8.5.9) are applicable only if the boundary layer developed on a flat plate is either turbulent or laminar all through. When the length of plate is such that both types of boundary layers exist and each one of them contributes appreciably to the total friction drag, one would require a combined relation.

Assuming that the length of transition region (i.e., the region in which boundary layer changes from laminar to turbulent) is negligible and that the boundary layer changes from laminar to turbulent at distance x_t from the leading edge, the total frictional drag on the plate would be

$$F_D = \frac{1}{2}\rho U^2 b\left[\frac{0.074L}{\left(R_{el}\right)^{1/5}} - \frac{0.074x_t}{\left(R_{et}\right)^{1/5}} + \frac{1.328x_t}{\left(R_{et}\right)^{1/2}}\right] \qquad \text{...(8.5.18)}$$

in which $R_{et} = \dfrac{Ux_t}{\nu}$ and $R_{el} < 10^7$. Since $\dfrac{R_{et}}{R_{el}} = \dfrac{x_t}{L}$, Eq. (8.5.18) can be written as

$$F_D = \frac{1}{2}\rho U^2 bL\left[\frac{0.074}{\left(R_{el}\right)^{1/5}} - \frac{0.074 R_{et}^{4/5}}{R_{el}} + \frac{1.328 R_{et}^{1/2}}{R_{el}}\right] \qquad \text{...(8.5.19)}$$

or

$$C_f = \left[\frac{0.074}{\left(R_{el}\right)^{1/5}} - \frac{A}{R_{el}} \right] \qquad \qquad ...(8.5.20)$$

in which $A = 0.074\,R_{et}^{4/5} - 1.328\,R_{et}^{1/2}$. This means that A depends on R_{et}. For example, if $R_{et} = 5 \times 10^5$, $A = 1743$. If R_{el} is between 10^7 and 10^9, one should use the following equation, instead of Eq. (8.5.20):

$$C_f = \left[\frac{0.455}{\left(\log_{10}^{R_{el}}\right)^{2.58}} - \frac{A}{R_{el}} \right] \qquad \qquad ...(8.5.21)$$

Values of A for different R_{et} are listed in Table 8.5.3.

Table 8.5.3: Values of A for Eqs. (8.5.20) and (8.5.21)

R_{et}	3×10^5	5×10^5	10^6
A	1055	1743	3340

It may be noted that at large R_{el} ($>10^7$), the contribution of the laminar drag is negligible and $\dfrac{A}{R_{el}}$

may be taken as approximately equal to zero. Variation of the drag coefficient C_f with Reynolds number for a smooth flat plate aligned parallel to the flow is as shown in Fig. 8.5.1.

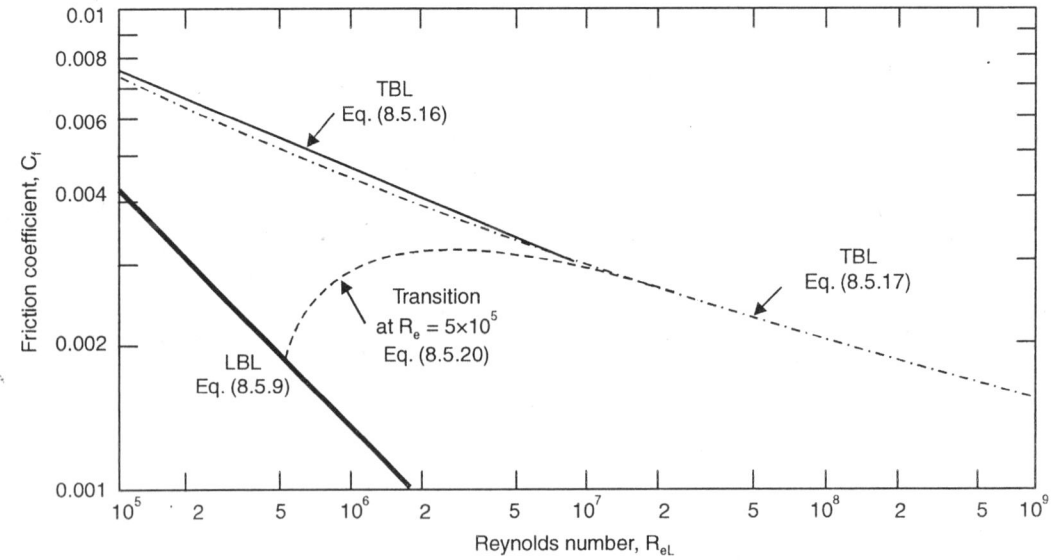

Fig. 8.5.1. Friction coefficient C_f for a smooth flat plate aligned parallel to the flow

8.6 BOUNDARY LAYER SEPARATION

In Art. 8.5, it was observed (Tables 8.5.1 and 8.5.2) that at high Reynolds number, the boundary layers are so thin that their displacement effect on the outer inviscid ambient flow is negligible. For slender bodies, such as flat plate and airfoils aligned parallel to the oncoming ambient stream, the assumption of negligible interaction between the boundary layer and the outer pressure distribution is an excellent approximation. However, for a blunt body, such as sphere or cylinder with its axis normal to the oncoming ambient stream, there is strong interaction between viscous and inviscid regions of flow in the rear of the blunt body, Fig. 8.6.1.

The analysis of the boundary layer flow over a flat plate makes one aware of the behaviour of both laminar and turbulent boundary layer except for what is known as flow separation giving rise to wake formation behind blunt bodies placed in an ambient stream, Fig. 8.6.1 (b).

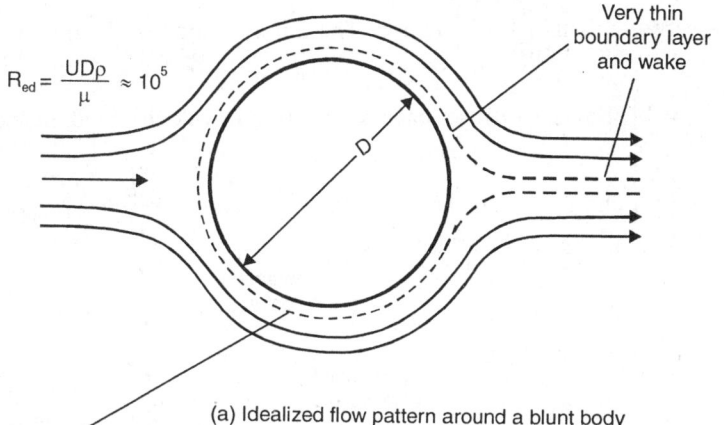

(a) Idealized flow pattern around a blunt body

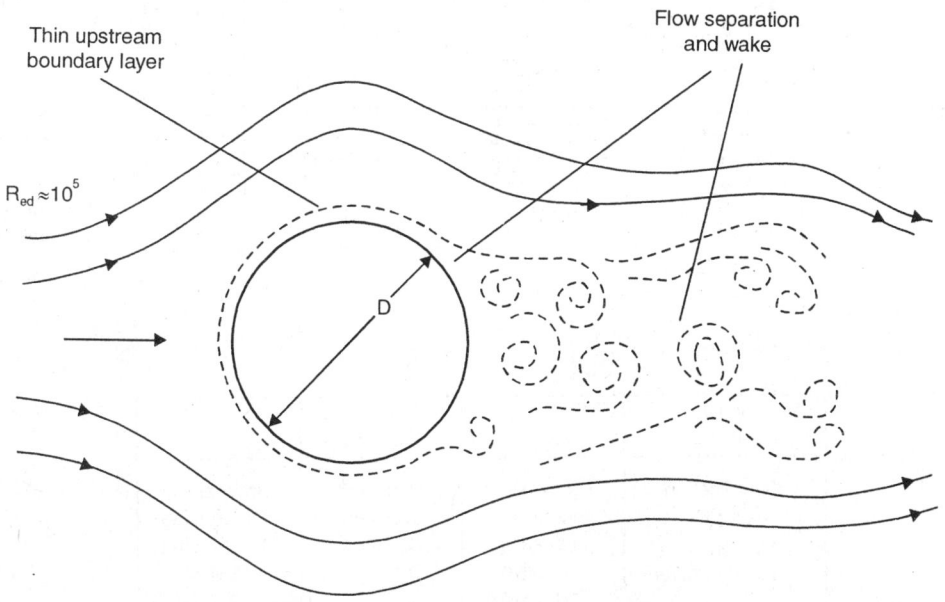

(b) Actual flow pattern around a blunt body

Fig. 8.6.1. Viscous and inviscid flow regions at the downstream of a blunt body

Separation of flow is caused by excessive momentum loss adjacent to the wall in a boundary layer fluid trying to move downstream against rising pressure ($\frac{dp}{dx} > 0$, i.e., adverse pressure gradient). If the pressure decreases in the direction of flow, the pressure gradient $\frac{dp}{dx}$ is negative and is termed favourable pressure gradient and in such a situation, flow separation can never occur.

Consider flow in a duct consisting of a nozzle, throat, and diffuser, Fig. 8.6.2. The nozzle part of the duct is a favourable pressure gradient $\left(\frac{dp}{dx} < 0\right)$ region. The throat section of the duct is the region of zero pressure gradient $\left(\frac{dp}{dx} = 0\right)$. But, the diffuser causes reduction in velocity (and, hence, momentum) of the flowing fluid particles and is also the adverse pressure gradient $\left(\frac{dp}{dx} > 0\right)$ region.

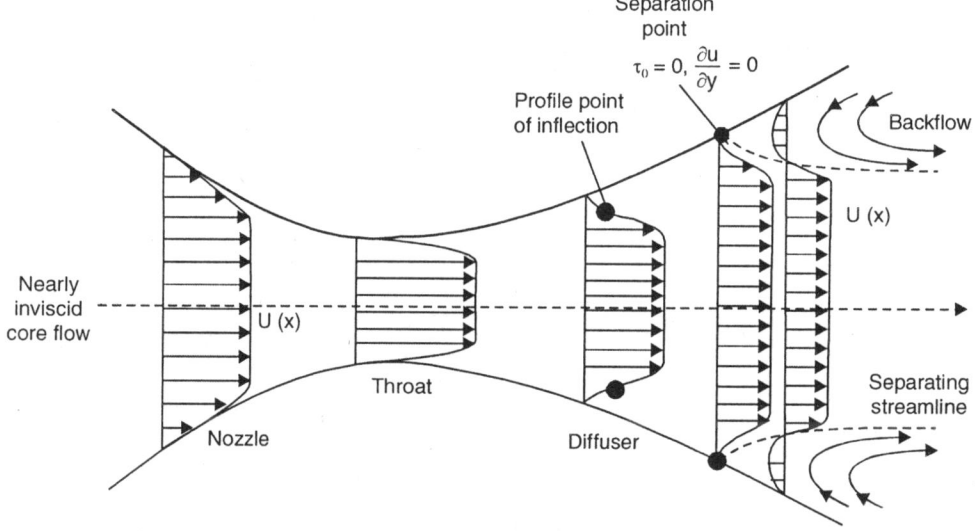

Flow Characteristics	Nozzle	Throat	Diffuser
Flow Section	Decreasing	Constant	Increasing
Velocity	Increasing	Constant	Decreasing
Momentum	Increasing	Constant	Decreasing
Pressure Gradient	Negative (Favourable)	Zero	Positive (Adverse)
Pressure	Decreasing	Constant	Increasing

Fig. 8.6.2 Separation in a nozzle-diffuser assembly (11)

Fluid particles close to the solid wall of the duct always experience retarding shear force irrespective of the sign of the pressure gradient. In the nozzle region $\left(\dfrac{dp}{dx}<0\right)$, the pressure behind a fluid particle close to the nozzle wall is greater than the pressure opposing the motion ahead of the particle; the particle is, therefore, sliding down a "pressure hill" and as such there is no possibility of the particle being slowed to zero velocity.

For the throat region, $\dfrac{\partial p}{\partial x}=0$ and there is a decrease in momentum which, however, is insufficient to bring the fluid particles to rest.

In the diffusion region $\left(\dfrac{dp}{dx}>0\right)$, the already-slowed (due to reduced velocity and viscous shear) fluid particle near the solid wall encounters a "pressure hill" that it must climb (1). The fluid particle near the wall, in such situation, could come to rest. This phenomenon is called *separation* or *flow separation* and, as a result, the boundary streamline leaves the boundary at the separation point. Because of the continued action of pressure increasing in the downstream direction, the fluid particles near the wall may start moving in a direction opposite to the direction of the main motion. Thus, there exists back flow near the duct wall downstream of the point of separation. This region of back flow is known as *wake*. The separation causes reduction in the net amount of work done by a fluid element on its surrounding fluid at the expense of its kinetic energy, with the net result that the pressure recovery is incomplete and, therefore, the fluid drag increases (9). The losses too increase.

Since the separation results into increased flow losses and drag, one should avoid or delay the separation, say, by streamlining the boundary or by injecting high velocity fluid in the boundary layer thereby accelerating the retarded fluid particles or by sucking the slow-moving fluid particles from near the boundary continuously.

If one compares the velocity distributions of laminar and turbulent boundary layers, Fig. 8.6.3, one finds that for the same ambient stream, the velocity (and, hence, momentum) within the turbulent boundary layer is greater than that in the laminar boundary layer. As such, turbulent boundary layer resists (or delays) separation of flow more effectively.

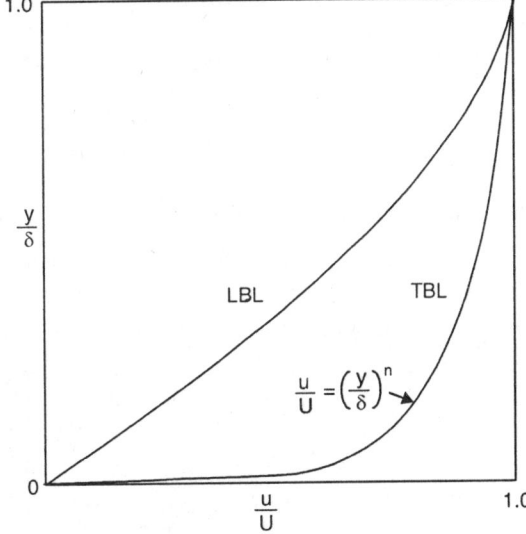

Fig. 8.6.3. Non-dimensional velocity profiles for boundary layer flow over a flat plate

8.7 VISCOUS (OR LAMINAR) SUBLAYER

Nikuradse's experimental analysis of velocity distribution in turbulent boundary layer has revealed that the parabolic velocity distribution exists near the boundary even when the flow elsewhere in the boundary layer is turbulent for which the velocity distribution is logarithmic. The distance of the intersection of the two velocity profiles from the boundary can be termed as *viscous (or laminar) sublayer thickness* δ'. From Nikuradse's experimental work, the value of δ' is given as (7)

$$\delta' = \frac{11.6\nu}{u_*}$$

in which $u_* = \sqrt{\dfrac{\tau_0}{\rho}}$ and is termed the shear velocity, and τ_0 is the wall shear stress.

Surface (or boundary) roughness is identified by the average height of irregular surface projections. In fluid mechanics, the boundaries are classified as smooth or rough depending upon the roughness height k as well as fluid and flow characteristics that are represented by the thickness of laminar sublayer. When δ' is much larger than k, the eddies present in the outer turbulent region are not able to penetrate the sublayer and reach the boundary surface because of inherent stability of laminar flow. The roughness elements are, therefore, not exposed to turbulent flow. Therefore, the boundary acts as smooth boundary and is termed *hydrodynamically smooth boundary*. The boundary resistance, therefore, depends on Reynolds number and not on the roughness.

If the flow conditions are such that the value of k is much greater than δ', the laminar sublayer is completely destroyed and the roughness projections are exposed to the external turbulent flow. Wakes are formed behind each element and the losses increase considerably. The resistance offered by the boundary now depends on roughness k. The boundary is, then, said to be *hydrodynamically rough boundary*.

Experiments have shown that the boundary acts as hydrodynamically smooth boundary when $\dfrac{k}{\delta'} < 0.25$. If $\dfrac{k}{\delta'} > 6.0$, the boundary is hydrodynamically rough (2). For $0.25 < \dfrac{k}{\delta'} < 6.0$, the boundary is said to be in transition when the boundary resistance depends on Reynolds number as well as roughness height.

8.8 EXTERNAL FLOWS

When a solid body of arbitrary shape is immersed in a fluid and there exists a relative motion between the two, the fluid exerts, on the body, forces and moments about all three coordinate axes, Fig. 8.8.1. It is customary and convenient to choose one axis parallel to the motion and positive downstream. The

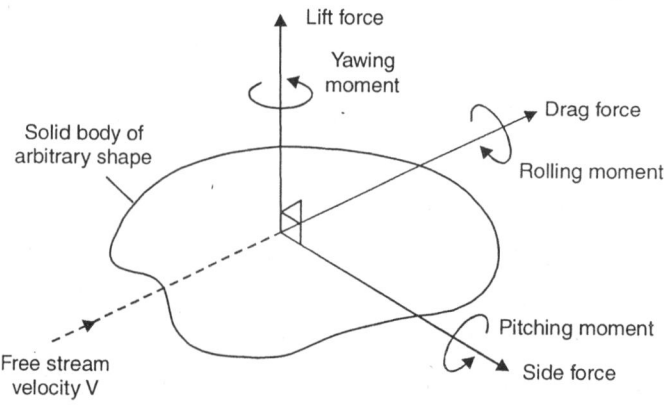

Fig. 8.8.1. Forces and moments on a body immersed in a flow of fluid

force acting on the body is on account of both pressure and viscous stresses present in the fluid flow. The component of the force, exerted on the body, in the direction of motion is known as *drag* that opposes the motion and is to be overcome if the motion is to exist. Another component that is perpendicular to the drag and performs a useful task of supporting the weight of the body is termed *lift*. The third component perpendicular to both drag and lift is called *side* or *transverse* force. It should be noted that the drag and lift are produced by the dynamic action of the fluid flowing past the body. Other forces (such as buoyant force or gravitational force) are not included in either drag or lift.

Consider two-dimensional flow around an airfoil, Fig. 8.8.2, of unit thickness perpendicular to the plane of the diagram. For a differential surface area dA, the drag force dF_D is given as,

$$dF_D = p\,dA\,\sin\theta + \tau_0\,dA\,\cos\theta$$

$\therefore$
$$F_D = \int_A (p\sin\theta + \tau_0\cos\theta)\,dA \qquad\qquad ...(8.8.1)$$

Similarly, the lift force,

$$F_L = \int_A (p\cos\theta - \tau_0\sin\theta)\,dA \qquad\qquad ...(8.8.2)$$

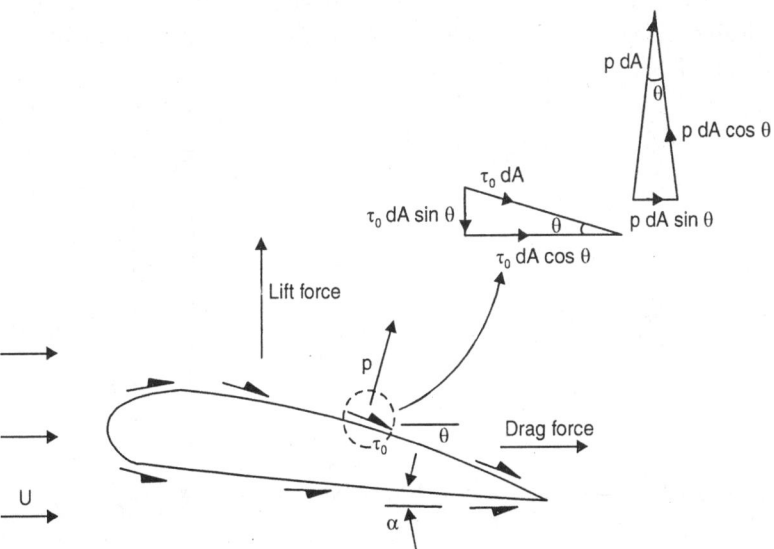

Fig. 8.8.2. Viscous and pressure forces on an airfoil

The flow pattern around the immersed body determines the magnitude of the drag and lift forces, and the boundary layer that develops on the body plays an important role. Conceptually speaking, the drag and lift forces are computable from Eqs. (8.8.1) and (8.8.2). However, there are only very few cases in which one can estimate drag and lift forces without recourse to experimentation. This is because of flow separation, caused by the presence of adverse pressure gradient, that prohibits the analytical determination of the drag and lift forces. Boundary layer theory can predict the location of the point of separation. But, it cannot predict the distribution of pressure (usually low) in the separated region. The difference between the high pressure in the front stagnation region and the low pressure in the rear separated region causes a large contribution to the total drag. This part of the drag is called *pressure drag* or *form drag* since the pressure difference depends on the form (i.e., shape) of the body (or boundary). This drag combined with the viscous (or friction) drag yields the total (or profile) drag.

Pressure drag depends on the magnitude of pressure as well as the orientation of the body. For example, the pressure force on a flat plate, aligned parallel to the flow, may be very large without contributing to the total drag since the pressure force acts in the direction normal to the flow direction.

For flows in which viscous effects are small and inertial effects are large (i.e., large Reynolds number), the pressure difference, $p - p_0$ is proportional to the dynamic pressure $\frac{1}{2}\rho U^2$. Therefore, the pressure coefficient (and also the drag coefficient) is relatively independent of the Reynolds number.

Consider a thin plate placed parallel to a fluid stream, Fig. 8.8.3 (a). Because of the flow symmetry, the boundary layers that develop on the two sides of the plate are the same and, hence, pressures above and below balance each other. Therefore, the pressure term in Eq. (8.8.1) drops and one can compute the drag force using the relevant equation (Art. 8.5). There is no lift on the plate because the flow is symmetrical. However, when the plate is placed at right angles to the flow direction, a positive pressure develops on the front side of the plate due to stagnation. But, a much lower pressure exists on the rear side on account of the flow separation at the edges of the plate. For this case, therefore, the first term of Eq. (8.8.1) is the sole contributor to the total drag force on the plate. Due to symmetry of flow, once again, the lift force on the plate, Fig. 8.8.3(b), is zero. Since pressure in the separated region cannot be predicted theoretically or analytically, one has to take recourse to experimentation to estimate the total drag on the plate, Fig. 8.8.3 (b).

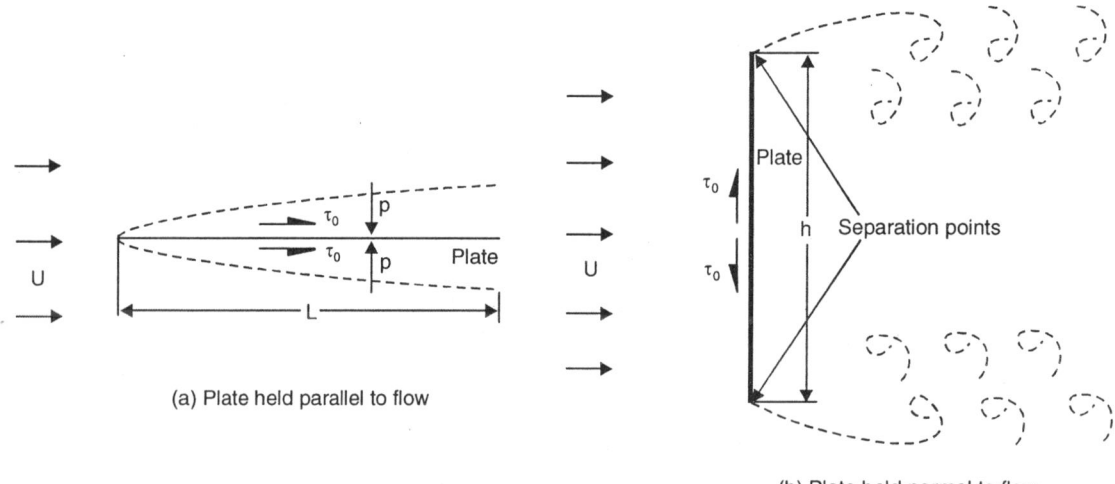

(a) Plate held parallel to flow

(b) Plate held normal to flow

Fig. 8.8.3. Flow past a flat plate

The drag coefficient for flow over a body immersed in a fluid flow is usually based on frontal area of the object. For flow over a flat plate aligned normal to the flow direction, the wall shear stress does not contribute to the total drag force. If the Reynolds number (with characteristic length as height of the plate) is higher than 1000, the drag coefficient, $C_D \left[= \dfrac{F_D}{\left(\frac{1}{2}\rho U^2 bh\right)} \right]$ of the plate held normal to the flow direction is essentially independent of the Reynolds number and depends on width (*b*) to height (*h*) ratio *b/h* and is as shown in Fig. 8.8.4 (3). The drag coefficient for all shapes of objects with sharp edges is independent of the Reynolds number (for $R_e \geq 1000$) because the separation points are fixed by the geometry of the object, Table 8.8.1.

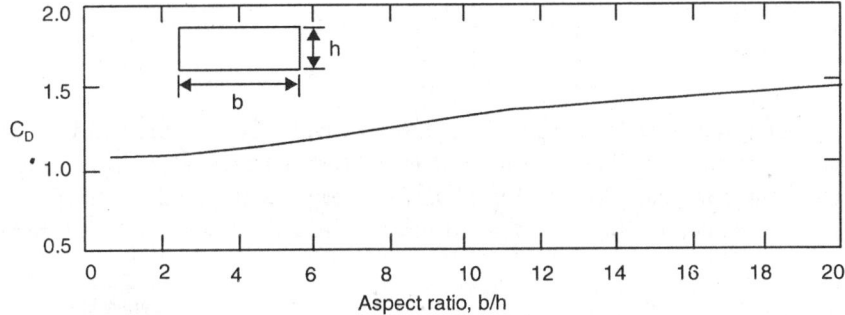

Fig. 8.8.4. Variation of the drag coefficient with aspect ratio for a flat plate of finite width normal to the flow with $R_{eh} > 1000$ (3)

Table 8.8.1: Typical values of the drag coefficient for selected objects (1, 11)

Object	Two- dimensional or three-dimensional	$C_D (R_e > 1000)$ (based on frontal area)
Rectangular plate (normal to flow)	3-D (b/h = 1)	1.18
	3-D (b/h = 5)	1.2
	3-D (b/h = 10)	1.3
	3-D (b/h = 20)	1.5
	2-D (b/h = ∞)	2.0
Disc (normal to flow)	3-D	1.17
Hemisphere (open end facing flow)	3-D	1.42
Hemisphere (open end facing downstream)	3-D	0.38
Square prism	3-D (b/h = 1)	1.05
	2-D (b/h = ∞)	2.05
Cube	3-D	1.07
60° Cone (flow along axis and towards base)	3-D	0.8
Cylinder (flow along axis)	3-D (l/D = 0.5)	1.15
	3-D (l/D = 1.0)	0.90
	3-D (l/D = 2)	0.85
	3-D (l/D = 4)	0.87
	3-D (l/D = 8)	0.99

8.8.1. Drag around Sphere and Cylinder

In rounded objects, the point at which the boundary layer separates from the object cannot be predicted and, as such, Eq. (8.8.1) cannot be used. In such cases too, therefore, one has to conduct experimental investigations.

Figure 8.8.5(a and b) illustrates the flow separation as a result of boundary layer encountering adverse pressure gradient in the flow around a cylinder and the consequent pressure variation. Based on potential flow theory, the theoretical inviscid pressure distribution around a cylinder is given as (11),

$$C_p = \frac{p - p_0}{\frac{1}{2}\rho V^2} = 1 - 4\sin^2\theta$$

...(8.8.3)

and is as shown in Fig. 8.8.5(c). Here, p_0 and V are, respectively, the pressure and velocity for the free stream. The actual pressure distributions for laminar and turbulent boundary layers, Fig. 8.8.5(c), are significantly different from that given by Eq. (8.8.3). At very low Reynolds number R_e, say less than about 0.5, the flow pattern is almost similar to that of ideal fluid flow and the pressure recovery

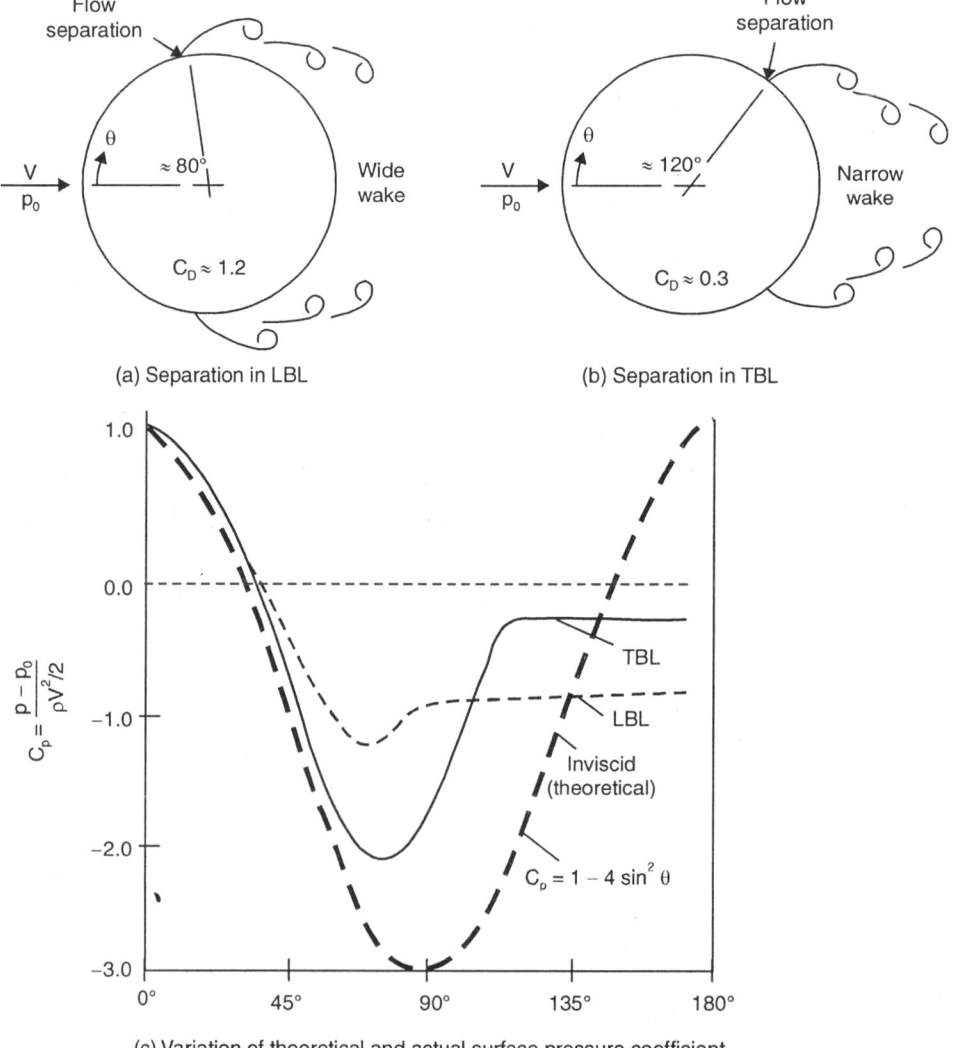

(a) Separation in LBL

(b) Separation in TBL

(c) Variation of theoretical and actual surface pressure coefficient

Fig. 8.8.5. Pressure distribution around a cylinder (11)

downstream is almost complete. Hence, the total drag is due to friction alone. At higher Reynolds number, laminar boundary layer forms and the flow separates at $\theta \approx 80°$ on account of the adverse pressure gradient prevailing on the rear side of the cylinder, Fig. 8.8.5(a). This prediction could not have been made by inviscid flow theory. Relatively low pressure in the wake (region between separated boundary layer and the boundary), in comparison to p_0, causes large pressure (or form) drag and $C_D \approx 1.2$. Besides, two symmetrical eddies (or vortices) that rotate in opposite direction are formed. At Reynolds number between 2 and 30, the vortices remain fixed in position and the main flow closes downstream of the vortices. With increases in R_e, the fixed eddies elongate and begin to oscillate. When R_e is around 90, the elongated vortices break away from the cylinder alternately and are carried downstream by the main flow while other vortices are similarly formed, elongated, and carried alternately. With further increase in R_e, this phenomenon is intensified and becomes continuous. This results in the formation of two discrete rows of vortices in the wake region as shown in Fig. 8.8.5. These two rows or vortices are known as *vortex street* or von Karman's vortex street.

Shedding of each vortex produces circulation and gives rise to a lateral force on the cylinder and, thus, subjecting the cylinder to a forced vibration. The familiar 'singing' of telephone wires is caused by this phenomenon due to lateral movement of wind. If the natural frequency of a suspension bridge coincides with the frequency of the forced vibration due to vortex shedding, the bridge may fail.

At high Reynolds number, the boundary layer becomes turbulent and is less vulnerable to the adverse pressure gradient due to the increased momentum of the flowing fluid near the surface of the cylinder. Therefore, the separation is delayed, i.e., the point of separation shifts downstream at $\theta \approx 120°$, Fig.8.8.5(b). The wake size, therefore, is smaller and pressure on the rear side of the cylinder is higher (compared to that when the boundary layer was laminar) resulting into a lower $C_D \approx 0.3$ which is almost 25 percent of the drag coefficient for laminar boundary layer conditions, Fig. 8.8.6(a).

For three-dimensional spherical bodies too, similar difference between the more vulnerable laminar boundary layer separation and more resistant turbulent boundary layer separation has been observed in experiments and typical values of C_D are 0.5 (when boundary layer is laminar) and 0.2 when boundary layer is turbulent, Fig. 8.8.6 (b). Figure 8.8.6 shows the sharp reduction in the drag coefficient C_D for cylinder as well as sphere at the transition (critical) Reynolds number.

For flow around a sphere, both viscous (or friction) drag and pressure drag contribute significantly to the total drag. The drag coefficient for a smooth sphere C_D, therefore, depends on the Reynolds number as shown in Fig. 8.8.6(b), which is based on experimental analysis. When $R_e \leq 1$, there is no flow separation and the drag F_D is predominantly viscous drag the expression for which was first obtained analytically by Sir G.G. Stokes (1819 – 1903) as

$$F_D = 3\,\pi\,\mu\,UD \qquad \qquad ...(8.8.4)$$

in which D is the diameter of sphere and U is velocity of flow of fluid (or velocity of sphere in stationary fluid) of viscosity μ.

Equation (8.8.4), known as Stokes' law, holds good when inertial forces are negligible. This requires that Reynolds number, $\dfrac{UD\rho}{\mu}$ must be less than 0.1 and the sphere is falling in an infinite expanse of fluid. Therefore, for ordinary fluids like air and water, the sphere must be of a very small size. Stokes law, however, does have practical applications. When a small solid spherical particle is falling through a fluid under its own weight, the particle initially accelerates and, then, attains constant (or terminal) velocity when net downward force on the particle is zero, i.e., when the weight of the particle is exactly balanced by the sum of the buoyant and resisting forces caused by the flow of fluid (relative to the

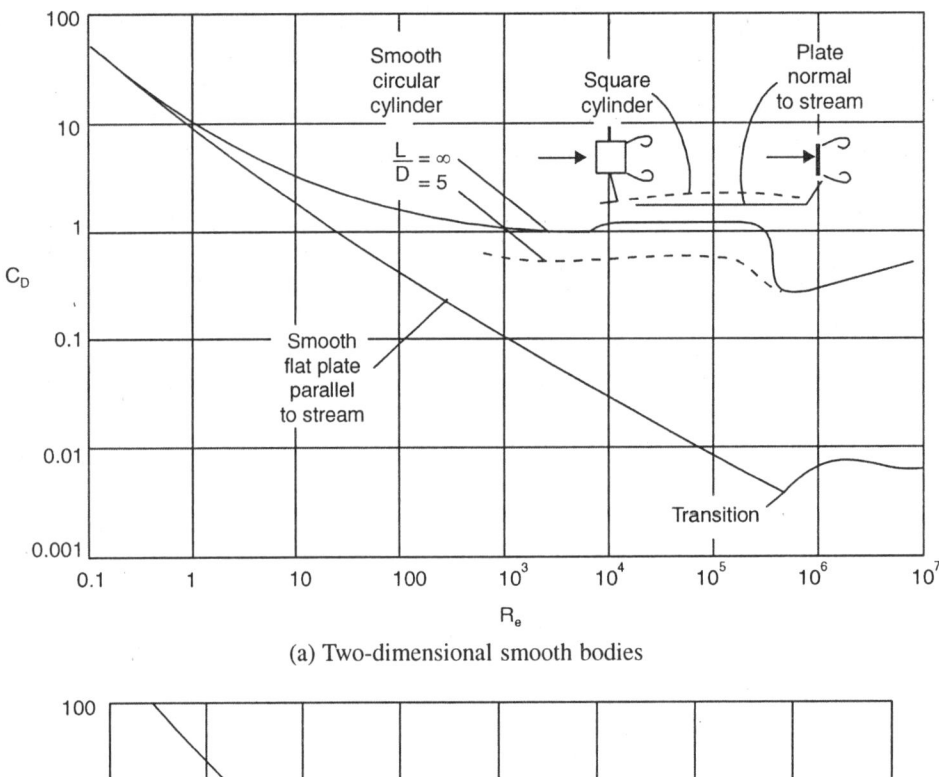

(a) Two-dimensional smooth bodies

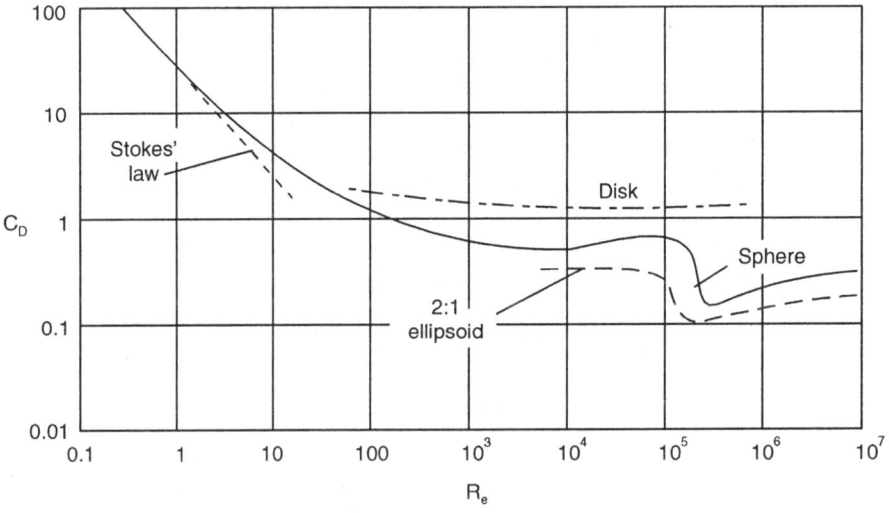

(b) Three-dimensional smooth bodies

Fig. 8.8.6. Variation of drag coefficient with Reynolds number(11)

falling particle) round the particle. The terminal velocity of such a particle (like tiny drops of rain water falling on earth or fine dust particles settling down through atmospheric air or flowing water) can be estimated using Stokes' law, Eq. (8.8.4), assuming that the Reynolds number is small enough for the Stokes' law to be applicable. If the particle is of diameter D, and the mass density of the material is ρ_s and that of the fluid is ρ, and the terminal velocity U has been attained, then,

Downward force (i.e., weight of the particle)

= total upward force = Buoyant force + Viscous force

or
$$\frac{\pi}{6}D^3 \rho_s g = \frac{\pi}{6}D^3 \rho g + 3\pi D\mu U$$

or
$$U = \frac{D^2(\rho_s - \rho)g}{18\mu} \qquad ...(8.8.5)$$

Likewise, the rate of rise of gas bubbles in a liquid can also be determined using the Stokes' law provided that the Reynolds number is less than 0.1. In this case,

Buoyant force = weight of bubble + Viscous force

and, therefore,

$$U = \frac{D^2(\rho_l - \rho_g)g}{18\mu} \qquad ...(8.8.6)$$

in which, ρ_l and ρ_g are the mass densities of liquid and gas, respectively.

Further, on substituting the Stokes' law in the expression for coefficient of drag, one gets

$$C_D = \frac{F_D}{\left(\frac{1}{2}\rho U^2 A\right)}$$

$$= \frac{3\pi\mu U D}{\left(\frac{1}{2}\rho U^2\right)\left(\frac{\pi D^2}{4}\right)} = \frac{24\mu}{UD\rho}$$

$$\therefore \qquad C_D = \frac{24}{R_e} \qquad ...(8.8.7)$$

Equation (8.8.7) has been found to be in agreement with experimental values (Fig. 8.8.6) at Reynolds number less than or equal to one. For $R_e > 1$, Eq. (8.8.7) deviates significantly with the experimental values (Fig. 8.8.6). When Reynolds number exceeds 1, flow separation occurs and the total drag is the sum of the friction drag and pressure drag. The friction drag decreases with increasing Reynolds number and at $R_e \approx 1000$, the friction drag is only about 5 percent of the total drag.

For Reynolds number between 1000 and about 3×10^5, there is no change in the pattern of flow separation as a result of which C_D remains almost constant. At $R_e \cong 3 \times 10^5$, the boundary layer formed on the sphere turns turbulent and, therefore, has relatively more momentum than that of laminar boundary layer. Hence, the effect of adverse pressure gradient is better resisted and the separation is delayed, i.e., the point of separation shifts downstream of the mid-section of sphere. Size of the wake zone too is reduced and the pressure difference between upstream and downstream of sphere is less. Therefore, the drag and, hence, C_D reduces considerably.

Transition in the boundary layer from laminar to turbulent is affected by the turbulence in the free stream and the roughness of the surface over which the boundary layer develops. Accordingly, the transition does not occur at a unique value of the Reynolds number. Reynolds number at which the laminar boundary layer turns to turbulent boundary layer may range between 50,000 (for rough body surface and/or highly turbulent free stream) and 4×10^5 (for smooth objects placed in free stream with low turbulence), Fig. 8.8.7. Golf balls fly in this range of Reynolds number. The golf balls are, therefore, deliberately 'dimpled' to 'trip' the boundary layer so that the boundary layer formed around the ball is necessarily turbulent and the resulting drag is minimum. Table tennis balls are not dimpled as their motion is at much lower Reynolds number.

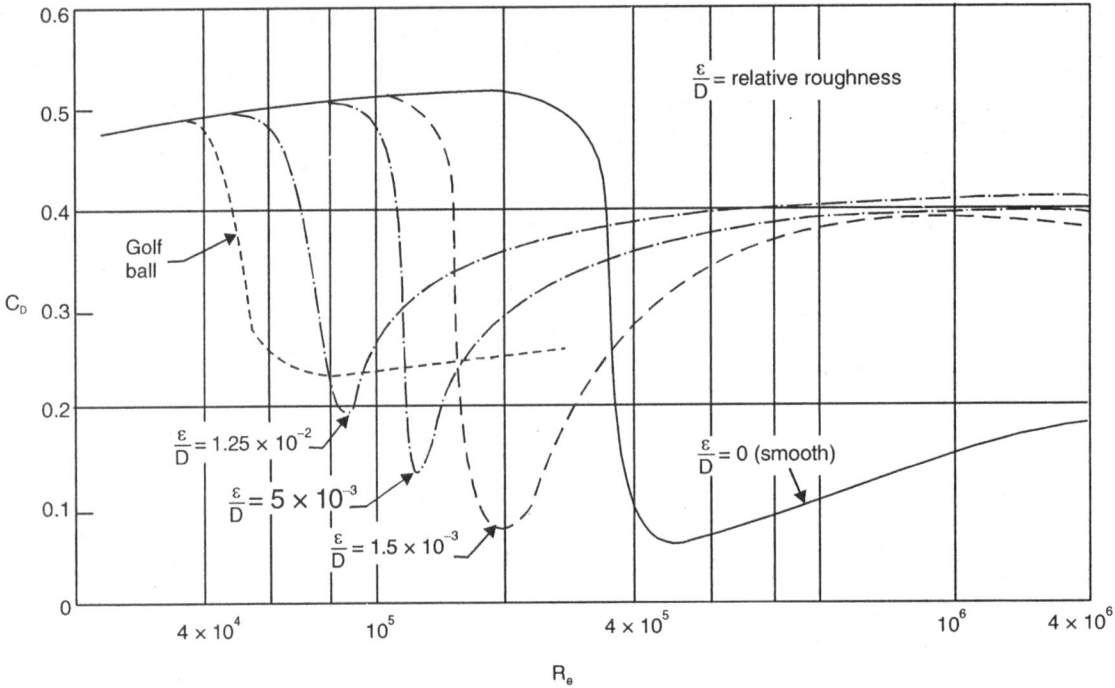

Fig. 8.8.7. Effect of surface roughness on transition to turbulent boundary layer (12)

The extent of separated flow region behind a body can be reduced or eliminated by streamlining the body shape so as to reduce the adverse pressure gradient. Streamlining not only delays the separation but, also, decreases the wake size and, thus, reduces the pressure drag. However, addition of tail section increases the surface area and, therefore, the friction drag. The optimum streamlined shape is the one that causes minimum total drag.

Reduction of aerodynamic drag is important for automobiles to effect fuel economy. However, there is limit to the overall length of an automobile and, hence, completely streamlined tails are impractical. Therefore, the contours of an automobile are optimized within the limits of overall length of the automobile so as to have minimum drag.

8.9 LIFT FORCE

Consider flow of an inviscid fluid around a cylinder, Fig. 8.9.1(a). Since there is no viscosity, there is no boundary layer formation and, hence, no friction drag on the cylinder. Further, the pressure distribution, around the cylinder, is symmetrical about the axis of the cylinder and, hence, pressure drag too is zero. Based on inviscid flow theory, it can be shown that the drag is zero for any shape of object (symmetrical or otherwise) placed in a uniform stream. However, this prediction does not match with real fluid flows and an object always experiences drag, when placed in a real fluid. This has been termed *d'Alembert's paradox* – that the drag on an object in an inviscid fluid is zero, but the drag on an object in a real fluid is not zero, even if the viscosity of the fluid is negligibly small. This paradox was pointed out by d 'Alembert in 1752 and this made everyone to overreact and reject all inviscid flow theory until 1904, when Prandtl first pointed out the profound effect of the thin viscous boundary layer on the flow pattern (11).

The pressure distribution with respect to the direction of motion is symmetrical for steady irrotational flow (with uniform velocity, U_0) of an inviscid fluid around a two-dimensional cylinder of large length L with its axis normal to the flow. This means that the pressure distributions on the upper and lower halves of the cylinder are symmetrical. As a result, there is no force on the cylinder in the direction normal to the flow direction. That is, there is no lift force. If one superimposes on this flow pattern a constant clockwise circulation (i.e., flow with concentric streamlines) of magnitude Γ, Fig. 8.9.1(b), the resulting flow pattern, Fig. 8.9.1(c), would have unsymmetrical streamline pattern with respect to the irrotational flow direction. As a result, the streamlines are now closer in the upper part and farther in the lower part causing the velocities higher in the upper part relative to the velocities in the lower part. Application of the Bernoulli's equation along a streamline would indicate that the pressures are lower in the upper part and higher in the lower part as a consequence of which the cylinder experiences an upward force. This force is termed lift force that is the component of the resultant force, acting on the body, perpendicular to the fluid motion in the gravitational direction. The component that is not in the gravitational direction but is at right angles to the fluid motion is termed transverse force. This effect would be witnessed when one repeats the experiment with real fluid when drag force too would be present. This phenomenon was first investigated by a German scientist Heinrich Magnus (1802-1870) and, hence, is named as *Magnus effect*.

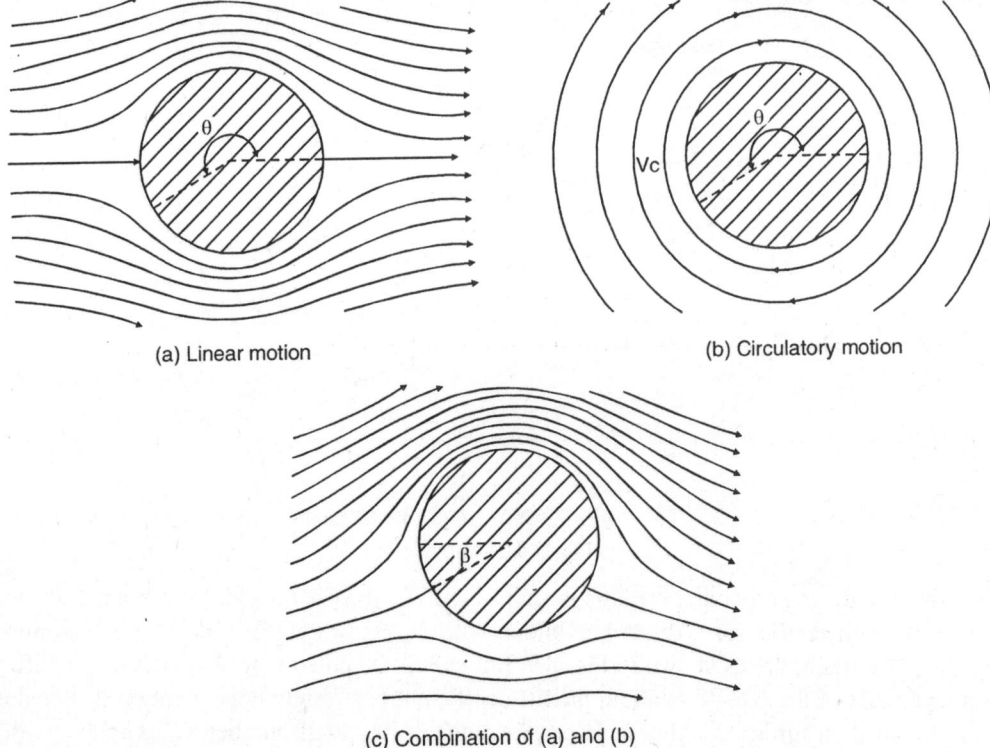

(a) Linear motion (b) Circulatory motion

(c) Combination of (a) and (b)

Fig. 8.9.1. Circulation superimposed upon irrotational flow past a cylinder (7)

According to inviscid flow theory, the lift force F_L per unit length of any cylinder of any shape immersed in a uniform stream equals $\rho U_0 \Gamma$ where Γ is the total net circulation around the body. This was first derived by W.M. Kutta in 1902 and, again, independently, by N. Joukowski in 1906 and, therefore, termed *Kutta-Joukowsky's law* and is applicable to any two-dimensional object. It is emphasized that circulation is a must for lift force to exist.

8.9.1. Coefficient of Lift

Coefficient of lift, C_L, like coefficient of drag, is defined as

$$C_L = \frac{F_L}{\frac{1}{2}\rho V^2 A_p} \qquad ...(8.9.1)$$

In designing lifting bodies such as airfoils, Fig. 8.9.2, the objective of the designer is to create a larger lift force and minimize drag at the same time. The lift and drag coefficients of an airfoil depend primarily on the Reynolds number and angle of attack α that is the angle between the airfoil chord (i.e., the straight line joining the leading edge and trailing edge) and the free stream velocity vector. The shape of airfoil is obtained by combining a mean line and thickness (of airfoil) distribution along the mean line. If the airfoil is of symmetric section, the mean line and chord line both are straight lines and they coincide. An airfoil with curved mean line is termed cambered airfoil. The area of an airfoil at right angles to the flow direction changes with angle of attack. Consequently, the planform area (i.e., the projected area of airfoil on the plane containing the chord line), A_p is used to define lift and drag coefficients of an airfoil.

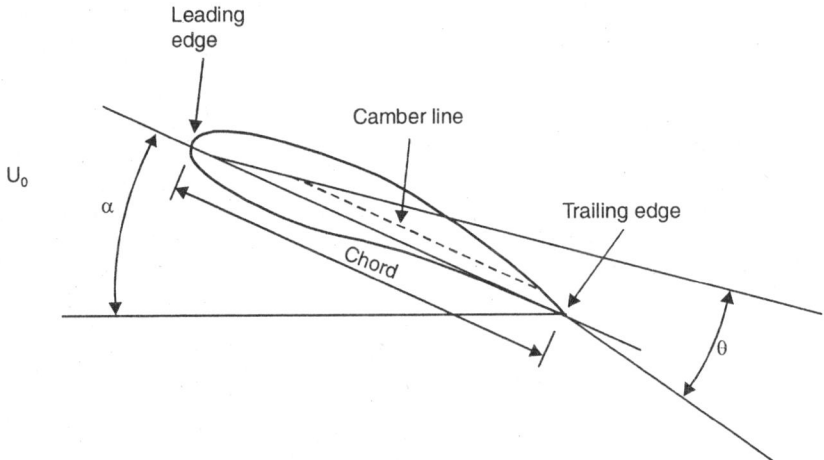

Fig. 8.9.2. An airfoil

Theoretical and experimental variations of C_L and C_D (based on planform area, (i.e., length L multiplied by diameter D), for a rotating cylinder, with U_c/U_0 [$= (\pi DN)/(60U_0)$ with N as number of revolutions per minute] are as shown in Fig. 8.9.3. It should be noted that while the drag coefficient is fairly independent of the rate of rotation, the lift coefficient is strongly dependent on it. Besides, both C_L and C_D depend on surface roughness. In certain range of Reynolds number, C_D is known to decrease with increase in surface roughness. But, increase in surface roughness helps in dragging more fluid

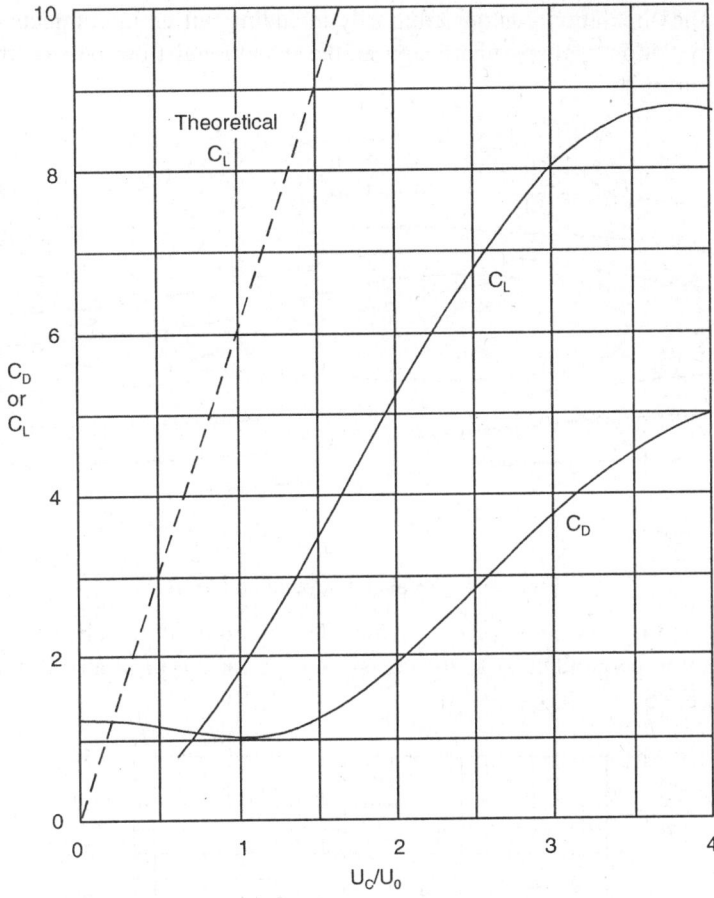

Fig. 8.9.3. Coefficients of lift and drag for a rotating cylinder (7)

around a sphere and, hence, increase the circulation (for a given angular velocity) that may increase the lift coefficient. Therefore, a rotating rough golf ball would travel farther than a smooth one due to the reduced drag and increased lift.

8.9.2 Lift and Drag of an Airfoil

Existence of circulation is essential for the lift to exist. A rotating cylinder placed in a real fluid produces circulation due to viscous action of its surface on the surrounding fluid. In case of airfoils that do not rotate, there should be a different mechanism by which circulation around the airfoil is caused. As explained in Art 8.2, the flow inside the boundary layer is rotational (or viscous) while the flow outside may be treated as irrotational (or non-viscous). Therefore, there would exist circulation and, hence, vorticity on the upper as well as lower surface of an airfoil. For an inclined airfoil, the velocities over the upper and lower surfaces are unequal and, therefore, the circulation around the airfoil, i.e.,

$\oint v\,ds$, will be clockwise if the velocities over the upper surface are greater than the velocities over the lower surface. For the circulation contour, shown in Fig. 8.9.4, the circulation would be clockwise if

$$\int_1^2 vds > \int_3^1 vds$$

One may draw the circulation contour arbitrarily including within it, complete boundary layer so that the value of circulation remains unaffected as the irrotational flow outside the boundary layer makes no contribution to it.

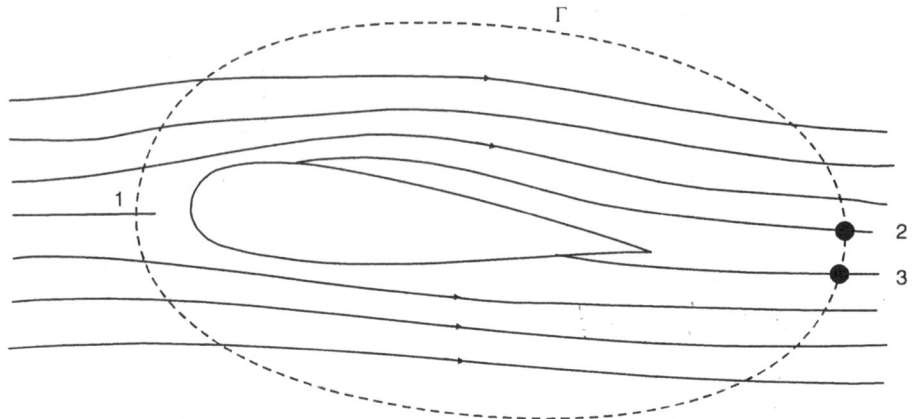

Fig. 8.9.4. Circulation around an airfoil

Variations of C_L and C_D (based on plan form area, i.e., chord multiplied by span; span is the length of airfoil in the direction perpendicular to the cross-section), for a typical airfoil, with angle of attack α are shown in Fig. 8.9.5.

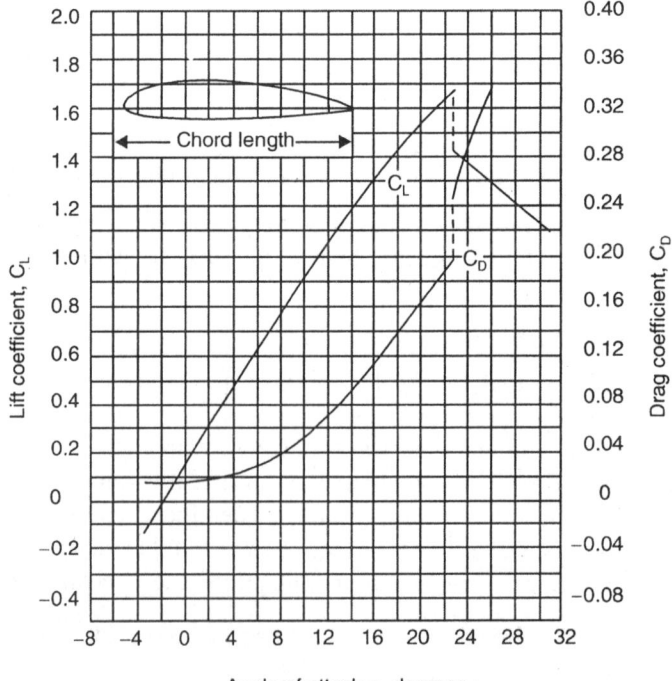

Fig. 8.9.5. Typical variation of lift and drag coefficients (based on maximum projected wing area) for an airfoil (9)

At small angles of attack, near the trailing edge of an airfoil, the pressure gradient is adverse but not large enough to cause significant boundary layer separation. Therefore, the boundary layer adheres to the surface of the airfoil. The lack of symmetry of flow produces lift at small angles of attack. As the angle of attack is increased, the adverse pressure gradient on the upper surface increases and the separation point moves forward (i.e., towards the leading edge). At a certain angle of attack $\alpha \approx 15$ to 20°, depending on the shape of the airfoil profile, maximum lift is achieved. Further increase in angle of attack, α causes flow to separate completely from the upper surface of the airfoil. (i.e., separation starts from the leading edge). The airfoil is, then, said to be *stalled*. There is, then, sudden decrease in lift coefficient and increase in drag coefficient. The airfoil is no longer flyable.

EXAMPLES

E8.1. The sinusoidal velocity profile for laminar boundary layer on a flat plate is generally given as

$$u = A \sin (By) + C$$

Evaluate constants A, B, and C.

Solution: The boundary conditions for a boundary layer are

    (i)   at $y = 0$, $u = 0$

    (ii) at $y = \delta$, $u = U$

    (iii) at $y = \delta$, $\dfrac{\partial u}{\partial y} = 0$

Substituting these conditions in the given equation

$$u = A \sin (By) + C$$

gives

$0 = A \sin (0) + C$	$\therefore\ C = 0$
$U = A \sin (B\delta) + C$	$\therefore\ U = A \sin (B\delta)$
and $0 = AB \cos (B\delta) + C$	$\therefore\ 0 = AB \cos (B\delta)$
$AB \cos (B\delta) = 0$ gives $B\delta = \pi/2$	$\therefore\ B = \pi/2\delta$

Accordingly,

$$U = A \sin \left(\frac{\pi \delta}{2\delta} \right) = A \sin \frac{\pi}{2} = A$$

$\therefore$ $A = U$, $B = \pi/2\delta$, and $C = 0$

Thus, $u = U \sin \left(\dfrac{\pi}{2} \dfrac{y}{\delta} \right)$

or $\dfrac{u}{U} = \sin \left(\dfrac{\pi}{2} \dfrac{y}{\delta} \right)$

E8.2. Obtain expressions for δ, δ^*, and frictional force on a flat plate of length L and width B if the boundary layer velocity profile is as obtained in E 8.1.

Solution: For a steady, incompressible flow over a flat plate, U = constant,

$$\frac{dp}{dx} = 0 \text{, and from Eqs. (8.4.8) and (8.5.4)}$$

$$\tau_0 = \rho U^2 \frac{d\theta}{dx}$$

or

$$\tau_0 = \rho U^2 \frac{d\delta}{dx} \int_0^1 \frac{u}{U}\left(1 - \frac{u}{U}\right) d\eta$$

Here,

$$\eta = \frac{y}{\delta} \text{ so that } \frac{u}{U} = \sin\left(\frac{\pi}{2}\frac{y}{\delta}\right) = \sin\left(\frac{\pi}{2}\eta\right)$$

Therefore,

$$\tau_0 = \rho U^2 \frac{d\delta}{dx} \int_0^1 \sin\left(\frac{\pi}{2}\eta\right)\left[1 - \sin\left(\frac{\pi}{2}\eta\right)\right] d\eta$$

$$= \rho U^2 \frac{d\delta}{dx} \int_0^1 \left[\sin\left(\frac{\pi}{2}\eta\right) - \sin^2\left(\frac{\pi}{2}\eta\right)\right] d\eta$$

$$= \rho U^2 \frac{d\delta}{dx} \frac{2}{\pi}\left[\left\{-\cos\frac{\pi}{2}\eta - \frac{1}{2}\frac{\pi}{2}\eta + \frac{1}{4}\sin\pi\eta\right\}\right]_0^1$$

$$= \rho U^2 \frac{d\delta}{dx}\frac{2}{\pi}\left\{0 - \frac{\pi}{4} + 0 + 1 + 0 + 0\right\}$$

$$= 0.137\rho U^2 \frac{d\delta}{dx}$$

Also,

$$\tau_0 = \mu\frac{\partial u}{\partial y}\bigg|_{y=0} = \mu U \frac{\pi}{2\delta}\cos\frac{\pi}{2}\frac{y}{\delta}\bigg]_{y=0} = \frac{\pi\mu U}{2\delta}$$

Thus,

$$\frac{\pi\mu U}{2\delta} = 0.137\rho U^2 \frac{d\delta}{dx}$$

or

$$\delta\, d\delta = \frac{\pi\mu}{2(0.137)\rho U}dx = 11.5\frac{\mu}{\rho U}dx$$

Hence, on integration,

$$\frac{\delta^2}{2} = 11.5\frac{\mu}{\rho U}x + C$$

Since at $x = 0$, $\delta = 0$, $\therefore C = 0$

Hence,

$$\delta = \sqrt{\frac{23x\mu}{\rho U}}$$

or
$$\frac{\delta}{x} = \frac{4.8}{\sqrt{R_{ex}}}$$

Now
$$\delta* = \delta \int_0^1 \left(1 - \frac{u}{U}\right) d\eta$$

$$= \delta \int_0^1 \left(1 - \sin\frac{\pi}{2}\eta\right) d\eta$$

$$= \delta \left[\eta + \frac{2}{\pi}\cos\frac{\pi}{2}\eta\right]_0^1$$

$$= \delta \left[1 + 0 - 0 - \frac{2}{\pi}\right]$$

$$= \delta \left[1 - \frac{2}{\pi}\right]$$

$$= 0.3634\,\delta$$

$$= (0.3634)\frac{4.8\,x}{\sqrt{R_{ex}}}$$

$$\therefore \qquad \frac{\delta*}{x} = \frac{1.74}{\sqrt{R_{ex}}}$$

Frictional force on one side of the plate is given by

$$F = \int_0^L \tau_0\, b\, dx$$

$$= \int_0^L \frac{\pi\mu U \sqrt{\rho U}}{2\sqrt{23\,x\mu}} b\, dx$$

$$= \frac{\pi\sqrt{\mu\rho}\, U^{3/2} b}{9.59} \int_0^L x^{-1/2} dx$$

$$= \frac{\pi\sqrt{\mu\rho}\, U^{3/2} b}{9.59}\left(2L^{1/2}\right)$$

$$F = \frac{0.655\rho U^2 bL}{\sqrt{R_{el}}}$$

E8.3. Water flows at U = 1.2 m/s past a flat plate with length of plate in the flow direction as 1.5m. The boundary layer is tripped so that it is turbulent at the leading edge. Compute δ, $\delta*$ and τ_0 at the trailing edge (i.e., x = L). Assume velocity distribution in the boundary layer to be one-seventh power law type.

Solution: At $x = L = 1.5\ m$; $R_{el} = \dfrac{(1.2)(1.5)}{10^{-6}} = 1.8 \times 10^6$.

From Eq. (8.5.14)

$$\frac{\delta}{x} = \frac{0.382}{(R_{ex})^{0.2}}$$

Hence, $\delta = \dfrac{(0.382)(1.5)}{(1.8 \times 10^6)^{0.2}} = 0.032\ m = 32\ mm$

As obtained in Art. 8.5.2, for turbulent boundary layer with one-seventh power law type velocity distribution,

$$\delta* = \frac{\delta}{8} = \frac{32}{8} = 4\ mm$$

From Eq. (8.5.15)

$$c_f = \frac{0.059}{(R_{ex})^{0.2}} = \frac{0.059}{(1.8 \times 10^6)^{0.2}} = 3.31 \times 10^{-3}$$

∴

$$\tau_0 = c_f \left(\frac{1}{2} \rho U^2\right) = (3.31 \times 10^{-3})\left(\frac{1000}{2} \times 1.2 \times 1.2\right)$$

$$= 2.38 N/m^2$$

E8.4. The thickness of a turbulent boundary layer at the trailing edge of a 3.2 m long plate is 40 mm. If the flowing fluid is water, determine free stream velocity, average shear stress, and shear stress at the trailing edge. Assume that the boundary layer is turbulent at the leading edge.

Solution: For turbulent boundary layer,

$$\frac{\delta}{x} = \frac{0.382}{(R_{ex})^{0.2}}$$

At the trailing edge, $x = L = 3.2\ m$

$$R_{ex} = R_{el} = \frac{U(3.2)}{10^{-6}} = 3.2 \times 10^6\ U$$

∴

$$\frac{0.040}{3.2} = \frac{0.382}{(3.2 \times 10^6\ U)^{0.2}}$$

or
$$U^{0.2} = \frac{0.382 \times 3.2}{0.04 (3.2 \times 10^6)^{0.2}} = \frac{30.56}{(3.2 \times 10^6)^{0.2}}$$

$\therefore$
$$U = \frac{26.65 \times 10^6}{3.2 \times 10^6}$$

$$= 8.33 \text{ m/s}$$

Hence, $R_{el} = 8.33 \times 3.2 \times 10^6 = 26.65 \times 10^6 = 2.7 \times 10^7$

$\therefore$
$$C_f = \frac{0.455}{\left[\log_{10} 2.7 \times 10^7 \right]^{2.58}} = 2.57 \times 10^{-3}$$

$\therefore$ Average shear stress, $\overline{\tau}_0 = C_f \frac{1}{2} \rho U^2$

$$= 2.57 \times 10^{-3} \left(\frac{1}{2} \right) 1000 \times (8.33)^2$$

$$= 89.17 \text{ N/m}^2$$

Also,
$$c_f = \frac{0.059}{(R_{ex})^{0.2}} = \frac{0.059}{(2.7 \times 10^7)^{0.2}} = 1.93 \times 10^{-3}$$

$\therefore$ Local shear stress at the trailing edge

$$\tau_0 = c_f \frac{1}{2} \rho U^2$$

$$= 1.93 \times 10^{-3} \left(\frac{1}{2} \right) 1000 \times 8.33^2$$

$$= 66.96 \text{ N/m}^2$$

E8.5. Wind with uniform velocity of 40 km/h flows past a 20 m high cylindrical chimney having diameter of 1.2m. Compute the bending moment at the base of the chimney due to wind forces.

Solution: For air, $\rho = 1.2 \text{ kg/m}^3$ and $\nu = 1.5 \times 10^{-5} \text{ m}^2/\text{s}$

$$U = \frac{40 \times 1000}{60 \times 60} = 11.11 \text{ m/s}$$

$\therefore$
$$R_e = (11.11)(1.2) \left(\frac{1}{1.5 \times 10^{-5}} \right) = 8.89 \times 10^5$$

Hence, $C_D = 0.34$ from Fig. 8.8.6(a)

$\therefore$
$$F_D = C_D \frac{1}{2} \rho U^2 A_p$$

$$= (0.34)\frac{1}{2}(1.2)(11.11)^2 (1.2 \times 20)$$

$$= 604.32 \text{ N}$$

This is the resultant force that will act at the mid-height section of the chimney. Hence, the bending moment at the base of the chimney, M_0 is obtained as

$$M_0 = 604.32 \times 10$$

$$= 6043.2 \text{ Nm}$$

$$= 6.04 \text{ kNm}$$

E8.6.   A ship 50 m long and having a wetted area of 500 m^2 requires 200 kW of power to overcome the resistance (due to friction and wave effects) to ship's motion at 7.0 m/s velocity. Determine the frictional and wave resistances. Assume $\rho = 1020$ kg/m^3, $\nu = 9.3 \times 10^{-7}$ m^2/s, and critical Reynolds number = 5×10^5.

Solution: Power $= F_{Dt} U$

$\therefore$
$$F_{Dt} = \frac{200 \times 1000}{7} = 28571 \text{ N}$$

$$\text{Re}_L = \frac{7.0 \times 50}{9.3 \times 10^{-7}} = 3.76 \times 10^8$$

From Eq. (8.5.21) and Table 8.5.3,

$$C_f = \frac{0.455}{\left[\log_{10} 3.76 \times 10^8\right]^{2.58}} - \frac{1743}{3.76 \times 10^8}$$

$$= 1.78 \times 10^{-3} - 464 \times 10^{-8}$$

$$\cong 1.78 \times 10^{-3}$$

Therefore, frictional resistance $= 1.78 \times 10^{-3} \ (1/2) \ 1020 \times 7 \times 7 \times 500$

$$= 22.241 \text{ kN}$$

Hence, wave (or residual) resistance $= 28.571 - 22.241$

$$= 6.33 \text{ kN}$$

E8.7.   A small sphere ($D = 6$ mm) is observed to fall through caster oil ($\rho = 969$ kg/m^3 and $\mu = 0.9$ Ns/m^2) at a terminal speed of 60 mm/s. Compute the drag coefficient for the sphere and also the density of the sphere.

Solution: Reynolds number, $R_e = \dfrac{0.060 \times 0.006 \times 969}{0.9}$

$$= 0.388$$

Since $R_e < 0.1$, Eq. (8.8.7) gives

$$C_D = \frac{24}{0.388} = 61.86$$

Since the sphere is falling at terminal speed, its submerged weight equals the drag force. Therefore,

$$\frac{\pi}{6}D^3(\rho_s - \rho)g = C_D \frac{\pi}{4}D^2 \frac{1}{2}\rho U^2$$

or

$$\rho_s = \rho + C_D \frac{3}{4}\frac{\rho U^2}{gD}$$

$$= 960 + 61.86\left(\frac{3}{4}\right)\frac{969 \times 0.06 \times 0.06}{9.8 \times 0.006}$$

$$= 3721.45 \text{ kg/m}^3$$

E8.8.   A dust particle falling in air is observed to settle at 2 mm/s. The specific gravity of the particle is 2.65. Estimate its diameter.

Solution: Reynolds number, $R_e = \dfrac{(2 \times 10^{-3})D}{1.5 \times 10^{-5}}$

Therefore, $C_D = \dfrac{24}{R_e} = \dfrac{24 \times 1.5 \times 10^{-5}}{(2 \times 10^{-3})D} = \dfrac{0.18}{D}$ (Assuming that $R_e < 1$)

$$\therefore \quad \frac{\pi}{6}(D^3)(2.65 \times 1000 - 1.2)(9.8) = \frac{0.18}{D}\left(\frac{1}{2} \times 1.2 \times 0.002 \times 0.002\right)\frac{\pi}{4}D^2$$

Hence, $D = 5 \times 10^{-6}$ m = 5 μm

Alternatively, using Eq. (8.8.5)

$$U = \frac{D^2(\rho_s - \rho)g}{18\mu}$$

$$\therefore \qquad D^2 = \frac{18\mu U}{(\rho_s - \rho)g} = \frac{18(1.8 \times 10^{-5})(2 \times 10^{-3})}{(2.65 \times 1000 - 1.2)9.80}$$

$$\therefore \qquad D = 5 \times 10^{-6} \text{ m} = 5 \text{ μm}$$

$$R_e = \frac{UD\rho}{\mu} = \frac{(2 \times 10^{-3})(5 \times 10^{-6})1.2}{1.8 \times 10^{-5}} = 6.67 \times 10^{-4} < 1$$

Therefore, assumption of validity of the Stokes' law is correct.

E8.9.   A sphere of steel (specific gravity = 7.96) and another sphere of magnesium (specific gravity = 1.82) fall freely in water and glycerin, respectively. What should be the ratio of their diameters for attaining dynamic similarity between them?

Solution: The dynamic similarity for the given problem would require equality of the Reynolds number R_e so that the drag coefficients C_D for the two cases would be equal. That is

$$R_{es} = R_{em}$$

and $$C_{Ds} = C_{Dm}$$

Here, subscripts s and m correspond to the two cases of steel sphere and magnesium sphere falling, respectively, in water and glycerin.

$$\left(\frac{UD}{\nu}\right)_s = \left(\frac{UD}{\nu}\right)_m$$

$\therefore$ $$\frac{U_s}{U_m} = \frac{D_m\,\nu_w}{D_s\,\nu_g}$$

and $$\left[\frac{F_D}{\left(\frac{1}{2}\rho U^2\right)\left(\frac{\pi}{4}D^2\right)}\right]_s = \left[\frac{F_D}{\left(\frac{1}{2}\rho U^2\right)\left(\frac{\pi}{4}D^2\right)}\right]_m$$

or $$\frac{\frac{\pi}{6}D_s^3\left(\gamma_s - \gamma_w\right)}{\left(\frac{1}{2}\rho_w U_s^2\right)\left(\frac{\pi}{4}D_s^2\right)} = \frac{\frac{\pi}{6}D_m^3\left(\gamma_m - \gamma_g\right)}{\left(\frac{1}{2}\rho_g U_m^2\right)\left(\frac{\pi}{4}D_m^2\right)}$$

Here, the subscripts w and g correspond to the fluids water and glycerin. Thus,

$$\frac{D_s\left(\gamma_s - \gamma_w\right)}{\rho_w U_s^2} = \frac{D_m\left(\gamma_m - \gamma_g\right)}{\rho_g U_m^2}$$

or $$\left(\frac{U_s}{U_m}\right)^2 = \frac{D_s\left(\gamma_s - \gamma_w\right)}{D_m\left(\gamma_m - \gamma_g\right)}\frac{\rho_g}{\rho_w} = \left(\frac{D_m\,\nu_w}{D_s\,\nu_g}\right)^2$$

i.e., $$\left(\frac{D_s}{D_m}\right)^3 = \left(\frac{\nu_w}{\nu_g}\right)^2\left(\frac{\rho_w}{\rho_g}\right)\frac{\left(\gamma_m - \gamma_g\right)}{\left(\gamma_s - \gamma_w\right)}$$

$$= \left(\frac{10^{-6}}{1.18\times10^{-3}}\right)^2\left(\frac{1000}{1.26\times1000}\right)\left(\frac{1.82-1.26}{7.96-1.0}\right)$$

$$\therefore \qquad \left(\frac{D_s}{D_m}\right)=3.58\times10^{-3}$$

or
$$\frac{D_m}{D_s}=279.4$$

E 8.10. Compute the diameter of a parachute (in the form of a hemispherical shell) to be used for dropping an object of 880 N so that it touches the earth at a velocity less than 6 m/s. The drag coefficient for a hemispherical shell with its concave side upstream may be taken as 1.32 (for $R_e > 1000$).

Solution: The drag force experienced by the parachute and the weight it drops must be equal. Therefore,

$$C_D\left(\frac{\pi}{4}D^2\right)\left(\frac{1}{2}\rho U^2\right)=880$$

or
$$1.32\left(\frac{\pi}{4}D^2\right)\left(\frac{1}{2}\times1.2\times6^2\right)=880$$

Hence, $D = 6.27$ m

PROBLEMS

P8.1 Consider boundary layer development in parallel flow past a flat plate for which the characteristic length dimension for the Reynolds number is x, the distance measured from the leading edge. Prepare a plot of boundary layer thickness versus distance for $0.01 \le x \le 8$ m.

P8.2 Velocity profiles in laminar boundary layer are sometimes expressed by either linear or cubic equation:

$$\text{Linear}: \frac{u}{U}=\frac{y}{\delta} \qquad \text{Cubic:}\ \frac{u}{U}=\frac{3}{2}\left(\frac{y}{\delta}\right)-\frac{1}{2}\left(\frac{y}{\delta}\right)^3$$

Compare these velocity profiles by plotting y/δ (on the ordinate) versus u/U (on the abscissa). Evaluate θ/δ and δ^*/θ for each of the two profiles.

P8.3 The velocity profile in a turbulent boundary layer is often approximated by the "$\frac{1}{7}$-power-law" equation

$$\frac{u}{U}=\left(\frac{y}{\delta}\right)^{1/7}$$

Compare this profile with the cubic laminar boundary layer velocity profile (P 8.2) by plotting y/δ (on the ordinate) versus u/U (on the abscissa) for both profiles.

P8.4 Consider a laminar boundary layer on a flat plate with velocity profile given by the cubic expression of P 8.2. For this profile

$$\frac{\delta}{x}=\frac{4.64}{\sqrt{R_{ex}}}$$

Find expressions for δ^*/x and θ/x.

P8.5 The velocity profile in a laminar boundary layer flow at zero pressure gradient is approximated by the linear expression given in P 8.2. Use the momentum integral equation with this profile to obtain expression for δ/x and C_f.

P8.6 Air at 20°C, 100-kPa (absolute) flows along a smooth plate with a velocity of 140 km/h. How long should the plate be to obtain a boundary layer thickness of 10 mm? Assume the boundary layer to be turbulent right from the leading edge.

P8.7 Compute the skin-friction drag on an airplane 120 m long, average diameter 20 m, with velocity of 120 km/h travelling through air at 100 kPa (absolute) and 20°C.

P8.8 A small airplane at a velocity of 40 m/s tows an advertising signboard. The dimensions of the signboard are 40 m long and 1.5 m high. Assume $p = 1$ atm, and $t = 20°C$. Assuming the signboard to be a flat plate, calculate the power required to tow the signboard.

P8.9 Estimate the settling velocity of a small metal sphere, specific gravity 7.6 and diameter 0.1 mm, in crude oil at 20°C, specific gravity 0.86 and kinematic viscosity $= 8 \times 10^{-5}$ m^2/s.

P8.10 What is the maximum diameter for a spherical particle of dust, specific gravity 2.65, that settles in atmospheric air at 20°C in accordance with the Stokes' law? What is its settling velocity?

P8.11 At what speed should a 100 mm sphere travel through water at 20°C to have a drag of 4 N?

P8.12 How many parachutes ($C_D = 1.2$) of diameter 30 m are required to drop a bulldozer weighing 50 kN at a terminal speed of not more than 10 m/s through air at 100 kPa absolute at 20°C.

REFERENCES

(1) Fox, RW and McDonald, AT: *"Introduction to Fluid Mechanics"* John Wiley & Sons, New York, 1994.

(2) Garde, R.J. and Mirajgaonkar, A.G: *"Engineering Fluid Mechanics"*, Scitech Publications (India) Pvt. Ltd., Chennai (India), 2005).

(3) Hoerner, SF: *"Fluid-dynamic Drag"*, Published by author, 1965.

(4) Massey, B: *"Mechanics of Fluids"*, 7th ed., Nelson Thornes Ltd., U.K., 2001.

(5) Munson, BR Young, DF and Okiishi, TH: *"Fundamentals of Fluid Mechanics"*, John Wiley & Sons, 2002.

(6) Prandtl, L: *"Essentials of Fluid Dynamics"*, Hafner, New York, 1952 [Streeter et al. (9)].

(7) Rouse, H: *"Elementary Mechanics of Fluids*, Wiley Eastern Pvt. Ltd., New Delhi, 1970.

(8) Schlichting, H: *"Boundary Layer Theory"*, McGraw-Hill, New York, 1979.

(9) Streeter, VL, Wylie, EB, and Bedford, KW: *"Fluid Mechanics"*, McGraw-Hill Book Co,, Singapore, 1998.

(10) Von Karman, T: *"On Laminar and Turbulent Friction"*, Z. Angew, Math. Mech. Vol. 1, 1921 [Massey (4)].

(11) White, FM: *"Fluid Mechanics"*, McGraw-Hill, 2003.

(12) Blevins, R.D: *"Applied Fluid Dynamics Handbook"*, Van Nostrand Reinhold, New York, 1984. [Munson et al. (5)].

CHAPTER 9

Incompressible Viscous Flow in Pipes

9.1 FLOW DEVELOPMENT

Fluid flows have been categorized in various ways (Chapter-3). Viscous flows, i.e., both laminar and turbulent flows, may be either external (i.e., unbounded; for example, the growth of boundary layer over a flat plate or motion of an airplane in the atmospheric air) or internal (i.e., bounded by walls; for example, flow in pipes, ducts, nozzles, diffusers, sudden contractions, valves, fittings etc.). In case of external flows, the flow pattern around a body immersed in fluid is of interest as was discussed in Chapter-8. Internal flows are bounded by the surrounding walls and the viscous effects grow so as to encompass the entire flow.

Figure 9.1.1 illustrates an internal flow in a long pipe. The fluid enters the pipe through well-rounded entrance with uniform velocity U_o and the flow at the entrance is, therefore, inviscid. Because of the no-slip condition at the pipe wall, the fluid particles in contact with the wall remain at rest, i.e., the velocity at the wall must be zero along the entire length of the pipe. The wall shear stress retards the fluid near the wall. Therefore, in order to meet continuity (or mass conservation) requirement, the velocity must increase in the central region. At successive sections along the pipe, the effect of the pipe wall is felt farther out into the flow and the shape of the velocity profile continuously changes. The flow in the central region is inviscid (or irrotational or potential) while flow near the pipe wall is viscous due to boundary layer development.

Sufficiently far downstream from the pipe entrance (say, at $x = L_e$), the developing boundary layer reaches the centreline of the pipe and the flow becomes entirely viscous and the axial velocity at $x \geq L_e$, no longer changes with increase in x. When the shape of the velocity profile does not change with distance x along the pipe, the flow is said to be *fully-developed*. The distance between the entrance and the location (or section) at which the entire flow has become viscous (i.e., fully developed flow) is known as *entrance length* or *transitional length* or *length of establishment*. For $x > L_e$, the velocity profile remains the same, the wall shear is constant, and the pressure drops linearly with x for both laminar and turbulent flows.

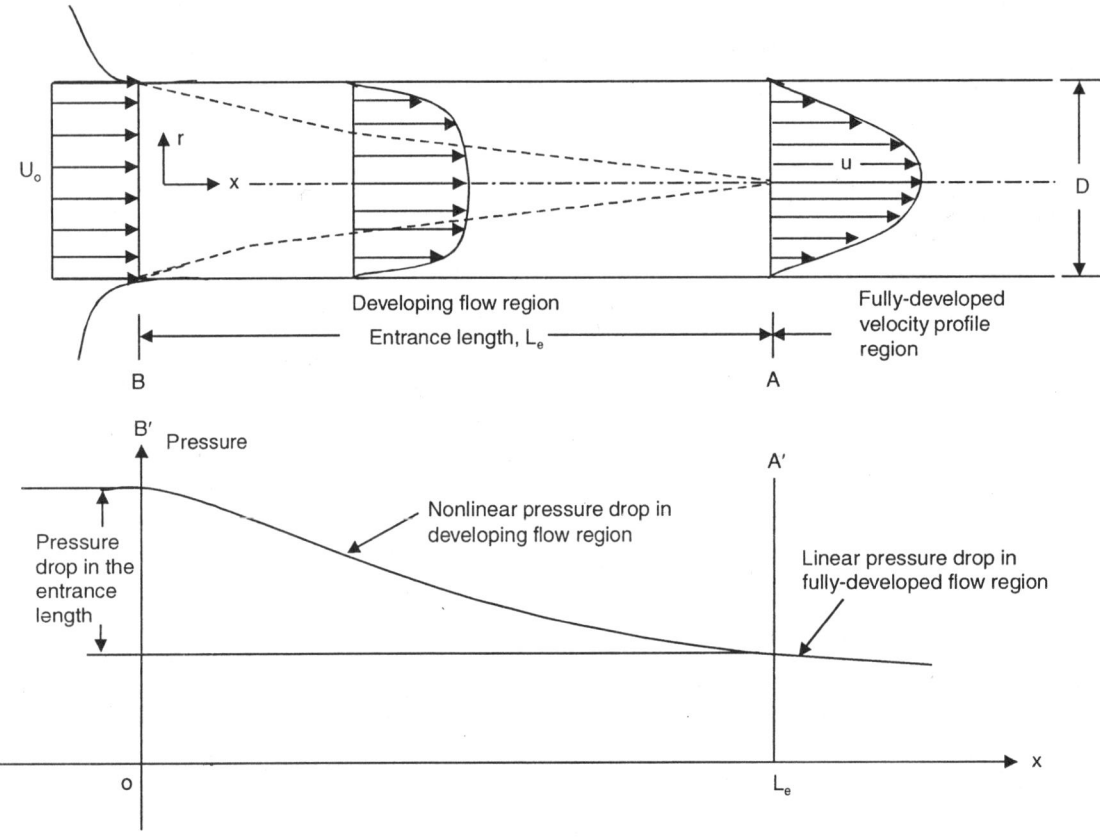

Fig. 9.1.1. Velocity profiles and pressure variations in the entrance region of pipe flow

The length of establishment L_e is expected to depend on pipe diameter D, average velocity U, and fluid properties ρ and μ. Therefore, dimensional analysis indicates that

$$\frac{L_e}{D} = \phi\left(\frac{UD\rho}{\mu}\right) = \phi\left(R_e\right)$$

For laminar flow, Langhaar (6) developed the theoretical formula

$$\frac{L_e}{D} = 0.058\,R_e \cong 0.06\,R_e \qquad \qquad ...(9.1.1)$$

This formula agrees well with the observations. Accepting lower critical Reynolds number for flow in pipes as 2000, the maximum entrance length for laminar flow is 120 D. For turbulent flow, the boundary layer grows much faster and, hence, entrance length is relatively shorter and is given as (12)

$$\frac{L_e}{D} \cong 4.4\,R_e^{1/6} \qquad \qquad ...(9.1.2)$$

This means that the transition length would be about 44 times the diameter for turbulent flow having Reynolds number equal to 10^6. For typical pipe flow applications, length of pipeline is considerably greater than 120 times its diameter. Therefore, the entrance effects may be neglected for both laminar and turbulent flows and a simple analysis, based on fully-developed flow, is carried out.

9.2 STEADY LAMINAR FLOW IN CIRCULAR PIPES

Steady, uniform, laminar flow in a circular pipe or annulus can be analyzed in the same manner as explained in Chapter-7 for laminar flow between parallel plates. Consider an annular element of internal radius r, length δx, and radial thickness δr in an inclined (at an angle θ with horizontal) circular pipe of uniform radius R, Fig. 9.2.1. A fluid with viscosity μ and mass density ρ flows in the pipe such that the flow conditions correspond to the fully developed steady laminar flow. Since the flow is steady and uniform, the acceleration is zero. Therefore, sum of all the forces (due to pressure p, viscous shear τ, and gravity) acting on the element is zero. Therefore,

$$\left\{ (p)2\pi r\,\delta r - \left(p + \frac{\partial p}{\partial x}\delta x \right)2\pi r\,\delta r \right\} + \left\{ \tau 2\pi r\,\delta x - \right.$$

$$\left. -(\tau 2\pi r\,\delta x + \frac{\partial}{\partial r}(\tau 2\pi r\,\delta x)\delta r \right\} + \rho g\,2\pi r\,\delta r\,\delta x \sin\theta = 0 \qquad \qquad ...(9.2.1)$$

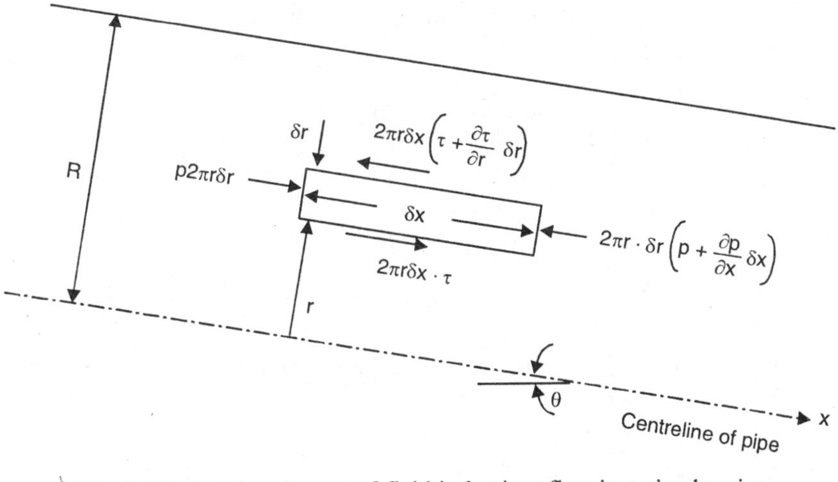

Fig. 9.2.1. Annular element of fluid in laminar flow in a circular pipe

Here, p is, obviously, the static pressure. This equation, after simplification and dividing by the volume of the chosen annular element (i.e., $2\pi\,r\,\delta r\,\delta x$), reduces to

$$-\frac{\partial p}{\partial x} - \frac{1}{r}\frac{\partial(\tau r)}{\partial r} - \rho g\frac{\partial z}{\partial x} = 0 \qquad \qquad ...(9.2.2)$$

where, z is the elevation of the pipe above a chosen horizontal datum and, hence, $\sin\theta = (-)\partial z/\partial x$. Equation (9.2.2) can also be written as

$$\frac{\partial(p + \rho gz)}{\partial x} + \frac{1}{r}\frac{\partial(\tau r)}{\partial r} = 0 \qquad \qquad ...(9.2.3)$$

The term $(p + \rho gz)$ is, obviously, the piezometric pressure, say p^*, that is independent of r and varies only in x-direction. Therefore, Eq. (9.2.3) can be written as

$$\frac{dp*}{dx} + \frac{1}{r}\frac{d(\tau r)}{dr} = 0 \qquad \qquad ...(9.2.3)$$

On integration,

$$\frac{dp*}{dx}\frac{r^2}{2} + \tau r + C_1 = 0 \qquad \qquad ...(9.2.4)$$

where, C_1 is a constant of integration. Equation (9.2.4) is valid for circular pipe as well. Therefore, it must be satisfied at $r = 0$ too. This means that the constant of integration C_1 is zero for flow in a circular pipe.

Recognizing that the direction of measurement of distance r is from the centre of the pipe and not from the pipe wall, the shear stress τ is related to the velocity gradient as

$$\tau = -\mu\frac{du}{dr} \qquad \qquad ...(9.2.5)$$

Note that du/dr is negative. On substituting the value of τ, Eq. (9.2.5), in Eq. (9.2.4), one obtains

$$\frac{r^2}{2}\frac{dp*}{dx} - r\mu\frac{du}{dr} + C_1 = 0$$

$$\therefore \qquad \qquad du = \frac{r}{2\mu}\frac{dp*}{dx}\,dr + \frac{C_1}{\mu r}\,dr$$

Hence, $\qquad \qquad u = \frac{r^2}{4\mu}\frac{dp*}{dx} + \frac{C_1}{\mu}\ln r + C_2 \qquad \qquad ...(9.2.6)$

Here, C_2 is another constant of integration. Constants of integration C_1 and C_2 are to be evaluated using the boundary conditions.

For flow in a circular pipe, C_1 is zero and C_2 can be determined using the boundary condition at $r = R$, $u = 0$. Therefore, Eq. (9.2.6) gives

$$C_2 = -\frac{R^2}{4\mu}\frac{dp*}{dx}$$

and, hence,

$$u = -\frac{1}{4\mu}\left(\frac{dp*}{dx}\right)\left(R^2 - r^2\right) \qquad \qquad ...(9.2.7)$$

Here, it should be noted that $dp*/dx$ is negative. Equation (9.2.7) gives the velocity distribution for fully-developed steady, laminar flow in a circular pipe of uniform diameter. The velocity distribution is a paraboloid of revolution, Fig. 9.2.2. Therefore, its volume is one-half the volume of its circumscribing cylinder. Hence, the average velocity, U is one-half the maximum velocity, u_m. The maximum velocity would occur at $r = 0$. Thus,

$$u_m = -\frac{R^2}{4\mu}\frac{dp*}{dx} \qquad \qquad ...(9.2.8)$$

and
$$U = -\frac{R^2}{8\mu}\frac{dp^*}{dx}$$
...(9.2.9)

Therefore, the discharge Q is given as

$$Q = -\frac{\pi R^4}{8\mu}\frac{dp^*}{dx} = \int_0^R 2\pi r\,u\,dr$$
...(9.2.10)

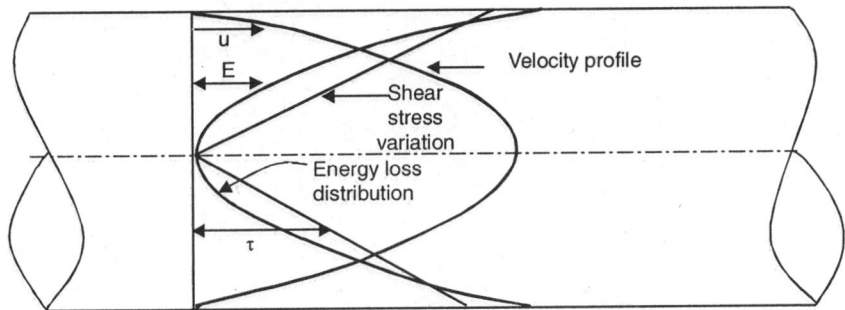

Fig. 9.2.2. Distribution of velocity, shear and energy losses for a circular pipe

For a horizontal pipe, z = constant and the piezometric pressure equals static pressure. Substituting the pressure drop Δp in a length L of a circular pipe of diameter D, Eq. (9.2.10) yields

$$Q = \frac{\pi D^4 \Delta p}{128\,\mu L}$$
...(9.2.11)

Therefore, the average velocity

$$U = \frac{Q}{\pi D^2/4} = \frac{D^2\,\Delta p}{32\mu L}$$
...(9.2.12)

Equations (9.2.11) and (9.2.12) can, alternatively, be written as

$$\Delta p = \frac{128\mu LQ}{\pi D^4} = \frac{32\mu UL}{D^2}$$
...(9.2.13)

In terms of head loss h_f,

$$h_f = \frac{\Delta p}{\rho g} = \frac{128\mu LQ}{\pi \rho g D^4} = \frac{32\mu UL}{\rho g D^2}$$
...(9.2.14)

Thus, it is seen that for steady laminar flow in a circular pipe of uniform diameter, the losses vary directly with the viscosity, the length, and the discharge and inversely as the fourth power of the diameter, Eq. (9.2.14). Equation (9.2.14) is known as Hagen-Poiseullie's equation having been determined experimentally by Hagen in 1839 and independently by Poiseullie in 1840. The analytical derivation was made by Wiedemann in 1856 (10).

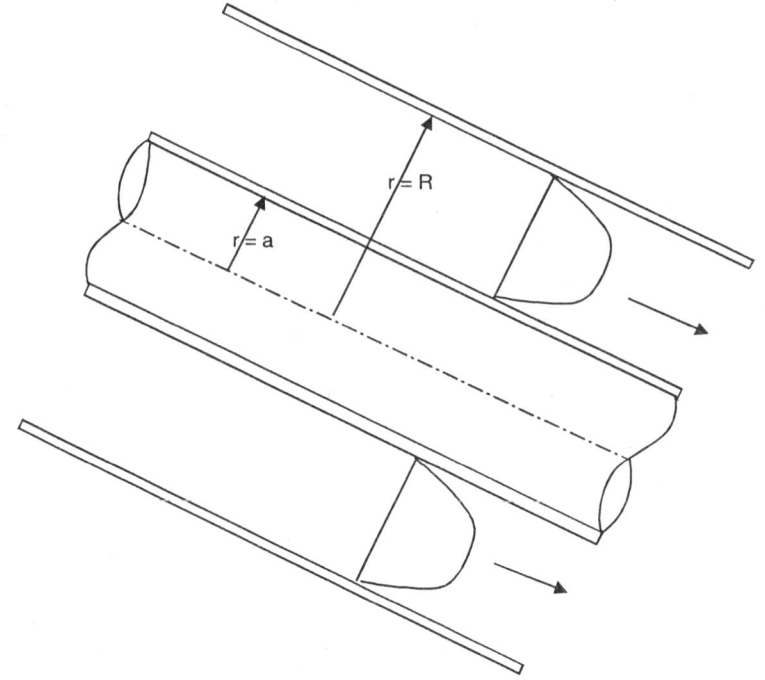

Fig. 9.2.3. Flow through an annulus

For the annular flow, Fig. 9.2.3, the boundary conditions are

$$u = 0 \text{ at } r = a, \text{ the inner tube radius}$$

and

$$u = 0 \text{ at } r = R.$$

Therefore, substitution of these boundary conditions in Eq. (9.2.6) yields

$$0 = \frac{a^2}{4\mu} \frac{dp^*}{dx} + \frac{C_1}{\mu} \ln a + C_2 \qquad \qquad ...(9.2.15)$$

and

$$0 = \frac{R^2}{4\mu} \frac{dp^*}{dx} + \frac{C_1}{\mu} \ln R + C_2 \qquad \qquad ...(9.2.16)$$

$$\therefore \qquad C_1 = -\frac{dp^*}{dx} \frac{(R^2 - a^2)}{4 \ln R/a} \qquad \qquad ...(9.2.17)$$

Hence, Eq. (9.2.6) becomes

$$u = \frac{r^2}{4\mu} \frac{dp^*}{dx} - \frac{R^2 - a^2}{4\mu \ln R/a} \frac{dp^*}{dx} \ln r + C_2 \qquad \qquad ...(9.2.18)$$

and Eq. (9.2.18) becomes

$$0 = \frac{R^2}{4\mu} \frac{dp^*}{dx} - \frac{R^2 - a^2}{4\mu \ln R/a} \frac{dp^*}{dx} \ln R + C_2 \qquad \qquad ...(9.2.19)$$

Subtracting Eq. (9.2.19) from Eq. (9.2.18), one gets

$$u = \frac{1}{4\mu} \frac{dp^*}{dx} \left[(r^2 - R^2) + \frac{R^2 - a^2}{\ln R/a} \ln R/r \right]$$

$$= -\frac{1}{4\mu} \frac{dp^*}{dx} \left[(R^2 - r^2) + \frac{R^2 - a^2}{\ln R/a} \ln R/r \right] \qquad \text{...(9.2.20)}$$

Discharge through annulus is, therefore, given as

$$Q = \int_a^R 2\pi r u \, dr = -\frac{\pi}{8\mu} \frac{dp^*}{dx} \left[R^4 - a^4 - \frac{(R^2 - a^2)^2}{\ln R/a} \right] \qquad \text{...(9.2.21)}$$

9.3 SHEAR STRESS IN A CIRCULAR PIPE

Equation (9.2.4) for a circular pipe (i.e., $C_1 = 0$) yields

$$\tau = -\frac{r}{2} \frac{dp^*}{dx} \qquad \text{...(9.3.1)}$$

In deriving Eq. (9.2.4) and, therefore, Eq. (9.3.1) too, no assumption about the type of flow (viz., laminar or turbulent) has been made. Hence, Eq. (9.3.1) gives the variation of shear stress for fully-developed steady uniform flow in a circular pipe, Fig. 9.2.2. Equation (9.3.1) is valid for both laminar as well as turbulent flow and indicates that the shear stress varies linearly with radial distance r from a value of zero at the centreline of the pipe, where $r = 0$, to a maximum value τ_0 at the wall, where $r = R$, Fig. 9.2.2. At the pipe wall, the shear stress τ_0 is given as

$$\tau_0 = -\frac{R}{2} \frac{dp^*}{dx} \qquad \text{...(9.3.2)}$$

and for horizontal pipe (i.e., z = constant),

$$\tau = -\frac{r}{2} \frac{dp}{dx} \qquad \text{...(9.3.3)}$$

and

$$\tau_0 = -\frac{R}{2} \frac{dp}{dx} \qquad \text{...(9.3.4)}$$

Combining Eqs. (9.3.1) and (9.3.2) or Eqs. (9.3.3) and (9.3.4), one gets

$$\frac{\tau}{\tau_0} = \frac{r}{R} \qquad \text{...(9.3.5)}$$

$$\therefore \qquad\qquad \tau = \tau_0 \frac{r}{R} \qquad \text{...(9.3.6)}$$

9.3.1 Energy Dissipation in Circular Pipes

In Art. 7.3, it was shown that the rate of energy dissipation for steady, uniform, and laminar flow, the rate of energy dissipation per unit volume equals $\mu \left(\dfrac{du}{dy} \right)^2$. Therefore, for horizontal laminar flow in a circular pipe, the rate of energy dissipation is equal to

$$\int_o^R \tau \left(\frac{du}{dy} \right) 2\pi r dr$$

$$= \int_o^R \frac{\tau^2}{\mu} 2\pi r \, dr$$

$$= \int_o^R \frac{r^2}{4\mu} \left(-\frac{dp}{dx} \right)^2 2\pi r \, dr \text{ using Eq. (9.3.3)}$$

$$= \frac{\pi}{2\mu} \left(-\frac{dp}{dx} \right)^2 \int_o^R r^3 dr$$

$$= \frac{\pi R^4}{8\mu} \left(-\frac{dp}{dx} \right)^2$$

$$= \frac{\pi D^4}{128\mu} \left[\frac{32\mu U}{D^2} \right] \left(-\frac{dp}{dx} \right) \text{ using Eq. (9.2.13)}$$

$$= \frac{\pi D^2}{4} U \left(-\frac{dp}{dx} \right)$$

$$= Q \left(\frac{p_1 - p_2}{L} \right)$$

Further, the term $\dfrac{p}{\rho}$ in the Bernoulli's equation, Eq. (4.6.5), represents the work done on unit mass of the flowing fluid by the unbalanced pressure forces. For horizontal conduit flow, the pressure force per unit volume of the fluid equals pressure gradient [see Art. 9.2], i.e., $\left(-\dfrac{\partial p}{\partial x} \right)$. The power P (i.e., the rate at which work must be done) required to maintain a viscous flow through a horizontal conduit must, then, be equal to the product of the pressure gradient, volume of the fluid (i.e., AL) and the velocity of flow, U. That is,

$$P = -\frac{\partial p}{\partial x} (AL)(U)$$

$$= -\frac{\partial p}{\partial x} LQ$$

$$= (p_1 - p_2) Q$$

$$= \rho g Q h_f$$

in which h_f represents head loss due to friction. Therefore, power required for viscous flow per unit

length of a horizontal conduit equals $\left(\dfrac{p_1 - p_2}{L}\right) Q$ which is the expression for the rate of energy dissi-

pation per unit volume of the fluid flowing in a horizontal pipe.

Hence, in steady, uniform, laminar flow in a horizontal circular pipe, the product $Q(p_1 - p_2)$ represents not only the average rate of energy dissipation or the rate at which work is done, but also the average rate at which heat is generated in a circular pipe. That is, energy used in maintaining such flow is completely transformed into heat.

It should be noted that pressure p would be replaced by piezometric pressure p^* if the pipe is not horizontal.

9.4 VELOCITY DISTRIBUTION IN TURBULENT FLOW

Equation (7.8.4) derived for turbulent flow over a surface agrees well with the experiments and gives satisfactory results. This equation is

$$\frac{u}{u_*} = \frac{1}{K}\ln y + C \qquad \qquad ...(9.4.1)$$

This means that the velocity distribution for turbulent flow is logarithmic. The constant C cannot be evaluated from the boundary condition, $u = 0$ at $y = 0$ as $\ln 0 = -\infty$. However, the law of the wall, Eq. (7.8.2) is

$$\frac{u}{u_*} = \frac{u_* y}{v} \qquad \qquad ...(9.4.2)$$

which, for the boundary condition $u = u_w$ at $y = \delta'$, yields (1)

$$\frac{u_w}{u_*} = \frac{u_* \delta'}{v} = N \qquad \qquad ...(9.4.3)$$

From Eq. (9.4.3), one can conclude that $u_* \delta'/v$ should have a critical value N (sort of Reynolds number) at which flow changes from laminar to turbulent. Substitution of the boundary condition, $u = u_w$ at $y = \delta'$ in Eq. (9.4.1) yields

$$\frac{u_w}{u_*} = N = \frac{1}{K}\ln \delta' + C = \frac{1}{K}\ln Nv/u_* + C$$

Hence, $$C = N - \frac{1}{K}\ln Nv/u_* \qquad \qquad ...(9.4.4)$$

Substituting the value of constant C from Eq. (9.4.4) in Eq. (9.4.1), one obtains

$$\frac{u}{u_*} = \frac{1}{K}\ln y + N - \frac{1}{K}\ln Nv/u_*$$

or
$$\frac{u}{u_*} = \frac{1}{K}\ln u_* y/v + \left(N - \frac{1}{K}\ln N\right) \qquad \ldots(9.4.5)$$

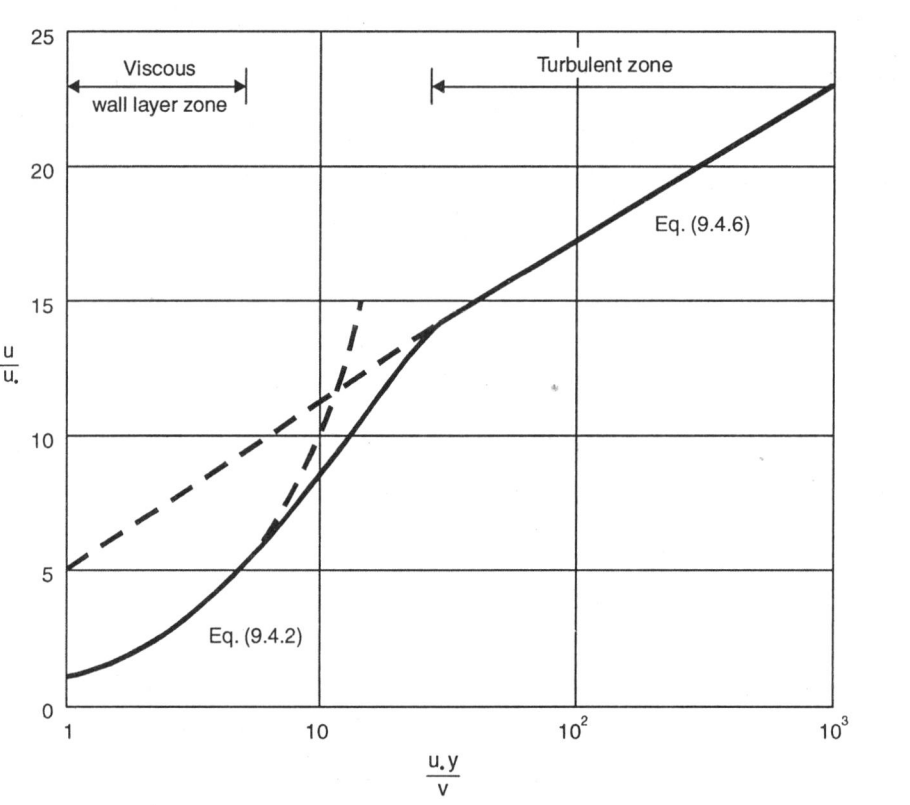

Fig. 9.4.1. Velocity variation for turbulent flow in smooth pipes

The term, $\left(N - \dfrac{1}{K}\ln N\right)$ has been determined experimentally by plotting $\dfrac{u}{u_*}$ against $\ln u_* y/v$, Fig. 9.4.1. For hydrodynamically smooth surfaces, Nikuradse's experimental data yielded $K = 0.40$ and $\left(N - \dfrac{1}{K}\ln N\right) = 5.5$. Thus, Eq. (9.4.5) reduces to

$$\frac{u}{u_*} = 5.75 \log u_* y / v + 5.5 \qquad \ldots(9.4.6)$$

This equation is known as the Karman-Prandtl equation for velocity distribution in turbulent flow near hydrodynamically smooth surfaces.

Experimental data indicate that Eq. (9.4.2) is suitable for $u_* y / \nu < 5$ and Eq. (9.4.6) is valid for $u_* y / \nu > 70$. When $5 < \dfrac{u_* y}{\nu} < 70$, von Karman has suggested the following equation for velocity variation in turbulent flow over hydrodynamically smooth surfaces:

$$\frac{u}{u_*} = 11.5 \log u_* y / \nu - 3.05 \qquad \qquad ...(9.4.7)$$

Similarly, for hydrodynamically rough surfaces in which the viscous sublayer is destroyed by roughness (of average height k), one can apply a boundary condition, $u = u_k$ at $y = k$. Therefore,

$$\frac{u_k}{u_*} = \frac{1}{K} \ln k + C$$

Again, treating u_k / u_* as another dimensionless parameter N', one gets

$$C = N' - \frac{1}{K} \ln k$$

Substituting the value of C in Eq. (9.4.1), one obtains

$$\frac{u}{u_*} = \frac{1}{K} \ln y + N' - \frac{1}{K} \ln k$$

or

$$\frac{u}{u_*} = \frac{1}{K} \ln y / k + N' \qquad \qquad ...(9.4.8)$$

Based on Nikuradse's experimental data, Eq. (9.4.8) reduces to

$$\frac{u}{u_*} = 5.75 \log y / k + 8.5 \qquad ...(9.4.9)$$

This is Prandtl-Karman equation for velocity distribution in turbulent flow near hydrodynamically rough surfaces.

Alternatively, one may assume the boundary condition $u = 0$ at $y = y'$, Fig. 9.4.2, so that Eq. (9.4.1) reduces to

$$\frac{u}{u_*} = \frac{1}{K} \ln y / y' \qquad ...(9.4.10)$$

Based on experimental evidence, $y' = \dfrac{\delta'}{107}$

$$= \frac{11.6 \nu}{107 u_*} = \frac{0.108 \nu}{u_*} \text{ for smooth surfaces and}$$

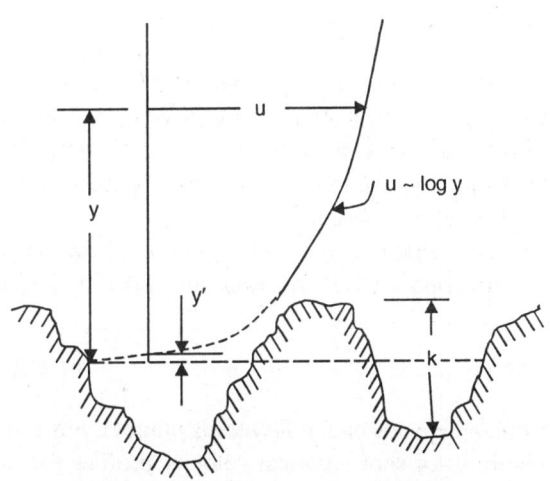

Fig. 9.4.2. Definition sketch for roughness boundary

$y' = \dfrac{k}{30}$ for rough surfaces. On substituting these values of y' in Eq. (9.4.10), one at a time, one would, once again, obtain Eqs. (9.4.6) and (9.4.9).

Although Eqs. (9.4.6) and (9.4.9) have been developed from the data near boundaries, they have been found applicable farther away from the boundaries too. Both these equations are valid for turbulent flow in pipes as well. Therefore, discharge Q in a pipe of radius R can be determined by integrating the velocity profile, i.e.,

$$Q = \int_0^R 2\pi r u \, dr \qquad \qquad ...(9.4.11)$$

By substituting appropriate expression for u alongwith replacing y (distance from the wall) with $(R - r)$ and, then, solving Eq. (9.4.11), one can obtain the value of discharge flowing through a pipe. On dividing the discharge Q by area of cross-section πR^2, one can obtain the following expressions for the average value U:

For smooth pipes,
$$\dfrac{U}{u_*} = 5.75 \log u_* R / \nu + 1.75 \qquad \qquad ...(9.4.12)$$

For rough pipes,
$$\dfrac{U}{u_*} = 5.75 \log y/k + 4.75 \qquad \qquad ...(9.4.13)$$

On subtracting either Eq. (9.4.12) from Eq. (9.4.6) or Eq. (9.4.13) from Eq. (9.4.9), one obtains the following equation:

$$\dfrac{u - U}{u_*} = 5.75 \, \log y/R + 3.75 \qquad \qquad ...(9.4.14)$$

By substituting $u = u_m$ at $y = R$ (i.e., the centreline of pipe) in Eq. (9.4.14), one gets

$$\dfrac{u_m - U}{u_*} = 3.75 \qquad \qquad ...(9.4.15)$$

Equations (9.4.14) and (9.4.15) are known as velocity-deficit laws that are valid for both rough as well as smooth pipes.

The logarithmic laws, Eqs. (9.4.6) and (9.4.9), are widely applicable to turbulent flows in pipes. The viscous sublayer thickness is very small and covers only a very small portion of the flow section even in hydrodynamically smooth turbulent flows. The viscous sublayer is almost non-existent in rough turbulent flows since roughness size k exceeds the thickness of viscous sublayer and, hence, the sublayer is of no consequence.

The velocity variation for turbulent flow in a smooth pipe can also be represented by the following empirical power-law equation developed by Prandtl:

$$\dfrac{u}{u_m} = \left(\dfrac{y}{R}\right)^{1/n} = \left(1 - \dfrac{r}{R}\right)^{1/n} \qquad \qquad ...(9.4.16)$$

Here, n depends on the Reynolds number, Fig. 9.4.3. The value of n generally varies between 6 and 10. Figure 9.4.4 shows typical velocity profiles for fully developed flow in pipes.

Equation (9.4.16) gives infinite velocity gradient near the wall and, therefore, the power-law equation is not applicable close to the wall ($y/R < 0.04$). Likewise, at the centre, the power-law equation does

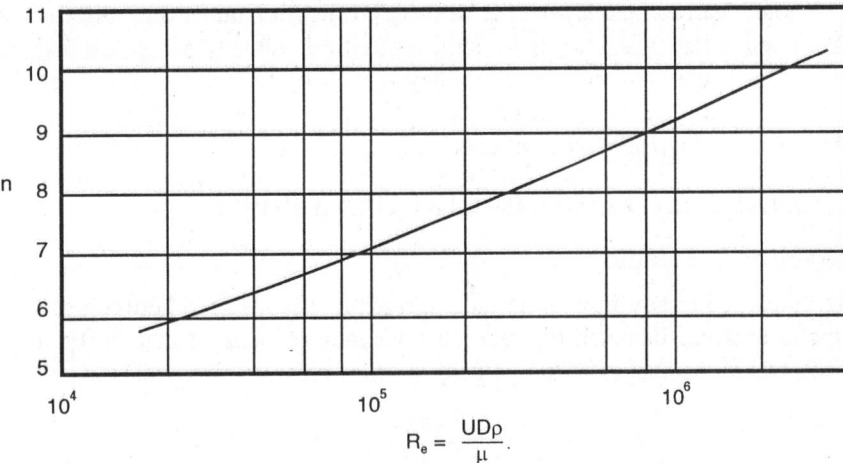

Fig. 9.4.3. Exponent n for power-law velocity profiles (5)

not yield zero slope although the experimental data confirm the validity there. The logarithmic law too does not yield non-zero value of the velocity gradient du/dy at the centreline of pipe (i.e., at $r = 0$). Being easy to use, the power-law velocity profile, Eq. (9.4.16) is used very often.

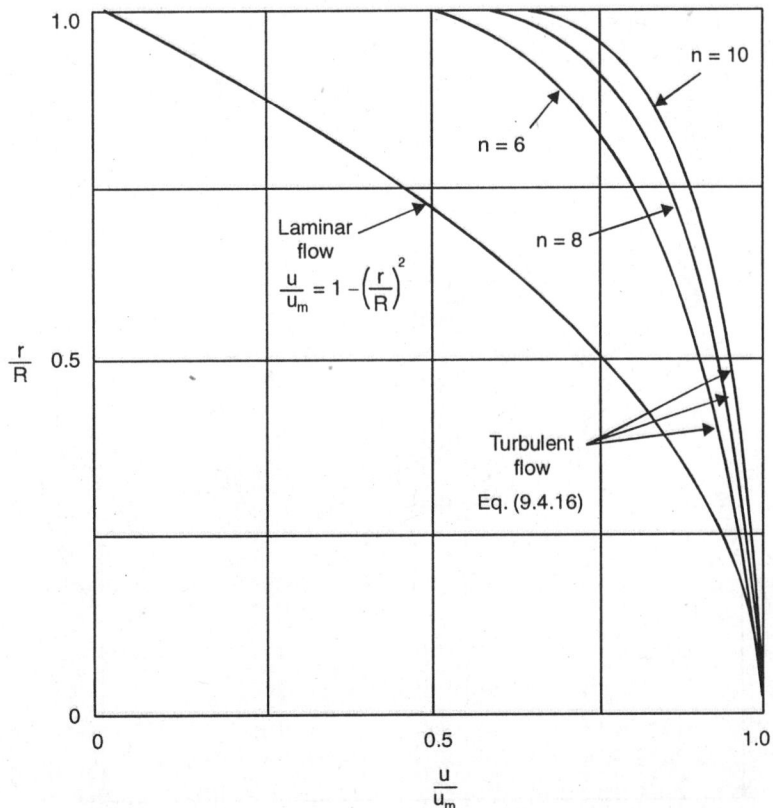

Fig. 9.4.4. Typical velocity profiles for laminar and turbulent flow in a circular pipe

It should be noted that, compared with the velocity profile for laminar flow, the velocity profiles for turbulent flow are much flatter and closer to fictitious uniform velocity profile for inviscid fluid. Most of the flows are of turbulent kind and, therefore, have nearly uniform velocity profiles. The usefulness of the inviscid flow theory is, therefore, obvious. Of course, there are several flows that cannot be analyzed without accounting for viscous effects.

9.5 FRICTIONAL HEAD LOSS IN CIRCULAR PIPES

9.5.1 Darcy-Weisbach Equation

Consider fully-developed steady flow of an incompressible viscous fluid between sections 1 and 2 of an inclined pipe of uniform diameter, Fig. 9.5.1. Application of energy equation, Eq. (4.6.9), between sections 1 and 2, where the shapes of the velocity profiles are the same, yields

$$\frac{p_1}{\rho g} + z_1 = \frac{p_2}{\rho g} + z_2 + h_f \qquad \qquad ...(9.5.1)$$

where, h_f represents the head loss due to friction. Thus,

$$h_f = \Delta z + \frac{\Delta p}{\rho g} = \Delta \left(\frac{p}{\rho g} + z \right) \qquad \qquad ...(9.5.2)$$

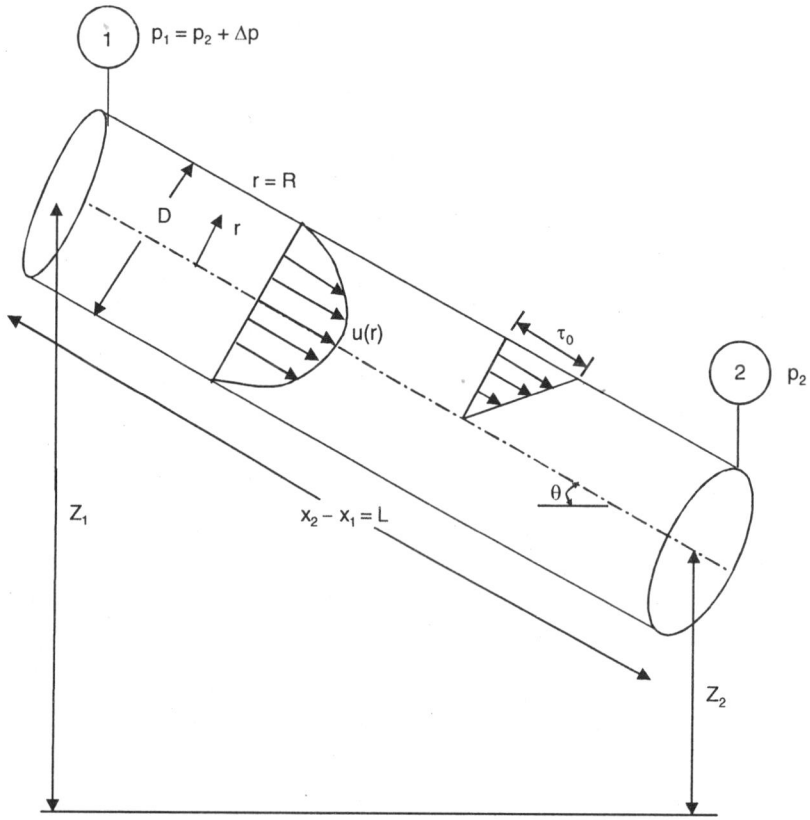

Fig. 9.5.1. Control volume for steady, fully-developed flow in an inclined pipe

Hence, the frictional head loss equals the change in the piezometric pressure head, i.e., the drop in the elevation of the hydraulic gradient line (HGL) or the total energy line (TEL). Application of momentum principle, Eq. (4.2.7) for the control volume between sections 1 and 2 of the pipe, Fig. 9.5.1, yields

$$\sum F_x = \Delta p(\pi R^2) + \rho g(\pi R^2 L)\sin\theta - \tau_o(2\pi RL) = \rho Q(U_2 - U_1) = 0$$

where, τ_0 is the wall shear stress. Also, $L \sin\theta = \Delta z$.

Therefore,
$$\frac{\Delta p}{\rho g} + \Delta z = \frac{2\tau_o L}{\rho g R} \qquad \qquad ...(9.5.3)$$

Hence, Eqs. (9.5.2) and (9.5.3) yield

$$h_f = \frac{2\tau_o L}{\rho g R} = \frac{4\tau_o L}{\rho g D} \qquad \qquad ...(9.5.4)$$

Equation (9.5.4) shows that the frictional head loss is proportional to the wall shear stress and also to the ratio L/D irrespective of whether the pipe is horizontal or inclined. It should be noted that Eq. (9.3.2) would also yield Eq. (9.5.4) that can, alternatively, be written as

$$h_f = \frac{4\tau_o(0.5U^2)L}{(0.5\rho U^2)gD}$$

or
$$h_f = f\frac{L}{D}\frac{U^2}{2g} = \frac{8fLQ^2}{\pi^2 g D^5} \qquad \qquad ...(9.5.5)$$

in which, f is dimensionless and equals $\dfrac{4\tau_0}{(1/2)\rho U^2}$, and is known as Darcy's *friction factor* or, simply,

friction factor, after Henry Darcy (1803-1858), a French engineer whose pipe flow experiments in 1857 first established the effect of roughness on pipe resistance. The friction factor f, in general, depends on the Reynolds number UD/ν and k/D for flow in circular pipes. Equation (9.5.5) was first proposed in 1850 by Julius Weisbach, a German professor. Equation (9.5.5) is known as Darcy-Weibach equation.

9.5.1 Exponential pipe-friction formulas

For convenience of computations for designing pipeline systems, sometimes exponential pipe-friction formula is proposed. One such formula in S.I. units, proposed by Hazen-Williams, is (10)

$$\frac{h_f}{L} = \frac{10.675 Q^n}{C^n D^m} \qquad \qquad ...(9.5.6)$$

Here, $\dfrac{h_f}{L}$ is the head loss per unit length of pipe, $n = 1.85$, $m = 4.87$, and C depends upon the condition of pipe surface as follows (10):

C	Pipe surface condition
140	Extremely smooth, straight pipes; asbestos-cement
130	Very smooth pipes; concrete; new cast iron
120	New welded steel
110	Vitrified clay; new riveted steel
100	Cast iron after years of use
95	Riveted steel after years of use
60-80	Old pipes in poor condition

An exponential formula of the kind of Eq. (9.5.6) is valid only for the fluid viscosity for which it has been developed and is valid for limited range of Reynolds number. Therefore, such formulas are empirical, i.e., developed from experimental results and are useful and convenient to work with for the range of data used for developing these formulas. Compared to such empirical formulas, Darcy-Weisbach equation, Eq. (9.5.5), is more rational and universal and, therefore, has received wide acceptance. However, empirical exponential formula may be preferred for specific applications when the conditions for specific application are within the range of the experimental data used for developing the formula.

9.6 VARIATION OF FRICTION FACTOR

Equation (9.5.5) would yield head loss correctly only if the value of the friction factor f is properly chosen. It is apparent that the friction factor f is not a constant and depends on U, D, ρ, and μ besides the characteristics of roughness represented by its size k, spacing of roughnesses k', and shape of roughness that can be represented by a dimensionless parameter such as form factor m (10). Thus,

$$f = f_1 (U, D, \rho, \mu, k, k', m) \qquad \qquad ...(9.6.1)$$

Using principles of dimensional analysis, Eq. (9.6.1) can be rewritten as

$$f = f_2 \left(\frac{UD\rho}{\mu}, \frac{k}{D}, \frac{k'}{D}, m \right) \qquad \qquad ...(9.6.2)$$

For hydrodynamically smooth pipes, $k = k' = m = 0$ and, thus

$$f = f_3 \left(\frac{UD\rho}{\mu} \right) = f_3 (R_e) \qquad \qquad ...(9.6.3)$$

The nature of the functions in Eqs. (9.6.2) and (9.6.3) can be confirmed only through experimentation. Nikuradse (9) used three pipes of different diameters. The interior walls of these pipes were coated with sand grains of almost constant diameter k so as to have the same values of k/D for different pipes. His experiments proved that for one value of k/D, the data (f versus R_e) plotted on a smooth curve regardless of the pipe diameter. His experimental study did not include variation of k'/D and m, however. Therefore, for one type (with respect to the shape and spacing) of roughness,

$$f = f_4 \left(R_e, \frac{k}{D} \right) \qquad \qquad ...(9.6.4)$$

The plot of friction factor f against R_e for different values of k/D called relative roughness on a log-log graph paper is called Stanton diagram, Fig. 9.6.1. Blasius correlated the experimental data for smooth pipes through an empirical equation (valid upto R_e = 100,000) known as Blasius equation,

$$f = \frac{0.316}{R_e^{0.25}} \qquad\qquad ...(9.6.5)$$

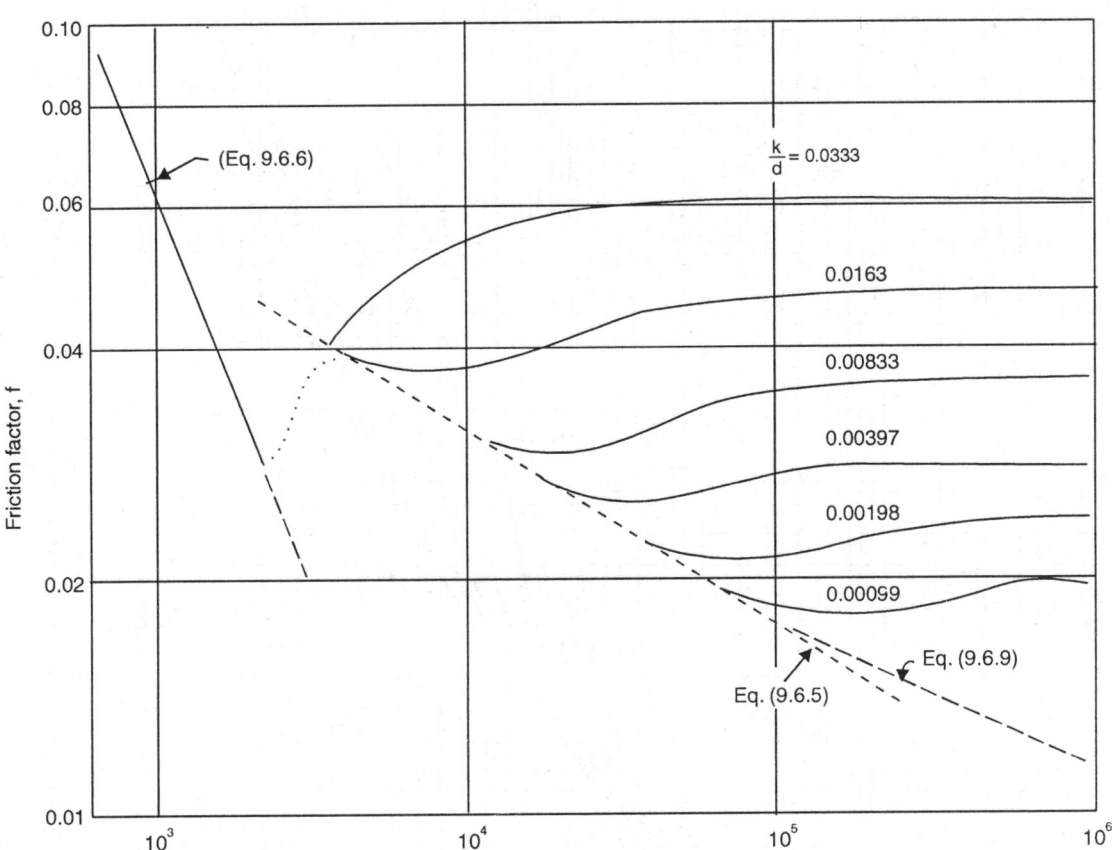

Fig. 9.6.1. Variation of friction factor f (Stanton diagram) for Nikuradse's test on sand-roughened pipes (9)

Compared to the sand-coated rough surfaces, rough surfaces of commercially-produced pipes are very complex. The sand-coated rough surface either gets fully covered by the laminar sublayer or protrudes uniformly as the thickness of the laminar sublayer decreases. The commercial roughness is non-uniform and, therefore, only some part of the roughness protrudes first and amount of protrusion increases with the decreases in the laminar sublayer thickness. Therefore, the relationship between f and R_e for sand-coated rough pipes is expected to be different than that for commercially-produced rough pipes. Hence, most of the advances made in the subject are due to the experiments on artificially roughened commercial pipes. Moody (7) developed a convenient chart, Fig. 9.6.2, for estimation of friction factor f in clean commercial pipes. From Moody's diagram, Fig. 9.6.2, it is seen that for Reynolds number less that about 2000, when flow in a pipe is laminar, f depends on R_e alone and relative roughness k/D has no effect on it. In fact, Eq. (9.2.14) can be rewritten as

$$h_f = \frac{32\mu UL}{\rho g D^2}\left(\frac{2U}{2U}\right) = \frac{64\mu}{UD\rho}\frac{L}{D}\frac{U^2}{2g} = \frac{64}{R_e}\frac{L}{D}\frac{U^2}{2g} = f\frac{L}{D}\frac{U^2}{2g}$$

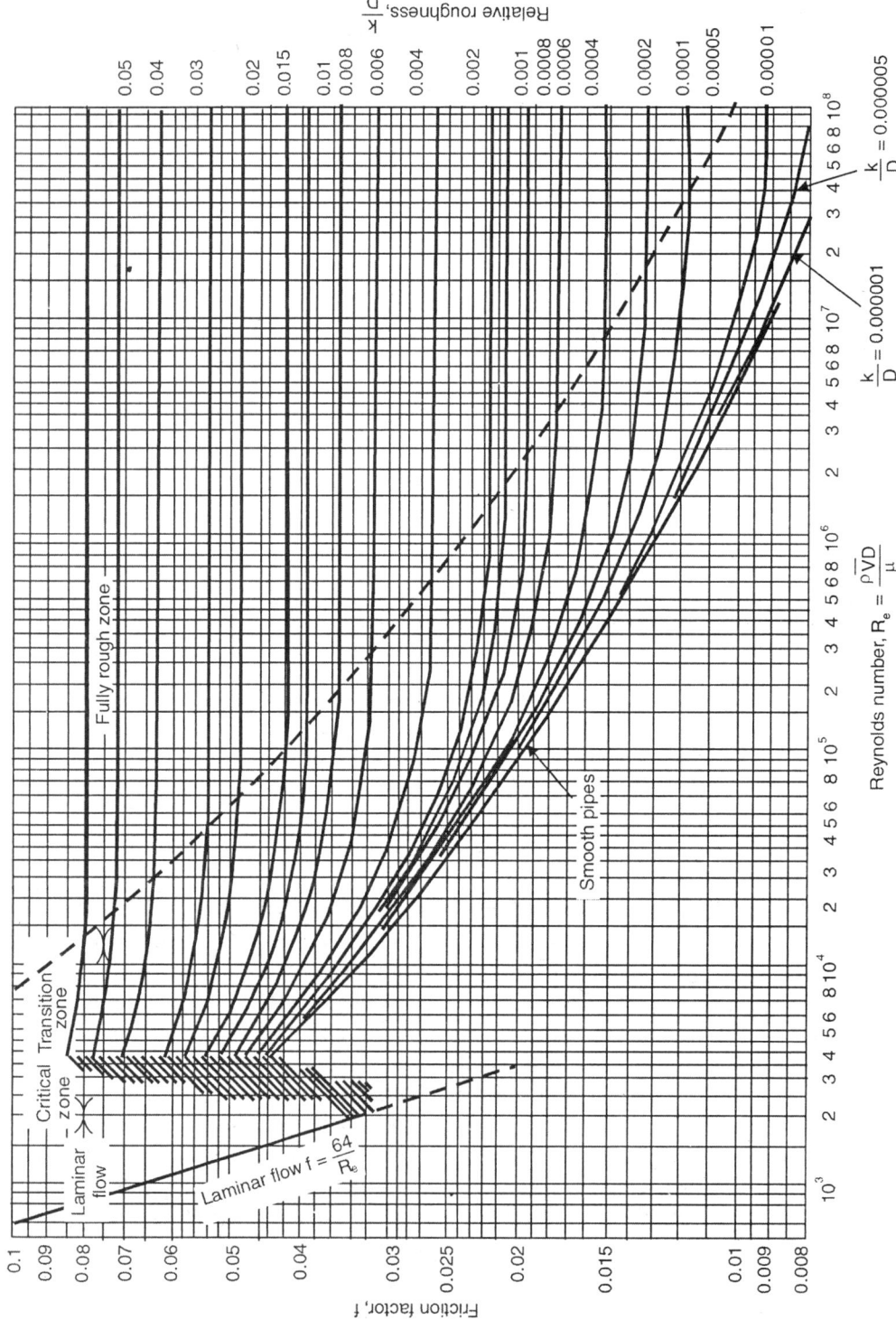

Fig. 9.6.2. Variation of friction factor f for fully-developed flow in commercial circular pipes (7)

This means that for laminar flow,

$$f = \frac{64}{R_e}$$

...(9.6.6)

Equations (9.6.6) agrees very well with the experimentation of both Nikuradse as well as Moody. Further, Eqs. (9.5.4) and (9.5.5) yield

$$f \frac{L}{D} \frac{U^2}{2g} = \frac{4\tau_0 L}{\rho g D} = \frac{4u_*^2 L}{g D}$$

Therefore,

$$u_* = U \sqrt{f/8}$$

...(9.6.7)

Here, u_* is the shear velocity. Substituting the value of u_* in Eq. (9.4.12), one obtains

$$\frac{U}{U\sqrt{f/8}} = 5.75 \log U \sqrt{f/8}\,(D/2)/v + 1.75$$

$$\therefore \qquad \frac{1}{\sqrt{f}} = \frac{5.75}{\sqrt{8}} \log \frac{UD}{v} \sqrt{f} + \frac{5.75}{\sqrt{8}} \log \sqrt{1/32} + \frac{1.75}{\sqrt{8}}$$

Hence,

$$\frac{1}{\sqrt{f}} = 2.03 \log R_e \sqrt{f} - 0.91$$

...(9.6.8)

The derived equation, Eq. (9.6.8), has been modified slightly in view of the experimental evidence. The corresponding equation that fits the experimental data on smooth pipes is

$$\frac{1}{\sqrt{f}} = 2.0 \log R_e \sqrt{f} - 0.8$$

...(9.6.9)

Equation (9.6.9) is the accepted formula for smooth pipes. The values of f computed from this equation are as follows:

R_e	4000	10^4	10^5	10^6	10^7	10^8
f	0.0399	0.0309	0.0180	0.0116	0.0081	0.006

Thus, f decreases by a factor of about 5 when Reynolds number increases from 4000 to 10^7.

The value of f for smooth pipes can be determined by Eq. (9.6.9) only by cumbersome trial procedure. Hence, one can use Blasius equation, Eq. (9.6.5), for smooth pipes if $R_e < 100,000$. For higher Reynolds number, one may use (13)

$$f = 0.0032 + \frac{0.221}{R_e^{0.237}}$$

...(9.6.10a)

or

$$f = \left[1.8 \log 0.145 R_e \right]^{-2}$$

...(9.6.10b)

if one wants to avoid trial.

It should also be noted that a simple substitution of Blasius' equation, Eq. (9.6.5), in Darcy-Weisbach equation, Eq. (9.5.5), would yield that the frictional head loss is not proportional to the square of flow velocity when the pipe surface is hydrodynamically smooth.

Likewise, substitution of the value of u_*, Eq. (9.6.7), in Eq. (9.4.13) and subsequent modification of the resulting equation on the basis of experimental evidence for rough pipes, yields

$$\frac{1}{\sqrt{f}} = 2.0 \log R/k + 1.74 \qquad \qquad ...(9.6.11)$$

Therefore, there is no Reynolds number effect and, hence, the frictional head loss varies exactly as the square of the velocity U, Eq. (9.5.5). The values of f computed from Eq. (9.6.11) are as follows:

k/D	0.00001	0.0001	0.001	0.01	0.05
f	0.00806	0.0120	0.0196	0.0379	0.0716

Thus, friction factor increases by about 9 times when k/D increases by a factor of 5000.

Equation (9.6.11) is valid only when flow is fully turbulent in hydrodynamically rough pipes for which condition f depends on k/D alone and this value of f is known as the *limiting friction factor*. Equation (9.6.9) and (9.6.11) can be rewritten as

$$\frac{1}{\sqrt{f}} - 2\log R/k = \phi\left(\frac{Re\sqrt{f}}{R/k}\right)$$

$$= 2\log Re\sqrt{f} - 0.8 \quad \text{for smooth pipes}$$

$$= 1.74 \qquad \qquad \text{for rough pipes.}$$

Since

$$\frac{R_e\sqrt{f}}{R/k} \equiv \frac{\dfrac{UD}{v}\sqrt{f/8}\sqrt{8}}{D/2k} \equiv \frac{u_* 2\sqrt{8}k}{v} \equiv \frac{11.6(2\sqrt{8})k}{11.6v/u_*}$$

$$\equiv 65.6\,\frac{k}{\delta'}$$

Since the nature of the surface (i.e., rough or smooth or in transition) is decided by the parameter k/δ', one can expect a single curve between $\left[\dfrac{1}{\sqrt{f}} - 2\log R/k\right]$ and $\left[\dfrac{R_e\sqrt{f}}{R/k}\right]$ or $\left[\dfrac{k}{\delta'}\right]$ as shown in Fig. 9.6.3 (13). Based on this plot, the transition range of surface roughness begins when $R_e\sqrt{f}/(R/k) \approx 17$, i.e., $k/\delta' = 0.25$ and ends when $R_e\sqrt{f}/(R/k) \approx 400$, i.e., $k/\delta' = 6.0$. Accordingly, surface behaves as hydrodynamically smooth if $k/\delta' < 0.25$ and hydrodynamically rough if $k/\delta' > 6.0$.

Colebrook (2) combined Eqs. (9.6.9) and (9.6.11) suitably to obtain a formula that is valid for all ranges of Reynolds number and relative roughness when flow is turbulent. The formula is

$$\frac{1}{\sqrt{f}} = -2.0 \log \left[\frac{k/D}{3.7} + \frac{2.51}{R_e \sqrt{f}} \right] \qquad \qquad ...(9.6.12)$$

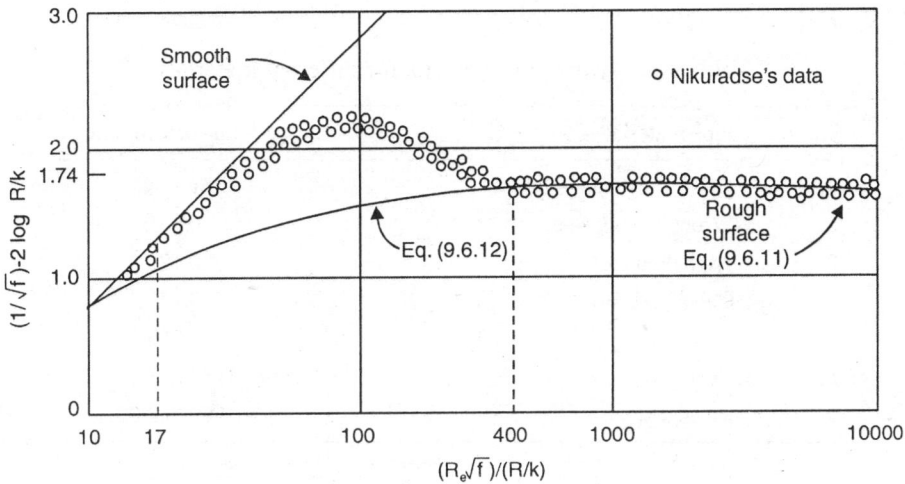

Fig. 9.6.3. Variation of $(1/\sqrt{f}) - 2\log R/k$ with $R_e \sqrt{f}/(R/k)$

This is the accepted design formula and was plotted by Moody in the form of Fig. (9.6.2). Equation (9.6.12) too is an implicit equation for the friction factor f and, therefore, it is cumbersome to estimate f from Eq. (9.6.12).

In place of the Moody's diagram, one may use the following explicit formulas (4,11) to obtain the value of f within about 2 percent of the value obtained from Eq. (9.6.12):

$$f = \frac{1.328}{\left[\ln\left(\dfrac{k/D}{3.7} + \dfrac{5.74}{R_e^{0.9}} \right) \right]^2} \qquad \qquad ...(9.6.13)$$

for $10^{-6} \le \dfrac{k}{D} \le 10^{-2}$ and $5000 \le R_e \le 10^8$

Alternatively,

$$f = \frac{1.636}{\left[\ln\left\{ \left(\dfrac{k/D}{3.7} \right)^{1.11} + \dfrac{6.9}{R_e} \right\} \right]^2} \qquad \qquad ...(9.6.14)$$

Surface roughness of a commercial pipe cannot be expressed easily in terms of sand grain diameter. Nevertheless, one can determine the size of the sand grains such that, when the surface of a pipe is coated with these sand grains, this pipe yields the same limiting value of friction factor (i.e., the value of friction factor for fully rough turbulent flow condition when f depends only on surface condition) as that given by the commercial pipe. This is known as *equivalent sand grain roughness* and its size can be determined by first determining experimentally the value of the friction factor for the given commercial pipe and, then, substituting this value of f in Eq. (9.6.11) to obtain the value of k, i.e., the size of the *equivalent sand grain roughness* or the *equivalent sand grain diameter*. Some typical values of k are given in Table 9.6.1.

Table 9.6.1 Average values of k for different materials

Sl.	Pipe material	Value of k in mm
1.	Glass, drawn glass tubing	Smooth
2.	Wrought iron, steel	0.045
3.	Asphalted cast iron	0.120
4.	Galvanised iron	0.150
5.	Cast iron	0.250
6.	Concrete	0.30 to 3.00
7.	Riveted steel	0.90 to 9.00

9.7 PROBLEMS ON FRICTION LOSS IN PIPELINES

Problems involving only friction losses in long pipeline systems are, relatively, simple pipe flow problems and can be solved by using Darcy-Weisbach equation, Eq. (9.5.5), and Moody's diagram, Fig. 9.6.2, or some suitable equation for prediction of the friction factor. For using Moody's diagram, however, one needs to know variables required to compute Reynolds number R_e and k/D. Simple pipe flow problems can be of the following four types:

Type I : μ, ρ, k, L, Q or U, and D are known, to determine Δp or h_f.

Type II : $\mu, \rho, k, \Delta p$ or h_f, Q or U, and D are known, to determine L.

Type III : $\mu, \rho, k, \Delta p$ or h_f, L, and D are known, to determine Q or U.

Type IV: $\mu, \rho, k, \Delta p$ or h_f, Q, and L are known, to determine D.

Problems of type I and type II are straight forward. One can compute R_e and k/D and obtain f from Moody's diagram or relevant equation. Thereafter, using Darcy-Weisbach equation, Eq. (9.5.5), one can compute h_f (or Δp) or L whichever is unknown.

Problems of type III require iteration procedure since f and R_e both are unknown as Q (or U) appears in both of the two ordinates of Moody's diagram. Iteration procedure, however, is quite fast as f varies rather slowly with R_e. Since k/D is known, trial value of f may be taken as the limiting value of friction factor (i.e. the value of f for given k/D for fully rough turbulent condition when it no longer depends on the Reynolds number) that can be read either from Moody's diagram or obtained from Eq. (9.6.11). With this value of f, first trial value of U or Q can be computed from Eq. (9.5.5) and corresponding Reynolds number estimated. Now obtain an improved value of f for given k/D and estimated R_e using Moody's diagram. This iteration is carried till the value of f has been found correct to two significant digits. The corresponding value of Q (or U) is the desired value.

Alternatively, if one eliminates the velocity of flow (or flow rate) from Eq. (9.5.5) by substituting the value of U (in terms of R_e) in Eq. (9.5.5), one gets

$$h_f = f \frac{L}{D} \left(\frac{R_e \nu}{D} \right)^2 \frac{1}{2g} = \frac{f L R_e^2 \nu^2}{2 g D^3} \qquad \text{...(9.7.1)}$$

or

$$f R_e^2 = \frac{2 g D^3 h_f}{L \nu^2} \qquad \text{...(9.7.2)}$$

Now, one can obtain another curve between $\dfrac{2 g D^3 h_f}{L \nu^2}$ $(i.e., f R_e^2)$ and R_e for given k/D using the

relevant curve f versus R_e of Moody's diagram. One can, then, obtain R_e (and, hence, U) for the given value of $2 g D^3 \, h_f / L \nu^2$.

Still, alternatively, using Eq. (9.7.2), one can get values of

$$R_e \sqrt{f} \left(= \sqrt{\frac{2 g D^3 h_f}{L \nu^2}} \right) \text{ and } \frac{1}{\sqrt{f}} \left(= \frac{{}^\bullet R_e \, \nu L^{1/2}}{\sqrt{2 g D^3 h_f}} \right)$$

and substitute in Colebrook equation, Eq. (9.6.12), to obtain (2)

$$R_e = -\left(8 \frac{g D^3 h_f}{L \nu^2} \right)^{1/2} \log \left(\frac{k/D}{3.7} + \frac{1.775 \nu}{\sqrt{g D^3 h_f / L}} \right) \qquad \text{...(9.7.3)}$$

Hence, R_e (and, hence, U or Q) can be obtained directly using Eq. (9.7.3) for all types of turbulent flows in pipes.

Pipe flow problems of type IV are of considerable significance as an engineer, dealing with flow of fluids in pipelines, is often required to determine the smallest (and, hence, least costly) pipe size that can deliver the desired flow rate, Q under specified conditions of head loss. Since the diameter D occurs in all three parameters (f, R_e, and k/D) of Moody's diagram, one has to adopt an iterative solution for this kind of pipe flow problem.

Using Eq. (9.5.5) the pipe diameter D can be expressed as

$$D^5 = \frac{8 L Q^2}{g \pi^2 h_f} f = C_1 f \qquad \text{...(9.7.4)}$$

in which C_1 can be determined from the given data. Again, substituting the value of U in terms of Q,

$$R_e = \frac{UD}{\nu} = \frac{4Q}{\pi \nu} \frac{1}{D} = C_2 D \qquad \text{...(9.7.5)}$$

in which C_2 is the known quantity for the given data. The following iterative procedure should yield the desired solution:

1. Choose an arbitrary value of f.
2. Calculate D from Eq. (9.7.4) and R_e from Eq. (9.7.5).
3. Obtain relative roughness k/D

4. Find f for computed R_e (step 2) and k/D (step 3).
5. Repeat steps 2, 3, and 4 with the value of f obtained in step 4 till the value of f does not change in the first two significant digits. The corresponding value of D is the required pipe size. Since pipes are available only in some selected standard sizes, the value of D obtained is raised to the next larger size available.

An alternative iteration procedure can be as follows:

1. Assume a trial pipe diameter D that is available commercially.
2. Compute R_e and k/D and obtain the friction factor f.
3. Compute head loss h_f using Eq. (9.5.5) and the value of f (step 2).
4. If the computed h_f (step 3) is too large, repeat steps 2 and 3 with another larger trial diameter D that is available commercially. If the computed h_f is too small, repeat steps 2 and 3 with smaller trial diameter that is available commercially.

9.8 FRICTIONAL HEAD LOSS IN NON-CIRCULAR CONDUITS

One can analyze fully-developed flow in non-circular conduits following the procedure followed for circular pipes. However, the analysis is very complicated and one may, instead, use a simpler, but approximate, method based on the concept of hydraulic radius that is frequently used for flow in open channel the cross-section of which is usually not circular.

For a non-circular conduit too, the control volume concept of Fig. 9.5.1 is still valid with the difference that A does not equal πR^2 and the cross-sectional perimeter (P) for the shear stress action does not equal $2\pi R$. The momentum principle yields

$$\Delta p \, A + \rho g \, AL \, sin\theta - \tau_0 \, PL = 0$$

$\therefore$
$$h_f = \frac{\Delta p}{\rho g} + \Delta z = \frac{\tau_0}{\rho g} \frac{L}{A/P} \qquad ...(9.8.1)$$

Equation (9.8.1) is identical to Eq. (9.5.3) except that the shear stress τ_0 is not constant over the wetted perimeter but an integrated average value around the wetted perimeter and that $R/2$ is replaced with the term A/P known as hydraulic radius R_h and defined as the ratio of cross-sectional area A and wetted perimeter P. For a circular pipe,

$$R_h = \frac{A}{P} = \frac{\frac{\pi}{4}D^2}{\pi D} = \frac{D}{4}$$

$\therefore$
$$D = 4R_h$$

Thus, replacing D with $4R_h$ in Eq. (9.5.5), one gets

$$h_f = f \frac{L}{(4R_h)} \frac{U^2}{2g}$$

It is emphasized that wetted perimeter includes all surfaces acted upon by the shear stress. Further, the Reynolds number would now be $\dfrac{U(4R_h)}{\nu}$ and relative roughness would be $k/(4R_h)$. Thus, the friction factor f is

$$f = F\left(\frac{U(4R_h)}{\nu}, \frac{k}{4R_h}\right) \qquad \qquad ...(9.8.2)$$

One should, however, not expect to use Moody's diagram and yet obtain accurate results. However, this approach is known to yield results for friction factor within ±15 percent for turbulent flow. The error increases to about ±40 percent for laminar flow. Hence, the approximate method suggested is relatively much less suitable for laminar flows.

9.9 MINOR LOSSES IN PIPE SYSTEMS

A pipeline system may not have only uniform straight pipe and may have, additionally, a variety of pipe fittings such as valves, bends, joints, elbows etc. or abrupt or gradual changes in flow cross-section for meeting specific requirements. Therefore, besides head loss due to friction in the pipeline system, there would be additional loss due to these fittings and cross-sectional changes. These additional losses are termed *minor losses* or *local losses* or *secondary losses* that are caused primarily due to flow separation giving rise to large scale turbulence on account of which considerable energy is dissipated by turbulent mixing of fluid in the separated region. The 'minor loss' is a misnomer because in relatively short pipeline system with a variety of fittings, the minor losses may be more significant than the losses due to friction in the pipeline and, therefore, may not remain minor or secondary.

While the source of the minor loss is contained in a very short length of a pipeline, its effect in the form of turbulence may persist for considerable distance downstream thereby affecting there the frictional phenomenon too. However, in the analysis, it is always assumed that the process of frictional loss remains unaffected by the presence of the fittings causing minor loss and the minor loss is concentrated at the location of the source (i.e., fitting) causing it. The total loss in the pipeline system, then, equals the sum of frictional loss and minor loss.

The flow pattern in the vicinity of the pipe fittings is quite complex due to flow separation and is, therefore, not amenable to theoretical solution. The losses are, therefore, measured experimentally and correlated with the pipe flow parameters. Since the frictional loss in a pipe having rough turbulent flow condition is proportional to square of the mean flow velocity, the minor loss h_L (usually, the pressure head difference across the fitting) is commonly expressed as

$$h_L = K_L\left(\frac{U^2}{2g}\right) \qquad \qquad ...(9.9.1)$$

in which, K_L is termed the *loss coefficient* that depends upon geometry of the fitting, and fluid and flow properties. That is,

$$K_L = \frac{h_L}{\left(U^2/2g\right)} = \frac{\Delta p}{\frac{1}{2}\rho U^2} = \phi \text{ (geometry, } R_e) \qquad \qquad ...(9.9.2)$$

Fittings cause flow cross-section to vary along the flow direction. The flowing fluid, therefore, experiences relatively large accelerations and/or decelerations. Thus, the flow is dominated by inertial effects rather than viscous effects. Hence, the pressure drops and head losses correlate directly with the dynamic pressure. This is particularly true when the Reynolds number is rather large. It is for this reason that the friction factor for fully developed pipe flow with large Reynolds number is independent of the Reynolds number, Fig. (9.6.2). Hence, in most cases of practical interest, the loss coefficients for pipe fittings are functions of geometry of the fitting alone. That is, $K_L = f\,(geometry)$. The value of K_L is often determined by experiments except, however, for sudden expansion in a pipeline, Art. 5.9. Almost all data on loss coefficient, K_L are reported for turbulent flow conditions.

Minor loss is sometimes expressed in terms of *equivalent length* l_e of a pipe of specified uniform diameter and friction factor such that the frictional head loss in the given pipe equals the minor loss of the fitting. That is,

$$h_L = K_L \frac{U^2}{2g} = f\,\frac{l_e}{D}\frac{U^2}{2g} \qquad \qquad ...(9.9.3)$$

A single pipeline system may have many straight lengths of uniform diameter pipes causing friction losses and many fittings that cause minor losses. Since both friction and minor losses are correlated with $U^2/2g$, these losses can be summed together to obtain total loss, h_{Lt}

$$h_{Lt} = \sum h_f + \sum h_L \qquad \qquad ...(9.9.4)$$

If the pipe is of uniform diameter, then

$$h_{Lt} = \frac{U^2}{2g}\left(f\frac{L}{D} + \sum K_L \right) \qquad \qquad ...(9.9.5)$$

9.9.1 Transition Loss

Many pipeline systems require various transition sections by which the pipe diameter is changed from one size to another. Such changes may be effected either abruptly or gradually and, in both cases, contribute head loss that is not accounted for in the computations for frictional head loss. The extreme examples of the transition include sudden expansion (e.g., flow from a pipe into a reservoir, i.e., an exit) and sudden contraction (e.g., flow into a pipe from a reservoir, i.e., an entry).

9.9.1.1 Expansion loss

For flow through sudden expansion, the shear stress in the corner region of separated flow, i.e., dead water region, is negligible. Control volume analysis between the inlet end of the expansion and the end of the separation zone gives theoretical loss in sudden expansion as (Art. 5.9)

$$h_L = \frac{(U_1 - U_2)^2}{2g} = \frac{U_1^2}{2g}\left[1 - \left(\frac{D_1}{D_2}\right)^2 \right]^2 = K_L \frac{U_1^2}{2g} \qquad \qquad ...(9.9.6)$$

$$\therefore \qquad K_{Lse} = \left[1 - \left(\frac{D_1}{D_2}\right)^2 \right]^2 \qquad \qquad ...(9.9.7)$$

It should be noted that the minor loss is based on the velocity head in the smaller pipe. Further, the head loss due to sudden expansion is not due to wall shear stress (i.e., the direct viscous effect) but due to the dissipation of kinetic energy (i.e., another type of viscous effect) on account of inefficient (due to flow separation) deceleration of fluid. In case of flow from a pipe into a reservoir, the ratio $D_1/D_2 \approx 0$ and $K_{Lse} \approx 1$. The resulting head loss is termed *exit loss* that equals the velocity head. Equation (9.9.7) is in agreement with experimental evidence.

If the expansion is gradual, the head loss (including loss due to pipe friction for the length of the expansion) is quite different. Figure 9.9.1(3) shows a gradual conical expansion (usually called a *diffuser* that is so shaped as to decelerate a fluid) and the variation of loss coefficient K_L with the diffuser angle θ and the diameter ratio D_2/D_1. Diffuser is intended to decelerate a fluid and, thus, raise the static pressure of the flow. Accordingly, efficiency of diffuser can be decided on the basis of *pressure recovery coefficient C_p* defined as

$$C_p = \frac{p_2 - p_1}{\frac{1}{2}\rho U_1^2} \qquad \qquad ...(9.9.8)$$

The head loss h_{Ld} is given as

$$\frac{U_1^2}{2g} + \frac{p_1}{\rho g} = \frac{U_2^2}{2g} + \frac{p_2}{\rho g} + h_{Ld}$$

or

$$h_{Ld} = \left(\frac{U_1^2}{2g} - \frac{U_2^2}{2g}\right) + \left(\frac{p_1}{\rho g} - \frac{p_2}{\rho g}\right)$$

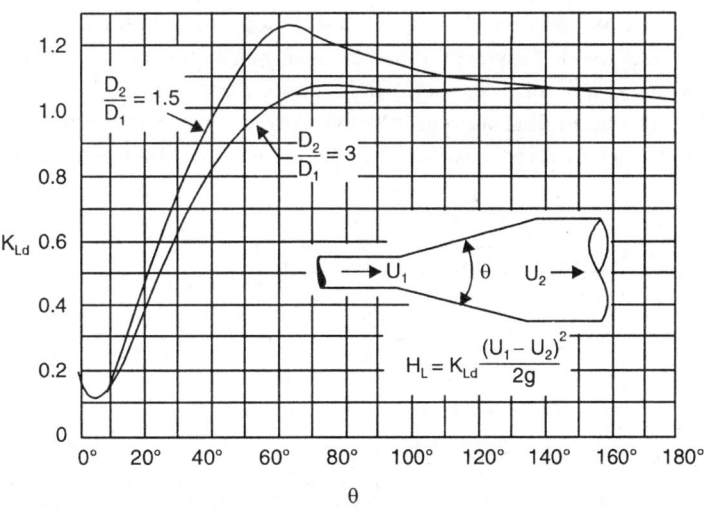

Fig. 9.9.1. Loss coefficient for conical expansion (3)

or

$$h_{Ld} = \frac{U_1^2}{2g}\left(1 - \frac{D_1^4}{D_2^4}\right) - \frac{p_2 - p_1}{\frac{1}{2}\rho U_1^2}\left(\frac{U_1^2}{2g}\right)$$

or

$$h_{Ld} = \frac{U_1^2}{2g}\left[1 - \frac{D_1^4}{D_2^4} - C_p\right] = K_{Ld}\left(\frac{U_1^2}{2g}\right) \qquad ...(9.9.9)$$

$\therefore$
$$K_{Ld} = 1 - \frac{D_1^4}{D_2^4} - C_p$$
...(9.9.10)

For a given diameter ratio, therefore, a higher C_p would mean lower head loss. This means that an efficient diffuser would have larger pressure recovery coefficient.

From Fig. 9.9.1, it is obvious that a diffuser angle (θ) of about 5° results in minimum loss and, therefore, maximum pressure recovery. For very small diffuser angles θ (less than about 8° for the diffuser shown), the diffuser length increases and most of the head loss may be due to wall shear (i.e., friction) as in fully developed pipe flow. For diffuser angles greater than about 40°, the loss is so excessive that it would be advantageous to use sudden expansion rather than diffuser. This unexpected effect is due to flow separation from the diffuser walls and the losses are mainly due to uncontrolled dissipation of kinetic energy of the jet leaving the small diameter pipe.

Actual loss coefficient of a gradual expansion (and, therefore, the pressure rise in the direction of flow) depends not only on diameter ratio and diffuser angle but also on characteristics of flow at the inlet and outlet sections of the diffuser.

9.9.1.2 Contraction loss

In case of sudden (or abrupt) contraction, Fig. 9.9.2, curvature of the streamlines and the accompanying acceleration of the fluid causes the pressure at the upstream end (section 1) of the junction to vary in an unknown way. Therefore, it is not possible to analyze the flow using momentum equation as was possible in case of sudden expansion. However, it has been shown experimentally that the process of converting pressure head into kinetic head is very efficient and, therefore, the head loss from section 1 to vena contracta (section 0) is negligible compared with the loss, due to flow expansion, between sections 0 and 2 where the velocity head gets reconverted into pressure head. Making use of Eq. (9.9.6), the head loss h_{LC}

$$h_{LC} = \frac{(U_0 - U_2)^2}{2g} = \frac{U_2^2}{2g}\left(\frac{U_0}{U_2} - 1\right)^2 = \frac{U_2^2}{2g}\left(\frac{1}{C_c} - 1\right)^2$$
...(9.9.11)

Since,
$$U_0 A_0 = U_0 C_c A_2 = U_2 A_2$$

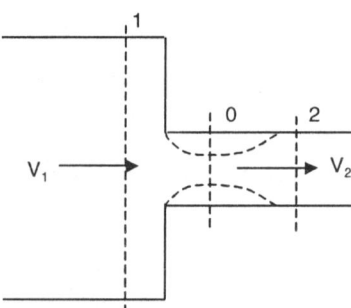

Fig. 9.9.2. Sudden contraction in a pipeline

Here, C_c is the coefficient of contraction. Equation (9.9.11) can, alternatively, be written as

$$h_{LC} = K_{LC}\left(\frac{U_2^2}{2g}\right)$$
...(9.9.12)

Here, K_{LC} is the loss coefficient for sudden contraction and its values, based on values of C_c for water as determined by Weisbach, are as given in Table 9.9.1. For Table 9.9.1, $(A_2/A_1)=(D_2/D_1)^2$.

Table 9.9.1 Values of loss coefficient for sudden contraction (10)

$\dfrac{A_2}{A_1}$	≈ 0	0.1	0.2	0.3	0.4	0.5	0.6	0.7	0.8	0.9	1.0
C_c	0.617	0.624	0.632	0.643	0.659	0.681	0.712	0.755	0.813	0.892	1.0
K_{LC}	0.5	0.46	0.41	0.36	0.30	0.24	0.18	0.12	0.06	0.02	0

The head loss at the entrance to a pipeline from a reservoir is taken as $0.5\left(U^2/2g\right)$ in which U is the velocity of flow in the pipeline and the entry is square-edged, Fig. 9.9.3. For well-rounded entrance, the loss coefficient varies between 0.01 to 0.05 and is, therefore, considered negligible.

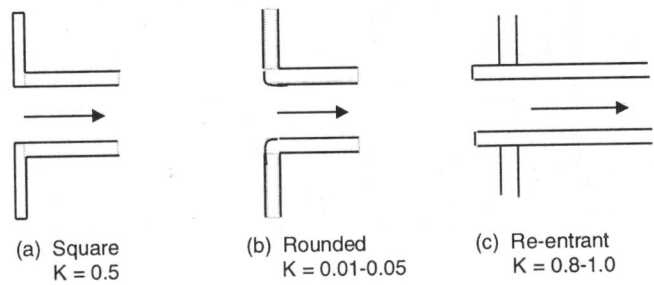

(a) Square
K = 0.5

(b) Rounded
K = 0.01-0.05

(c) Re-entrant
K = 0.8-1.0

Fig. 9.9.3. Head-loss coefficient K for velocity heads for a pipe entrance

For a gradual contraction, the head loss is very small and the loss coefficient values, as determined experimentally, are 0.02, 0.04, and 0.07 for contraction angles of 30°, 45°, and 60°, respectively (12).

9.9.2 Losses due to bends and elbows in a pipeline

The head loss of a bend in a pipeline is always larger than the frictional head loss for fully-developed flow through a straight pipeline of the same length. This is due to the flow separation on the curved walls and a swirling secondary flow arising from the centripetal acceleration.

Consider a pipe bend, Fig. 9.9.4. Because of the curved streamlines along the pipe bend, there exists a force acting radially inwards on the fluid due to inward acceleration. Therefore, there is an increase of pressure near the outer wall of the pipe bend starting from, say, P and rising to a maximum value at Q. On the inner wall of the pipe bend, the pressure reduces and attains a minimum value at R and, therefore, it rises till upto, say, S whereafter the pipeline is straight. Thus, between P and Q and also between R and S, there is an adverse pressure gradient that gives rise to flow separation and associated losses due to accompanying turbulence. These losses can be reduced by making the radius of the pipe bend relatively large.

Because of relatively higher pressures along the outer wall of the pipe bend compared to the pressures along the inner wall of the pipe bend the secondary flow is set up. The fluid particles in the vicinity of the pipe boundary, which are moving with smaller velocities due to boundary layer effects, start moving from outer wall to the inner wall in the transverse direction and along the pipe wall. To satisfy the continuity, the fluid particles from the inner wall start moving radially towards the outer wall. This secondary flow too causes additional head loss.

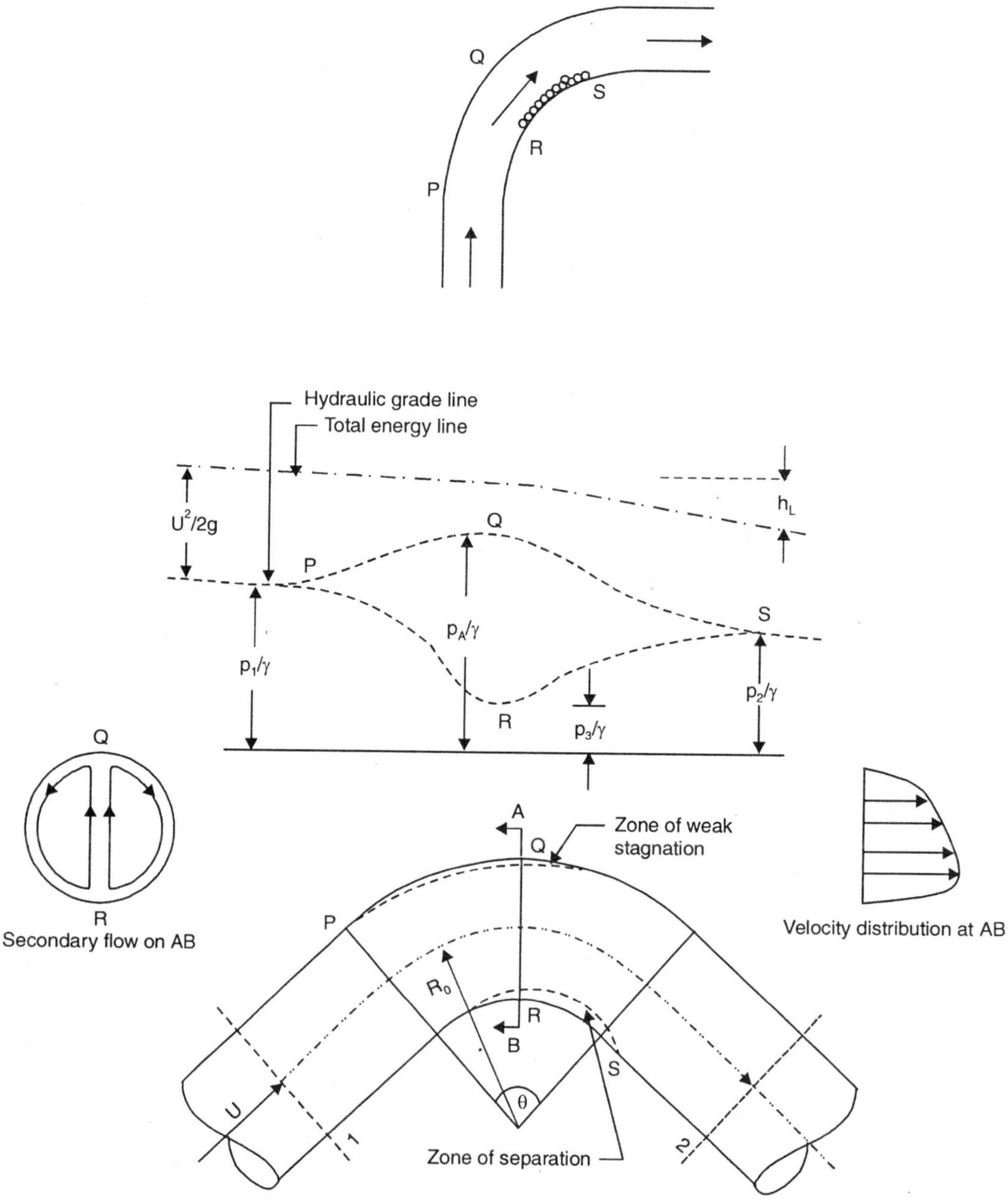

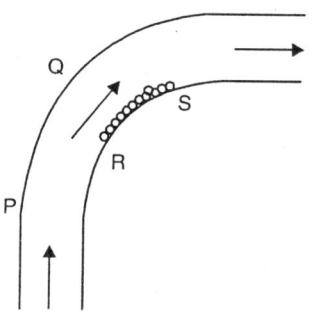

Fig. 9.9.4. Flow in a pipe bend (13)

The head loss due to a bend is, once again, conveniently expressed as K_{Lb} $(U^2/2g)$ in which K_{Lb} is the loss coefficient for the bend. The value of K_{Lb} depends upon the roughness, the bend angle θ and the ratio of radius of the pipe bend R_b to the pipe diameter D and is independent of Reynolds number when its value is relatively high. Typical values of K_L for different types of bends and elbows are given in Table 9.9.2.

Table 9.9.2 Typical values for loss coefficients for pipe components (8)

Component	K_L
Elbows	
Regular 90° (flanged)	0.3
Regular 90° (threaded)	1.5
Long radius 90° (flanged)	0.2
Long radius 90° (threaded)	0.7
Long radius 45° (flanged)	0.2
Regular 45° (threaded)	0.4
180° return bends	
180° return bend (flanged)	0.2
180° return bend (threaded)	1.5
Tees	
Line flow (flanged)	0.2
Line flow (threaded)	0.9
Branch flow (flanged)	1.0
Branch flow (threaded)	2.0
Union (threaded)	0.08
Valves	
Globe valve (fully open)	10
Angle valve (fully open)	2
Gate valve (fully open)	0.15
Gate valve (¼ closed)	0.26
Gate valve (½ closed)	2.1
Gate valve (¾ closed)	17
Ball valve (fully open)	0.05
Ball valve (1/3 closed)	5.5
Ball valve (2/3 closed)	210

The difference in pressures or piezometric pressures, Δp at the outer and inner walls of a given pipe bend depends on discharge if the Reynolds number is sufficiently large. This characteristic of a pipe bend enables it to work as discharge measuring device (appropriately named as *bend meter* or *elbow meter*) requiring no additional fitting. The discharge Q through a pipe bend is related to piezometric pressure head difference Δh (= $\Delta p/\rho g$) as follows:

$$Q = C_d \left(\frac{\pi}{4} D^2 \right) \sqrt{2g\,\Delta h}$$ ...(9.9.13)

Here, D is the diameter of pipe and C_d is the coefficient of discharge for the bend.

9.9.3 Losses due to valves in a pipeline

There are several kinds of valves that are in commercial use. The loss coefficient for valves would primarily depend upon the geometry of the valve and the extent of its opening. Typical values of the loss coefficient for use in Eq. (9.9.1), to determine the head loss due to valve, are included in Table 9.9.2.

EXAMPLES

E9.1   Determine the direction of flow through a circular pipe of diameter 12 mm and length 10 m such that its end A is 5 m above the end B of the pipe. The weight density and viscosity of the flowing fluid are 9000 N/m³ and 0.05 kg/ms, respectively.

Pressures at A and B are, respectively, 250 kPa and 300 kPa. Also, determine the flow rate.

Solution: At section A, the piezometric pressure,

$$p + \rho gh = 250 \times 10^3 + (9000)\,(5) = 295000 \text{ N/m}^2$$

and at section B that is chosen as datum,

$$p + \rho gh = 300 \times 10^3 = 300000 \text{ N/m}^2$$

Since the velocity head at sections A and B are equal and flow takes place in the direction of decreasing energy (head), the flow is from B to A.

Since the diameter of the pipe is rather small and the viscosity of the flowing fluid is relatively higher, the flow is expected to be laminar. Therefore, using Eq. (9.2.13),

$$300000 - 295000 = \frac{(128)(0.05)(10)Q}{\pi(12 \times 10^{-3})^4}$$

$$\therefore \qquad Q = 5.09 \times 10^{-6} \text{ m}^3/\text{s}$$

$$= 5.09 \times 10^{-3} \, l/s$$

The corresponding Reynolds number,

$$R_e = \frac{4Q\rho}{\pi D\mu} = \frac{4 \times 5.09 \times 10^{-6} \times 9000}{\pi(0.012)(0.05)(9.8)} = 9.92$$

Hence, the flow is laminar and use of Hagen – Poiseullie's equation, Eq. (9.2.13) is justified.

E9.2   A pipe carries water such that the local velocities at the centre of the pipe (i.e., $y = R$) and at $y = R/2$ were, respectively, 1.5 m/s and 1.36 m/s, respectively. If the diameter of the pipe is 25 cm, determine the discharge, friction factor, and height of pipe wall roughness.

Solution: On substituting the values of u for given values of y/R in Eq. (9.4.14), one gets

$$\frac{1.5 - U}{u_*} = 3.75 \qquad \therefore \; 1.5 - U = 3.75\,u_*$$

and $$\frac{1.36-U}{u_*} = 5.75 \log 0.5 + 3.75 = 2.019 \qquad \therefore 1.36 - U = 2.019 u_*$$

These two equations yield

$$U = 1.20 \text{ m/s and } u_* = 0.08 \text{ m/s}$$

Hence, $$Q = \frac{\pi}{4}(0.25)^2 (1.2) = 0.0589 \text{ m}^3/\text{s} = 58.9 \text{ l/s}$$

and $$u_* = U\sqrt{f/8} = 0.08$$

$\therefore$ $$f = \left[\frac{0.08\sqrt{8}}{1.2}\right]^2$$

$$= 0.036$$

and $$R_e = 266750$$

The data are in the fully rough range of turbulent flow. Therefore, f can also be obtained from Eq. (9.6.11)

$$\frac{1}{\sqrt{f}} = 2.0 \log R/k + 1.74$$

i.e., $$\frac{1}{\sqrt{0.036}} = 2.0 \log \frac{0.125}{k} + 1.74$$

$\therefore$ $$k = 2.15 \times 10^{-3} \text{ m}$$

$$= 2.15 \text{ mm}$$

E9.3 For turbulent flow in smooth as well as rough pipes, determine the value of y/R at which local velocity is equal to the mean velocity of flow over the cross-section of the pipe.

Solution: Substituting $u = U$ in the velocity–deficit equation (valid for both smooth as well as rough pipes), Eq. (9.4.14), one gets

$$5.75 \log y/R = -3.75$$

or $$\frac{y}{R} = 0.2228$$

E9.4 Obtain the following expression for the maximum velocity u_m for turbulent flow in smooth as well as rough pipes.

$$\frac{u_m}{U} = 1.326 \sqrt{f} + 1$$

Solution: On substituting the value of u_* from Eq. (9.6.7) in Eq. (9.4.15), one gets

$$\frac{u_m - U}{U\sqrt{f/8}} = 3.75$$

or

$$\frac{u_m - U}{U} = 3.75\sqrt{f/8} = 1.326\sqrt{f}$$

$\therefore$

$$\frac{u_m}{U} = 1.326\sqrt{f} + 1$$

E9.5   Shear stress at the bed of a channel can also be determined by observing the velocities at two different elevations at the relevant section. Obtain suitable expression for the shear stress.

Solution: On substituting $u = u_1$ at $y = y_1$ and $u = u_2$ at $y = y_2$ in the general logarithmic velocity variation, Eq. (9.4.1), one gets

$$\frac{u_1}{u_*} = \frac{1}{K}\ln y_1 + C$$

and

$$\frac{u_2}{u_*} = \frac{1}{K}\ln y_2 + C$$

Eliminating C from these equations, one gets

$$\frac{u_2 - u_1}{u_*} = \frac{1}{K}\ln y_2/y_1$$

or

$$u_* = \sqrt{\tau_0/\rho} = \frac{K(u_2 - u_1)}{\ln y_2/y_1}$$

Hence,

$$\tau_0 = \rho \frac{K^2(u_2 - u_1)^2}{(\ln y_2/y_1)^2}$$

E9.6   Determine the head (or energy) loss due to friction in a 500 m length of 200 mm diameter cast iron ($k = 0.25$ mm) pipe when an oil ($v = 10^{-5}$ m^2/s) flows in it at the rate of 150 l/s.

Solution: Reynolds number, $R_e = \dfrac{4Q}{\pi D v}$

$$= \frac{4 \times 150 \times 10^{-3}}{\pi(200 \times 10^{-3})(10^{-5})}$$

$$= 95493$$

Relative roughness, $\dfrac{k}{D} = \dfrac{0.25}{200}$

$$= 0.00125$$

From Moody's diagram, Fig. 9.6.2, friction factor, $f = 0.0235$

Alternatively, f can be obtained from Eq. (9.6.13) or Eq. (9.6.14). From Eq. (9.6.13)

$$f = \frac{1.328}{\left[\ln\left(\dfrac{k/D}{3.7} + \dfrac{5.74}{R_e^{0.9}}\right)\right]^2}$$

$$= 0.0233$$

and from Eq. (9.6.14)

$$f = \frac{1.636}{\left[\ln\left\{\left(\dfrac{k/D}{3.7}\right)^{1.11} + \dfrac{6.9}{R_e}\right\}\right]^2}$$

$$= 0.0229$$

whereas Colebrook's equation, Eq. (9.6.12), gives

$$f = 0.0231$$

Adopting $f = 0.0231$ and using Darcy-Weisbach equation, Eq. (9.6.2),

$$h_f = \frac{8(0.0231)(500)(150 \times 10^{-3})^2}{\pi^2(9.80)(200 \times 10^{-3})^5}$$

$$= 67.17 \text{ m}$$

E9.7   Determine the diameter of clean wrought iron ($k = 0.045$ mm) pipe required to carry 250 l/s of oil ($v = 10^{-5}$ m²/s) if the head loss is not to exceed 10 mm per metre length of pipe.

Solution: From Eq. (9.7.4)

$$D^5 = \frac{8LQ^2}{g\pi^2 h_f} f$$

$$= \frac{8(0.250)^2}{(9.8)(\pi)^2(0.01)} f$$

$$\therefore \qquad D^5 = 0.517 f$$

and from Eq. (9.7.5)

$$R_e = \frac{4Q}{\pi v}\frac{1}{D}$$

$$= \frac{4(0.25)}{\pi(10^{-5})D}$$

$$= \frac{31831}{D}$$

Further computations (see Art. 9.7) can be carried out in the following tabular form. Last three columns of the following Table give the values of f as obtained from Moody's diagram and computed from Eqs. (9.6.13) and (9.6.14), respectively.

Trial value of f	D	R_e	k/D	f (Fig.9.6.2)	f (Eq.(9.6.13))	f (Eq.(9.6.14))
0.02	0.4	79578	1.125×10^{-4}	0.0195	0.0194	0.0191
0.0195	0.399	79826	1.128×10^{-4}	0.0195	0.0194	0.0191

Thus, the required diameter should be equal to or greater than 0.399 m. One may, therefore, adopt a pipe of diameter, say 0.4 m that may be available in the market.

E9.8   Determine the head H for the 140 mm diameter clean cast iron ($k = 0.25$ mm) pipeline shown in Fig. E9.8 if the discharge through the pipeline is 50 l/s.

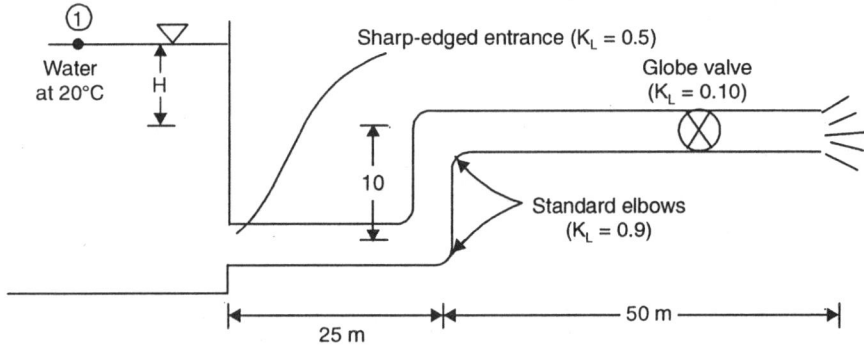

Fig. E9.8

Solution: Applying Bernoulli's equation, Eq. (5.1.5) between points 1 and 2,

$$(H+10)+0+0 = 10 + \frac{U_2^2}{2g} + 0 + 0.5\frac{U_2^2}{2g} + f\left(\frac{25+10+50}{0.140}\right)\frac{U_2^2}{2g} + 2(0.9)\frac{U_2^2}{2g} + 10\frac{U_2^2}{2g}$$

or

$$H = \frac{U_2^2}{2g}(13.3 + 566.7f)$$

since

$$Q = 50 \times 10^{-3} \text{ m}^3/\text{s}, \quad U_2 = \frac{50 \times 10^{-3}}{\frac{\pi}{4}(0.140)^2} = 3.248 \text{ m/s}$$

and

$$R_e = \frac{4Q}{\pi D \nu} = \frac{4 \times 50 \times 10^{-3}}{\pi(0.140)(10^{-6})} = 454728$$

and

$$\frac{k}{D} = \frac{0.25}{140} = 1.79 \times 10^{-3}$$

∴ from Moody's diagram, $f = 0.023$

and from Eq. (9.6.13), $f = 0.0232$

from Eq. (9.6.14), $f = 0.0232$

Adopting a value of f as 0.023

$$H = \frac{(3.248)^2}{2 \times 9.8}\left(13.3 + 566.7(0.023)\right)$$

$$= 14.174 \text{ m}$$

Thus, the required value of head, H is 14.174 m.

E9.9   Determine the equivalent length for the pipeline shown in Fig. E9.8.

Solution: The head loss in the pipeline is 14.174 m (including the exit loss at section 2). Therefore, from Eq. (9.9.3),

$$14.174 = 0.023\frac{l_e}{0.140}\left(\frac{(3.248)^2}{2 \times 9.8}\right)$$

∴   Equivalent length, $l_e = 160.3$ m.

E9.10   What would be the discharge in the pipeline of E9.8 if $H = 10$ m? Assume all other data (except Q which is unknown) to be the same as given in E9.8.

Solution: As obtained in E 9.8.

$$H = \frac{U_2^2}{2g}(13.3 + 566.7 f)$$

But

$$U_2 = \frac{Q}{\frac{\pi}{4}(0.140)^2} = 64.96 Q$$

∴ $H = 215\, Q^2\, (13.3 + 555.7\, f)$

∴

$$Q = \left[\frac{10}{215(13.3 + 566.7 f)}\right]^{1/2} = \frac{0.216}{(13.3 + 566.7 f)^{1/2}}$$

Hence, $R_e = \dfrac{4Q}{\pi Dv} = \dfrac{4 \times 0.042}{\pi(0.14)(10^{-6})} = 9.095 \times 10^6\, Q$

and $\dfrac{k}{D} = 1.79 \times 10^{-3}$

Trial f	Q	R_e	f
0.020	0.0435	39581	0.023
0.023	0.042	381972	0.023

Hence, $Q = 0.042$ m³/s $= 42$ l/s

PROBLEMS

P9.1   Determine the maximum wall shear stress for laminar flow of a fluid of known properties in a circular tube of given diameter.

P9.2   Compute head loss per meter of a circular tube (diameter = 0.6 mm) in which mercury flows at Reynolds number equal to 1500.

P9.3   Compute the pressure drop per meter of a horizontal tube (diameter = 3 mm) in which a liquid (μ = 50 cP, specific gravity = 0.80) flows at Reynolds number equal to 150.

P9.4   Determine the diameter of a vertical pipe needed for flow of liquid (ν = 1.5 $\times$ 10^{-6} m^2/s) at Reynolds number equal to 1500 when the pressure remains constant.

P9.5   At what distance from the centre of a circular pipe of diameter D does the average velocity occur if the flow in the pipe is laminar?

P9.6   Fluid flows through a pipe (diameter = 8 mm) at a Reynolds number equal to 1600. The head loss in the pipe is 15 m in a pipe length equal to 60 m. Determine the discharge in litres per minute.

P9.7   What should be the minimum value of Reynolds number so that the flow through a riveted steel pipe (diameter = 3 m and k = 3mm) remains unaffected by the viscosity of the flowing liquid?

P9.8   What size of new cast iron pipe is required to transport 350 l/s of water for 1 km with a total head loss of 2m?

P9.9   An old pipe (diameter = 2 m and k = 30mm) is coated with a lining so that the roughness reduces to k = 1mm. What would be the saving (in percent) in annual pumping costs per km for water discharge of 5 m^3/s? The pumps are 80% efficient.

P9.10 Water is being carried from a reservoir 1 at higher elevation to another reservoir 2 at lower elevation. The reservoirs are connected by a pipeline. The pipeline has 40 m of 50mm pipe followed by 25 m of 150 mm pipe followed by yet another 50 m of 75 mm pipe. All these pipes are of cast iron. There are two 90^0 elbows in the first pipe and an open globe valve in the third pipe and all of these fittings are flanged. What would be the discharge in the pipeline if the difference in the water surface elevation of the reservoir is 30 m.

REFERENCES

(1)   Bakhmeteff, BA: "*The Mechanics of Turbulent Flow*", Princeton University Press, Princeton, 1941 [Streeter et al. (10)].

(2)   Colebrook, CF: "*Turbulent flow in pipes, with particular reference to the transition between the smooth and rough pipe laws*", J. Inst. Civil Engg., London, Vol.11, 1938-39 [Streeter et al. (10)].

(3)   Gibson, AH: "*The conversion of kinetic energy to pressure energy in the flow of water through passages having divergent boundaries*", Engineering, Vol. 93, 1912. [Streeter et al. (10)].

(4)   Haaland, SE: "*Simple and explicit formulas for the friction factor in turbulent pipe flow*", J. Fluids Engg., March 1983.

(5)   Hinze, JO: "*Turbulence*", McGraw-Hill, New York, 1975 [Munson et al. (8)].

(6)   Langhaar, HL: "*Steady flow in transitional length of a straight tube*", J. App. Mech., Vol. 9, 1942.

(7)   Moody, LF: "Friction factors for pipe flow", Trans. ASME, Nov. 1944 [Streeter et al. (10)].

(8)   Munson, BR, Young, DF, and Okiishi, TH:"*Fundamental of Fluid Mechanics*", John Wiley & Sons, 2002.

(9)   Nikuradse, J: "*Stromungsgesetze in rauhen rohren*", Ver. Dtsch. Ing. Forschungsh, Vol. 361, 1933 [Streeter et al. (10)].

(10) Streeter, VL, Wylie, EB, and Bedford, KW: "*Fluid Mechanics*", McGraw-Hill, 1998.

(11) Swamee PK and A.K. Jain: "*Explicit equations for pipe-flow problems*", J. Hyd. Div., Proc. ASCE, May 1976.

(12) White, FM: "*Fluid Mechanics*", McGraw-Hill, 2003.

(13) Garde, RJ, Mirajgaonkar, AG: "*Engineering Fluid Mechanics*", Scitech Publications (India) Pvt. Ltd., Chennai (India), 2003.

CHAPTER 10

Pipe Flow Systems

10.0 GENERAL

The basic principles for solving problems related to steady incompressible flow, including minor losses, in conduits were developed in Chapter 9. A large majority of practical problems relating to pipe flow systems would have turbulent flow. Therefore, Darcy-Weisbach equation for estimation of Q or head loss due to friction, and Moody's diagram (or relevant equations) for the estimation of f would often be used.

10.1 HYDRAULIC AND ENERGY GRADE LINES FOR A PIPELINE

If one plots values of $\{(p/\rho g)+(U^2/2g)\}$ as vertical ordinate above the centreline of a pipe in which a fluid is flowing, the resulting plot is known as total energy (or head) line (or energy grade line), Art. 5.4, and its elevation, at any section of pipe, with respect to a chosen datum (horizontal line) would represent the total head of the flowing fluid with respect to the datum. For a reservoir with still water surface, the total head line coincides with the free surface. In case of a flow in a pipe of uniform diameter, the energy dissipation is due to friction only and is uniform along the pipeline. Therefore, the total head line falls uniformly in the direction of flow. When there is concentrated dissipation of energy (as at an abrupt transition or valve etc.) or addition of mechanical energy to the fluid (say, by means of a pump), the total head line would have an abrupt step downward or upward, respectively. While drawing total head line, it is always assumed that (i) minor losses are concentrated at one point, and (ii) the flow is fully developed. Although not essential, but it is usually assumed that the energy correction factor equals unity.

Like total energy (or head) line, one can draw *pressure (or piezometric) line* or *hydraulic grade line* by plotting values of $p/\rho g$ as vertical ordinate above the centreline of a pipe in which a fluid is flowing. Obviously, the pressure line would be $U^2/2g$ below the total head line. If pressure line, at any section of the pipeline, coincides with the centreline of the pipe, the pressure there is zero (i.e., atmospheric).

A pipe that rises above its pressure (or hydraulic grade) line is known as siphon (Art. 10.2). The pressure in this part of pipeline is, therefore, negative gauge pressure. Under no circumstances, the pressure can fall below − 100 kPa (if the atmospheric pressure is 100 kPa), i.e., the perfect vaccum condition at which the flow would stop. In reality, the flow of a liquid would stop much before the pressure fell to absolute zero value (i.e., vacuum) as either the dissolved air or other gases would get released from the liquid or vapourization of the liquid would begin causing the released air/gases/ vapours to collect in the highest region of siphon to form an air-lock.

Figure 10.1.1 shows a typical pattern of hydraulic grade line (HGL) and total energy line (TEL) for a pipeline with a pump and siphon.

The *hydraulic gradient* is the slope of the hydraulic gradient line with respect to the axis of the pipeline, i.e., $d[z + (p/\rho g)]/dL$. Similarly, the *energy gradient* is the slope of the total energy line with respect to the axis of the pipeline, i.e., $d[z + (p/\rho g) + (U^2/2g)]/dL$.

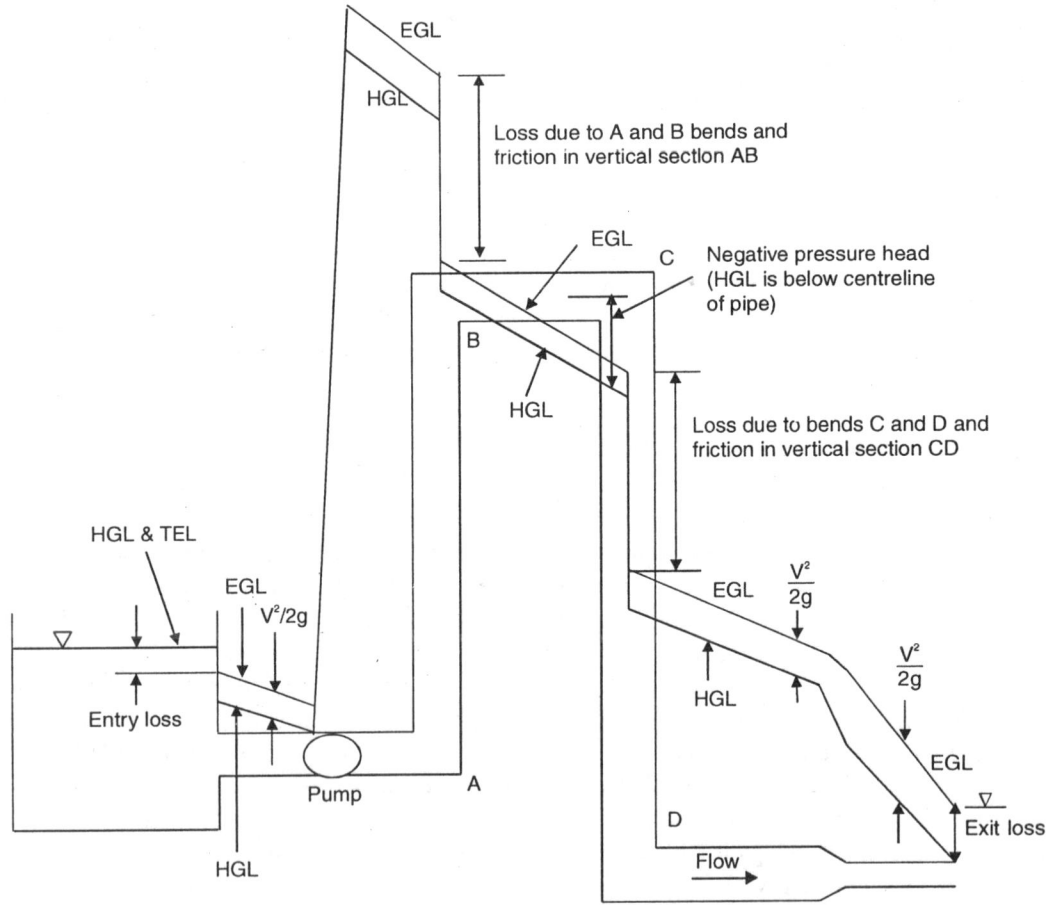

Fig. 10.1.1 Hydraulic and energy grade lines for a pipeline system with pump and siphon (3)

In case of long pipelines, minor losses may be less than 5 percent of the friction losses in the pipeline. One may, therefore, consider the minor losses negligible or, alternatively, these may be in-cluded by adding their corresponding equivalent lengths to the actual length of the pipeline and, then, considering only friction loss.

10.2 SIPHON

Siphon is a closed conduit, Fig. 10.2.1, that lifts the liquid to an elevation higher than the free surface in the reservoir feeding the conduit and, then, discharges it at a lower elevation. Assuming that the fluid flowing in siphon occupies the whole space (i.e., siphon flows full), one may apply energy equation between locations 1 and 2 to obtain

$$H = \frac{U^2}{2g} + K_L \frac{U^2}{2g} + f \frac{L}{D} \frac{U^2}{2g}$$

or

$$H = \frac{U^2}{2g}\left(1 + K_L + f \frac{L}{D}\right) \qquad \qquad ...(10.2.1)$$

Here, K_L is the sum of all relevant minor loss coefficients. For known discharge or velocity in a given pipeline, one can solve Eq. (10.2.1) directly to obtain the value of H. However, one would need trial solution for obtaining the velocity U for known H, since f depends on U.

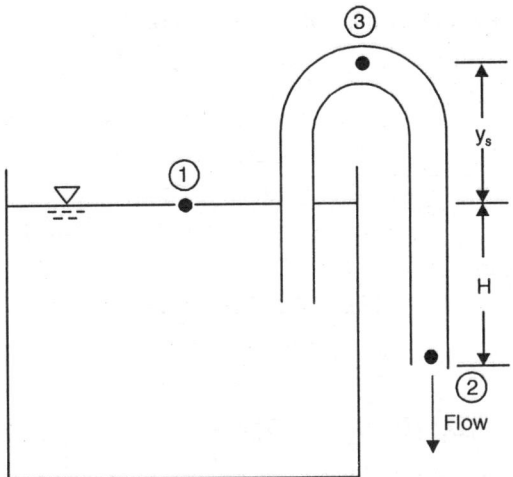

Fig. 10.2.1. Siphon

Similarly, on applying energy equation between locations 1 and 3, one gets

$$0 = \frac{U^2}{2g} + \frac{p_3}{\rho g} + y_s + K'_L \frac{U^2}{2g} + f \frac{L'}{D} \frac{U^2}{2g} \qquad \qquad ...(10.2.2)$$

in which K'_L is the sum of all minor loss coefficients for the pipeline 1-3, and L' is the length of the pipeline 1-3. Equation (10.2.2), when solved for pressure at 3, gives

$$\frac{p_3}{\rho g} = -y_s - \frac{U^2}{2g}\left(1 + K'_L + f \frac{L'}{D}\right) \qquad \qquad ...(10.2.3)$$

Equation (10.2.3) shows that the pressure at the summit (section 3) of the siphon is always negative and it decreases with increase in y_s as well as $U^2/2g$. Obviously, for values of p_3 less than vapour pressure of the flowing liquid, Eq. (10.2.3) is not valid because the vapourization of the liquid could make the flowing liquid compressible and the assumption of incompressibility used in deriving the energy equation is invalidated.

In reality, a siphon stops working or does not work satisfactorily when the pressure intensity at the summit is close to the vapour pressure of the flowing liquid. Large siphons, required to work continuously, therefore, have vacuum pumps to remove the gases/vapours that collect at the summit.

10.3 MULTIPLE PIPE SYSTEMS

In many practical applications, such as water supply or fire protection systems, a hydraulic engineer has to deal with multiple pipe systems in different configurations. The governing principles for the flow in such multiple pipe systems are the same as those for the single pipe systems discussed so far in this chapter. However, the number of unknowns in multiple pipe systems are relatively more and, hence, such systems require additional rules for analysis.

10.3.1 Pipes in Series

When two (or more) pipes of different sizes or roughnesses are so connected, Fig. 10.3.1, that every particle of the flowing fluid passes through each pipe of the pipe system, the pipes are said to be connected in series. Since flow rate is the same in all pipes,

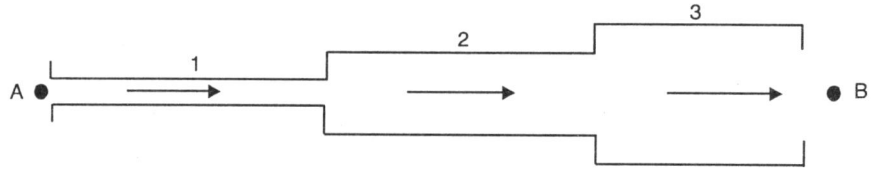

Fig. 10.3.1. Pipes in series

$$Q_1 = Q_2 = Q_3 = Q \qquad \qquad ...(10.3.1a)$$

or
$$U_1 D_1^2 = U_2 D_2^2 = U_3 D_3^2 = Q \qquad \qquad ...(10.3.1b)$$

when pipes are circular. Further, the total head loss through the system equals the sum of the head loss in each pipe of the pipe system. That is,

$$\Delta h_{A-B} = \Delta h_1 + \Delta h_2 + \Delta h_3 \qquad \qquad ...(10.3.2a)$$

$$= \frac{U_1^2}{2g}(f_1 \frac{L_1}{D_1} + K_{L1}) + \frac{U_2^2}{2g}(f_2 \frac{L_2}{D_2} + K_{L2}) + \frac{U_3^2}{2g}(f_3 \frac{L_3}{D_3} + K_{L3}) \qquad ...(10.3.2b)$$

For a typical system of pipe in series, Fig. 10.3.2, one can obtain the following expression by applying energy equation between A and B,

$$H + 0 + 0 = 0 + 0 + 0 + K_{Le} \frac{U_1^2}{2g} + f_1 \frac{L_1}{D_1} \frac{U_1^2}{2g} + \frac{(U_1 - U_2)^2}{2g} + f_2 \frac{L_2}{D_2} \frac{U_2^2}{2g} + \frac{U_2^2}{2g} \qquad ...(10.3.3)$$

in which non-zero terms on the right hand side of Eq. (10.3.3) are, respectively, entry loss, friction loss in pipe 1, sudden expansion loss, friction loss in pipe 2, and exit loss from pipe 2. Substituting the value of U_2 from the continuity equation

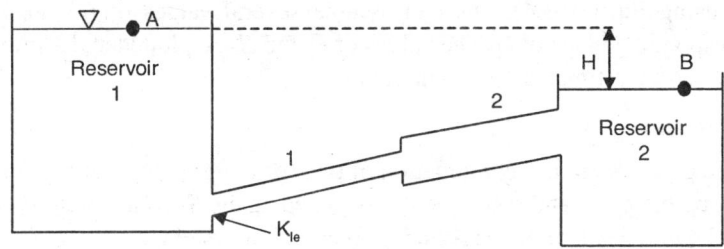

Fig. 10.3.2. Pipes connected in series

$$U_1 D_1^2 = U_2 D_2^2$$

i.e.,

$$U_2 = U_1 \frac{D_1^2}{D_2^2}$$

in Eq. (10.3.3), one gets

$$H = \frac{U_1^2}{2g} \left\{ K_{Le} + f_1 \frac{L_1}{D_1} + \left[1 - \left(\frac{D_1}{D_2} \right)^2 \right] + f_2 \frac{L_2}{D_2} \left(\frac{D_1}{D_2} \right)^4 + \left(\frac{D_1}{D_2} \right)^4 \right\} \qquad ...(10.3.4a)$$

or

$$H = \frac{U_1^2}{2g} \left\{ C + C_1 f_1 + C_2 f_2 \right\} \qquad ...(10.3.4b)$$

in which

$$C = K_{Le} + \left[1 - \left(\frac{D_1}{D_2} \right)^2 \right] + \left(\frac{D_1}{D_2} \right)^4$$

$$C_1 = \frac{L_1}{D_1}$$

and

$$C_2 = \frac{L_2}{D_2} \left(\frac{D_1}{D_2} \right)^4$$

The values of C, C_1, and C_2 are determinable from the known characteristics of pipes.

For known discharge flowing in a given pipe system, H can be directly computed as R_e and, hence, f are known quantities. However, with H known and discharge unknown, friction factors too are unknown. For such problems, one has to resort to trial solution by assuming friction factors (assumed equal for all pipes) and obtaining a trial U_1 and Q. Therefore, trial R_{e1} and R_{e2} are determined, and f_1 and f_2 are obtained from Moody's chart or relevant equation. With these values, one can obtain improved trial values for f_1 and f_2. The procedure is repeated till the trial solution converges. Since friction factor f varies slightly with R_e, the trial solution converges rapidly. This solution procedure is also applicable to a pipe system with more than two pipes in series.

Alternatively, using Eq. (10.3.4), one can assume several values of Q and, for each of these, determine the corresponding value of H. The values of Q and H are plotted and a smooth curve is fitted. From this curve, Q can be estimated for given value of H.

10.3.2 Pipes in parallel

The second multiple pipe system is a combination of two (or more) pipes that are connected so that the flow is divided among the pipes and subsequently is joined again. Such a system is called parallel pipe system, Fig. (10.3.3), that has two nodes A and B. For such a system,

$$\Delta h_{A-B} = \Delta h_1 = \Delta h_2 = \Delta h_3 \qquad \qquad ...(10.3.5)$$

and
$$Q = Q_1 + Q_2 + Q_3 \qquad \qquad ...(10.3.6)$$

In analyzing flows in parallel pipe systems, it is often assumed that minor losses are either negligible or added into the lengths of each pipe using the concept of equivalent length.

For known head loss Δh_{A-B}, the estimation of Q requires simple pipe solution for each of the pipes in the system. The discharges in each of the pipes are, then, added together to obtain total discharge.

The second type of problem is determination of distribution of flow (i.e., Q_1, Q_2 etc.) and the head loss Δh, when the total discharge Q is known. This is relatively complex and can be solved using the following procedure:

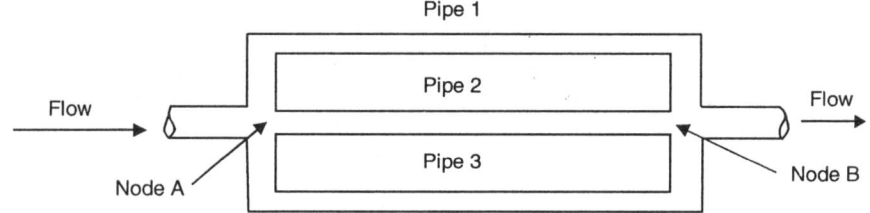

Fig. 10.3.3. Parallel pipe flow system

1. Assume a discharge Q_1' through pipe 1.
2. Obtain $\Delta h_1'$ for the assumed Q_1'.
3. Using $\Delta h_1' = \Delta h_2' = \Delta h_3'$ etc., calculate Q_2', Q_3' etc.
4. Assuming that the given discharge Q is divided among the pipes in the same proportions as Q_1', Q_2', Q_3' etc., the discharges, Q_1, Q_2, Q_3 etc. can be calculated as follows:

$$Q_1 = Q \frac{Q_1'}{\sum Q'} \qquad \qquad ...(10.3.7a)$$

$$Q_2 = Q \frac{Q_2'}{\sum Q'} \qquad \qquad ...(10.3.7b)$$

$$Q_3 = Q \frac{Q_3'}{\sum Q'} \qquad \qquad ...(10.3.7c)$$

etc. Here, $\sum Q' = Q_1' + Q_2' + Q_3'$.

5. Compute $\Delta h_1, \Delta h_2, \Delta h_3$ etc. for the computed values of Q_1, Q_2, Q_3 etc. and check that

$\Delta h_1 = \Delta h_2 = \Delta h_3$ etc.

This method of estimating the distribution of flow yields solution within a few percent. Alternatively, one can write, for the pipe system of Fig. 10.3.3, the following equations:

$$\frac{8 f_1 L_1 Q_1^2}{\pi^2 g D_1^5} - \frac{8 f_2 L_2 Q_2^2}{\pi^2 g D_2^5} = 0 \qquad\qquad ...(10.3.8)$$

$$\frac{8 f_2 L_2 Q_2^2}{\pi^2 g D_2^5} - \frac{8 f_3 L_3 Q_3^2}{\pi^2 g D_3^5} = 0 \qquad\qquad ...(10.3.9)$$

and $$Q - Q_1 - Q_2 - Q_3 = 0 \qquad\qquad ...(10.3.10)$$

Equations (10.3.8 – 10.3.10) can be solved suitably to obtain Q_1, Q_2, and Q_3 and, thereafter, Δh_1 or Δh_2 or Δh_3. The solution procedure would be iterative if f depends on R_e and is not constant (as in fully rough turbulent flow when f depends on k/D alone).

10.3.3 Branching Pipes (Three Reservoir Network)

A simple branching pipe system, Fig. 10.3.4, can be represented by three reservoirs connected together through pipes 1, 2, and 3 meeting at a junction, say J. Besides meeting the continuity requirement at the junction J, the pressures in each pipe must change so as to yield the same static pressure at the junction J. Reservoir 1 is at the highest elevation. Hence, flow in pipe 1 will always be towards the junction. Similarly, the flow in pipe 3 would always be towards the reservoir 3 as this reservoir is at the lowest elevation. The direction of flow in pipe 2 would depend upon the relative elevation of the HGL at the junction J and the water level in the reservoir 2. Therefore, the continuity requirement at J can be expressed as

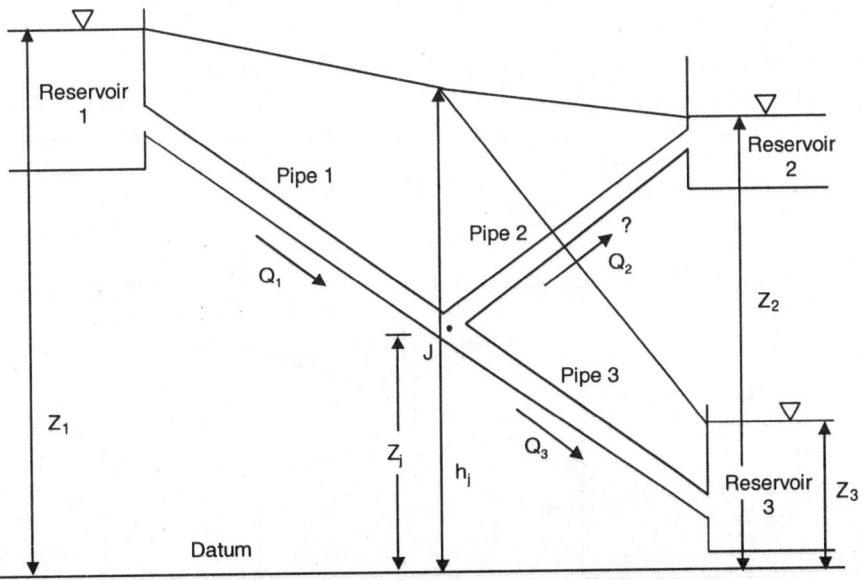

Fig. 10.3.4. Three reservoir connections

$$Q_1 \pm Q_2 - Q_3 = 0 \qquad \text{...(10.3.11)}$$

If the HGL at the junction has an elevation of h_j, then

$$h_j = z_j + \frac{p_j}{\rho g} \qquad \text{...(10.3.12)}$$

where p_j (gauge pressure) is the static pressure at J. Since gauge pressures at water levels in reservoirs 1, 2, and 3 are all zero, the head loss Δh in each pipe must be as follows:

$$\Delta h_1 = z_1 - h_j = \frac{8 f_1 L_1 Q_1^2}{\pi^2 g D_1^5} \qquad \text{...(10.3.13)}$$

$$\Delta h_2 = \pm (z_2 - h_j) = \frac{8 f_2 L_2 Q_2^2}{\pi^2 g D_2^5} \qquad \text{...(10.3.14)}$$

$$\Delta h_3 = h_j - z_3 = \frac{8 f_3 L_3 Q_3^2}{\pi^2 g D_3^5} \qquad \text{...(10.3.15)}$$

In analyzing flows in branching pipes, like the parallel pipe systems, minor losses are either considered negligible or accounted for by adding equivalent pipe length to the actual pipe length. Solution procedure for branching pipes too is an iterative one and involves assuming h_j (and, hence, the direction of flow in pipe 2) and solving Eqs. (10.3.13 – 10.3.15) for Q_1, Q_2, and Q_3 and checking the accuracy of the solution using Eq. (10.3.11) with the assumed flow direction. The procedure is repeated till Eq. (10.3.11) is satisfied.

To reduce the number of trials, one may first assume $h_j = z_2$ and, hence, $Q_2 = 0$ and compute Q_1 and Q_3. If $Q_1 > Q_3$, the flow in pipe 2 must be towards the reservoir which means that h_j should be higher than z_2 and $Q_1 - (Q_2 + Q_3) = 0$. On the other hand, if $Q_1 < Q_3$, h_j should be lower than z_2 and $(Q_1 + Q_2) - Q_3 = 0$.

One can also plot values of algebraic sum of computed Q's against the corresponding assumed values of h_j and fit a smooth curve to these points. Intersection of this curve with h_j axis would yield the desired value of h_j for which the algebraic ΣQ is zero. For this value of h_j, the values of Q_1, Q_2, and Q_3 can now be calculated using Eqs. (10.3.13 – 10.3.15).

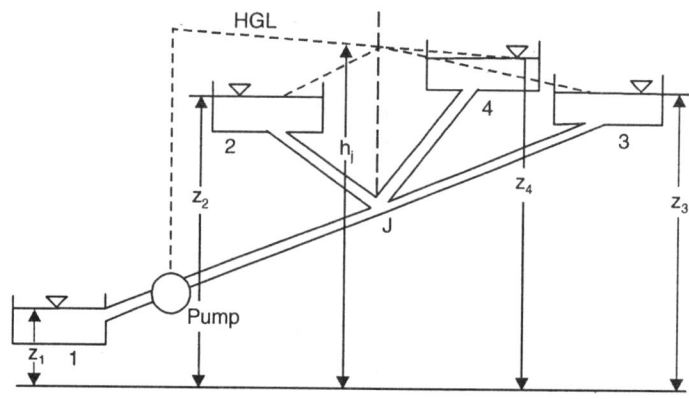

Fig. 10.3.5. Pumping from one reservoir to three other reservoirs

Another simple branching pipe system, Fig. 10.3.5, may include a pump that pumps water from one reservoir to two or more reservoirs. If the pump is assumed to run at constant speed and its head depends on discharge, the following iterative procedure can be adopted to compute the discharges in pipes connected to reservoirs (3):

1. Assume a suitable value of discharge through the pump.
2. Compute the pressure at the upstream (i.e., suction side) of the pump.
3. For the assumed value of discharge through pump, determine the head produced by the pump. One has to use characteristic curves of the pump for this purpose. Add this pump head to the suction pressure head (step 2).
4. Compute the head loss in the pipe between the pump and the junction J and thus, determine the pressure head at J (corresponding to the assumed value of discharge through the pump).
5. For the computed pressure head at J, calculate the discharges in pipes connected to reservoirs 2, 3, and 4.
6. Verify if the assumed pump discharge equals the net outflow discharge from J. If so, the problem is solved. Otherwise, repeat the procedure with another suitable value of discharge through the pump.

10.3.4 Pipe Networks

The ultimate example of multi-pipe system is the pipe network, Fig. 10.3.6, that are used for city water supply distribution systems. Once again, the basic principles of continuity and uniqueness of pressure head at any point form the basis of analyzing a pipe network. In fact, the consequence of the uniqueness of pressure head at any point (or junction) is that the net (i.e., algebraic sum with due regard to flow

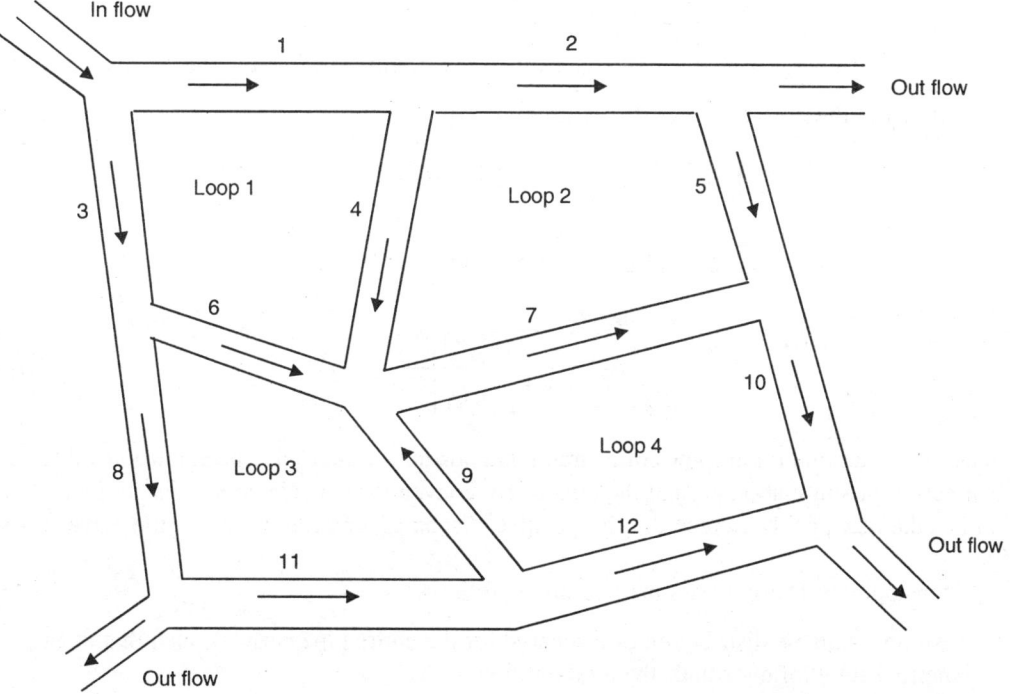

Fig. 10.3.6. Line sketch for a pipe network

direction) head loss round any closed loop in any network must be zero. Ordinary iterative procedure would be very time-consuming as flow directions in many of the pipes of the network may not be known in the beginning. Hardy Cross [1], in 1936, has suggested a systematic method that yields results of acceptable accuracy with relatively lesser number of iterations. Computers are being increasingly used to deal with such problems. As in other multi-pipe systems, equivalent single pipes are substituted for the sets of specified pipes that may be effectively either in series or in parallel or containing additional fittings causing additional losses. This simplifies the analysis.

Hardy Cross method expresses Darcy-Weisbach equation, Eq. (9.5.5) in the following form:

$$h_f = rQ^n \qquad\qquad ...(10.3.16)$$

in which r depends on pipe characteristics (i.e., diameter, length, surface roughness) and also fluid and flow characteristics. Therefore, the value of r is a suitable constant for each pipe. The value of n is around 2.0 since the flow is turbulent.

Hardy Cross method begins with assumed flow distribution that satisfies the continuity requirement at each junction, but, may not satisfy the requirement of net head loss round all the closed loops of the pipe network to be zero. Therefore, the assumed discharges Q_0 would be suitably corrected for the subsequent iteration. The corrective term ΔQ is obtained as follows:

For any pipe of a circuit such as ABDF, the corrected discharge Q is

$$Q = Q_0 + \Delta Q \qquad\qquad ...(10.3.17)$$

Hence, for every pipe in the loop

$$h_f = r(Q_0 + \Delta Q)^n \qquad\qquad ...(10.3.18)$$

$$h_f = r(Q_0^n + nQ_0^{n-1}\Delta Q + ...) \qquad\qquad ...(10.3.19)$$

If the corrective discharge ΔQ is small, then all terms of the series after the second term can be neglected. If ΔQ is the correct value of correction, $\sum h_f = 0$.

Hence,

$$\sum h_f = \sum rQ_0^n + \sum nrQ_0^{n-1}\Delta Q = 0$$

or

$$\sum rQ_0|Q_0|^{n-1} + \Delta Q \sum nr|Q_0|^{n-1} = 0$$

∴

$$\Delta Q = -\frac{\sum rQ_0|Q_0|^{n-1}}{\sum nr|Q_0|^{n-1}} \qquad\qquad ...(10.3.20)$$

Note that ΔQ is the same for all pipes of a circuit and absolute values have been introduced to account for the direction of summation around the circuit. By convention, the clockwise discharge and, hence, loss too is taken as positive. Accordingly, positive discharge correction ΔQ would have clockwise direction.

Steps for applying Hardy Cross method are as follows:

1. Assume a suitable distribution of discharge for the entire pipe network such that at each of the junctions total inflow equals the total outflow.

2. For an elementary circuit, compute the head loss $h_f (= rQ_0^n)$ in each of its pipes and obtain the algebraic sum of head loss for the circuit, i.e., $\sum rQ_0 |Q_0|^{n-1}$. Also obtain arithmetic sum $\sum nr |Q_0|^{n-1}$ for the circuit. Calculate the value of ΔQ i.e., the ratio of $\sum rQ_0 |Q_0|^{n-1}$ and $\sum nr |Q_0|^{n-1}$. The value of ΔQ is algebraically added to each of the pipe discharges in the circuit.

3. Repeat step 2 for all the remaining elementary circuits of the network.
4. Repeat steps 2 and 3 as many times as required until the corrections (ΔQ's) are negligible.

10.4 COMPRESSIBLE FLOW

The analysis of flow presented so far is applicable to the flow of incompressible fluids whose mass density remains invariant. However, all real fluids are compressible to some extent. Hence, the flow of a real fluid is truly incompressible only when the pressure is invariant (i.e., uniform) throughout the flow field. Such a flow would rarely exist. However, there may exist such flows in which $\delta\rho/\rho$ is much less than unity and in which case the flow may be approximated as incompressible even if the flowing fluid is a highly compressible gas.

Consider steady flow of an inviscid fluid. The continuity equation, Eq. (3.8.4) can be written in vectorial form as

$$\nabla.\left(\rho\bar{V}\right)=0 \qquad\qquad ...(10.4.1)$$

If one considers coordinate s along a streamline, Eq. (10.4.1) can be written as

$$\frac{\partial}{\partial s}\left(\rho V\right)=0 \qquad\qquad ...(10.4.2)$$

in which V is the velocity of flow in s-direction. Hence,

$$\rho\frac{\partial V}{\partial s}+V\frac{\partial \rho}{\partial s}=0 \qquad\qquad ...(10.4.3)$$

For the flow to be incompressible, the change in mass density should be negligibly small. In other words,

$$\left|V\frac{\partial\rho}{\partial s}\right|<<\left|\rho\frac{\partial V}{\partial s}\right|$$

or

$$\left|\frac{1}{\rho}\frac{\partial\rho}{\partial s}\right|<<\left|\frac{1}{V}\frac{\partial V}{\partial s}\right| \qquad\qquad ...(10.4.4)$$

Neglecting the gravitational term (i.e., the body force term) from Eq. (4.5.7),

$$V\frac{\partial V}{\partial s}=-\frac{1}{\rho}\frac{\partial p}{\partial s}$$

$$\frac{1}{V}\frac{\partial V}{\partial s}=-\frac{1}{\rho V^2}\frac{\partial p}{\partial s} \qquad\qquad ...(10.4.5)$$

Combining Eqs. (10.4.4) and (10.5.5) one gets

$$\left|\frac{1}{\rho}\frac{\partial\rho}{\partial s}\right| << \left|\frac{1}{\rho V^2}\frac{\partial p}{\partial s}\right|$$

Alternatively, this means that

$$\left|V^2\right| << \left|\frac{\partial p}{\partial\rho}\right| \qquad \qquad ...(10.4.6)$$

The changes in density and pressure of a fluid are related through its bulk modulus of elasticity K, Eq. (1.4.12), as

$$K = \rho\frac{dp}{d\rho} \qquad \qquad ...(10.4.7)$$

Combining Eqs. (10.4.6) and (10.4.7), one gets

$$V^2 << \left|\frac{K}{\rho}\right|$$

or

$$\frac{V}{\sqrt{K/\rho}} << 1$$

The term $\sqrt{K/\rho}$ is the velocity of sound a in the fluid under consideration. Therefore, the condition of incompressibility is

$$\frac{V}{a} << 1 \qquad \qquad ...(10.4.8)$$

The ratio V/a is known as Mach number M_a (Art. 6.4.5). Velocity of sound in air is about 330 m/s. This means that a velocity of air as high as 100 m/s would give $V/a \cong 0.3$ which is much less than 1. And, hence, flow of air with velocity as high as 100 m/s can be considered incompressible. If M_a exceeds 0.4, the effects of compressibility must be considered.

10.5 WATER HAMMER IN PIPES

Barring the derivation of few equations such as Eq. (5.5.10), the analysis of flow presented so far is applicable to steady flow in which fluid and flow characteristics (such as density, velocity, pressure etc.) do not change with time. Most of the flows that one counters are of turbulent kind and are, therefore, unsteady in the strictest sense of the term. However, if the time-averaged values of various flow parameters do not vary with time, the flow can be considered steady. Unsteadiness adds to the difficulties of solving problems of unsteady flow. Nevertheless, there are some unsteady flow situations that are amenable to analytical solution. One such problem is the continuous emptying (or filling) of a reservoir (with slow changes in reservoir level and, hence, velocity of flow) for which Eq. (5.5.10) was derived. For such flow problems, the forces causing the temporal acceleration are negligible compared to other forces that affect the flow. Other category of unsteady motion includes such problems (like flow in reciprocating hydraulic machines and in hydraulic and pneumatic servo-mechanisms) in which the flow changes rapidly and the forces producing temporal accelerations are important. Oscillatory motions (such as tidal movements) are also classified as unsteady. In yet another type of unsteady flow, the flow parameters change so rapidly that the elastic forces become important. One classical example of this type of unsteady flow is *water hammer* in pipelines that has been briefly dealt with in this section.

When an incompressible fluid flowing with certain velocity in a rigid pipe is brought to rest by closure of a valve at the downstream end, then all the fluid particles in the pipe would decelerate together as the fluid is incompressible and the pipe is rigid. Further, from Newton's second law of motion, the faster deceleration gives rise to greater force and, hence, with instantaneous stoppage of a moving fluid, the force would be infinite. However, neither the fluid (even if it is liquid) is entirely incompressible nor the walls of the pipe are perfectly rigid. Hence, an instantaneous (i.e., very rapid) closure of the valve would not cause instantaneous stoppage of the entire fluid. Instead, only the fluid particles in contact with the valve would be stopped at once and all other fluid particles would stop later at a time that would increase with the distance from the valve. Therefore, the pressure rise (due to the stoppage of fluid) would also travel with finite speed. In other words, the closing of the valve first stops only the fuid that is in direct contact with it; thereafter, the fluid particles in contact with the halted fluid particles come to rest and so on, Fig. 10.5.1. That is, the entire fluid of the pipe comes to rest by a "message" passed along the pipe from the halted fluid particles to the moving fluid particles (in contact with the halted ones) that they too must stop (2). The travel discontinuity is called *pressure* (or *elastic*) *wave* that acts as a message. It must be noted here that pressure waves travel through the material of the pipe walls too. However, the effect of this travel on pressure changes is negligible and, hence, ignored in the analysis.

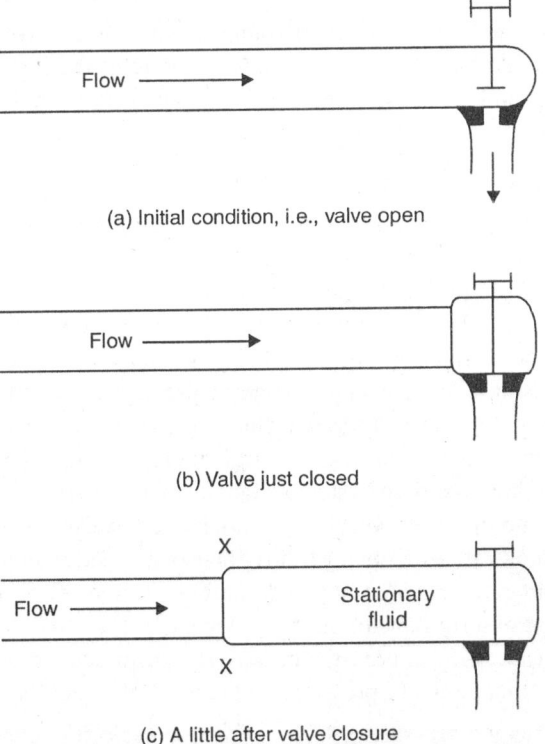

(a) Initial condition, i.e., valve open

(b) Valve just closed

(c) A little after valve closure

Fig. 10.5.1. Closing a valve in a pipeline (2)

Thus, one can appreciate that one can neglect the compressibility of the fluid (i.e., all the fluid particles are assumed to change velocity simultaneously) only if the time of travel of pressure wave is negligibly small in comparison with the time during which the velocity changes.

The behaviour of a suddenly stopped fluid is similar to that of a train, having loosely-coupled wagons, that has been suddenly stopped (2). The wagon immediately behind the engine compresses the buffer spring between itself and the engine. The force in the buffer spring, thus, increases and stops the wagon immediately behind the engine. In a similar manner, the second wagon is stopped and so on. The compression of the buffer springs is analogous to the compression of the fluid in the pipe. Obviously, after a wagon has stopped, the force in the buffer spring at its two ends must be equal, otherwise there would be a net force acting on the wagon and the wagon would start moving. Likewise, a pressure wave in a fluid changes the pressure at a particular point. However, after the wave has passed the point, the pressure at the point remains at its new value. Blast waves from an explosion too travel through atmosphere in similar manner.

Water hammer may occur in a closed pipe that is flowing full when there is a rapid change in velocity of flow caused by rapid change in the opening of a valve in the pipeline. Water hammer phenomenon is often experienced in domestic pipeline, particularly in the pipeline connected to a geyser, when a water tap is turned off very rapidly as a result of which a heavy knocking sound is heard and the pipeline vibrates. Due to rapid closure of tap, the velocity of flow decreases and pressure increases. Due to the elasticity of the water as well as that of the pipe material, the pressure changes do not occur instantaneously throughout the pipeline, but are propagated in the form of pressure wave. This phenomenon is known as *water hammer* and may occur in any fluid that flows in a pipeline. The pressure rise in a pipe as a result of sudden or rapid closure of a valve is tremendous in case of penstocks of a hydropower station and may even cause fracture of the pipe. Therefore, pipelines such as penstocks must be designed keeping in mind the likely pressure rise in the event of rapid closure of flow in the penstock due to sudden reduction in the demand of power.

10.5.1 Propagation of Pressure Wave in a Rigid Pipe

On changing the pressure at a point in a real fluid, the density of the fluid also changes – even if only marginally – and, hence, individual fluid particles undergo small changes in their position at the point under consideration. As a consequence of this, adjacent fluid particles also change their position and thus the new pressure is progressively, yet rapidly, transmitted throughout the fluid system. For incompressible fluids, however, there would not be any change in the density and, hence, no change in their position. Therefore, any disturbance, causing change in pressure, would be propagated with infinite velocity, i.e., instantaneously. For most of the real fluids, the pressure, changes may be transmitted so rapidly (compared to the time taken in causing the original change) that one can safely assume that the pressure has changed throughout the fluid system instantaneously. However, when the pressure changes are sudden or the fluid is moving with very high velocity relative to a solid body (or solid body is moving with very high velocity in a stationary fluid), the exact speed with which the pressure changes are transmitted, is of considerable significance and compressibility effects become important.

Consider the movement of a rigid piston in a rigid cylinder, Fig. 10.5.2(a), as a result of which the pressure change has been caused. This pressure change is transmitted through an abrupt discontinuity of pressure, Fig. 10.5.2(b), that is called pressure wave ahead of which is the original pressure p and behind which is the new changed pressure $p + \delta p$. The existence of the pressure difference δp across the wave (or wave front) results into an unbalanced force causing the fluid to accelerate. Likewise, other fluid ahead of the wave gets accelerated progressively. The fluid is, obviously, in unsteady state and to

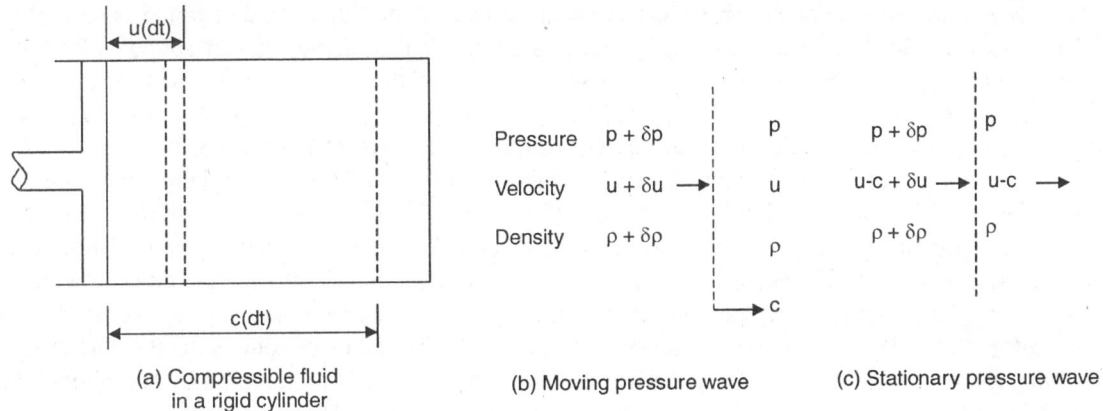

(a) Compressible fluid in a rigid cylinder

(b) Moving pressure wave

(c) Stationary pressure wave

Fig. 10.5.2. Movement of pressure wave

make the analysis simpler, a velocity c (velocity of pressure wave with respect to stationary set of coordinate axes) is superimposed in the direction opposite to that of the wave. Figure 10.5.2(c) shows the values of pressure, velocity, and mass density on the two sides of the stationary wave. For a wave cross-section of area ΔA, the continuity equation requires that

$$(\rho + \delta\rho)(u - c + \delta u)\Delta A = \rho(u - c)\Delta A$$

or

$$(u - c)\delta\rho + \delta u(\rho + \delta\rho) = 0$$

or

$$(c - u)\delta\rho = (\rho + \delta\rho)\delta u \qquad \qquad ...(10.5.1)$$

For a control volume enclosing the area of the pressure wave and neglecting the friction, the momentum equation yields

$$(p + \delta p)\Delta A - p\Delta A = \rho(u - c)\Delta A\{(u - c) - (u - c + \delta u)\} = \rho(u - c)\Delta A(-\delta u)$$

$\therefore$

$$\delta p = \rho(c - u)\delta u \qquad \qquad ...(10.5.2)$$

On eliminating δu from Eqs. (10.5.1) and (10.5.2), one gets

$$(c - u)^2 = \left(\frac{\rho + \delta\rho}{\rho}\right)\frac{\delta p}{\delta\rho} \qquad \qquad ...(10.5.3)$$

As $\delta p \to 0$, $\delta\rho \to 0$, hence, Eq. (10.5.3) becomes

$$(c - u) = \sqrt{dp/d\rho} \qquad \qquad ...(10.5.4)$$

The term $(c-u)$ represents the velocity of the pressure wave with respect to the fluid ahead of it. Thus, a small pressure change propagates at a velocity equal to $\sqrt{dp/d\rho}$ relative to the fluid in which it propagates. For any fluid, the bulk modulus of elasticity K, defined by Eq. (1.4.12) is $\rho\dfrac{dp}{d\rho}$ and, hence, Eq. (10.5.4) becomes

$$a = c - u = \sqrt{K/\rho} \qquad \qquad ...(10.5.5)$$

in which a is the velocity of the pressure wave with respect to the fluid ahead of it and is termed the *celerity* of wave. Thus, celerity of pressure wave in a fluid equals the velocity of sound in the fluid. Sound too is propagated by means of a succession of very small pressure waves. Human ear can detect the faintest sound that corresponds to a pressure fluctuation of about 3×10^{-5} Pa. Human ear can tolerate, without pain, the loudest sound that corresponds to a pressure fluctuation of about 100 Pa.

If the velocity of a fluid at a given location is less than the velocity of sound there, small pressure waves can be propagated in all directions. But, if the velocity of the fluid exceeds the local sonic velocity, the pressure wave cannot be propagated upstream. Thus, the sonic velocity ($=a$) distinguishes two different types of flow: one in which pressure wave can travel both upstream and downstream, i.e., *subsonic flow* and another in which pressure wave cannot travel upstream, i.e., *the supersonic flow*. It is, therefore, advantageous to express the velocity of the fluid in terms of sonic velocity. The ratio of the fluid velocity and sonic velocity is known as Mach number and is represented by the symbol M_a. For an entirely incompressible fluid, a would be infinite and, hence, M_a always zero.

It should be noted that the velocity of wave propagation (or transmission) does not involve movement of particles of any matter. It merely represents the speed with which a message of change in pressure is conveyed through the matter. It is for this reason that the word 'celerity' is used in place of velocity when one deals with the propagation of waves.

Further, the celerity of a pressure (or elastic) wave is dependent on the ratio of the bulk modulus of elasticity K and the mass density ρ rather than K alone. Hence, the celerity of a pressure wave in liquids and gases are nearly of the same order (for example, the velocity of sound, i.e., a pressure wave is 1414 m/s for water and 330 m/s for air) although the bulk modulus of elasticity for liquids and gases are vastly different.

As far as liquids are concerned, K and ρ – and, hence, the celerity a too – are found to be almost independent of the pressure intensity p. In case of gases, however, both E and ρ vary considerably with the change in pressure. But, the net effect of such variation on the celerity a is nil provided that the temperature does not change. For example, from Boyle's law for isothermal changes in the pressure and volume of a gas,

$$pv = \text{constant} \qquad \qquad \text{...(10.5.6)}$$

Here, p is the pressure intensity and v is the specific volume that equals $1/\rho$. Thus,

$$pdv + vdp = 0$$

or

$$-\frac{dp}{dv/v} = p$$

or

$$K = p$$

For an ideal (or perfect) gas

$$p = \rho gRT \qquad \qquad \text{...(10.5.7)}$$

in which R is the gas constant, T is the absolute temperature, and p is the absolute pressure. Thus,

$$a = \sqrt{K/\rho} = \sqrt{p/\rho} = \sqrt{gRT} \qquad \qquad \text{...(10.5.8)}$$

Likewise, for adiabatic changes in gases,

$$pv^k = \text{constant} \qquad \qquad \text{...(10.5.9)}$$

$\therefore$

$$v^k dp + pkv^{k-1}dv = 0$$

or
$$-\frac{dp}{dv/v} = kp$$

Hence,
$$K = kp$$

Therefore, for an ideal gas,

$$a = \sqrt{kgRT} \qquad \qquad ...(10.5.10)$$

Here, k is known as adiabatic index.

Equations (10.5.8) and (10.5.10) indicate that the net effect of change in pressure on the celerity of an elastic (or pressure) wave is nil provided that the temperature does not change.

10.5.2 Propagation of Pressure Wave in an Elastic Pipe

The analysis for the propagation of a pressure (or elastic) wave, as given in Art. 10.5.1, is applicable when the fluid medium is infinite or the surrounding boundaries containing the fluid are perfectly rigid. This section analyses the propagation of a pressure wave in a liquid (or gas) flowing within non-rigid (i.e., elastic) boundaries.

Consider a pipe in which the liquid flowing from left to right, is brought to rest (as in Fig. 10.5.1(c)) by a pressure wave travelling from right to left. Since the pipe is non-rigid, not only p and ρ, in the undisturbed liquid, change, respectiveiy, to $p + \delta p$ and $\rho + \delta\rho$, but area of cross-section of pipe A also changes to $A + \delta A$ for the disturbed liquid. It may be noted that δA and $\delta\rho$ are very small in comparison with δp and so is the change in head due to friction loss compared with the loss of head due to the pressure wave. In the following analysis, frictional effects are ignored and the velocity, u is, therefore, considered uniform over the cross-section of the pipe.

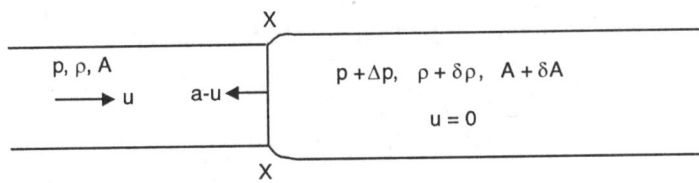

(a) Coordinate axes fixed relative to pipe

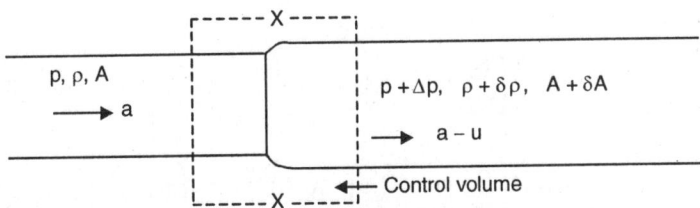

(b) Coordinate axes moving with wave

Fig. 10.5.3. Pressure wave in an elastic pipe (2)

Consider that a pressure wave, Fig. 10.5.3(a), travels upstream with a celerity a ($= c$-u) so that speed of the wave relative to the stationary pipe is a-u. Therefore, if one superposes a velocity a-u in the direction opposite to that of wave travel, the conditions would appear steady, Fig. 10.5.3(b). The continuity equation across the wave, then, gives

$$\rho A a = (\rho + \delta\rho)(A + \delta A)(a - u) \qquad \qquad ...(10.5.11)$$

$$= \rho(A + \delta A)(a - u) + \delta p(A + \delta A)(a - u)$$

$$= \rho A a + \rho a \delta A - \rho A u - \rho u \delta A + \delta\rho(A + \delta A)(a - u)$$

$$\therefore \qquad \frac{\delta\rho}{\rho} = \frac{u(A + \delta A) - a\delta A}{(A + \delta A)(a - u)} \qquad \qquad ...(10.5.12)$$

From Eqs. (1.4.12) and (10.5.12)

$$\frac{\delta\rho}{\rho} = \frac{\delta p}{K} = \frac{u(A + \delta A) - a\delta A}{(A + \delta A)(a - u)} \qquad \qquad ...(10.5.13)$$

Difference of pressure across the wave, δp causes a pressure force δpA towards the left. Applying momentum equation for the control volume shown in Fig. 10.5.3(b),

$$-A\delta p = \text{(mass flow rate) (change in velocity)}$$

$$= (\rho A a)\{(a - u) - a\}$$

$$= -\rho A a u$$

or

$$u = \frac{\delta p}{\rho a} \qquad \qquad ...(10.5.14)$$

On substituting the value of u from Eq. (10.5.14) in Eq. (10.5.13), one gets

$$\frac{\delta p}{K} = \frac{(\delta p/\rho a)(A + \delta A) - a\delta A}{(A + \delta A)\{a - (\delta p/\rho a)\}}$$

or

$$\frac{\delta p}{K} = \frac{\delta p(A + \delta A) - \rho a^2 \delta A}{(A + \delta A)(\rho a^2 - \delta p)}$$

$$\therefore \qquad \rho a^2 \left(\frac{\delta p}{K}\right)(A + \delta A) - \delta p\left(\frac{\delta p}{K}\right)(A + \delta A) + \rho a^2 \delta A - \delta p(A + \delta A) = 0$$

or

$$\rho a^2 \left(\frac{\delta p}{K}\right)\left(1 + \frac{\delta A}{A}\right) + \rho a^2 \frac{\delta A}{A} = \delta p\left(1 + \frac{\delta A}{A}\right) + \delta p\left(\frac{\delta p}{K}\right)\left(1 + \frac{\delta A}{A}\right)$$

Hence,

$$\rho a^2 = \frac{\delta p\left(1 + \dfrac{\delta A}{A}\right)\left(1 + \dfrac{\delta p}{K}\right)}{\dfrac{\delta p}{K}\left(1 + \dfrac{\delta A}{A}\right) + \dfrac{\delta A}{A}} \qquad \qquad ...(10.5.15)$$

Usually, the bulk modulus of elasticity K is very large compared to change in pressure δp. For example, K for water at moderate pressures is about 2 GPa, i.e, 2×10^9 Pa. Therefore, $\delta p/K$ is negligible in comparison with 1. That is,

$$1 + \frac{\delta p}{K} \approx 1.0$$

and likewise,

$$1 + \frac{\delta A}{A} \approx 1.0$$

Therefore, Eq. (10.5.15) reduces to

$$\rho a^2 = \frac{\delta p}{\dfrac{\delta p}{K} + \dfrac{\delta A}{A}} \qquad \qquad ...(10.5.16)$$

For a perfectly rigid pipe, $\delta A = 0$, and, hence, Eq. (10.5.16) gives

$$\rho a^2 = K$$

or

$$a = \sqrt{K/\rho} \qquad \qquad ...(10.5.17)$$

That is, the velocity with which a wave travels relative to the liquid flowing in a rigid pipe is the same as the velocity of sound in an infinite expansion of the liquid. In case of very small diameter pipes, frictional effects would reduce this value slightly.

For non-rigid pipes, however, one needs to estimate the value of $\delta A/A$ that would depend primarily on the material of the pipe and also on the freedom of movement of the pipe. For a circular pipe of diameter d and wall thickness t, an increase in pressure δp of the flowing liquid would cause hoop stress ($= \delta pd/2t$; see Eq. (2.5.7)) that tends to burst the pipe. If E represents the modulus of elasticity of the pipe material, the resulting hoop strain would be $\delta pd/2tE$ giving rise to a longitudinal strain. However, for the present problem, this strain is not of much significance as the pipe merely slides past the bulk of liquid inside it and the pipe, therefore, has no longitudinal stress. Since

$$A = \frac{\pi}{4} d^2$$

$\therefore$
$$\frac{\delta A}{A} = \frac{2\delta d}{d}$$

$$= 2 \text{ (hoop strain)}$$

$$= \frac{\delta pd}{tE}$$

Hence, Eq. (10.5.16) yields

$$\rho a^2 = \frac{\delta p}{\dfrac{\delta p}{K} + \dfrac{\delta pd}{tE}} = \frac{1}{\dfrac{1}{K} + \dfrac{d}{tE}}$$

$$a = \sqrt{\frac{1}{\rho}\left\{\frac{1}{\frac{1}{K}+\frac{d}{tE}}\right\}}$$

∴   ...(10.5.18)

or $$a = \sqrt{\frac{K'}{\rho}}$$ ...(10.5.19)

where K' is given as

$$\frac{1}{K'} = \frac{1}{K} + \frac{d}{tE}$$ ...(10.5.20)

One can regard K' as the effective bulk modulus of elasticity of the liquid flowing in a non-rigid or elastic pipe. It may be noted that the effect of transmission of elastic waves through the material of the pipe is generally negligible and has, therefore, been ignored in the analysis.

For water ($K = 2.05$ GPa $= 2.05 \times 10^9$ N/m^2) flowing in a rigid pipe, the value of a would be $= 1432$ m/s. If the pipe ($d = 100$ mm) is of steel ($t = 5$ mm) and considered elastic (E for steel $= 200$ GPa), the wave speed can be worked out as follows:

From Eq. (10.5.20),

$$K' = 6.72 \times 10^8 \text{ Pa}$$

and, hence, $$a = \sqrt{\frac{6.72 \times 10^8}{10^3}}$$

or $$a = 820 \text{ m/s}$$

Normally, the flow velocity, u in a pipe is much smaller than the value of a and hence, $c = a - u \approx a$. That is, one may consider a itself to be relative to the pipe. Since K (and, therefore, K' too) depends only slightly on pressure and temperature, it is usually treated as constant for a particular pipeline without much loss of accuracy in the computed values.

10.5.3 Magnitude of Pressure Change in a Pressure Wave

The rise in pressure caused by reducing the velocity of flow from u to zero can be estimated by using Eq. (10.5.14) that yields change in pressure

$$\delta p = \rho a u$$ ...(10.5.21)

If the velocity changes from u_1 to u_2 then the corresponding change in pressure is

$$\delta p = \rho a (u_1 - u_2)$$ ...(10.5.22)

Obviously, the pressure rises if the velocity decreases and vice versa. In terms of head, h, Eq. (10.5.22) becomes

$$\delta h = \frac{a(u_1 - u_2)}{g}$$

or $$\Delta h = -\frac{a}{g}\Delta u$$ ...(10.5.23)

Equation (10.5.23) shows that a mere reduction of 3 m/s in the velocity of water flowing in a rigid pipe ($a = 1432$ m/s) will increase head by about 440 m (i.e., about 4.32 MPa) which is too large to be neglected in the design of a pipe system.

It should be noted that the loss in kinetic energy due to reduction of velocity has been converted into strain energy that remains stored in the compressed material and gets released when the strain is removed (2).

10.5.4 Reflection of Pressure Waves

Consider a pipeline AB of length L, connected to a large reservoir, having a valve at its outlet end, Fig. 10.5.4(i). For normal flow condition, with valve open, the hydraulic gradient line CD represents the variation of piezometric pressure along the pipeline. If the valve is instantaneously closed, Fig. 10.5.4(ii), the fluid immediately upstream of the valve comes to rest and gets compressed giving rise to increased

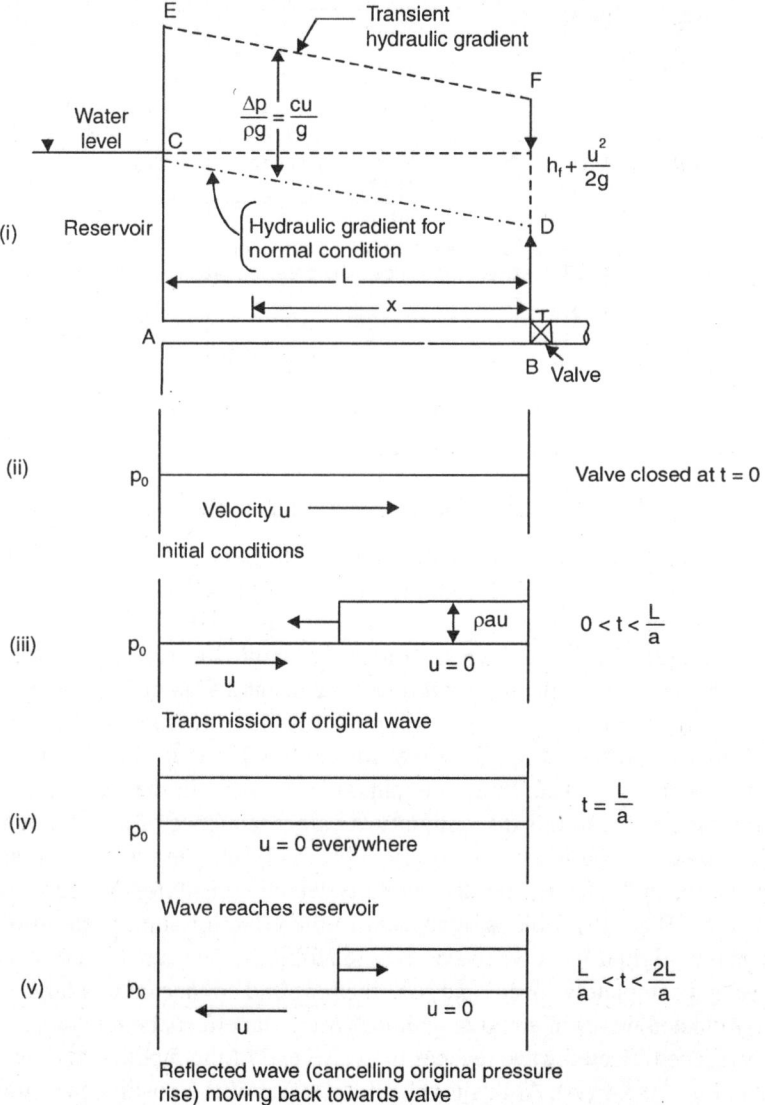

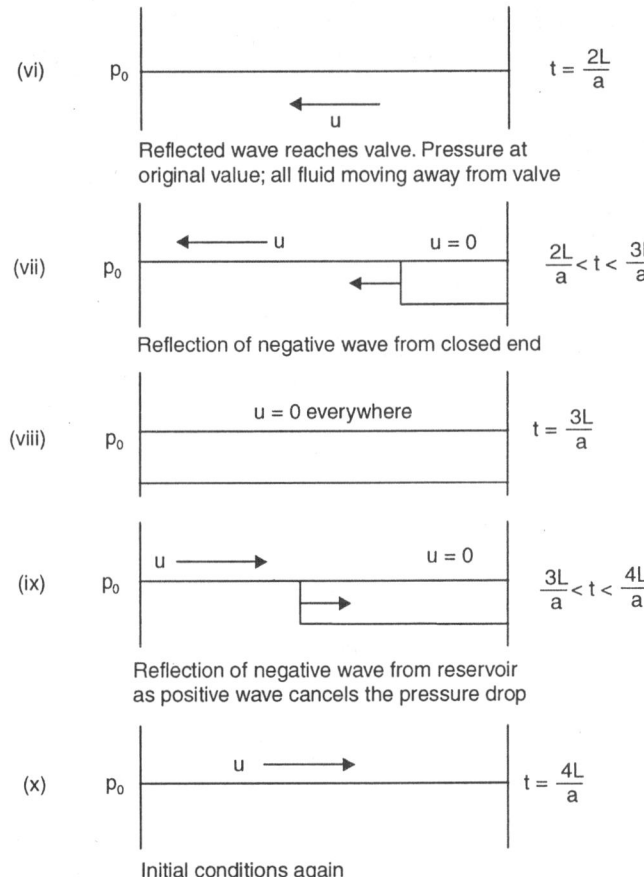

Fig. 10.5.4. Propagation of pressure wave in an elastic pipe (2)

pressure that propagates, through the fluid in the pipeline, towards the reservoir with celerity a as a positive wave, Fig. 10.5.4(iii). At time t (= L/a) since the closure of the valve, the wave reaches the reservoir, Fig. 10.5.4(iv), and the entire fluid in the pipeline has come to rest and has been compressed (i.e., the pressure has increased). Since the reservoir is very large, the velocity of flow in the reservoir is negligible and, hence, pressure in the reservoir remains constant. This means that as soon as the wave reaches the reservoir, (t = L/a), there exists a discontinuity; the pressure in the pipe (at its inlet or reservoir end) is higher (represented by E on the transient HGL) than the pressure in the reservoir represented by its water surface level. Therefore, although the fluid in the entire pipe is at rest (at t = L/a), it is not in equilibrium due to this discontinuity. As a consequence of this, the fluid starts flowing (the velocity of flow, u equals $\delta p/\rho a$ as given by Eq. (10.5.14)) into the reservoir so as to equalize the pressures at the reservoir end. Hence, the discontinuity which constitutes pressure wave now travels towards the valve, Fig. 10.5.4 (v). This is equivalent to the effect of superposition of a 'negative' or 'unloading' wave on the original 'positive' wave so as to nullify it. One can, therefore, conclude that the reflection of a wave at a completely 'open' end (i.e., the pipe end connected to a large reservoir) causes another wave (i.e., reflected wave) of equal magnitude (frictional effects being ignored) but opposite in sign (Fig. 10.5.4 (v)). The reflected wave reaches the valve end of the pipeline at time = $2L/a$ since the closure of the valve, Fig. 10.5.4 (vi). At this moment, the effect of the positive pressure wave has been nullified in the entire pipeline and, therefore, everywhere in the pipeline, the pressure is the same as its original value at $t = 0$, and the fluid density and pipe diameter are also at their original values. But, the fluid movement continues towards the reservoir from the valve even though the pressure gradient favours the flow from reservoir towards the valve. This reverse (towards the reservoir) movement of the fluid decompresses the fluid immediately next to the valve resulting in fall of pressure, Fig. 10.5.4 (vii). In the absence of frictional effects, the magnitude of this change of pressure is, again, the same as the magnitude of the pressure changes because the corresponding change in the velocity of fluid flow is the same u. Thus, a negative wave now travels towards the reservoir. Fig. 10.5.4(vii). At $t = 3L/a$, the negative wave reaches the reservoir (i.e., the open end of the pipeline) where the pressure is now higher than that at the reservoir end of the pipeline and the velocity of fluid flow is zero everywhere, Fig. 10.5.4 (viii). Thus, again, there is discontinuity or unbalanced state and fluid starts moving from the reservoir towards the valve through the pipeline and a positive wave travels towards the valve to nullify the negative wave, Fig. 10.5.4(ix). This positive wave reaches the valve at time $t = 4L/a$ since the closure of the valve, Fig. 10.5.4(x). Now, the conditions in the pipeline are the same as those which existed at $t = 0$ when the valve was closed; all the fluid in the pipeline moving towards the valve with velocity u, and the pressure too is back to its original value represented by the hydraulic gradient line CD, Fig. 10.5.4(i). The complete cycle of propagation of pressure wave and its reflection at open and closed ends is, therefore, repeated and, in the absence of friction, would be repeated indefinitely. Each cycle would complete in a time duration of $4L/a$. In reality, however, energy is continuously, although gradually, dissipated due to friction. Therefore, the magnitude (or intensity) of wave diminishes continuously and ultimately it vanishes.

If the valve is closed gradually in a time that is longer than $2L/a$, the rise of pressure would be different (in fact, it would be lesser) than ρau, and its estimation would be rather too complex. If one

assumes that the pressure rise is proportional to $\left(\dfrac{2L}{c}\right)\left(\dfrac{1}{t_c}\right)$ then,

$$\delta p = \left(\frac{2L}{a}\right)\left(\frac{1}{t_c}\right)\rho au = \frac{2\rho Lu}{t_c}$$

Here, t_c is the time of closure and is less than $2L/a$.

EXAMPLES

E10.1: Draw hydraulic and energy grade lines for the pipeline (diameter = 150 mm; length = 60 m; friction factor = 0.02) having a valve ($K_L = 10$) as shown in Fig. E10.1. Assume loss coefficient for nozzle (diameter = 75mm) as 0.1.

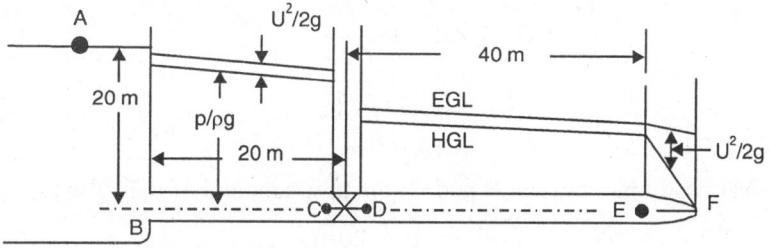

Fig. E10.1

Solution: Choosing the centreline of the pipeline as the datum, and writing energy equation between the reservoir (say, point A) and the exit F,

$$0+0+20=0+\frac{U_F^2}{2g}+0+0.5\frac{U^2}{2g}+0.02\left(\frac{60}{0.15}\right)\frac{U^2}{2g}+10\frac{U^2}{2g}+0.1\frac{U_F^2}{2g}$$

From the continuity equation,

$$U_F\left(\frac{\pi}{4}\right)(0.075)^2=U\left(\frac{\pi}{4}\right)(0.150)^2$$

i.e., $\qquad U_F = 4U$

Therefore, the energy equation reduces to

$$20=\frac{U^2}{2g}\left[16+0.5+8+10+16(0.1)\right]$$

$$=\frac{U^2}{2g}(36.1)$$

Therefore, $\qquad \dfrac{U^2}{2g}=0.554\,\text{m}$

Elevation of HGL (as well as TEL since the velocity head in the reservoir is zero) in the reservoir coincides with the reservoir level, i.e., its elevation is 20 m above the centreline of the pipe.

The elevation of TEL at B (immediately after the entrance)

$$(\text{TEL})_B = 20 - \text{entry loss at B}$$

$$= 20 - 0.5\frac{U^2}{2g}$$

$$= 20 - 0.5\,(0.554)$$

$$= 20 - 0.277$$

$$= 19.723 \text{ m}$$

and the elevation of HGL at B (immediately after the entrance)

$$(HGL)_B = 19.723 - \frac{U^2}{2g}$$

$$= 19.723 - 0.554$$

$$= 19.169 \text{ m}$$

Frictional head loss between B and C (the upstream end of the valve)

$$= 0.02\left(\frac{20}{0.15}\right)(0.554)= 1.477\,\text{m}$$

Therefore, the elevation of TEL at C, $(TEL)_C = (TEL)_B - 1.477$ m

$$= 19.723 - 1.477$$

$$= 18.246 \text{ m}$$

and, the elevation of HGL at C, $(HGL)_C = (TEL)_C - \dfrac{U^2}{2g}$

$$= 18.246 - 0.554$$

$$= 17.692 \text{ m}$$

Likewise, $\qquad (TEL)_D = (TEL)_C - 10\dfrac{U^2}{2g}$

$$= 18.246 - 5.540$$

$$= 12.706 \text{ m}$$

and $\qquad (HGL)_D = (TEL)_D - \dfrac{U^2}{2g}$

$$= 12.706 - 0.554$$

$$= 12.152 \text{ m}$$

$$(TEL)_E = (TEL)_D - 0.02 \left(\frac{40}{0.15}\right)(0.554)$$

$$= 12.706 - 2.955$$

$$= 9.751 \text{ m}$$

and $\qquad (HGL)_E = 9.751 - 0.554$

$$= 9.197 \text{ m}$$

Further, $\qquad (TEL)_F = (TEL)_E - 0.1\dfrac{U_F^2}{2g}$

$$= 9.751 - 0.1\,(16)\,(0.554)$$

$$= 8.865 \text{ m}$$

and $\qquad (HGL)_F = 8.865 - \dfrac{U_F^2}{2g}$

$$= 8.865 - 16(0.554)$$

$$= 8.865 - 8.864$$

$$= 0.001$$

The elevation of HGL at F is, obviously, zero and the value of 0.001, instead, is due to rounding off of the values at different steps of the calculations. The TEL and HGL are shown plotted in Fig. E10.1.

E10.2: Determine the location and magnitude of minimum pressure for the siphon shown in Fig. E10.2. Neglect minor losses and assume that the length of pipe is equal to the horizontal distance between the reservoirs. Equation of pipeline axis is $y = 0.002x^2$ for the coordinate system shown in Fig. E10.2.

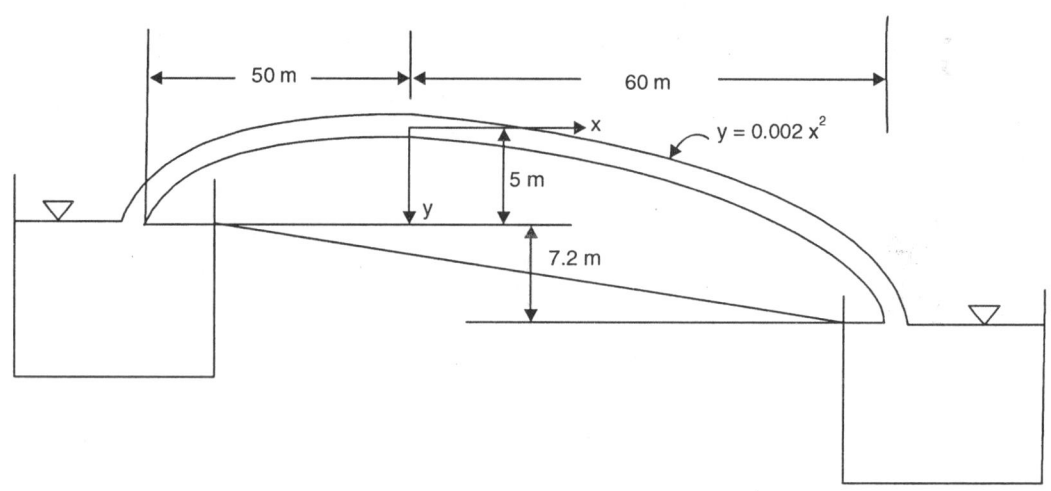

Fig. E10.2

Solution: When minor losses (proportional to $U^2/2g$) are negligible, the term $U^2/2g$ too is small and, therefore, can be neglected. Hence, the hydraulic grade line (HGL) would be the straight line connecting the water levels in the two reservoirs.

Choosing the coordinate system as shown in Fig. E10.2, the coordinates of the two points on the HGL are (-50, 5) and (60, 12.2). Hence, equation of HGL is

$$y = mx + C$$

in which m and C can be determined by substituting the coordinates of the two points on HGL

$$5 = -50\,m + C$$
$$12.2 = +60\,m + C$$

Hence, $\quad 110\,m = 7.2$

$\therefore \quad m = 0.0655$

and $\quad 12.2 = 60 \times 0.0655 + C$

$$C = 8.27$$

Hence, equation of HGL is

$$y = 0.0655x + 8.27$$

The HGL is below the pipeline. Hence, pressure in the entire length of pipe is negative. The pressure head at any section in the pipe is equal to the elevation difference between the pipe axis and the HGL and is negative.

Thus, the magnitude of pressure head at any section of the pipeline is

$$\frac{p}{\rho g} = 0.002\,x^2 - (0.0655x + 8.27)$$

assuming that the distance between the inlet (or outlet) of the siphon and the adjacent reservoir wall is negligible.

The minimum pressure would occur at a section where the distance between the pipeline axis and HGL is maximum and for this,

$$d\left(\frac{p}{\rho g}\right)\Big/dx = 0.004x - 0.0655 = 0$$

$$\therefore \qquad x = 16.375\,\text{m}$$

and the corresponding value of $\dfrac{p}{\rho g}$ is

$$\frac{p}{\rho g} = 0.002\,(16.375) - 0.0655(16.375) - 8.27$$

$$= 0.536 - 1.073 - 8.27$$

$$= -8.807\ \text{m}$$

E10.3:   For the two reservoir system, Fig. 10.3.2, determine the discharge of water flowing from one reservoir to the other. Data given are: $H = 65$ m and $K_{Le} = 0.5$

	Pipe 1	Pipe 2
Length L (m)	300	250
Diameter D (m)	0.5	1.0
Roughness height k (mm)	2.0	0.3

Solution: From the energy equation

$$6.5 = 0.5\frac{U_1^2}{2g} + f_1\frac{300}{0.5}\frac{U_1^2}{2g} + \frac{(U_1 - U_2)^2}{2g} + f_2\frac{250}{1.0}\frac{U_2^2}{2g} + \frac{U_2^2}{2g}$$

and from the continuity equation,

$$U_1\frac{\pi}{4}D_1^2 = U_2\frac{\pi}{4}D_2^2$$

$$\therefore \qquad U_2 = \left(\frac{D_1}{D_2}\right)^2 U_1 = (0.5)^2\,U_1 = 0.25U_1$$

Hence,

$$6.5 = 0.5\frac{U_1^2}{2g} + f_1(600)\frac{U_1^2}{2g} + \frac{U_1^2}{2g}(1 - 0.25)^2 + f_2(250)(0.25)^2\frac{U_1^2}{2g} + (0.25)^2\frac{U_1^2}{2g}$$

or
$$6.5 = \frac{U_1^2}{2g}\left[0.5 + 600\,f_1 + 0.5625 + 15.625\,f_2 + 0.0625\right]$$

$$= \frac{U_1^2}{2g}\left[1.125 + 600\,f_1 + 15.625\,f_2\right]$$

Since
$$\frac{k_1}{D_1} = 0.004 \text{ and } \frac{k_2}{D_2} = 0.0012$$

$\therefore$ Assuming the flow to be in fully rough turbulent region,

$$f_1 = 0.028 \text{ and } f_2 = 0.0197 \quad \text{(from Fig. 9.6.2)}$$

Hence,
$$6.5 = \frac{U_1^2}{2g}\left[1.125 + (600 \times 0.028) + (15.625 \times 0.0197)\right]$$

$\therefore$
$$U_1 = 2.643 \text{ m/s and } U_2 = 0.661 \text{ m/s}$$

and
$$R_{e1} = \frac{2.64 \times 0.5}{10^{-6}} = 1320000 \text{ and } R_{e2} = \frac{6.661 \times 1}{10^{-6}} = 661000$$

$\therefore$ Revised values of friction factor: $f_1 = 0.028$ and $f_2 = 0.02$ (from Fig. 9.6.2)

Hence,
$$6.5 = \frac{U_1^2}{2g}\left[1.125 + (600 \times 0.028) + (15.625 \times 0.02)\right]$$

$\therefore$
$$U_1 = 2.643 \text{ m/s and } U_2 = 0.661 \text{ m/s}$$

These values do not differ from the previous values.

Therefore,
$$Q = \frac{\pi}{4} D_1^2 (U_1)$$

$$= \frac{\pi}{4}(0.5)^2 (2.643)$$

$$= 0.519 \text{ m}^3/\text{s}$$

E10.4:   Compute the total discharge flowing in the pipe system of Fig. 10.3.3. The total head loss between the points A and B is 20 m. The pipe data are

Pipe	L(m)	D(cm)	k(mm)	k/D	f (for fully rough condition)
1	100	8	0.24	0.003	0.0262
2	150	6	0.12	0.002	0.0234
3	80	4	0.20	0.005	0.0304

Solution: From Eqs. (10.3.5) and (9.5.5)

$$20 = f_1 \frac{L_1}{D_1} \frac{U_1^2}{2g} = f_2 \frac{L_2}{D_2} \frac{U_2^2}{2g} = f_3 \frac{L_3}{D_3} \frac{U_3^2}{2g}$$

or $\quad 20 = (0.0262)(1250)\dfrac{U_1^2}{2g} = (0.0234)(2500)\dfrac{U_2^2}{2g} = (0.0304)(2000)\dfrac{U_3^2}{2g}$

$$U_1 = 3.46 \text{ m/s}, \ U_2 = 2.59 \text{ m/s, and } U_3 = 2.54 \text{ m/s}$$
$$R_{e1} = 276800, \ R_{e2} = 155400 \text{ , and } R_{e3} = 101600$$

∴ Revised values of friction factor are

$$f_1 = 0.0268, \quad f_2 = 0.0247 \quad \text{and} \quad f_3 = 0.0315$$
$$U_1 = 3.42 \text{ m/s}, \ U_2 = 2.52 \text{ m/s, and } U_1 = 2.49 \text{ m/s}$$

Accepting these values to be true values,

$$Q_1 = 0.0172 \text{ m}^3/\text{s}, \ Q_2 = 0.0071 \text{ m}^3/\text{s and } Q_3 = 0.0031 \text{ m}^3/\text{s}$$

Total discharge $= Q_1 + Q_2 + Q_3 = 0.0172 + 0.0071 + 0.0031$

$$= 0.0274 \text{ m}^3/\text{s}$$
$$= 98.64 \text{ m}^3/\text{hr}$$

E10.5: Consider the three-reservoir system shown in Fig. 10.3.4 with the following data:

$$L_1 = 125 \text{ m}, L_2 = 160 \text{ m and } L_3 = 95 \text{ m}$$
$$z_1 = 115 \text{ m}, z_2 = 85 \text{ m and } z_3 = 25 \text{ m}$$
$$D_1 = D_2 = D_3 = 25 \text{ cm}; k_1 = k_2 = k_3 = 1 \text{ mm}$$

Compute steady flow rate in all pipes for water at 20°C. What would be the effect on discharges if the diameter of pipes is reduced?

Solution: For $\dfrac{k}{D} = \dfrac{1}{250} = 0.004$, $f = 0.028$ assuming flow to be fully rough turbulent flow. Assuming the piezometric head $h_j = z_2$, and $Q_2 = 0$

$$\therefore \qquad 115 - 85 = 30 = \frac{8(0.028)(125)Q_1^2}{\pi^2 (9.80)(0.25)^5}$$

$$\therefore \qquad Q_1 = 0.318 \text{ m}^3/\text{s}$$

Also, $\qquad 85 - 25 = 60 = \dfrac{8(0.028)(95)Q_3^2}{\pi^2 (9.80)(0.25)^5}$

$$\therefore \qquad Q_3 = 0.516 \text{ m}^3/\text{s}$$

Since, $Q_3 > Q_1$, the flow in pipe 2 should be from the reservoir 2 towards the junction J and the continuity equation would be

$$Q_1 + Q_2 - Q_3 = 0$$

and the Darcy-Weisbach equation would be

$$\Delta h = \frac{8(0.028)LQ^2}{\pi^2(9.8)(0.25)^5}$$

∴ $$Q = 0.6494\sqrt{\Delta h/L}$$

The remaining computations can be carried out in the following tabular form:

Assumed h_j	Pipe 1 $z_1 = 115m, L_1 = 125m$		Pipe 2 $z_2 = 85m, L_2 = 160m$		Pipe 3 $z_3 = 25m; L_3 = 95m$		$Q_1 + Q_2 - Q_3$
	Δh_1	Q_1	Δh_2	Q_2	Δh_3	Q_3	
70	45	0.3896	15	0.1988	35	0.3942	+0.1942
80	35	0.3436	5	0.1148	55	0.4941	-0.0357
78	37	0.3533	7	0.1358	53	0.4851	+0.0040
78.2	36.8	0.3523	6.8	0.1339	53.2	0.4860	+0.0002

The continuity equation is satisfied with $h_j = 78.2$ m and the corresponding discharges are

$$Q_1 = 0.3523 \text{ m}^3/\text{s}$$
$$Q_2 = 0.1339 \text{ m}^3/\text{s}$$
and $$Q_3 = 0.4860 \text{ m}^3/\text{s}$$

On reduction of diameter of pipes, k/D and, hence, f would increase. Also L/D would increase. Therefore, the discharges in all pipes would reduce.

E10.6: Determine the distribution of flow in the pipe network as shown in Fig. E10.6(i). Assume n in the equation $h_f = rQ^n$ as equal to 2.0

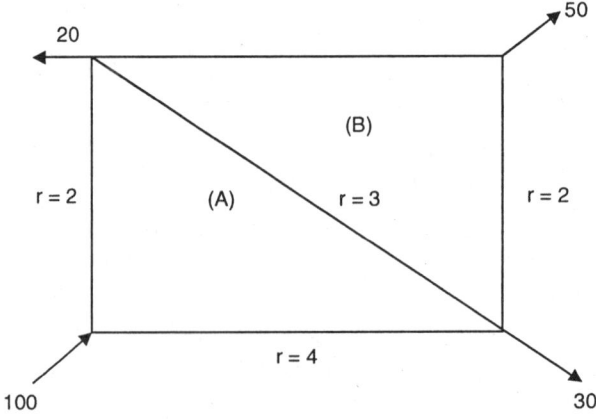

Fig. E10.6 (i)

Solution: Assume the distribution of discharges in different pipes of the network as shown in Fig. E10.6(ii).

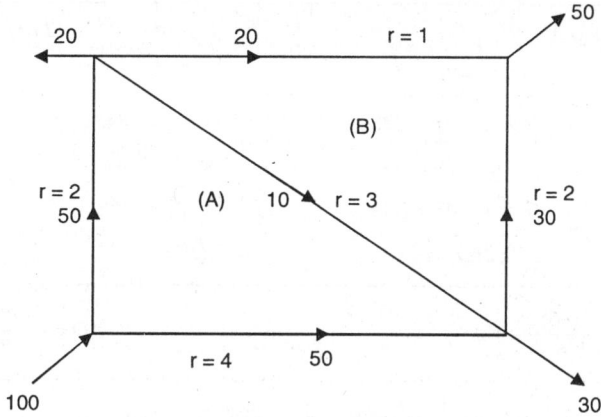

Fig. E10.6 (ii)

Hardy Cross method is applied for both loops A and B in the following tabular form:

Loop A		Loop B	
rQ^2	$2rQ$	rQ^2	$2rQ$
$2 \times 50 \times 50 = 5000$	$2 \times 2 \times 50 = 200$	$1 \times 20 \times 20 = 400$	$2 \times 1 \times 20 = 40$
$3 \times 10 \times 10 = 300$	$2 \times 3 \times 10 = 60$	$-3 \times 10 \times 10 = -300$	$2 \times 3 \times 10 = 60$
$-4 \times 50 \times 50 = -10000$	$2 \times 4 \times 50 = 400$	$-2 \times 30 \times 30 = -1800$	$2 \times 2 \times 30 = 120$
$\Sigma rQ^2 = -4700$	$\Sigma 2rQ = 660$	$\Sigma rQ^2 = -1700$	$\Sigma 2rQ = 220$

$$\Delta Q = -\frac{-4700}{660} \cong 7.0 \qquad\qquad \Delta Q = -\frac{-1700}{220} \cong 8.0$$

The revised distribution of discharges is as shown in Fig. E10.6(iii).

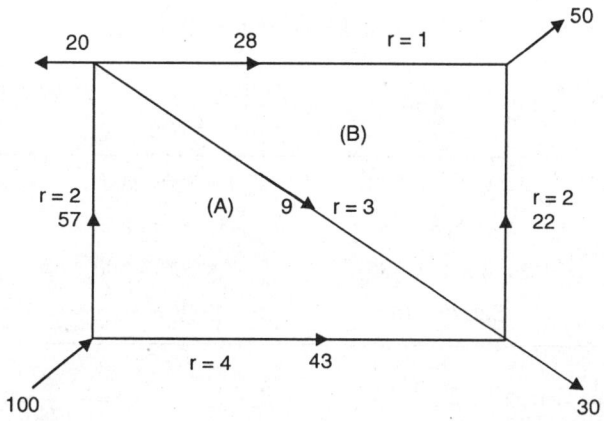

Fig. E10.6 (iii)

Applying Hardy Cross method to loops A and B of Fig. E10.6(iii),

Loop A		Loop B	
rQ^2	$2rQ$	rQ^2	$2rQ$
$2 \times 57^2 = 6498$	$2 \times 2 \times 57 = 228$	$1 \times 28^2 = 784$	$2 \times 1 \times 28 = 56$
$3 \times 9^2 = 243$	$2 \times 3 \times 9 = 54$	$-3 \times 9^2 = -243$	$2 \times 3 \times 9 = 54$
$-4 \times 43^2 = -7396$	$2 \times 4 \times 43 = 344$	$-2 \times 22^2 = -968$	$2 \times 2 \times 22 = 88$
$\Sigma rQ^2 = -655$	$\Sigma 2rQ = 626$	$\Sigma rQ^2 = -427$	$\Sigma 2rQ = 198$

$$\Delta Q = -\frac{-655}{626} \cong 1.0 \qquad\qquad \Delta Q = -\frac{-427}{198} \cong 2.0$$

The revised distribution of discharges is as shown in Fig. E10.6(iv).

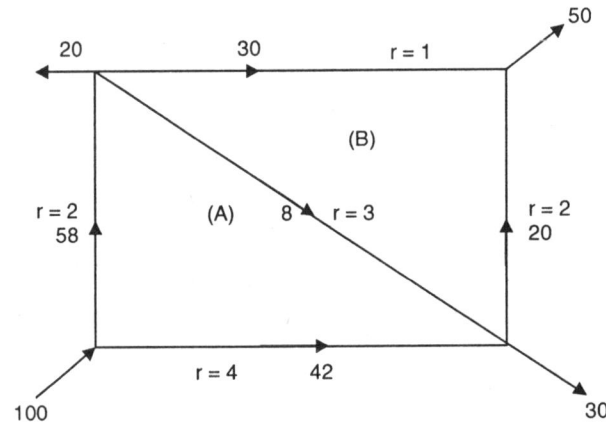

Fig. E10.6 (iv)

Applying Hardy Cross method to loops A and B of Fig. E10.6(iv),

Loop A		Loop B	
rQ^2	$2rQ$	rQ^2	$2rQ$
$2 \times 58^2 = 6728$	$2 \times 2 \times 58 = 232$	$1 \times 30^2 = 900$	$2 \times 1 \times 30 = 60$
$3 \times 8^2 = 192$	$2 \times 3 \times 8 = 48$	$-3 \times 8^2 = -192$	$2 \times 3 \times 8 = 48$
$-4 \times 42^2 = -7056$	$2 \times 4 \times 42 = 336$	$-2 \times 20^2 = -800$	$2 \times 2 \times 20 = 80$
$\Sigma rQ^2 = -136$	$\Sigma 2rQ = 616$	$\Sigma rQ^2 = -92$	$\Sigma 2rQ = 128$

$$\Delta Q = -\frac{-136}{616} = 0.22 \qquad\qquad \Delta Q = -\frac{-92}{128} = 0.72$$

The revised distribution of discharges is as shown in Fig. E10.6(v).

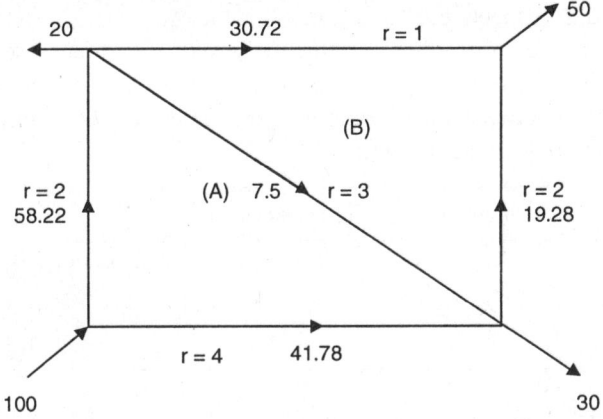

Fig. E10.6 (v)

Applying Hardy Cross method to loops A and B of Fig. E10.6(v),

Loop A		Loop B	
rQ^2	$2rQ$	rQ^2	$2rQ$
$2 \times 58.22^2 = 6779.14$	$2 \times 2 \times 58.22 = 232.88$	$1 \times 30.72^2 = 943.72$	$2 \times 1 \times 30.72 = 61.44$
$3 \times 7.5^2 = 168.75$	$2 \times 3 \times 7.5 = 45.00$	$-3 \times 7.5^2 = -168.75$	$2 \times 3 \times 7.5 = 45.00$
$-4 \times 41.78^2 = -6982.27$	$2 \times 4 \times 41.78 = 339.24$	$-2 \times 19.28^2 = -743.44$	$2 \times 2 \times 19.28 = 77.12$
$\Sigma rQ^2 = -34.38$	$\Sigma 2rQ = 612.6$	$\Sigma rQ^2 = -31.53$	$\Sigma 2rQ = 183.56$

$$\Delta Q = -\frac{-34.48}{612.6} = 0.06 \qquad\qquad \Delta Q = -\frac{-31.53}{183.56} = 0.17$$

Since ΔQ for both loops is negligibly small, the computations can be terminated and the computed distribution is as shown in Fig. E10.6(v).

PROBLEMS

P10.1 In Fig. 10.3.2, $K_{le} = 0.5$, $L_1 = 250$ m, $D_1 = 600$ mm, $k_1 = 1.5$ mm, $L_2 = 210$ m, $D_2 = 900$ mm, $k_2 = 0.5$ mm, $v = 2 \times 10^{-6}$ m²/s, and $H = 5$ m. Compute the discharge that would flow in the pipe system.

P10.2 Solve P10.1 by means of equivalent pipe concept.

P10.3 Determine the head loss and flow distribution for the following system of horizontal parallel pipes in which water flows at 20^0 C when the total discharge is 100 l/s and the upstream pressure is 0.5 MPa.

Pipe	Length, m	Diameter, mm	k, mm
1	500	100	5
2	1000	150	3
3	1250	200	1

P10.4 Solve pipe network of Fig. E10.6 if the values of *r* and n in Eq. (10.3.16) for all pipes in the network are 3 and 2, respectively.

P10.5 Compute the flow rate in each of the three pipes of the three reservoir system shown in Fig. 10.3.4. The difference between the elevations of the water surface in reservoirs 1 and 2 is 20 m. The difference between the elevations of the water surface in reservoirs 1 and 3 is 30 m. The pipe characteristics are as follows:

Pipe	Length, m	Diameter, mm	k, mm
1	250	150	0.020
2	120	100	0.015
3	70	120	0.018

P10.6 The velocity of water in a 600 mm diameter, 15 mm thick, and 300 m long cast iron pipe ($E_m = 1.039 \times 10^{11}$ N/m^2) is changed from 3.0 m/s to zero in 0.3 seconds by closing a valve. Determine the pressure rise. What would be the pressure rise if this closure takes place in 5 seconds? Also, determine the pressure rise if the closure takes place in 1 second.

REFERENCES

(1) Cross, H: "*Analysis of flow in networks of conduits or conductors*", Bull. Illinois Univ. Engg. Exp. Station, Report 286, 1936.

(2) Massey, B and Ward-Smith, J: "*Mechanics of Fluids*", Nelson Thornes Ltd., 1998.

(3) Streeter, VL, Wylie, EB, and Bedford, KW: "*Fluid Mechanics*", 9th Edition, McGraw-Hill Companies, Singapore, 1998.

CHAPTER 11

Basic Concepts of Fluid Flow
in Open Channels

11.1 OPEN CHANNEL FLOW

An open channel is a conduit in which a liquid flows with a free surface that is essentially an interface between the flowing liquid and the atmospheric air. The free surface of the flowing liquid is one of the boundaries. The free surface and the flow are interdependent. Flow in a conduit may be either open channel flow having a free surface or pipe flow that has no free surface since the flowing fluid fills the whole conduit. A free surface is subject to atmospheric pressure. Pipe flow and open channel flow are compared in Fig. 11.1.1. If the liquid flowing in an open channel of small slope is water and the flow is parallel and has uniform velocity distribution, then, the water surface (i.e., the free surface) is the hydraulic grade line, and the depth of water, h corresponds to the pressure head, and the elevation of the water surface with respect to an arbitrarily chosen datum is the piezometric head, i.e., $(h+z)$. The flow is caused by gravitational effects and the pressure distribution within the flowing fluid is generally hydrostatic. Open channel flows are almost always turbulent and are generally not affected by surface tension. The difference between the total energy line and hydraulic grade line equals the velocity head. Flow in rivulets flowing across a field to big rivers, canals, and partially-filled sewers are some of the familiar examples of open channel flow. Other practical problems of open channel flow include design of irrigation, power, and navigation channels, preparation of stage-discharge curves, propagation of flood wave in a stream, dispersion of pollutants in rivers, hydraulic design of stilling basin, spillway, weir etc.

Compared to pipes, channels are generally suited for carrying large discharges and, hence, are generally used for irrigation supplies. However, channel flows incur large evaporation loss and are more susceptible to contamination. Also, channels cannot be laid underground.

Although pipe flow and open channel flow are similar, Fig. 11.1.1, it is often much more difficult to analyze channel flow problems for the following reasons:

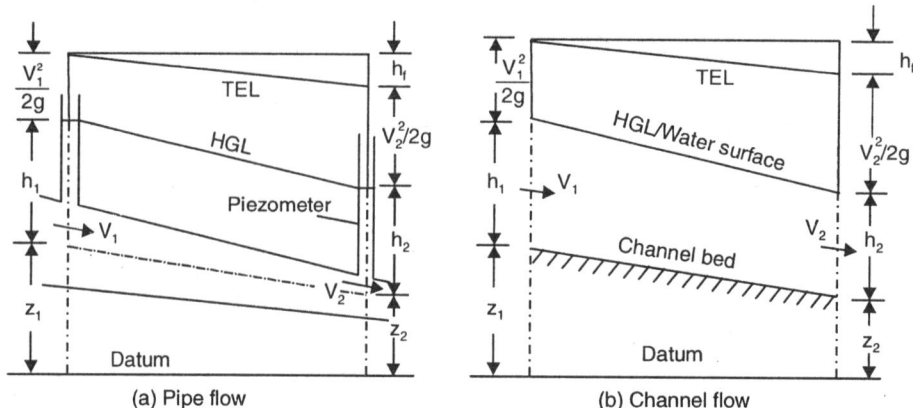

Fig. 11.1.1. Schematic comparison between pipe flow and channel flow (1)

(i) Free surface and, hence, the depth varies with time and space. This means that the flow cross-section is not constant. In pipes, the flow cross-section does not change with time even for unsteady flow.

(ii) For channel flows, the quantities like bh or h, discharge Q, bed slope S_o, and energy gradient S_f are interdependent. Here, b is the channel width.

(iii) It is relatively difficult to obtain reliable experimental data.

(iv) Physical condition of a channel varies over a wide range.

(v) Cross-section of a channel may be completely irregular.

(vi) Surface roughness too varies over a wider range and depends upon depth of flow h too, besides other parameters and, hence, its estimation is relatively more uncertain.

Owing to these difficulties, analysis of channel flow is more empirical than that of pipe flow. However, the empirical methods, if applied judiciously, do yield good results.

11.2 TYPES OF CHANNELS

Open channels (or, simply, channels) can be broadly categorized as either *natural channel* or *artificial channel*. Natural channels are the ones that have been developed by natural processes and have not been significantly improved by human efforts. Creeks, large and small rivers, and tidal estuaries are the examples of natural channels. Artificial channels are, however, man-made and include irrigation channels, power channels, navigation channels, drainage ditches, gutters etc. Artificial channels are designed to have sections of regular geometric shapes. Although, the basic principles governing the flow in natural as well as artificial channels remain the same, the analysis of flow in natural channels is relatively more difficult compared to that for artificial channels. It should be noted that the terms 'canal' and 'channel' are interchangeably used for artificial channels.

Canal (or *channel*) is usually long, mild-sloped, and built on or into the ground such that its free surface is exposed to the atmosphere. Part of a canal may have to be constructed below ground when the alignment of the canal is through a hill. In such a case, canal will be constructed like a tunnel but the free surface would still be exposed to the atmosphere. Similarly, a channel culvert meant for carrying a flow under roads, railway embankments, or runways too would be covered from top but its free surface would be exposed to the atmosphere.

A canal (or channel) may either be prismatic (i.e., having constant cross-sectional shape and bed slope) or non-prismatic. Also, a canal may be either unlined or lined with concrete, cement, bituminous materials, earthen material, or artificial membrane. Lined channels would have fixed bed slope and cross-section and, hence, the flow in such channels is much more amenable to analysis. Such channels are termed rigid boundary channels. Unlined channels have boundaries (bed and banks) of loose material that may get scoured. Therefore, the cross-section and bed slope of an unlined channel may change. Such channels, therefore, have mobile boundaries and are called mobile boundary channels. Alluvial channels are mobile boundary channels and analysis of flow in alluvial channels is more complex.

Based on the purpose served by a canal (or channel) it may be called as water supply canal, irrigation canal, power channel, navigation channel, sewer, road-side drain, or culvert etc.

The term *flume* refers to that part of a channel that has been built above the ground surface to convey a flow across a depression. A channel constructed in a laboratory for basic and applied research too is termed flume.

A channel having steep bed slope is often called *chute*. A *drop* too is a steep channel but of short length.

Channels can also be classified according to the geometrical shape of the channel section. Accordingly, a channel can be a rectangular channel, trapezoidal channel, triangular channel, or circular channel.

11.3 GEOMETRIC ELEMENTS OF A CHANNEL SECTION

Properties of a channel section are determined by the geometric shape of the channel, the depth of flow h, and the bed slope angle θ with respect to a horizontal line.

Flow section is the plane perpendicular to the main flow direction and bounded by the channel boundaries and the free surface of the flowing fluid.

The *depth of flow h* is the vertical distance from the lowest point (i.e., at the bed) of a channel (or flow) section to the water surface and is used interchangeably with the term *depth of flow section d* defined as the depth of flow measured perpendicular to the channel bed. Obviously,

$$d = h\cos\theta$$

and when θ is small, $d \approx h$, Fig.11.1.1.

Stage of a flow is the elevation of the water surface with respect to a suitable datum chosen arbitrarily. If the selected datum is the channel bed, then the stage and depth of flow are, obviously, equal.

Top width T is the width of the channel (or flow) section at the water surface.

Area of flow (or *flow area*) A is the area of the flow section.

Wetted perimeter P is the length of the line of intersection of the flow section and the channel boundary.

Hydraulic radius R is the ratio of the flow area A and the corresponding wetted perimeter P, i.e., $R = A/P$.

Hydraulic depth D is the ratio of the flow area A and the corresponding top width T, i.e., $D = A/T$.

The section factor for critical flow computation Z is the product of the water area and the square root of the hydraulic depth, i.e.,

$$Z = A\sqrt{D} = A\sqrt{A/T}$$

The section factor for uniform flow computation Z_n is defined as

$$Z_n = AR^{2/3}$$

Table 11.3.1 lists the values of geometric elements of different channel sections.

Table 11.3.1: Geometric Elements of Channel Sections (2)

Section	Area A	Wetted perimeter P	Hydraulic radius R	Top width T	Hydraulic depth D
Rectangular (Bed width = b)	bh	$b + 2h$	$\dfrac{bh}{b+2h}$	b	h
Trapezoidal (Base width = b) [Side slopes $z(H):1(V)$]	$(b+zh)h$	$b+2h\sqrt{1+z^2}$	$\dfrac{(b+zh)h}{b+2h\sqrt{1+z^2}}$	$b+2zh$	$\dfrac{(b+zh)h}{b+2zh}$
Triangular [Side slopes $z(H):1(V)$]	zh^2	$2h\sqrt{1+z^2}$	$\dfrac{zh}{2\sqrt{1+z^2}}$	$2zh$	$\dfrac{1}{2}h$
Circular (Diameter $= D_o$)	$\dfrac{1}{8}(\phi-\sin\phi)D_o^2$	$\dfrac{1}{2}\phi D_o$	$\dfrac{1}{4}\left(1-\dfrac{\sin\phi}{\phi}\right)D_o$	$(\sin\phi/2)D_o$ or $2\sqrt{h(d_o-h)}$	$\dfrac{1}{8}\left(\dfrac{\phi-\sin\phi}{\sin\phi/2}\right)D_o$
Parabolic (Top width $= T$)	$\dfrac{2}{3}Th$	$T+\dfrac{8}{3}\dfrac{h^{2*}}{T}$	$\dfrac{2T^2h^*}{3T^2+8h^2}$	$\dfrac{3}{2}\dfrac{A}{h}$	$\dfrac{2}{3}h$

* Satisfactory approximation for the interval $0<(4h/T)\leq 1$,

ϕ is the angle (towards the bed) between the radial lines to the points of intersection of the free surface and the channel boundary.

11.4 CLASSIFICATIONS OF CHANNEL FLOWS

Channel flows can be classified on the basis of many different criteria. One of the major criteria of classification is the variation of the depth of flow h with respect to time t and space coordinate x in the direction of flow. The flow is termed either *steady* if the depth of flow h, at any location, does not change with time t, i.e., $\partial h / \partial t = 0$ or unsteady when $\partial h / \partial t \neq 0$. This classification is, obviously, with respect to a stationary observer. Example of unsteady flow is propagation of a flood wave which phenomenon would appear steady to an observer moving with the same speed as that of the wave and in the same direction.

Using spatial variation of the depth of flow at a given instant, the channel flow can be classified as either uniform when $\dfrac{\partial h}{\partial x} = \dfrac{\partial U}{\partial x} = \dfrac{\partial Q}{\partial x} = 0$ or nonuniform when any of h, Q, and U vary with x. Here, h, Q, and U are, respectively, the depth of flow, discharge, and average velocity in the channel. Truly uniform flow in a channel requires, apart from $\dfrac{\partial h}{\partial x} = \dfrac{\partial Q}{\partial x} = \dfrac{\partial U}{\partial x} = 0$, that the velocity profiles along the length of the channel are identical which condition would not be satisfied in the entrance region of the channel, Fig. 11.4.1, due to the boundary layer formation. It is only when the thickness of the boundary layer becomes equal to the depth of flow, the flow gets established (or fully developed) and the velocity profiles attain identical shape downstream of the developing flow region. The length of the developing flow region is of the order of about $50h$ to $100h$ for subcritical flows. From practical viewpoint, existence of an unsteady uniform flow is impossible.

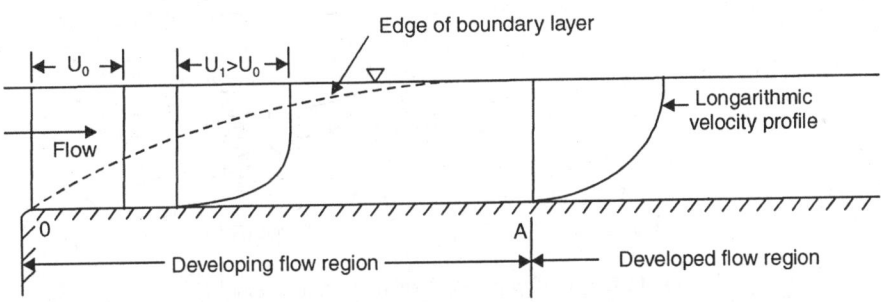

Fig. 11.4.1. Development of flow in a channel

Nonuniform flow is also called varied flow that is further classified as either rapidly varied flow (RVF) (when the depth of flow changes rapidly over a relatively short distance such as in the case of hydraulic jump) or gradually varied flow (GVF) when the depth of flow changes rather slowly over a relatively longer distance as in the case of a reservoir upstream of a dam, Fig. 11.4.2.

If water (or a liquid flowing in a channel) is added or withdrawn along the channel reach under consideration, as in gutters and side channel spillways, the flow is known as *spatially varied flow*.

It should be noted that classifications of flow in open channel are on the basis of variation of gross flow characteristics with respect to time and space and not on the basis of variation of individual streamlines or velocity vectors.

Depending upon the relative magnitudes of inertial and viscous forces, the channel flow can be either laminar, transitional, or turbulent. The basis for this classification is the familiar Reynolds number

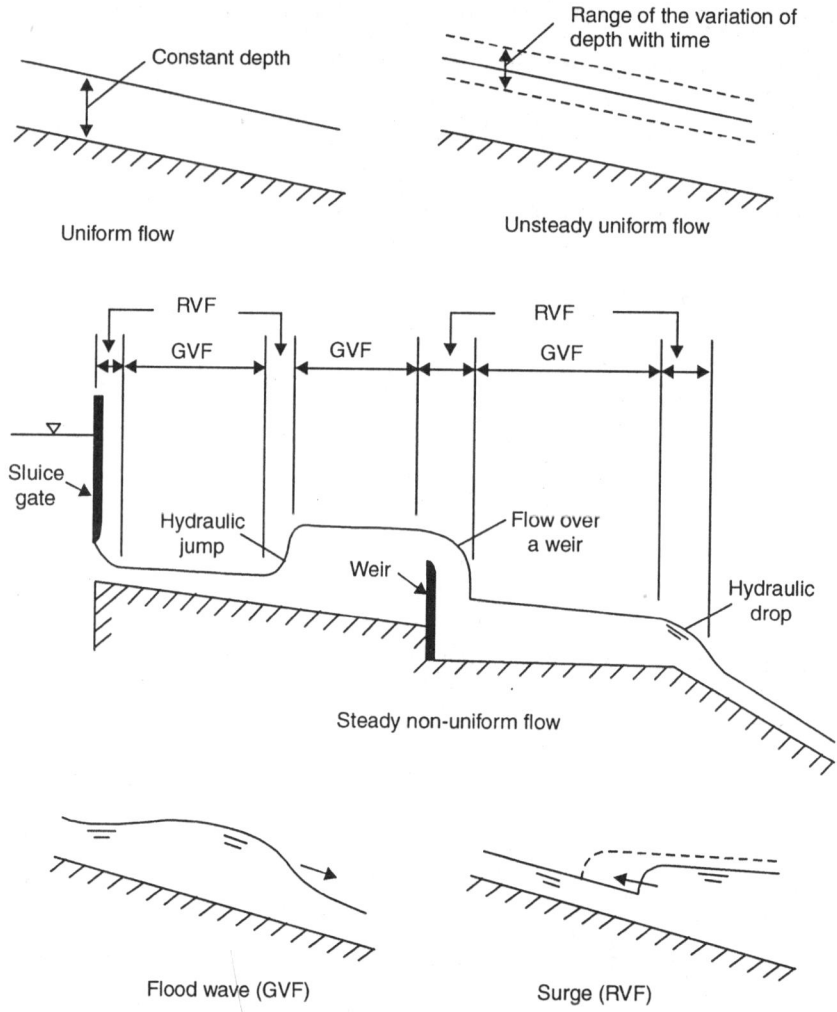

Fig. 11.4.2. Different types of flow in a channel (1)

R_e with the average velocity of flow U as the characteristic velocity and the hydraulic radius R as the characteristic length. Thus,

$$R_e = UR/\nu$$

The flow in a channel is laminar if $R_e < 500$, transitional if $500 < R_e < 2000$, and turbulent if $R_e > 2000$(1). The flow in a channel is usually turbulent.

Another useful classification of channel flows is on the basis of Froude number, F_r, i.e. the ratio of inertial and gravity forces defined as

$$F_r = \frac{U}{\sqrt{gD}}$$

The characteristic length for Froude number is often taken as the hydraulic depth D. The flow in a channel is classified as subcritical if $F_r < 1$, critical if $F_r = 1$, and supercritical if $F_r > 1$.

Flow in a channel can also be classified as either homogeneous if the density of the flowing fluid is constant in all spatial dimensions or stratified if the density of the flowing fluid varies in any direction. For stratified flows, the densimetric (or internal) Froude number is defined as (2)

$$F_r = \frac{U}{\sqrt{g(\Delta\rho/\rho)L}}$$

in which $\Delta\rho$ is the difference in mass densities of the two fluids, ρ is the mass density of the fluid in the layer under consideration (usually, the lower one) and U and L are characteristic velocity and depth of this layer.

Channel flows can also be defined as one-dimensional, two-dimensional, and three-dimensional depending upon the dependence of the velocity on space coordinates. In narrow channels, the velocity at a point depends on its distance from the bed as well as side walls of the channel and, hence, the flow is three-dimensional. However, in a wide channel (i.e., the width of channel greater than 5 – 10 times the depth of flow), the velocity at any elevation in a channel section would be practically constant (barring the region very close to the side walls or banks), i.e., independent of the distance of any point, at that elevation, from the side walls. This flow is called two-dimensional flow. Analysis of channel flow becomes simpler, if one considers only the average velocity over a cross-section of the channel ignoring the velocity variations in the cross-section and considering the variation of the average velocity in the main flow direction. Such an idealized flow is one-dimensional flow and the analysis is one-dimensional analysis.

11.5 BASIC EQUATIONS OF MOTION FOR OPEN CHANNEL FLOW

The three basic equations of any fluid motion are continuity equation (based on the law of mass conservation), energy equation (based on the law of energy conservation), and momentum equation (based on the Newton's second law of motion).

11.5.1 Continuity Equation

The concept of stream tube is commonly used by applying it to the whole region of flow such that the boundaries of the stream tube are also the physical boundaries of the pipe, or the river, or the canal etc. In other words, the waterway, i.e., the entire flow cross-section, is the stream tube. Consider unsteady flow in a channel, Fig. 11.5.1. In open channels, unsteadiness implies the changing surface level although unsteadiness may be due to the velocity variation or discharge variation also both of which would change, in general, the water surface level. Consider a stream tube which embraces the whole waterway of the channel and is of short length δx between sections 1 and 2 of Fig. 11.5.1. Let Q represent the discharge at the centre of the stream tube at any time t. Also, let A and T be, respectively, the area of cross-section and water surface width of the stream tube. The net mass inflow into the stream tube in time δt can be written as

$$\left[\left\{\rho Q - \frac{\partial(\rho Q)}{\partial x}\frac{\delta x}{2}\right\} - \left\{\rho Q + \frac{\partial(\rho Q)}{\partial x}\frac{\delta x}{2}\right\}\right]\delta t$$

$$= -\frac{\partial(\rho Q)}{\partial x}\delta x \, \delta t$$

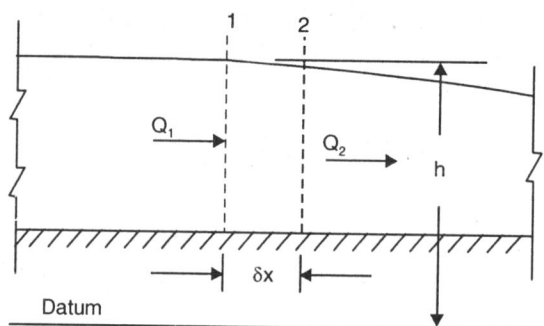

Fig. 11.5.1. Stream tube of length δx in an open channel flow

and this, according to the mass conservation law, must be equal to the increase in the mass of the stream tube in time δt, $\dfrac{\partial}{\partial t}(\rho A \delta x)\delta t$. Hence,

$$-\frac{\partial(\rho Q)}{\partial x}\delta x \delta t = \frac{\partial}{\partial t}(\rho A \delta x)\delta t$$

On dividing this equation by $\delta x \delta t$, one gets

$$\frac{\partial(\rho Q)}{\partial x}+\frac{\partial}{\partial t}(\rho A)=0 \qquad\qquad ...(11.5.1)$$

In case of withdrawal from or addition of water into the stream tube, from the lateral direction (volumetric rate being q_x per unit length), such as in cases of flow over a bottom rack or flow over a side channel spillway, Eq. (11.5.1) becomes

$$\frac{\partial(\rho Q)}{\partial x}+\frac{\partial}{\partial t}(\rho A)=\pm\rho q_x \qquad\qquad ...(11.5.2)$$

Obviously, the positive sign is to be used for lateral additions into the stream tube. If the flowing fluid is incompressible, Eq. (11.5.2) becomes

$$\frac{\partial Q}{\partial x}+\frac{\partial A}{\partial t}=\pm q_x \qquad\qquad ...(11.5.3)$$

and when there is no lateral addition or withdrawal, i.e., $q_x = 0$, Eq. (11.5.3) reduces to

$$\frac{\partial Q}{\partial x}+\frac{\partial A}{\partial t}=0 \qquad\qquad ...(11.5.4)$$

For steady flow, the depth of flow h and area of flow cross-section remain constant, i.e., $\dfrac{\partial A}{\partial t}=0$ and, hence, Eq. (11.5.4) gives

$$Q = \text{constant} = Q_1 = Q_2 \qquad\qquad ...(11.5.5)$$

Expressing the discharge Q in terms of the product of the area of flow section and average velocity of flow U, one may write Eq. (11.5.5) as

$$Q = A_1 U_1 = A_2 U_2 = \ldots\ldots = \text{constant} \qquad\qquad ...(11.5.6)$$

Here, the subscripts refer to the corresponding sections. For a rectangular channel of width B and depth of flow h, Eq. (11.5.6) yields,

$$Q = B_1 \, h_1 \, U_1 = B_2 \, h_2 \, U_2 = \ldots\ldots \text{ constant} \qquad \ldots(11.5.6)$$

If the width is constant, then

$$q = \frac{Q}{B} = h_1 U_1 = h_2 U_2 \qquad \ldots(11.5.7)$$

Here, q is termed unit discharge, i.e., the discharge per unit width of channel.

11.5.2 Energy Equation

Energy of a flowing fluid can also be expressed in terms of the 'head' of the flowing fluid. Therefore, the total head H_A at the point A (on a streamline) contained in the cross-section of flow at O, Fig.11.5.2, may be expressed as

$$H_A = z_A + d_A \cos\theta + \frac{U_A^2}{2g} \qquad \ldots(11.5.8)$$

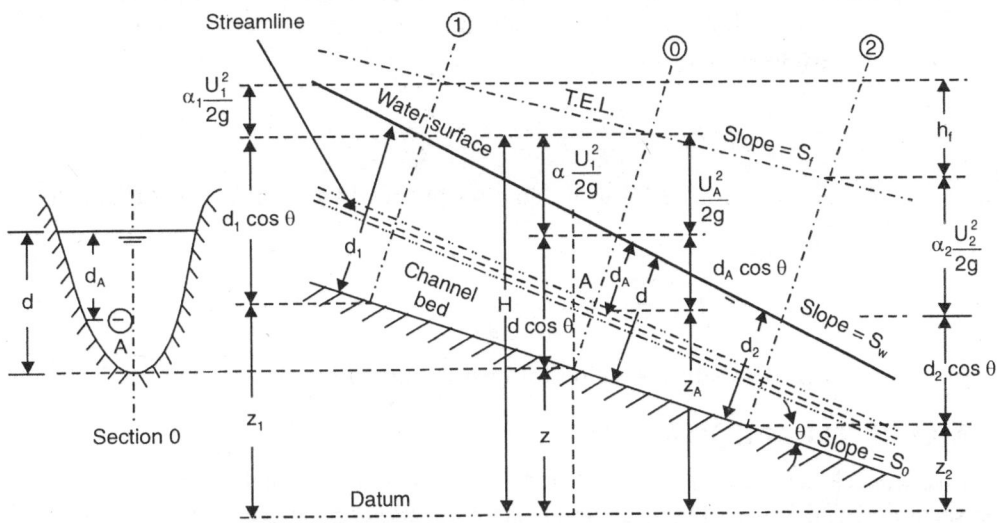

Fig. 11.5.2. Gradually varied flow in a channel (1)

Here, z_A is the elevation of the point A above the chosen datum plane, d_A is the depth of the point A below the free surface measured along the channel section (i.e., perpendicular to the flow direction), and θ is the angle of inclination of the channel bed with respect to the horizontal, Fig. 11.5.2. U_A is the velocity of flow at the point A of the selected streamline.

Only in case of an ideal parallel flow of uniform velocity distribution, one would have the same velocity at all points on a cross-section of the flow. An actual flow would always have non-uniform velocity distribution. However, one may assume, for all practical purposes, that the velocities (or velocity heads) at all points of a channel section are equal and that the effect of non-uniformity of the velocity distribution can be accounted for by using the energy correction factor (or energy coefficient) α. Hence, Eq. (11.5.8) can be written for a flow section as

$$H = z + d\cos\theta + \alpha\frac{U^2}{2g}$$ ...(11.5.9)

The line representing the elevation of total head, Fig. 11.5.2, is the total energy line (T.E.L.) and its slope is termed the energy gradient, S_f. Slope of the water surface is S_w and that of the channel bed is $S_0 = \sin\theta$. For uniform flow, $S_0 = S_w = S_f$.

For $\theta \approx 0$, $\cos\theta \approx 1.0$ and $d \approx h$. Hence, Eq. (11.5.9) reduces to

$$H = z + h + \alpha\frac{U^2}{2g}$$ ...(11.5.10)

Making use of the energy conservation law, one can write the energy equation between sections 1 and 2 of a prismatic channel of large slope as follows:

$$z_1 + d_1\cos\theta + \alpha_1\frac{U_1^2}{2g} = z_2 + d_2\cos\theta + \alpha_2\frac{U_2^2}{2g} + h_f$$ ...(11.5.11)

Here, h_f is the head loss between sections 1 and 2. The subscripts refer to the corresponding sections. Equation (11.5.11) applies to both parallel and gradually varied flow. For a channel of small slope, $\theta \approx 0$, Eq. (11.5.11) reduces to

$$z_1 + h_1 + \alpha_1\frac{U_1^2}{2g} = z_2 + h_2 + \alpha_2\frac{U_2^2}{2g} + h_f$$ ...(11.5.12)

If $\alpha_1 = \alpha_2 = 1.0$ (i.e., uniform velocity distribution in the flow) and $h_f = 0$ (as in ideal fluid), Eq. (11.5.12) reduces to

$$z_1 + h_1 + \frac{U_1^2}{2g} = z_2 + h_2 + \frac{U_2^2}{2g}$$ ...(11.5.13)

which is the well-known Bernoulli's equation for open channel flow.

11.5.3 MOMENTUM EQUATION

Applying Newton's second law of motion (Art. 4.2) to a flow in a channel of large slope, Fig. 11.5.2, between sections 1 and 2, one obtains the momentum equation for open channel flow as

$$\rho Q(\beta_2 U_2 - \beta_1 U_1) = P_1 - P_2 + W\sin\theta - F_f$$ ...(11.5.14)

in which β is the momentum correction factor to account for non-uniform velocity distribution, W is the weight of water enclosed between the two sections of the channel, and F_f is the total external force (due to friction or any other resistance) acting along the surface of contact between the flowing fluid and the channel surface. The weight W can be taken as $\rho g B\bar{h}L$ where $\bar{h} = \frac{1}{2}(h_1 + h_2)$ and the discharge Q can be calculated as $\frac{1}{2}(U_1 + U_2)B\bar{h}$. For parallel or gradually varied flow, the value of P (i.e., the resultant pressure force acting on a section) is computed as $\frac{1}{2}\rho g h^2 B$ for a rectangular channel of small slope and width B.

11.6 EFFECT OF CURVATURE OF STREAMLINES ON PRESSURE DISTRIBUTION IN OPEN CHANNELS

The pressure at any point in a channel cross-section is proportional to the height of the liquid column in a piezometer tube installed at the point. In parallel flow (with uniform velocity distribution) and, hence, no acceleration component in the plane of cross-section, this height would be equal to the depth of the point below the free surface. Therefore, the distribution of pressure over the flow cross-section of a channel is linear, i.e., hydrostatic and, hence, can be represented by a straight line, Fig. 11.6.1(a). The hydrostatic law of pressure distribution is valid only if either the fluid is at rest or the flow is parallel, i.e., the streamlines of the flow have neither substantial curvature nor appreciable divergence. Therefore, the flow filaments have no acceleration components in the plane of cross-section.

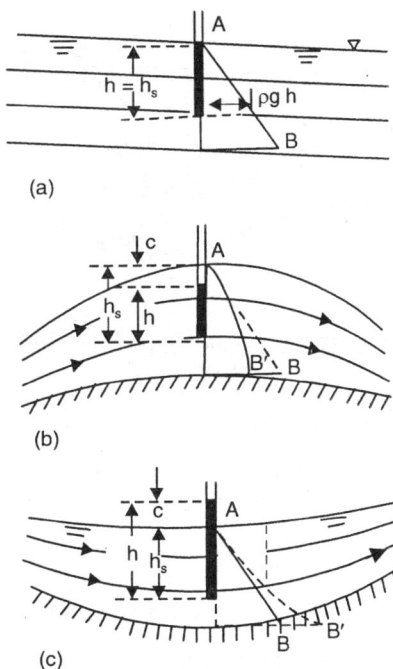

In actual channel flows, uniform flow is practically parallel flow. Gradually varied flow may also be regarded as parallel flow, since the change in the depth of flow is so gradual that the streamlines do not have either the substantial curvature or appreciable divergence. Hence, the effect of the acceleration components in the cross-sectional plane becomes negligible. Therefore, the hydrostatic law of pressure distribution is considered valid for uniform as well as gradually varied flows.

In curvilinear flow, as in flow over a spillway, the streamlines are curved and, hence, produce appreciable acceleration components and centrifugal forces normal to the flow direction. Such curvilinear flow may be either convex or concave, Fig. 11.6.1(b) and 11.6.1(c). In the convex flow, the forces, resulting from the streamline curvature, act against the gravitational force and, hence, the resulting pressure is less than the hydrostatic pressure that would exist if the flow were parallel. In concave flow, the forces, resulting from the streamline curvature, point downward and reinforce the gravitational force and, hence, the resulting pressure is greater than the corresponding hydrostatic pressure if the flow were parallel.

Similarly, when the divergence of streamlines is large enough to cause appreciable acceleration components normal to the flow direction, the pressure is non-hydrostatic.

For simplicity, the pressure head in a curvilinear flow may be obtained by multiplying the hydrostatic pressure head (say, h) with a pressure coefficient β' that would, obviously, be greater than 1 for concave flows and less than 1 for convex flows. For curvilinear flows, therefore, the energy and momentum equations must be suitably

AB – Hydrostatic pressure distribution, $\rho g h_s$

AB′ – Non-hydrostatic pressure distribution, $\rho g h_s \pm \rho g c$

Fig. 11.6.1. Pressure distribution in straight and curved channels (a) Parallel flow, (b) Convex flow, (c) Concave flow (h = pressure head; h_s= hydrostatic pressure head; c = pressure head correction for curvature (1)

modified to account for the effect of curvature of streamlines on the pressure variation. The forces P_1 and P_2 in Eq. (11.5.14), for curvilinear flows, would be replaced by $\beta'_1 P_1$ and $\beta'_2 P$ respectively, where β is the pressure correction factor (or pressure co-efficient).

11.7 EFFECT OF BED SLOPE ON PRESSURE DISTRIBUTION IN OPEN CHANNELS

Considering a straight channel of unit width and slope angle θ, Fig. 11.7.1, the weight of the fluid element AO is $\rho g \delta x d = \rho g \delta x h \cos\theta$ (as $d = h\cos\theta$) and the pressure at O due to this weight is $\rho g \delta x d \cos\theta / \delta x = \rho g d \cos\theta = \rho g h \cos^2\theta$. Hence, the pressure head at O is

$$h' = h\cos^2\theta \qquad \qquad ...(11.7.1)$$

or

$$h' = d\cos\theta \qquad \qquad ...(11.7.2)$$

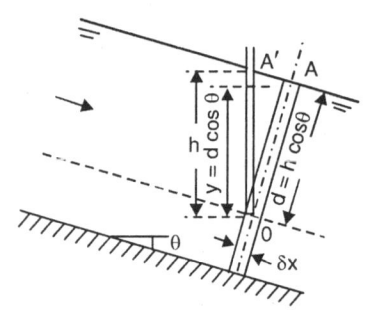

Fig. 11.7.1. Parallel flow in channels of large slope

Equation (11.7.1) means that the pressure head at any point O that is h below the free surface (h is the vertical depth at the point O) is equal to this depth h multiplied by a correction factor $\cos^2\theta$. Obviously, for small $\theta(\approx 0)$, the factor $\cos^2\theta$ will not differ much from unity. For example, for a slope of 1 in 10, i.e., $\tan\theta = 0.1$, $\cos^2\theta = 0.99$ and the pressure head, therefore, reduces by an amount less than one percent. Since the slope of field or laboratory channels is, usually, far less than 1 in 10, the correction for slope can safely be ignored and the pressure head at any point is taken as equal to its vertical depth below the free surface. It should also be noted that Eq. (11.7.1) is not applicable to varied flow particularly when θ is large, whereas, Eq. (11.7.2) would still be applicable. For channels of large slope, the pressure force P per unit width of a rectangular channel, Eq. (11.5.14), would be

$$\left(\rho g h \cos^2\theta\right)\left(\frac{1}{2}h\right) = \frac{1}{2}\rho g h^2 \cos^3\theta, \text{ Fig. 11.7.1. Also, the hydraulic grade line would be located below}$$

the free surface.

11.8 VELOCITY DISTRIBUTION IN A CHANNEL SECTION

Velocities in a channel section are not uniformly distributed due to the presence of free surface and friction along the channel boundary. The velocity distribution in a channel depends also on the shape of the channel section, Fig. 11.8.1, the roughness of the channel boundary, and presence of bends etc. On a bend, the velocity increases towards the convex side (i.e., outer bank) due to the centrifugal action of the flow. The measured maximum velocity in an ordinary channel usually occurs a little below the free surface at a distance of about 0.05 to 0.25 of the depth of flow. It has also been observed that in a narrower section the maximum velocity occurs at relatively deeper location.

The average velocity for the whole channel section equals the total discharge divided by the area of flow section. For this purpose, the channel section is divided into suitable number N of vertical strips by successive verticals, Fig. 11.8.1. By measuring the velocities at several depths along the centreline vertical of a strip, one can obtain the velocity profile for the strip. This profile for i^{th} strip of width b_i can be integrated to yield the discharge Q_i through the strip as

$$Q_i = \sum u b_i \Delta y \qquad \qquad ...(11.8.1)$$

and, thus, the total discharge in the channel Q is

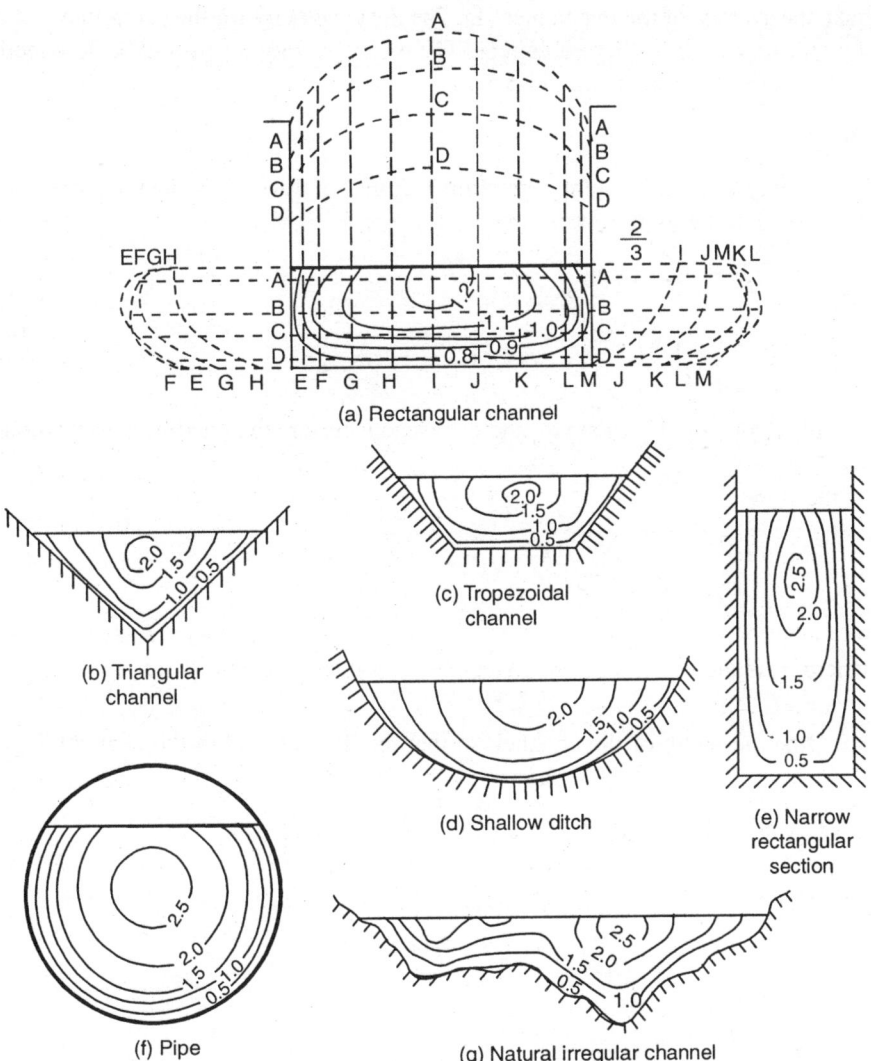

Fig. 11.8.1. Typical isovels (velocity contours) in channels of various shapes (1)

$$Q = \sum Q_i = \sum_{i=1}^{N} u \, b_i \, \Delta y \qquad \qquad ...(11.8.2)$$

Therefore, the average velocity for the whole channel section U is

$$U = \frac{Q}{A} = \frac{\sum Q_i}{A} \qquad \qquad ...(11.8.3)$$

where, A is the area of the flow cross-section.

Instead of measuring the entire velocity profile along a vertical, one can also obtain mean velocity for any vertical by measuring the velocity at 0.6 of the depth below the free surface in each vertical (1). For better accuracy, one can, instead, measure the velocities at 0.2 and 0.8 of the depth below the free surface and take the average of the two values (1). The mean velocity for the centreline vertical of a strip is multiplied by the area of the strip to obtain Q_i. The mean velocity for the entire flow section can now be determined using Eqs. (11.8.2) and (11.8.3).

EXAMPLE

E11.1: The energy and momentum correction factors for flow in an open channel of rectangular cross-section are as follows:

$$\alpha = 1 + 3\varepsilon^2 - 2\varepsilon^3$$

$$\beta = 1 + \varepsilon^2$$

in which $\varepsilon = \left(\dfrac{u_m}{U} - 1\right)$ where u_m and U, respectively, are the maximum and average velocities of the flow.

$$\frac{u}{u_m} = \left(\frac{h}{h_o}\right)^{1/n}$$

where h_0 is the depth of flow. Calculate the values of α and β for $n = 7$.

Solution: Let B be the width of the channel, so that the discharge Q in the channel is given as

$$Q = Bh_0 U = \int_0^{h_0} u\,B\,dh = \int_0^{h_0} B u_m \left(\frac{h}{h_0}\right)^{1/n} dh$$

or

$$U = \frac{u_m}{h_0^{(n+1)/n}} \left[\left(\frac{n}{n+1}\right) h^{(n+1)/n}\right]_0^{h_0}$$

$$= \left(\frac{n}{n+1}\right) u_m$$

$$\therefore \qquad \frac{u_m}{U} = \frac{n+1}{n}$$

$$\therefore \qquad \alpha = 1 + 3\left(\frac{n+1}{n} - 1\right)^2 - 2\left(\frac{n+1}{n} - 1\right)^3$$

$$= 1 + 3\left(\frac{1}{n^2}\right) - 2\left(\frac{1}{n^3}\right)$$

for $n = 7$, $\qquad n = 1 + \dfrac{3}{49} - \dfrac{2}{343} = 1.0554$

Similarly, $\qquad \beta = 1 + \left(\dfrac{n+1}{n} - 1\right)^2 = 1 + \dfrac{1}{n^2}$

for $n = 7$, $\qquad \beta = 1 + \dfrac{1}{49} = 1.0204$

PROBLEMS

P11.1 Mention the type of flow for the following flow conditions:
  (a) Flow in a straight prismatic channel.
  (b) Flow in a curved channel.
  (c) Flow downstream of a dam following the failure (breaking) of the dam.
  (d) Spreading of water on irrigated fields.
  (e) Flow in a channel upstream of a free overfall.

P11.2 Determine the state of flow (whether laminar or turbulent and subcritical or supercritical) in the following channels for $Q = 10$ m^3/s and depth of flow = 2.0 m

  (a) Rectangular channel having bed width of 10 m.
  (b) Triangular channel having side slopes 1V:2H.
  (c) Trapezoidal channel having base width of 10 m and side slope of 1V:2H.
  (d) Circular channel of diameter 5 m.
  (e) Rectangular channel having bed width of 1 m.

P11.3 The following pressures (in mm of water) were measured on a channel wall. Find the pressure coefficient at the base of the wall and determine the force per unit length of the wall.

y	0	50	100	150	200	250	300
p	0	55	110	162	221	281	342

Here, y is the distance (in mm) below the free surface and p is the pressure.

P11.4 The velocity in a wide river varies approximately as $u = 1 + 2(y/h)^2$. Obtain values of the energy and momentum correction factors when depth of flow in the river is 3.5 m.

REFERENCES

  (1) Chow, VT: "*Open-Channel Hydraulics*", McGraw-Hill Book Co., 1959.
  (2) French, RH: "*Open-Channel Hydraulics*", McGraw-Hill Book Co., 1994.

CHAPTER 12

Uniform Flow in Channels

12.1 INTRODUCTION

Uniform flow in open channels is the one in which (i) the depth of flow, area of flow cross-section, velocity of flow, and discharge at every section of the channel reach under consideration are constant at any instant of time, and (ii) the energy line, water surface line which is also the hydraulic grade line, and channel bed are parallel, i.e., their slopes are equal. The requirement of constant velocity at all sections, strictly speaking, means that the flow possesses a constant velocity at every point of any channel section within the uniform flow reach. For real fluids, however, this would mean that the velocity distribution across the channel section remains the same in the channel reach. This requirement of the invariant velocity distribution can be attained when the flow in the channel has been established or fully developed, i.e., the boundary layer has reached the free surface. For simplicity and practical purposes, however, the requirement of constant velocity is considered to have been satisfied if the flow possesses a constant mean velocity in the channel reach.

Uniform flow is considered to be steady only, as unsteady uniform flow, requiring flow parameter such as depth of flow changing simultaneously in the entire channel reach under consideration, is practically non-existent. Natural streams do not meet strict uniform flow conditions. Nevertheless, the uniform flow condition is frequently assumed for the analysis of flow in natural streams with the understanding that the results would be generally approximate.

The laws of flow resistance in open channels are essentially the same as those for pipe flow. That is, the resistance equation for open channel flow too can be obtained by equating the shear force at the boundary with the propulsive force acting in the direction of flow. In case of open channel flow, this propulsive force is due to the component of the weight of the flowing fluid in the direction of flow. One may recall that for pipe flow, the propulsive force is due to the pressure gradient.

The boundary conditions, however, are somewhat different in the two cases. In pipes, usually of circular cross-section, the shear stress opposing the fluid motion is uniformly distributed round the boundary of the flow section. But, for open channel flow, distribution of the boundary shear is non-uniform due to the existence of the free surface (on which the shear stress is negligibly small) and different cross-sectional shapes each of which with its own distribution of shear stress along the solid boundary.

Even if one considers a channel section to be equivalent to one-half of the corresponding pipe section, Fig. 12.1.1, the equivalence would not be exact. In a closed pipe running full, the point of maximum velocity would be at its centre A'. But, for an open channel flow, the maximum velocity often occurs not at the corresponding point A on the free surface, but at a point B little below A. This depression of the point of maximum velocity from A' to A is attributed to the action of secondary currents, i.e., circulating currents in the plane of the channel cross-section, Fig. 12.1.1. Such currents, present in all types of channels, tend to move the floating objects from near the banks towards the centre of the stream.

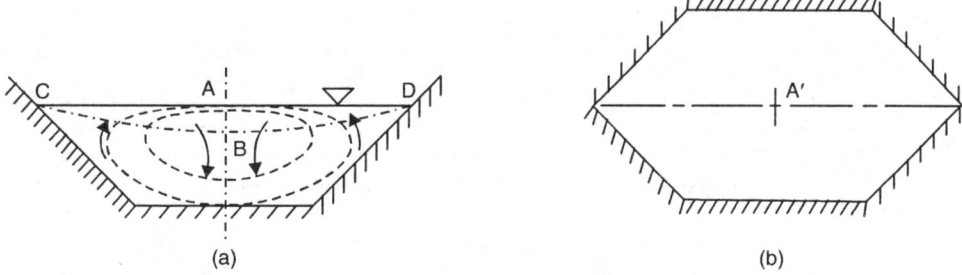

Fig. 12.1.1. Open channel flow and equivalent closed-conduit flow (a) Open channel flow, (b) Equivalent closed conduit flow

The distribution of shear stress on the boundaries of an open channel is affected not only by the asymmetry of the boundary but also by the presence of secondary flows. Therefore, for deriving the resistance equation, it is more convenient to use mean shear stress while balancing the propulsive forces causing motion and the resisting forces opposing the motion.

While it is possible to explain the occurrence of secondary currents or motion, for flow in a curved channel, there is no complete explanation yet for the occurrence of secondary flow in straight channels although it is generally attributed to the presence of the lateral component of turbulence (11). The secondary flow brings in the low momentum fluid from near the bank to the central region where the velocity is otherwise high, Fig. 12.1.2(a). The secondary flow thus reduces the velocity at and near the free surface. Therefore, the point of maximum velocity occurs not at the free surface but at about 0.05 h to 0.25 h below the free surface, Fig. 12.1.2(b). For wider channels, the flow becomes two-dimensional in the central region of the channel and, hence, the reduction in free surface velocity is likely to be small.

12.2 VELOCITY DISTRIBUTION IN OPEN CHANNEL FLOW

Flow in an open channel is mostly of turbulent nature. As such, only turbulent flow would be dealt in greater detail in this and the following chapters. However, laminar flow too has been discussed briefly in this section.

12.2.1 Uniform Laminar Flow in Open Channel

When water flows across a very wide surface such that the depth of flow is very small, the flow is specifically termed as *sheet flow*. In such a flow, the velocity and depth are very small so that viscous effects are more prominent and the flow becomes laminar. Therefore, the Newton's law of viscosity,

$\tau = \mu \dfrac{du}{dy}$, can be applied. For uniform laminar flow, the effective component of propulsive force (weight

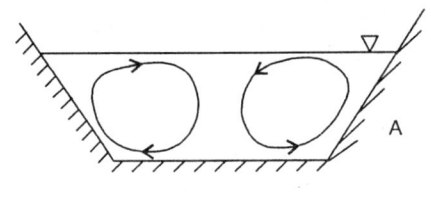

(a) Secondary circulation

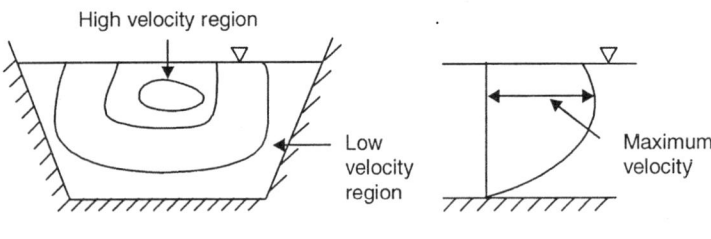

High velocity region

Low velocity region

Maximum velocity

(b) Velocity distribution

Fig. 12.1.2. Secondary circulation and velocity distribution in straight channels

of the flowing liquid) in the direction of flow must be balanced by the frictional force. Therefore, considering unit width of a channel, Fig. 12.2.1, one can write for flow along the layer PP

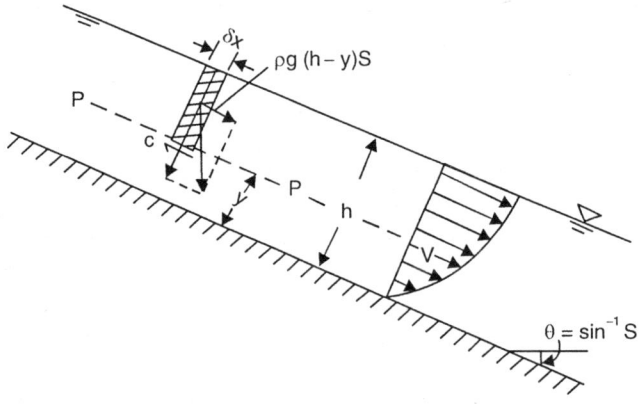

Fig. 12.2.1. Uniform laminar flow in a channel

$$\tau \delta x = \rho g (h - y) \delta x \sin \theta$$

$\therefore$
$$\tau = \rho g (h - y) S$$

where
$$S = \sin \theta$$

or
$$\mu \frac{du}{dy} = \rho g (h - y) S$$

or
$$du = \frac{gS}{v}(h - y)dy$$

Integrating and substituting the condition, $u = 0$ at $y = 0$,

$$u = \frac{gS}{v}\left(hy - \frac{y^2}{2}\right) \qquad \qquad ...(12.2.1)$$

Equation (12.2.1) is a quadratic equation. Hence, uniform laminar flow in a wide open channel has parabolic velocity distribution. The discharge per unit width of channel, q is obtained by integrating the velocity profile, Eq. (12.2.1) between $y = 0$ to $y = h$. Thus,

$$q = \int_o^h \frac{gS}{v}\left(hy - \frac{y^2}{2}\right)dy$$

or
$$q = \frac{gS}{v}\left[\frac{hy^2}{2} - \frac{y^3}{6}\right]_o^h$$

$$\therefore \qquad q = \frac{gSh^3}{3v} \qquad \qquad ...(12.2.2)$$

Hence, the average (or mean) velocity, U is

$$U = \frac{q}{h} = \frac{gSh^2}{3v} \qquad \qquad ...(12.2.3)$$

Uniform laminar flow turns to turbulent flow when the bed surface is relatively rough and the depth of flow is relatively large.

12.2.2 Uniform Turbulent Flow in Open Channel

Karman-Prandtl equations for velocity distribution in turbulent flow, Eqs. (9.4.6) and (9.4.9) are assumed to be applicable to the fully-developed open channel flow too. Keulegan (8) integrated Eqs. (9.4.6) and (9.4.9) to obtain the following expressions for mean velocity U, respectively, for smooth and rough channels:

$$\frac{U}{u_*} = 5.75 \log_{10}^{u_* R/v} + 3.25 \qquad \qquad ...(12.2.4)$$

$$\frac{U}{u_*} = 5.75 \log_{10}^{R/k} + 6.25 \qquad \qquad ...(12.2.5)$$

Here, R is the hydraulic radius and u_* is the shear velocity $(= \sqrt{\tau_o/\rho})$ and k is the height of surface roughness (Table 12.2.1). The numerical constants 3.25 and 6.25 remain constant for open channel flows with Froude number upto 4.0. Hence, Eqs. (12.2.4) and (12.2.5) can be used for most of the open channel flow problems. When Froude number for a flow exceeds 4, the roll waves are likely to develop rendering the flow to be non-uniform. If the boundary is in transition (i.e., $0.25 \le k/\delta' \le 6$), the following equation may be used (4):

$$\frac{U}{u_*} = 5.75 \log_{10} \frac{12.27\,Rx}{k} \qquad\qquad ...(12.2.6)$$

where x is a correction factor for viscous effects and is a function of k/δ' as shown in Fig. 12.2.2.

Table 12.2.1: Approximate values of roughness height k (2)

Material	k, mm
Brass, copper, lead, glass	0.03 – 0.91
Wrought iron, steel	0.06 – 2.44
Asphalted cast iron	0.12 – 2.13
Galvanized iron	0.15 – 4.57
Cast iron	0.24 – 5.49
Timber	0.18 – 0.91
Cement	0.40 – 1.22
Concrete	0.46 – 3.05
Drain tile	0.61 – 3.05
Riveted steel	0.91 – 9.14
Natural river bed	30.48 – 914.40

Application of any of the equations, Eqs. (12.2.4 – 12.2.6), requires the knowledge of u_* or the shear stress τ_0. Applying momentum principle for the control volume between sections 1 and 2 of the channel section shown in Fig. 12.2.3 and assuming that the shear stress τ_0 remains constant over the entire boundary of the channel, one obtains

$$P_1 - P_2 + W \sin\theta - \tau_0 P\delta x = (\beta\rho QU)_2 - (\beta\rho QU)_1 \qquad\qquad ...(12.2.7)$$

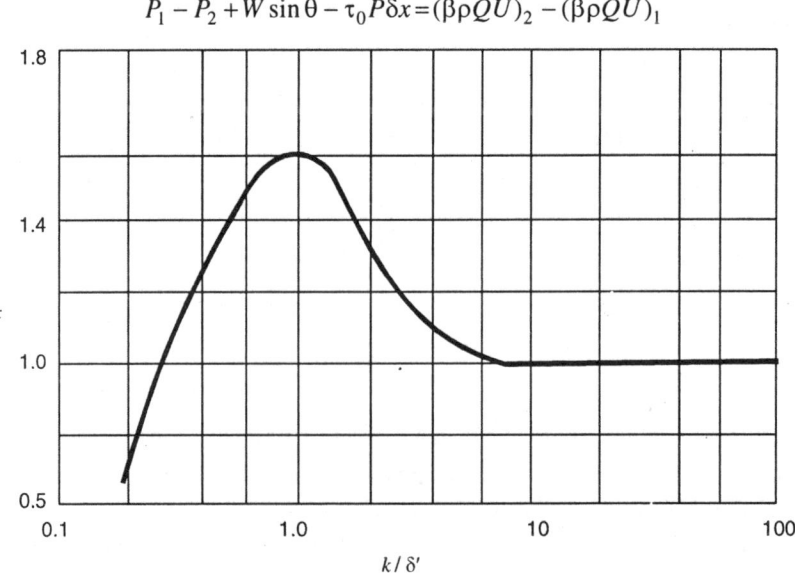

Fig. 12.2.2. Correction factor for use in Eq. (12.2.6) (4)

For uniform flow, however, $P_1 = P_2$, $W = \rho g\, A\delta x$, and the momentum flux $(= \beta \rho Q U)$ at sections 1 and 2 are equal. Also, for small channel bed slopes, $\sin\theta \approx \tan\theta = S$. Therefore, Eq., (12.2.7) reduces to

$$\rho g A\delta x S = \tau_0\, P\delta x$$

or
$$\tau_0 = \rho g\, RS \qquad \qquad ...(12.2.8)$$

∴
$$u_* = \sqrt{\frac{\tau_0}{\rho}} = \sqrt{gRS} \qquad \qquad ...(12.2.9)$$

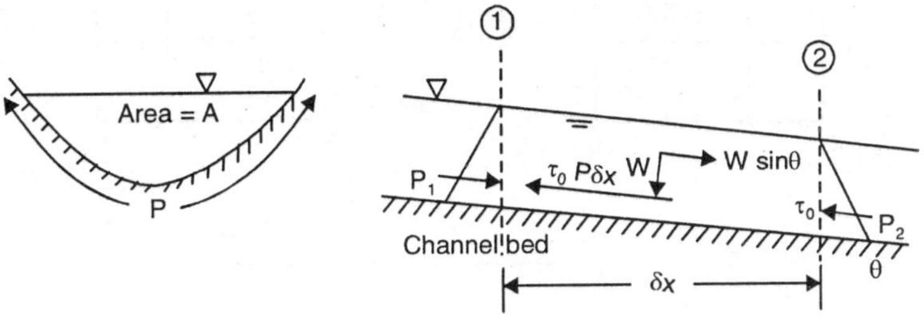

Fig. 12.2.3. Forces on a control volume of water

It should, however, be noted that the shear stress in channels, except for wide channels, is not uniformly distributed along the wetted perimeter, Fig. 12.2.4. The pattern of the shear stress distribution depends upon shape and surface roughness of the channel. Therefore, the local shear stress at any point of the channel boundary may be considerably different than the average value of the shear stress. Figure 12.2.5 shows the typical variation of the maximum shear stress on the bed and sides of smooth trapezoidal and rectangular channels (9). Therefore, for trapezoidal channels, the maximum shear stress on the channel bed is close to ρghs. For the sides of the channel, the maximum shear stress is approximately around $0.75\ \rho ghs$.

12.3 UNIFORM FLOW FORMULAS

Although the logarithmic equations for mean velocity of flow in open channels, Eqs. (12.2.4 – 12.2.6), have been obtained from theoretical considerations, these are less frequently used. Instead, the mean velocity U for turbulent uniform flow in open channels is usually expressed by some empirical power laws that are known as uniform flow formulas. These formulas can be expressed in the following general form:

$$U = C_1\, R^x\, S^y \qquad \qquad ...(12.3.1)$$

Here, x and y are suitable exponents, C_1 is a suitable factor of flow resistance that varies with the mean velocity, hydraulic radius, channel roughness, viscosity, and other factors, R is the hydraulic radius and S is the energy slope S_f. For uniform flow, however, S_f is the same as the water surface slope S_w or the bed slope S_0.

The flow in a natural stream is often approximated as uniform flow when there are no flood flows in the stream or significant variations in the flow caused by irregularities in the stream. It is, however, understood that the results of application of uniform flow formulas to natural channels would be approximate since the flow condition is subject to more uncertain factors than would be involved in a regular artificial channel.

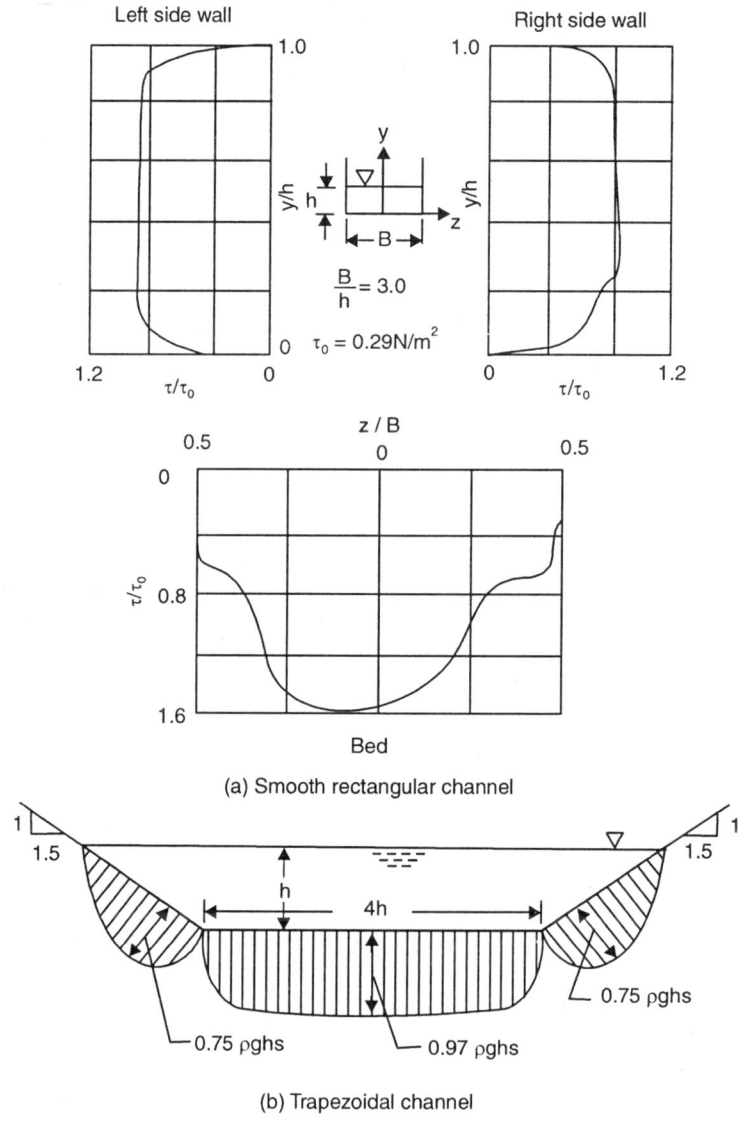

Fig. 12.2.4. Distribution of tractive force channel bed and banks (2, 9)

There are many empirical uniform flow formulas of the type of Eq. (12.3.1). But, none of these takes into account all the variables (flow cross-sectional area A, mean velocity U, velocity of flow at the free surface, wetted perimeter P, hydraulic radius R, the maximum depth of flow section, slope of the water surface, channel roughness, suspended sediment load, bed load, and viscosity of the flowing water) affecting the flow so as to qualify as a good formula. The best known and commonly used formulas for uniform flow are the Chezy's equation and Manning's equation. Therefore, only these two uniform flow equations have been discussed in this book.

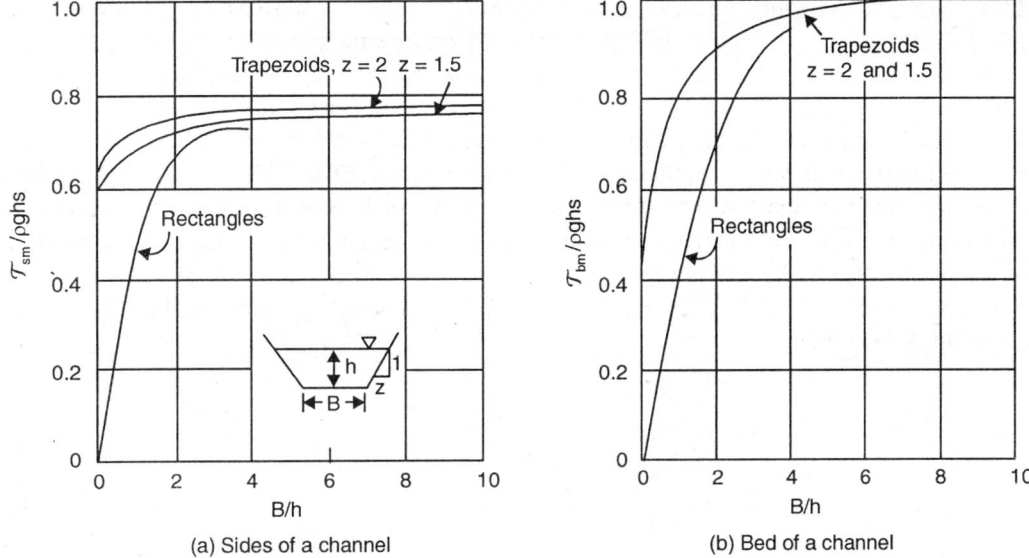

Fig. 12.2.5. Maximum tractive stress in a channel (5, 9)

12.3.1 Chezy's Equation

Using the expression for local friction coefficient for boundary layer over a flat plate (Art. 8.5), one may, for a channel boundary, write Eq. (12.2.8) as

$$c_f \frac{1}{2}\rho U^2 = \rho g R S$$

Therefore, $\qquad\qquad\qquad U = C\sqrt{RS}$ ...(12.3.2)

in which $C = \sqrt{\dfrac{2g}{c_f}}$ and is known as the Chezy's coefficient with dimensions of $[L^{1/2}/T]$. The Chezy's coefficient C is expected to depend on the Reynolds number UR/ν for smooth turbulent flow $(k/\delta' < 0.25)$ and k/R for rough turbulent flow $(k/\delta' > 6.0)$. For $0.25 \le k/\delta' \le 6.0$, C depends on both UR/ν as well as k/R. Equation (12.3.2) is known as the Chezy's equation. Several empirical expressions, based on systematic observations on rivers and large channels, were proposed for the Chezy's coefficient C. Based on these data, Hagen in 1881 stated that C varies as the sixth root of the hydraulic radius R.

12.3.2 Manning's Equation

An Irish engineer Robert Manning in 1889 presented a formula by incorporating Hagen's finding (i.e., $C \alpha R^{1/6}$) into the Chezy's formula. Manning's formula was later modified to its well-known form

$$U = \frac{1.486}{n} R^{2/3} S^{1/2}$$ ...(12.3.3)

in *FPS* system of units with R in ft, U in ft/s, and n and S dimensionless. Here, S is the slope of the energy gradient line and n, specifically known as the Manning's roughness (or rugosity) coefficient or, simply, the Manning's n, is the coefficient of roughness or rugosity and is dimensionless. This formula is based on large amount field data. It should be noted here that if the number 1.486, appearing in Eq. (12.3.3), were to be treated dimensionless, n would have the dimensions of $TL^{-1/3}$ which would mean that the roughness coefficient n would depend upon time parameter too. This, obviously, is unreasonable. Therefore, the possible way out is to treat the numerator of $1.486/n$ as dimensional number having the dimensions of $L^{1/3}T^{-1}$, thus leaving no dimensions for n. The equivalent of Eq. (12.3.3) in SI system of units can be obtained by finding the equivalent of 1.486 $(ft)^{1/3}/sec$ in S.I. system which would, obviously, be $1.486(0.3048)^{1/3}$ m$^{1/3}$/s = 1.0 m$^{1/3}$/s as 1 ft = 0.3048 m. Therefore, Eq. (12.3.3), in S.I. system of units, becomes

$$U = \frac{1}{n} R^{2/3} S^{1/2}$$

...(12.3.4)

Also,

$$Q = UA = \frac{1}{n} AR^{2/3} S^{1/2}$$

...(12.3.5)

Here, Q is the discharge in m^3/s, U is in m/s, R is in m and n and S are dimensionless. The numerical constant 1 has the units of m$^{1/3}$/s. And, the value of n, therefore, remains the same in both systems of units.

Henderson (6) has opined that the Manning's equation is suitable for all fully rough flows provided that

$$n^6 \sqrt{RS} \geq 3.44 \times 10^{-13}$$

Because of simplicity of the form of the Manning's equation, and satisfactory results it yields for practical problems of open channel flow, the Manning's equation has become the most widely used uniform flow resistance equation.

12.3.3 Estimation of the Manning's n

The application of the Manning's equation requires the estimation of the value of the Manning's n which is rather difficult as there is no exact method to select it. The difficulty in estimating the value of n can be easily appreciated by the fact that n depends on a number of factors. Some of these factors that exert considerable influence on the value of n are as follows (2):

(i) Roughness of the channel surface depends upon the size and shape of the grains of the material of the channel boundary. For finer material, such as sand, the value of n is less and remains unaffected by the changes in depth of flow. On the other hand, channel surface comprising of gravel and boulders would yield larger n that may vary with the depth of flow. In general, n would be relatively higher at low stage of the flow and relatively smaller at high stage of the flow.

(ii) Vegetation on the bed and sides of the channel can also be regarded as surface roughness. Besides, it reduces the channel capacity too. The value of n would depend upon the depth of flow as well as the height, density, distribution, and type of vegetation.

(iii) Irregularities in channel section may be due to variations in cross-section, size, and shape along the channel length. While a gradual change may not appreciably affect the value of n, abrupt change necessitates the adoption of a relatively larger n.

(iv) Channel alignment with curves of large radii without frequent changes in the curvature results in relatively low value of n. But, meandering alignment would result into higher value of n.

(v) Silting and scouring also affect the value of n. Silting, generally, tends to make an otherwise irregular channel into a regular channel, if the material being deposited on the channel bed is relatively fine. Therefore, the value of n is reduced. However, uneven deposits may result in increased value of n. Uniform scouring does not affect much the value of n. But, non-uniform and uneven scouring may result in more rough bed and, hence, increased value of n.

(vi) Obstructions such as fallen tress, bridge piers, debris or log jams can significantly increase the value of n depending upon the number, shape, and size of the obstruction.

(vii) Stage and discharge in the channel affect the value of n inversely. That is, increase in stage (and, hence, in discharge) decreases the value of n in most of the streams. For shallow depths, the bed roughness would have relatively more pronounced resistance to flow. If the banks at higher elevations are relatively rough and grassy, the value of n may increase with the increase in stage.

(viii) Suspended material and bed load would consume energy and, hence, cause additional head loss thereby implying increased value of n.

Cowan (3) has suggested the following expression for estimating the value of the Manning's n taking into account the main factors that affect the roughness coefficient:

$$n = (n_0 + n_1 + n_2 + n_3 + n_4) m_5 \qquad \qquad ...(12.3.6)$$

in which n_0 is the basic value (Table 12.3.1) for the Manning's n for a straight, uniform, and smooth channel in the natural materials and n_1, n_2, n_3, and n_4 are the values (Table 12.3.1) added to account for the effects of surface irregularities, variations of channel cross-section, obstructions, and vegetation and flow condition, respectively. The variable m_5 (Table 12.3.1) is a correction factor for the meandering of channel. Typical values of n for different channel surfaces are given in Table 12.3.2(2).

On equating the values of U obtained from Eq. (12.3.4) and Eq. (12.2.5), one gets

$$\frac{1}{n} R^{2/3} S^{1/2} = \sqrt{gRS} \; [5.75 \log_{10}^{R/k} + 6.25]$$

or

$$\frac{R^{1/6}}{n\sqrt{g}} = 5.75 \log_{10}^{R/k} + 6.25 \qquad \qquad ...(12.3.6)$$

By plotting $\dfrac{R^{1/6}}{n\sqrt{g}}$ versus $\dfrac{R}{k}$ on a logarithmic graph paper, Eq. (12.3.6) can be approximated as

$$\frac{R^{1/6}}{n\sqrt{g}} = 8.16 \left(\frac{R}{k} \right)^{1/6} \qquad \qquad ...(12.3.7)$$

for $5 < \dfrac{R}{k} < 700$. That is, for this range of R/k,

$$n = \frac{k^{1/6}}{25.6} \qquad \qquad .\;(12.3.8)$$

Table 12.3.1: Values for the computation of the Manning's *n* by Eq. (12.3.6) (2)

Channel conditions			Values
Channel bed material	Earth	n_0	0.020
	Rock cut		0.025
	Fine gravel		0.024
	Coarse gravel		0.028
Degree of irregularity	Smooth	n_1	0.000
	Minor		0.005
	Moderate		0.010
	Severe		0.020
Variations in channel cross-section	Gradual	n_2	0.000
	Alternating occasionally		0.005
	Alternating frequently		0.010-0.015
Relative effect of obstructions	Negligible	n_3	0.000
	Minor		0.010-0.015
	Appreciable		0.020-0.030
	Severe		0.040-0.060
Vegetation	Low	n_4	0.005-0.010
	Medium		0.010-0.025
	High		0.025-0.050
	Very high		0.050-0.100
Degree of meandering	Minor	m_5	1.000
	Appreciable		1.150
	Severe		1.300

in SI units (i.e., *k* is in meters). Equation (12.3.8) is often called as Strickler's equation after Strickler who developed the same equation using the data of gravel-bed streams in which *k* was the median size d_{50} of the bed material. Therefore, Eq. (12.3.8) can also be written as

$$n = \frac{d_{50}^{1/6}}{25.6} = 0.039\, d_{50}^{1/6} \qquad \qquad ...(12.3.9)$$

For bed materials having significant proportion of coarse–grained size, Meyer–Peter and Müller (10) suggested the following equation for *n* :

$$n = \frac{d_{90}^{1/6}}{26.3} = 0.038\, d_{90}^{1/6} \qquad \qquad ...(12.3.10)$$

Yet another method for estimation for the Manning's *n* is based on velocity measurements as suggested by Chow (2). The velocity profile for hydrodynamically rough turbulent flow depends on the

Table 12.3.2: Typical values of the Manning's *n* (12)

Channel surface	Manning's n
Artificial lined channels	
Glass	0.010 ± 0.002
Brass	0.011 ± 0.002
Steel, smooth	0.012 ± 0.002
Painted	0.014 ± 0.003
Riveted	0.015 ± 0.002
Cast iron	0.013 ± 0.003
Concrete, finished	0.012 ± 0.002
Unfinished	0.014 ± 0.002
Planed wood	0.012 ± 0.002
Clay tile	0.014 ± 0.003
Brickwork	0.015 ± 0.002
Asphalt	0.016 ± 0.003
Corrugated metal	0.022 ± 0.005
Rubble masonry	0.025 ± 0.005
Excavated earth channels	
Clean	0.022 ± 0.004
Gravelly	0.025 ± 0.005
Weedy	0.030 ± 0.005
Stony, cobbles	0.035 ± 0.010
Natural channels	
Clean and straight	0.030 ± 0.005
Sluggish, deep pools	0.040 ± 0.010
Major rivers	0.035 ± 0.010
Floodplains	
Pasture, farmland	0.035 ± 0.010
Light brush	0.05 ± 0.02
Heavy brush	0.075 ± 0.025
Trees	0.15 ± 0.05

surface roughness height, Eq. (9.4.10) with $y' = k/30$ and $K = 0.4$, and is given as

$$u = 5.75\, u_* \log_{10} \frac{30\,y}{k} \qquad \qquad ...(12.3.11)$$

If $u_{0.2}$ and $u_{0.8}$ are the velocities measured at, say, $0.2h$ and $0.8h$ below the free surface, respectively, so that,

$$u = u_{0.2} \text{ at } y = 0.8h$$

and
$$u = u_{0.8} \text{ at } y = 0.2h$$

On substituting these values of u in Eqs. (12.3.11), one gets

$$u_{0.2} = 5.75 u_* \log_{10} \frac{24h}{k}$$

and
$$u_{0.8} = 5.75 u_* \log_{10} \frac{6h}{k}$$

$\therefore$
$$x = \frac{u_{0.2}}{u_{0.8}} = \frac{\log \dfrac{24h}{k}}{\log \dfrac{6h}{k}} = \frac{1.38 + \log h/k}{0.778 + \log h/k}$$

$\therefore$
$$(1-x) \log^{h/k} = 0.778 x - 1.38$$

or
$$\log^{h/k} = \frac{0.778 x - 1.38}{1 - x} \qquad \qquad ...(12.3.12)$$

Substituting Eq. (12.3.12), with the assumption $h \approx R$, in Eq. (12.2.5), one gets

$$\frac{U}{u_*} = 5.75 \left[\frac{0.778 x - 1.38}{1 - x} \right] + 6.25$$

or
$$\frac{U}{u_*} = \frac{1.78 x + 1.685}{x - 1} \qquad \qquad ...(12.3.13)$$

Combining Eqs. (12.3.2), (12.2.9), (12.3.4), and (12.3.13) one gets

$$\frac{U}{u_*} = \frac{c}{\sqrt{g}} = \frac{R^{1/6}}{n\sqrt{g}} = \frac{1.78 x + 1.685}{x - 1}$$

or
$$n = \frac{(x-1) R^{1/6}}{5.58 x + 5.28} \approx \frac{(x-1) h^{1/6}}{5.58 x + 5.28} \qquad \qquad ...(12.3.14)$$

Equation (12.3.14), thus, enables the computation of n from two velocity measurements in a channel section at $0.2h$ and $0.8h$ below the free surface. Equation (12.3.14), however, needs further verification (5).

12.4 COMPUTATION OF UNIFORM FLOW

The discharge Q for uniform flow in any channel is obtained by multiplying the average velocity U obtained from uniform flow formula, Eq. (12.3.2) or Eq. (12.3.4) or Eq. (12.3.5), and the area of the flow section A. Thus,

$$Q = UA = C_1 A R^x \sqrt{S} \qquad \qquad ...(12.4.1)$$

in which $C_1 = C$ and $x = \dfrac{1}{2}$ for the Chezy's formula, and $C_1 = \dfrac{1}{n}$ and $x = \dfrac{2}{3}$ for the Manning's formula.

Equation (12.4.1) can, alternatively, be written as

$$Q = K\sqrt{S} \qquad \qquad \ldots(12.4.2)$$

so that the term K, known as *conveyance* of the channel, is given as

$$K = C A \sqrt{R} \qquad \qquad \ldots(12.4.3)$$

for the Chezy's formula, and

$$K = \frac{1}{n} AR^{2/3} \qquad \qquad \ldots(12.4.4)$$

for the Manning's formula.

Equation (12.4.1) indicates that the average velocity of flow, U or the volumetric flow rate, i.e., discharge Q depend on (i) the resistance coefficient Manning's n or Chezy's C, (ii) the channel shape, (iii) the longitudinal slope of the channel, S, and (iv) the depth of flow. Equation (12.4.4) can be combinted with Eq. (12.4.2) to get

$$AR^{2/3} = \frac{Qn}{\sqrt{S}} \qquad \qquad \ldots(12.4.5)$$

The term $AR^{2/3}$ is called the *section factor for uniform flow computation* and may be assigned the symbol z_1. Equation (12.4.5) is valid for uniform flow in any channel section. While the right side of Eq. (12.4.5) depends on Q, n, and S, the left side of the equation depends only on the geometry of the flow section. Hence, for specified condition of Q, n, and S in a channel, there would be only one possible depth for maintaining uniform flow in the channel provided that the value of $AR^{2/3}$ always increases with increase in depth which is true in most cases except in closed channels having a gradually closing top as in circular channel sections. This depth is called the *normal depth* and may be assigned the symbol h_n. Likewise, for given n and S for a channel section, there can be only one discharge, known as *normal discharge*, for maintaining a uniform flow through the section provided, again, that $AR^{2/3}$ always increases with increase in depth.

Manning's equation, Eq. (12.3.4) or Eq. (12.4.5), presents no computational difficulty for obtaining the value of either Q or S for known values of n, the depth of flow, channel shape and either S or Q. However, if the depth of flow, i.e., normal depth is to be determined for given channel shape, Q, n, and S, a trial solution would be required except in the simple case when the channel is very wide (so that $P \approx B$) or R varies exponentially with the depth of flow. The computation of the normal depth can be simplified by preparing (see Art. 12.4.1 and 12.4.2) dimensionless section factor curves for the relationship between the depth of flow h and section factor $AR^{2/3}$ for different channel shapes, Fig. 12.4.1. For given n, Q, and S, one can determine $AR^{2/3}$ using Eq. (12.4.5) and, thereafter, determine the depth h from Fig. 12.4.1. This value of h is, obviously, the normal depth of flow obtained without trial.

12.4.1 Section Factor Curves for Trapezoidal and Rectangular Channels

Considering a trapezoidal channel section of bed width B, side slopes z (horizontal): 1 (vertical), one can write

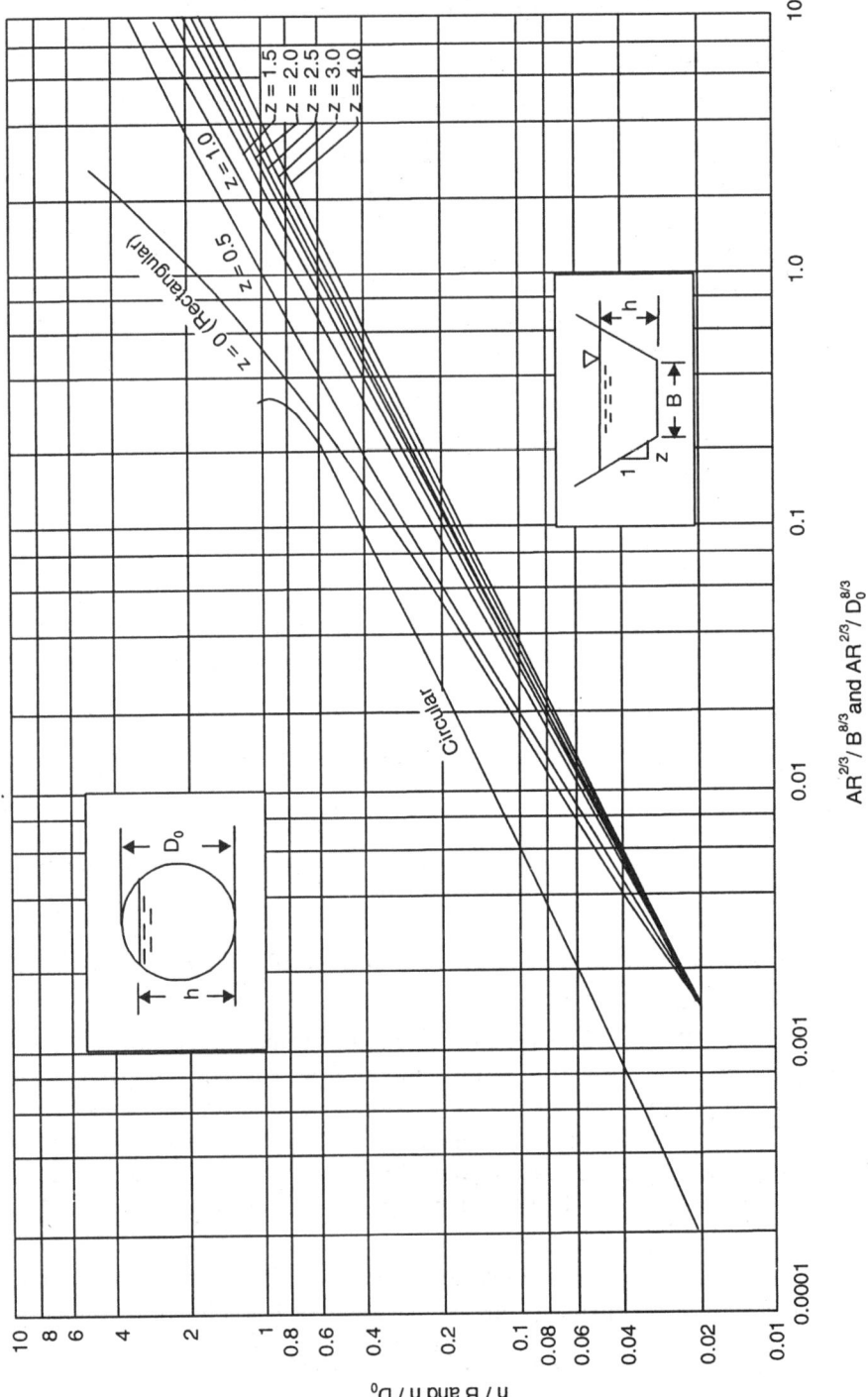

Fig. 12.4.1. Dimensionless section factor curves for computation of normal depth (2)

$$z_1 = AR^{2/3} = (B + zh)h \left[\frac{(B+zh)h}{B + 2h\sqrt{z^2 + 1}} \right]^{5/3} \qquad ...(12.4.6)$$

On dividing both sides by $B^{8/3}$, one gets

$$\frac{z_1}{B^{8/3}} = \frac{AR^{2/3}}{B^{8/3}} = \frac{[(h/B)(1 + zh/B)]^{5/3}}{[1 + (2h/B)\sqrt{z^2 + 1}]^{2/3}} \qquad ...(12.4.7)$$

For rectangular channel ($z = 0$), Eq. (12.4.7) becomes

$$\frac{z_1}{B^{8/3}} = \frac{AR^{2/3}}{B^{8/3}} = \frac{(h/B)^{5/3}}{\left(1 + \dfrac{2h}{B}\right)^{2/3}} \qquad ...(12.4.8)$$

Using Eq. (12.4.7), one can, therefore, obtain $z_1 / B^{8/3}$ for different values of h/B and z and obtain the relationship between $z_1 / B^{8/3}$ and h/B for different values of z, Fig. 12.4.1.

For given channel shape (i.e., B and z), Q, n, and S, one can now determine the normal depth h_n by the following procedure requiring no trial:

(i) Obtain $z_1 = \dfrac{nQ}{\sqrt{S}}$ and $z_1 / B^{8/3}$.

(ii) Read the value of $\dfrac{h}{B}$ for the known values of $z_1 / B^{8/3}$ and z from Fig. 12.4.1.

(iii) Determine the value of h which, obviously, is the required normal depth h_n.

One can, alternatively, write a computer program and obtain the value of h_n using the Manning's equation. Equation (12.4.6) could also be made dimensionless by dividing the equation with $h^{8/3}$ instead of $B^{8/3}$ and, thus, obtaining (11)

$$\frac{z_1}{h^{8/3}} = \frac{(1 + zh/B)^{5/3}}{\dfrac{h}{B}\left[1 + 2\sqrt{1 + z^2}\,\dfrac{h}{B}\right]^{2/3}} \qquad ...(12.4.9)$$

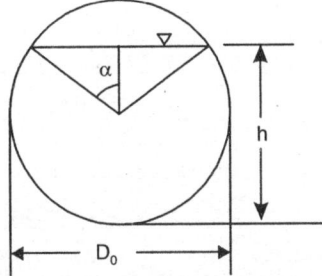

Fig. 12.4.2. Circular channel section

One can now prepare a plot between $\dfrac{h}{B}$ and $\dfrac{z_1}{h^{8/3}}$ for different values of z. This plot can be used to determine the value of B for known normal depth h_n (i.e., h in Eq. (12.4.9)) in a way similar to that used for determining the normal depth h_n using Fig. 12.4.1.

12.4.2 Section Factor for Circular Channels

For a circular channel of diameter D_0 and depth of flow equal to h, Fig. 12.4.2, the section factor z_1 is computed as follows (11):

$$\cos\alpha = \frac{h - \dfrac{D_0}{2}}{D_0/2} = \frac{2h - D_0}{D_0} = \frac{2h}{D_0} - 1$$

$$P = \pi D_0 - \frac{D_0}{2}(2\alpha) = D_0(\pi - \alpha)$$

$$A = \frac{\pi D_0^2}{4} - \frac{1}{2}\left(\frac{D_0}{2}\right)\left(\frac{D_0}{2} \times 2\alpha\right) + 2\left(\frac{1}{2} \times \frac{D_0}{2}\cos\alpha \times \frac{D_0}{2}\sin\alpha\right)$$

$$= \frac{\pi D_0^2}{4} - \frac{D_0^2}{4}(\alpha) + \frac{D_0^2}{8}\sin 2\alpha$$

$$\therefore \qquad A = \frac{D_0^2}{4}\left[\pi - \alpha + \frac{1}{2}\sin 2\alpha\right]$$

Hence,
$$R = \frac{A}{P} = \frac{D_0}{4}\frac{\left[\pi - \alpha + \dfrac{1}{2}\sin 2\alpha\right]}{(\pi - \alpha)}$$

$$AR^{2/3} = \frac{\left[\pi - \alpha + \dfrac{1}{2}\sin 2\alpha\right]^{5/3}}{(\pi - \alpha)^{2/3}}\frac{D_0^{8/3}}{4^{5/3}}$$

$$\therefore \qquad \frac{AR^{2/3}}{D_0^{8/3}} = \frac{\left[\pi - \alpha + \dfrac{1}{2}\sin 2\alpha\right]^{5/3}}{4^{5/3}(\pi - \alpha)^{2/3}} \qquad\qquad ...(12.4.10)$$

$$= f(\alpha)$$

$$= f\left(\frac{h}{D_0}\right) \text{ as } \cos\alpha = \frac{2h}{D_0} - 1$$

i.e.,
$$\frac{z_1}{D_0^{8/3}} = f\left(\frac{h}{D_0}\right)$$

Therefore, a plot of $\dfrac{z_1}{D_0^{8/3}}$ versus $\dfrac{h}{D_0}$, Fig. 12.4.1, can be prepared using Eq. (12.4.10) and can be used for estimating the value of the normal depth in a way similar to that used for determining the normal depth for trapezoidal channels.

12.4.2.1 Characteristics of flow in a circular channel

For given Q, n, and S, there would be only one possible depth for maintaining uniform flow provided that the value of $AR^{2/3}$ always increases with increase in depth of flow. In majority of the channels, this condition is satisfied. However, in channels of reducing top width, such as circular channels, this condition is not satisfied. Therefore, one can expect more than one normal depth of flow for some flow conditions in a circular channel.

If the subscript "o" indicates the full flow condition, one can obtain the following using the Manning's equation (11):

$$\frac{Q}{Q_0}=\frac{AR^{2/3}}{A_0R_0^{2/3}}=\frac{\left(\pi-\alpha+\dfrac{1}{2}\sin 2\alpha\right)^{5/3}}{\pi(\pi-\alpha)^{2/3}} \qquad\qquad ...(12.4.11)$$

and

$$\frac{U}{U_0}=\frac{R^{2/3}}{R_0^{2/3}}=\left[\frac{\pi-\alpha+\dfrac{1}{2}\sin 2\alpha}{\pi-\alpha}\right]^{2/3} \qquad\qquad ...(12.4.12)$$

for given n that is invariant with depth of flow. Obviously, the right side of both these equations would depend on h/D_o. Figure 12.4.3 shows the variation of $AR^{2/3}/A_0R_0^{2/3}$ (or Q/Q_0) and $R^{2/3}/R_0^{2/3}$ (or U/U_0 with h/D_0 for a circular channel. Here, Q_0 and U_0 are, respectively, the discharge and velocity when the conduit just becomes full, i.e. when $\alpha = 0$.

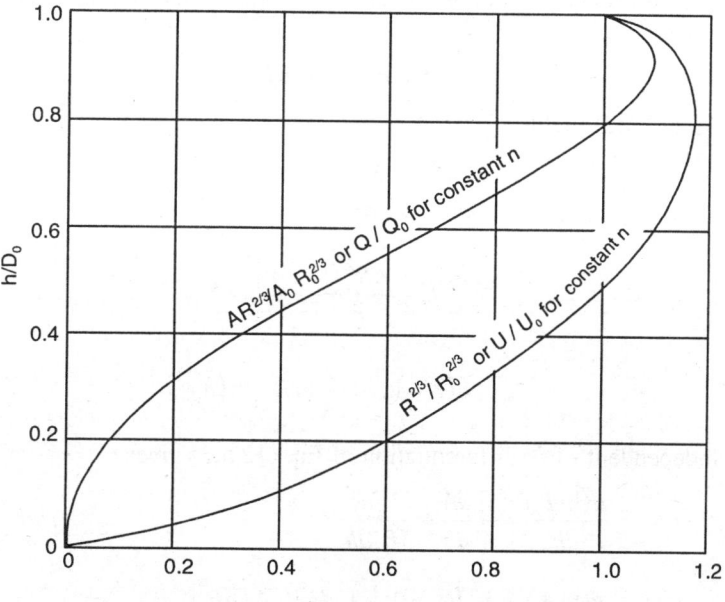

Fig. 12.4.3. Flow characteristics of a circular channel section(2)
(Subscript "o" indicates the full flow condition)

By equating the derivative of $AR^{2/3}$ with respect to h to zero, one can obtain the condition for the maximum discharge in a channel since, according to the Manning's formula, the discharge is proportional to $AR^{2/3}$ for given n and S. The condition for maximum discharge yields $h = 0.938\ D_0$. Similarly, the condition for maximum velocity yields $h = 0.81\ D_0$. Also, the dimensionless curves of Fig. 12.4.3 indicate that when the depth is greater than about $0.82\ D_0$, it is possible to have two different depths for the same discharge Q, one above and other below the value of $0.938\ D_0$ at which depth the channel would carry the maximum discharge. Likewise, when the depth is greater than $0.5\ D_0$, it is possible to have two different depths of flow for the same velocity, one above and other below the value of 0.81 D_0 at which depth occur the maximum velocity in a circular channel. One may, therefore, conclude that for some range of depths in any closed conduit with a gradually closing top, the section factor for uniform flow computation, i.e., $AR^{2/3}$ is not a single-valued function. However, the exact value of the depths for maximum discharge and velocity would depend upon the shape of the conduit.

Because of the two possible values of normal depths, as discussed above, circular conduits with gradually closing top are designed such that the depth of flow in the conduit does not exceed $0.8\ D_o$. In the event of the depth exceeding $0.8\ D_0$, waves and backwater effect may change the depth to other normal depth and eventually equal to the full depth, i.e., D_0 (2).

12.5 HYDRAULIC EXPONENT FOR UNIFORM FLOW COMPUTATIONS

It would be advantageous to obtain a relation between conveyance K of a channel and the depth of flow h by assuming that

$$K^2 = c_1 h^N \qquad \qquad ...(12.5.1)$$

where c_1 is a constant and N is called *hydraulic exponent for uniform flow computation*. In gradually varied flow, Eq. (12.5.1) is often used for convenience in computations. Taking logarithms on both sides of Eq. (12.5.1), one gets

$$2\ln K = \ln c_1 + N \ln h \qquad \qquad ...(12.5.2)$$

Differentiation of Eq. (12.5.2) yields

$$\frac{d(\ln K)}{dh} = \frac{N}{2h} \qquad \qquad ...(12.5.2)$$

Since

$$K = \frac{AR^{2/3}}{n}$$

∴

$$\ln K = \ln A + \frac{2}{3}\ln R - \ln(n) \qquad \qquad ...(12.5.3)$$

Assuming n to be independent of h, differentiation of Eq. (12.5.3) gives

$$\frac{d(\ln K)}{dh} = \frac{1}{A}\frac{dA}{dh} + \frac{2}{3R}\frac{dR}{dh}$$

or

$$\frac{d(\ln K)}{dh} = \frac{1}{A}\frac{dA}{dh} + \frac{2}{3R}\left[\frac{1}{P}\frac{dA}{dh} - \frac{A}{P^2}\frac{dP}{dh}\right] \qquad \qquad ...(12.5.4)$$

as

$$R = \frac{A}{P}$$

Equation (12.5.4) can further be simplified to

$$\frac{d(\ln K)}{dh} = \frac{1}{A}\frac{dA}{dh} + \frac{2}{3A}\frac{dA}{dh} - \frac{2}{3}\frac{R}{A}\frac{dP}{dh}$$

$$\therefore \qquad \frac{d(\ln K)}{dh} = \frac{1}{3A}\left[5\frac{dA}{dh} - 2R\frac{dP}{dh}\right] \qquad\qquad ...(12.5.5)$$

Considering change in flow section dA on account of change in depth of flow by dh, Fig. 12.5.1, one observes that

$$T\,dh = dA$$

Hence,

$$\frac{dA}{dh} = T$$

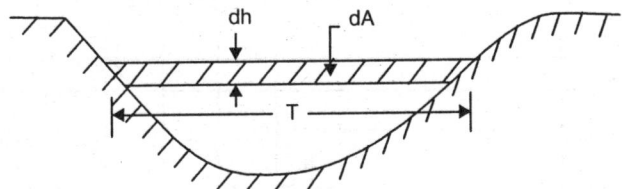

Fig. 12.5.1

Substituting the value of $\dfrac{dA}{dh}$ in Eq. (12.5.5), one gets

$$\frac{d(\ln K)}{dh} = \frac{1}{3A}\left[5T - 2R\frac{dP}{dh}\right] \qquad\qquad ...(12.5.6)$$

Comparing Eqs. (12.5.2) and (12.5.6), one gets

$$\frac{N}{2h} = \frac{1}{3A}\left[5T - 2R\frac{dP}{dh}\right] \qquad\qquad ...(12.5.7)$$

Equation (12.5.7) is the general equation for the hydraulic exponent N and is valid for all shapes of channels. For a trapezoidal channel with bed width B and side slopes of z(H):1(V), the expressions for A, P, T, R, and $\dfrac{dP}{dh}$ are

$$A = (B + zh)h$$

$$P = B + 2h\sqrt{1 + z^2}$$

$$T = B + 2zh$$

$$R = \frac{(B + zh)h}{B + 2h\sqrt{1 + z^2}}$$

and
$$\frac{dP'}{dh} = 2\sqrt{1+z^2}$$

Substituting these values in Eq. (12.5.7)

$$N = \frac{10}{3}h\frac{B+2zh}{(B+zh)h} - \frac{4h}{3}\frac{(2\sqrt{1+z^2})}{B+2h\sqrt{1+z^2}}$$

$$N = \frac{10}{3}\left[\frac{1+2zh/B}{1+zh/B}\right] - \frac{8}{3}\left[\frac{(h/B)(\sqrt{1+z^2})}{1+2(h/B)(\sqrt{1+z^2})}\right] \qquad ...(12.5.8)$$

Equation (12.5.8) shows that the hydraulic exponent N for trapezoidal channels is a function of z and h/B. The curves for N versus h/B for different values of z, Fig. 12.5.2, indicate that the value of N for

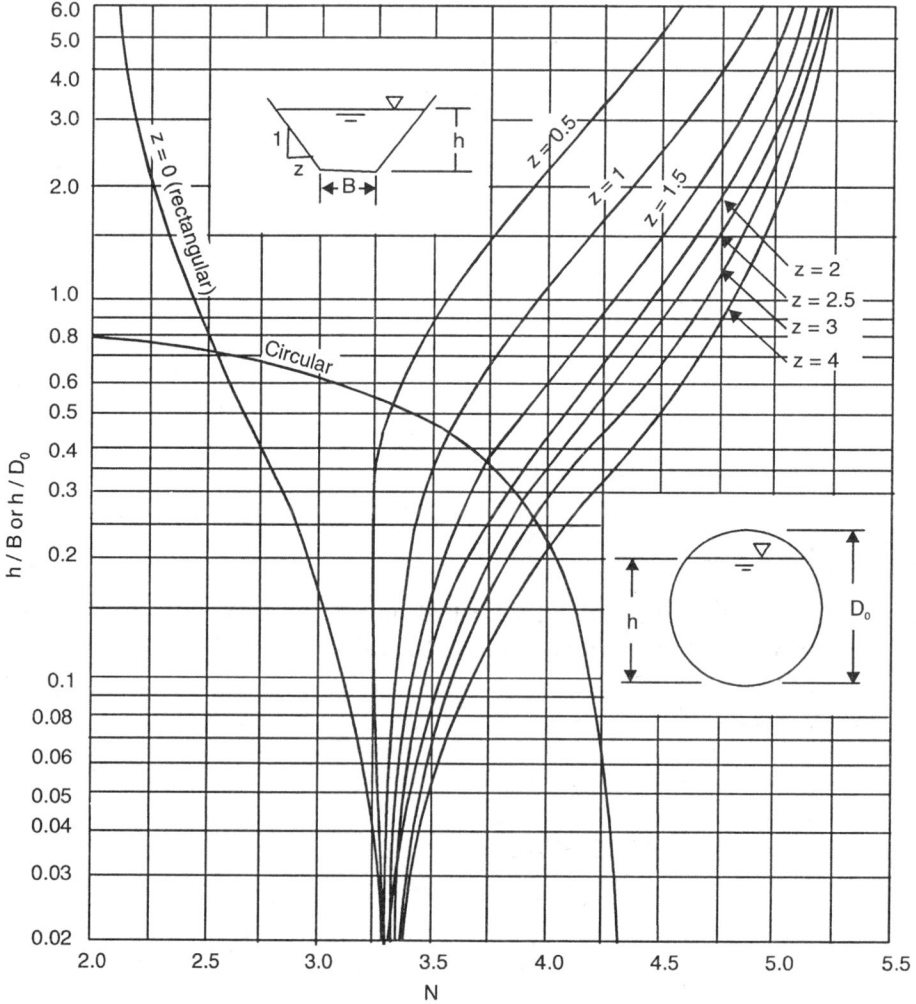

Fig. 12.5.2. Variation of hydraulic exponent, N (2)

trapezoidal channels vary within a range of 2.0 to 5.3 for the values of z between zero (i.e., rectangular channel) and 4.

Figure 12.5.2 also shows the variation of N with h/D_0 for a circular channel. One may observe that the value of N decreases rapidly as h/D_0 approaches unity. In fact, mathematical analysis indicates (2) that N will be equal to *zero* at $h/D_0 = 0.938$ (at which the discharge would be maximum, Art. 12.4) and becomes negative for h/D_0 exceeding 0.938.

For channel sections other than the rectangular, trapezoidal, and circular shapes, one can, similarly, prepare curves for the variation of N with depth h using Eq. (12.5.7) provided that dP/dh can be evaluated. For most of the channel shapes (including the irregular ones which cannot be described by standard geometrical shapes and excluding the channels with abrupt changes in their cross-sectional form and the closed conduits with gradually closing top), a logarithmic plot of $K(=(1/n)AR^{2/3})$ as ordinate against the depth h as abscissa, in accordance with Eq. (12.5.2), can be approximated as straight line. Using the coordinates of any two points on the straight line, one can compute the value of N from the equation (2),

$$N = 2\frac{\log(K_1/K_2)}{\log(h_1/h_2)} = 2 \times \text{slope of the line}$$

Alternatively, one may prefer, in accordance with Eq. (12.5.1), a logarithmic plot of K^2 as ordinate against the depth h as abscissa. This plot is, again, approximated by a straight line. The slope of this line would be equal to the average value of N for the range of depths for which the curve has been approximated as a straight line.

For the curved portion of the plot, however, the value of N for a given depth can be computed using the slope of the tangent to the curve at that depth.

12.6 FLOW IN A CHANNEL SECTION WITH VARYING ROUGHNESS

Even in a simple channel section, the roughness along the wetted perimeter may be significantly different in different parts of the perimeter. For example, a rectangular flume of concrete bed and glass walls must have different n values for the bed and walls. In applying the Manning's formula one should, therefore, compute value of an *equivalent roughness n* for the entire perimeter and use it for the uniform flow formula, i.e., the Manning's formula. For this purpose, the flow section is imaginatively divided into, say, J parts for which the wetted perimeters $P_1, P_2, \ldots, P_J$ and the coefficients of roughness $n_1, n_2, \ldots, n_J$ are known. Assuming that each of the J parts has the same mean velocity as that of the entire flow section (7), one may write

$$A = \frac{U^{3/2}n^{3/2}P}{S^{3/4}} = \sum_{i=1}^{J}\frac{U^{3/2}n_i^{3/2}P_i}{S^{3/4}}$$

$\therefore$

$$n = \left[\frac{1}{P}\sum_{i=1}^{J}P_i n_i^{3/2}\right]^{2/3}$$

or

$$n = \frac{(P_1 n_1^{1.5} + P_2 n_2^{1.5} + \ldots\ldots P_J n_J^{1.5})^{2/3}}{P^{2/3}} \qquad \ldots(12.6.2)$$

Here, P is the wetted perimeter for the entire flow section and S is the longitudinal slope of the channel.

12.7 CHANNELS OF COMPOUND SECTIONS

The cross-section of a channel, at times, may be composed of many distinct subsections such that each sub-section has Manning's *n* that is different from others, Fig. 12.7.1. For example, in an alluvial channel, consisting of a main channel and two side channels, the side channels are usually more rough than the main channel. As a result, the mean velocity in the main channel is greater than that in the side channels. For such case, the Manning's formula may be applied separately to each subsection and thus determine the discharge in each subsection. The sum of all these discharges equals the total discharge which may be divided by the area of the entire flow section to obtain the mean velocity for the whole section.

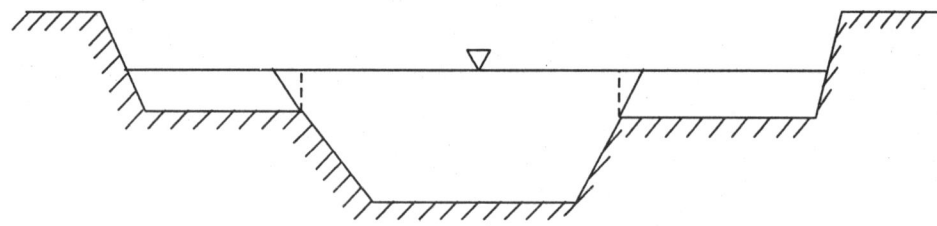

Fig. 12.7.1. A channel of compound section

12.8 EFFICIENT UNIFORM FLOW CHANNELS

The design of a rigid (*i.e.*, non-erodible) surface open channel should be such that it is capable of carrying maximum discharge with minimum excavation, construction, and lining costs. This, obviously, suggests that the designed channel should have maximum hydraulic radius for a given area or minimum perimeter for the given area since hydraulic radius equals A/P. Therefore, a circular or semicircular section is the most efficient section. The adoption of circular channel section is ruled out in view of the involved difficulties of construction, stable earthen slopes and other practical considerations. Hence, the designer has to determine the most efficient section of a given shape.

Considering a trapezoidal channel section, Fig. 12.8.1, the area of flow section *A* and the wetted perimeter *P* are given as

$$A = Bh + zh^2 \qquad\qquad ...(12.8.1)$$

$$P = B + 2h\sqrt{1 + z^2} \qquad\qquad ...(12.8.2)$$

Here, $\qquad\qquad z = cot\ \theta$

On substituting the value of *B* obtained from Eq. (12.8.1) into Eq. (12.8.2), one gets

$$P = \frac{A}{h} - zh + 2h\sqrt{1 + z^2} \qquad ...(12.8.3)$$

To minimize *P*, evaluate *dP/dh* with *A* and *z* constant and set *dP/dh* equal to zero. Therefore,

$$\frac{dP}{dh} = -\frac{A}{h^2} - z + 2\sqrt{1 + z^2} = 0$$

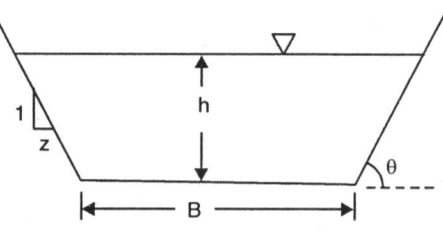

Fig. 12.8.1. Trapezoidal channel section

or
$$A = h^2[2\sqrt{1+z^2} - z]$$
...(12.8.4)

Substituting the value of A in Eq. (12.8.4), one gets

$$Bh + zh^2 = h^2[2\sqrt{1+z^2} - z]$$

∴
$$B = 2h[\sqrt{1+z^2} - z]$$
...(12.8.5)

Likewise, substituting the value of A from Eq. (12.8.4) in Eq. (12.8.3), one gets

$$P = h[2\sqrt{1+z^2} - z] - zh + 2h\sqrt{1+z^2}$$

or
$$P = 4h\sqrt{1+z^2} - 2zh = 2h[2\sqrt{1+z^2} - z]$$
...(12.8.6)

Thus,
$$R = \frac{A}{P} = \frac{h}{2}$$
...(12.8.7)

Equation (12.8.7) indicates that for any side slope θ, the most efficient uniform flow channel of trapezoidal shape would be the one for which hydraulic radius is half the depth of flow.

For a rectangular section, $\theta = 0$ or $z = 0$. Therefore, the most efficient rectangular channel section would be the one for which

$$A = 2h^2$$
...(12.8.8)

$$P = 4h$$
...(12.8.9)

$$R = h/2$$
...(12.8.10)

and
$$B = 2h$$
...(12.8.11)

To obtain the depth of flow for the most efficient section, these equations must be solved in conjunction with the Manning's equation.

Equations (12.8.4 – 12.8.7) are valid for any value of z. The best value of z for given h and A can be obtained by setting dP/dz, evaluated from Eq. (12.8.3), equal to zero.

$$\frac{dP}{dz} = -h + 2h\left[\frac{1}{2}\frac{1}{\sqrt{1+z^2}}\right]2z = 0$$

or
$$2z = \sqrt{1+z^2}$$

or
$$z = \frac{1}{\sqrt{3}} = \cot\theta$$

∴
$$\theta = 60°$$

Thus, the maximum-flow trapezoidal section would be half of a hexagon.

Similarly, one can show that a circular channel section running partially full would perform best when the depth of flow is half the diameter. Geometric elements of the most efficient channel sections of some selected shapes are given in Table 12.8.1. Obviously, the semi-circular open channel section is the best of all possible channel sections as it gives minimum wetted perimeter for a given area of flow section. However, the percentage increase in discharge for semicircular section, compared to that in semi-hexagonal section, is only marginal (see E12.4 and E12.5).

Table 12.8.1: Geometric elements of the most efficient hydraulic sections (5)

Cross-section	Normal depth, h_n	Area, A	Wetted perimeter, P	Hydraulic radius, R	Water surface width, T	Hydraulic depth, D
Rectangular	$0.917\,x$	$2\,h^2$	$4\,h$	$0.500\,h$	$2\,h$	h
Triangle (side slope 1:1)	$1.297\,x$	h^2	$2.83\,h$	$0.354\,h$	$2\,h$	$0.500\,h$
Trapezoid (half of hexagon)	$0.968\,x$	$1.73\,h^2$	$3.46\,h$	$0.500\,h$	$2.31\,h$	$0.750\,h$
Semicircle	$1.000\,x$	$0.5\,\pi h^2$	$\pi\,h$	$0.500\,h$	$2\,h$	$0.250\,\pi\,h$
Parabola $T = 2\sqrt{2h}$	$0.937\,x$	$1.89\,h^2$	$3.77\,h$	$0.500\,h$	$2.83\,h$	$0.667\,h$

$$\text{Here, } x = \left(Qn/\sqrt{s}\right)^{3/8}$$

In practice, however, the sharp corners in any chosen channel section are rounded so that these may not become zones of stagnation. Sometimes, the side slopes may also have to be adjusted depending upon the type of bank soil. In India, lined channels carrying discharges less than 55 m³/s are generally of triangular section of the permissible side slope and rounded bottom (Fig. 12.8.2)(11). A trapezoidal section with rounded corners (Fig. 12.8.3) is adopted for lined channels carrying discharges larger than 55 m³/s. The side slopes depend on the properties of the material through which the channel is to pass. Table 12.8.2 gives the recommended values of the side slopes for channels excavated through different types of material.

To avoid damage to the lining, the maximum velocity in lined channels is restricted to 2.0 m/s. Thus, the design is based on the concept of a limiting velocity.

Table 12.8.2: Suitable side slopes for channels excavated through different types of material (2)

Material	Side slopes (H:V)
Rock	Nearly vertical
Muck and peat soil	0.25:1
Stiff clay or earth with concrete lining	05:1 to 1:1
Earth with stone lining	1:1
Firm clay	1.5:1
Loose, sandy soil	2:1
Sandy loan	3:1

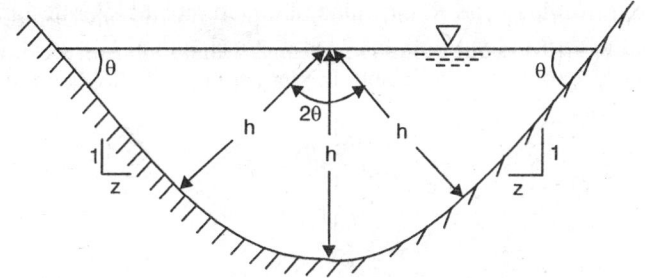

Fig. 12.8.2. Lined channel section for $Q < 55$ m³/s

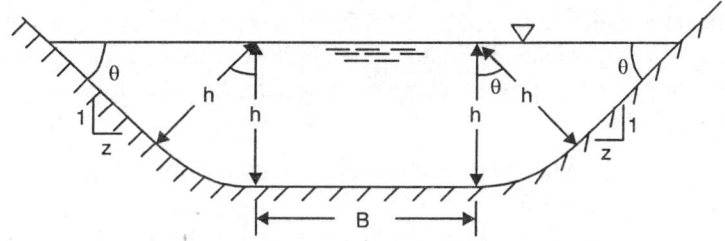

Fig. 12.8.3. Lined channel section for $Q > 55$ m³/s

The expressions for geometric elements of a lined channel section, Figs. 12.8.2 and 12.8.3, can be written as follows:

For triangular section, Fig. 12.8.2,

Area,
$$A = 2\left(\frac{1}{2}h^2 \cot\theta\right) + \frac{1}{2}h^2(2\theta)$$

$$= h^2(\theta + \cot\theta)$$

Wetted perimeter,
$$P = 2h\cot\theta + h(2\theta)$$

$$= 2h(\theta + \cot\theta)$$

Hydraulic radius,
$$R = \frac{A}{P} = \frac{h^2(\theta + \cot\theta)}{2h(\theta + \cot\theta)} = \frac{h}{2}$$

Similarly, for trapezoidal section, Fig. 12.8.3,

$$A = Bh + 2\left(\frac{1}{2}h^2\cot\theta\right) + 2\left(\frac{1}{2}h^2\theta\right)$$

$$= Bh + h^2(\theta + \cot\theta)$$

$$P = B + 2h(\theta + \cot\theta)$$

$$R = \frac{Bh + h^2(\theta + \cot\theta)}{B + 2h(\theta + \cot\theta)}$$

In all these expressions for A, P, and R, the value of θ is in radians. For designing a lined channel, one needs to solve these equations along with the Manning's equation. For given Q, n, S, and A and R expressed in terms of h for known Q, the Manning's equation will yield, for triangular section, an explicit relation for h as shown below:

$$Q = \frac{1}{n}[h^2(\theta + \cot\theta)]\left[\frac{h}{2}\right]^{2/3} S^{1/2}$$

$\therefore \qquad\qquad h = \left[\frac{nQ(2)^{2/3}}{\sqrt{S}\,(\theta + \cot\theta)}\right]^{3/8}$ $\qquad\qquad$ ...(12.8.11)

However, in case of trapezoidal section, Fig. 12.8.3, the design calculations would start with an assumed value of velocity (less than the maximum permissible velocity of 2.0 m/s) and the expression for h will be in the form of a quadratic expression as can be seen from the following (1):
From the Manning's equation,

$$R = \left(\frac{Un}{\sqrt{S}}\right)^{3/2}$$

$\therefore \qquad\qquad \left(\frac{Un}{\sqrt{S}}\right)^{3/2}[B + 2h(\theta + \cot\theta)] = A = \frac{Q}{U}$

$\therefore \qquad\qquad B = \frac{Q}{U}\left(\frac{\sqrt{S}}{Un}\right)^{3/2} - 2h(\theta + \cot\theta)$

On substituting this value of B in the expression for area of flow section A, one gets

$$\frac{Q}{U}\left(\frac{\sqrt{S}}{Un}\right)^{3/2} h - 2h^2(\theta + \cot\theta) + h^2(\theta + \cot\theta) = A = \frac{Q}{U}$$

$\therefore \qquad\qquad h^2(\theta + \cot\theta) - \frac{Q}{U}\left(\frac{\sqrt{S}}{Un}\right)^{3/2} h + \frac{Q}{U} = 0$

$\therefore \qquad h = \dfrac{\dfrac{Q}{U}\left(\dfrac{\sqrt{S}}{Un}\right)^{3/2} \pm \sqrt{\dfrac{Q^2}{U^2}\left(\dfrac{\sqrt{S}}{Un}\right)^3 - 4(\theta + \cot\theta)\dfrac{Q}{U}}}{2(\theta + \cot\theta)}$

Therefore, in order to have a feasible solution,

$$\frac{Q^2}{U^2}\left(\frac{\sqrt{S}}{Un}\right)^3 \geq 4(\theta + \cot\theta)\frac{Q}{U}$$

i.e.,
$$\frac{S^{3/2}}{U^4} \geq \frac{4n^3(\theta + \cot\theta)}{Q}$$

or
$$U \leq \left[\frac{QS^{3/2}}{4n^3(\theta + \cot\theta)}\right]^{1/4} \qquad ...(12.8.12)$$

This means that for designing a trapezoidal section for a lined channel, the velocity will have to be suitably chosen so as not to violate the above criterion in order to have a feasible solution (see E12.8 for illustration).

EXAMPLES

E12.1. A 2.5 m wide rectangular channel with a bed slope of 0.0004 has a depth of flow of 0.6 m. Assuming steady uniform flow, determine the discharge in the channel if the Manning's roughness coefficient is 0.015.

Solution: From Eq. (12.3.5)

$$Q = \frac{1}{n} A R^{2/3} S^{1/2}$$

$$= \frac{1}{n} \frac{A^{5/3}}{P^{2/3}} S^{1/2}$$

$$= \frac{1}{0.015} \frac{(2.5 \times 0.6)^{5/3}}{[2.5 + (2 \times 0.6)]^{2/3}} (0.0004)^{1/2}$$

$$= 1.096 \text{ m}^3/\text{s}$$

E12.2. Determine the normal depth of flow for the channel section of E12.1 if the discharge in the channel is 3.0 m³/s.

Solution: Substituting the given values in the Manning's equation, Eq. (12.3.5)

$$3.0 = \frac{1}{0.015} \frac{(2.5h)^{5/3}}{(2.5 + 2h)^{2/3}} (0.0004)^{1/2}$$

$$\therefore \qquad \frac{h^{5/3}}{(2.5 + 2h)^{2/3}} = 0.489$$

This equation can be solved by trial and error.

Assumed value of h (m)	1.0	1.2	1.25	1.225	1.235
L.H.S. of the equation	0.367	0.470	0.496	0.483	0.488

Hence, $h_n = 1.235$ m

Alternatively, one can use Fig. 12.4.1 as follows:

$$z_1 = AR^{2/3} = \frac{Qn}{\sqrt{S}} = \frac{0.015 \times 3}{\sqrt{0.0004}} = 2.25$$

$$\frac{z_1}{B^{8/3}} = \frac{2.25}{(2.5)^{8/3}} = 0.195$$

From Fig. 12.4.1, $\dfrac{h}{B} = 0.495$

Hence, $h_n = 0.495 \times 2.5$

∴ $h_n = 1.238\,\text{m}$

The difference in the two values of h_n is due to the limits in reading the graph.
Still, alternatively, one can prepare a computer program to solve the Manning's equation for normal depth of flow.

E12.3.   A trapezoidal channel (width $B = 10$ m, side slope $z = 2$, bed slope $S_0 = 0.0016$, and the Manning's $n = 0.02$) carries a discharge of 90 m³/s. Estimate the normal depth of flow.

Solution: Substituting the values of Q, n, A, R, S_0 for the given channel in the Manning's equation, Eq. (12.3.5), one gets

$$90 = \frac{1}{0.02}\left[(10 + 2h_n)h_n\right]^{5/3}\left[1/(10 + 2\sqrt{5}h_n)\right]^{2/3}(0.0016)^{1/2}$$

or $\left[(10 + 2h_n)h_n\right]^{5/3} = 45(10 + 2\sqrt{5}\,h_n)^{2/3}$

Solving this equation by trial, one obtains the desired normal depth

$$h_n = 2.25\,\text{m}$$

E12.4.   A rectangular brick-lined channel is to be built to carry a discharge of 5 m³/s of water in uniform flow condition. The channel slope is to be 0.001. Compute the value of B and h such that the channel behaves as the most efficient one.

Solution: For brick-lined channel, $n = 0.015$.

Substituting $A = 2h^2$ and $R = \dfrac{h}{2}$ in the Manning's equation, Eq. (12.3.5), one gets

$$5 = \frac{1}{0.015}(2h^2)\left(\frac{h}{2}\right)^{2/3}\cdot(0.001)^{1/2}$$

∴ $h^{8/3} = 1.8824$

and $h = 1.268$ m

Hence, $A = 2h^2$

$$= 3.216\,\text{m}^2$$

and $B = \dfrac{A}{h} = 2.536\,\text{m}$

E12.5.   Compute the value of discharge for (a) semi-hexagonal channel section and (b) semi-circular channel section with area of flow section the same as obtained in E12.4. Assume slope of the channel and the Manning's roughness also the same as in E12.4.

Solution: (a) With $z = \cot\theta = \dfrac{1}{\sqrt{3}} = 0.577$, Eq. (12.8.4) gives

$$3.216 = h^2\left(2\sqrt{1+(0.577)^2} - 0.577\right)$$

∴ $h = 1.363$ m

and $R = 0.681$ m

∴ $Q = \dfrac{1}{0.015}(3.216)(0.681)^{2/3}(0.001)^{1/2}$

∴ $Q = 5.248\,\text{m}^3/\text{s}$

(b)   For semi-circular channel section

$$A = pD_0^2/8$$

∴ $D_0 = \sqrt{\dfrac{8(3.216)}{\pi}}$

                 $= 2.862$ m

and $P = \dfrac{1}{2}\pi D_0$

                 $= 4.496$ m

∴ $R = 0.715$ m

∴ $Q = \dfrac{1}{0.015}(3.216)(0.715)^{2/3}(0.001)^{1/2}$

                 $= 5.421 \text{ m}^3/\text{s}$

It should be noted that this value is 8.42 percent more than that for the rectangular section and 3.3 percent more than that for the semi-hexagonal section.

E12.6.   A lined canal ($n = 0.015$) laid at a slope of 1 in 1600 is required to carry a discharge of 25 m³/s. The side slopes of the canal are to be kept at 1.25H: 1V. Determine the depth of flow.

Solution: Since $Q < 55$ m³/s, a triangular section with rounded bottom, Fig. 12.8.2, is considered suitable.

Here, cot $\theta = 1.25$

∴ $\theta = 38.66°$ or 0.675 radian

Thus, from Fig. 12.8.2,

          $A = h^2\,(\theta + \cot\theta) = h^2\,(0.675 + 1.25)$

          $= 1.92\,h^2$

and
$$P = 2h(\theta + \cot\theta) = 2h\,(0.675 + 1.25)$$
$$= 3.85\,h$$

∴
$$R = \frac{1.925h^2}{3.85h} = \frac{h}{2}$$

From the Manning's equation, $Q = \dfrac{1}{n} AR^{2/3} S^{1/2}$

Hence,
$$25 = \frac{1.925h^2}{0.015}\left(\frac{h}{2}\right)^{2/3}\left(\frac{1}{1600}\right)^{1/3}$$

∴
$$h^{8/3} = \frac{25 \times 0.015 \times 2^{2/3} \times 40}{1.925}$$

∴
$$h = 2.57 \text{ m}$$

E12.7.   The depth of flow in a hydraulically most efficient trapezoidal channel (side slopes 1.5 H:1V and the Manning's $n = 0.015$) is to be restricted to 1.6 m for its design discharge of 15 m³/s. Determine the bed slope of the channel.

Solution: From Eq. (12.8.5),

$$B = 2h[\sqrt{1 + z^2} - z]$$

$$= (2 \times 1.6)[\sqrt{1 + 2.25} - 1.5]$$

$$= 0.969 \text{ m}$$

∴
$$A = (B + zh)\,h$$

$$= [0.969 + (1.5 \times 1.6)](1.6)$$

$$= 5.39 \text{ m}^2$$

and
$$R = \frac{h}{2} = \frac{1.6}{2} = 0.8\,\text{m}$$

From the Manning's equation, Eq. (12.3.5)

$$15 = \frac{1}{0.015}(5.39)(0.8)^{2/3}\,S^{1/2}$$

∴
$$S^{1/2} = \frac{(15)(0.015)}{(5.39)(0.8)^{2/3}} = 0.0484$$

∴
$$S = 2.346 \times 10^{-3}$$

E12.8.   Design a lined channel to carry a discharge of 300 m³/s through an alluvium whose angle of repose is 31°. The bed slope of the channel is 7.75 x 10⁻⁵ and the Manning's n for the lining material is 0.016. [adapted from (1)]

Solution: Since $Q > 55$ m³/s, a trapezoidal section with rounded corners, Fig. 12.8.3, is to be designed.

Here, side slope $\theta = 31° = 0.541$ radians

$\therefore$ $\quad\quad\quad\quad$ $\cot \theta = 1.664$

$\therefore$ $\quad\quad\quad\quad$ $\theta + \cot \theta = 2.205$

$\therefore$ $\quad\quad\quad\quad$ $A = Bh + 2.205\, h^2$

$\quad\quad\quad\quad\quad$ $P = B + 4.41h$

Adopting $\quad\quad$ $U = 2$ m/s

$$A = \frac{300}{2} = 150\, \text{m}^2$$

$\therefore$ $\quad\quad\quad\quad$ $Bh + 2.205\, h^2 = 150$

and $\quad\quad\quad$ $R = \left(\dfrac{Un}{\sqrt{S}}\right)^{3/2} = \left(\dfrac{2 \times 0.016}{7.75 \times 10^{-5}}\right)^{3/2} = 6.93\,\text{m}$

$\therefore$ $\quad\quad\quad\quad$ $6.93\,(B+4.41h) = 150$

or $\quad\quad\quad\quad$ $B = 21.645 - 4.41\, h$

$\therefore$ $\quad\quad\quad\quad$ $21.645h - 4.41h^2 + 2.205\, h^2 = 150$

or $\quad\quad\quad\quad$ $2.205\, h^2 - 21.645\, h + 150 = 0$

$\therefore$ $\quad\quad$ $h = \dfrac{21.645 \pm \sqrt{(21.645)^2 - 4 \times 2.205 \times 150}}{4.41}$

Obviously, the roots of h are imaginary. Using the criterion, Eq. (12.8.12), one gets

$$U \leq \left[\frac{QS^{3/2}}{4n^3(\theta + \cot\theta)}\right]^{1/4}$$

$$\leq \left[\frac{300 \times (7.75 \times 10^{-5})^{3/2}}{4(0.016)^3(2.205)}\right]^{1/4}$$

$\quad\quad\quad\quad$ ≤ 1.543 m/s

$\therefore$ $\quad\quad\quad\quad$ Adopt $U = 1.5$ m/s

$\therefore$ $\quad\quad\quad\quad$ $A = 200$ m²

$$R = \left(\frac{1.5 \times 0.016}{\sqrt{7.75 \times 10^{-5}}}\right)^{3/2}$$

$\quad\quad\quad\quad$ $= 4.50$ m

$$4.50\,(B + 4.41h) = 200$$

$\therefore\qquad B = 44.44 - 4.41h$

Again, using $\quad Bh + 2.205\,h^2 = A$, one gets

$$44.44h - 4.41h^2 + 2.205\,h^2 = 200$$

or $\qquad 2.205\,h^2 - 44.44\,h + 200 = 0$

$\therefore\qquad h = \dfrac{44.44 \pm \sqrt{(44.44)^2 - 4 \times 2.205 \times 200}}{4.41}$

$$= 13.37 \text{ m or } 6.784 \text{ m}$$

$\therefore\qquad B = 44.44 - 4.41\,h$

$$= 14.52 \text{ m for } h = 6.784 \text{ m}$$

Other value of $h\ (= 13.37$ m) gives negative value of B which is meaningless.

$\therefore\qquad B = 14.52$ m and $h = 6.784$ m.

PROBLEMS

P12.1 Water flows with a depth of 1 m in a 2 m wide rectangular channel. If the slope of the channel is $1°$ and the lining is of brick work, estimate the discharge for uniform flow.

P12.2 A rectangular channel of width 6 m is laid at a slope of 6×10^{-3} and has the Manning's roughness coefficient for its bed and sides as 0.014. Determine the normal depth of flow and also average and maximum shear stresses on the channel bed when discharge in the channel is 5 m³/s.

P12.3 The cross-section of a river may be idealized as shown in Fig. P12.3. Determine the discharge carried by the river when its bed slope is 2×10^{-4} and the Manning's $n = 0.02$.

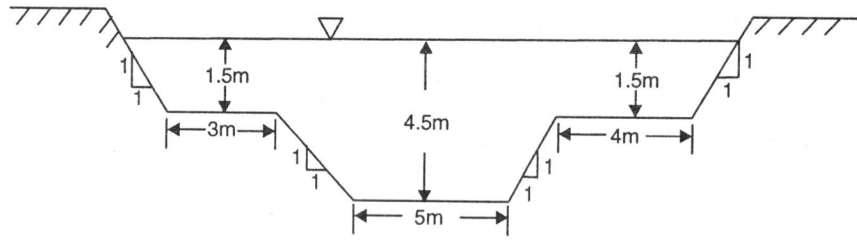

Fig. P.12.3

P12.4 A 3 m wide rectangular channel, lined with finished concrete, is designed for a flow rate of 5 m³/s at a normal depth of 1 m. Determine (i) the design slope of the channel, (ii) the percent change in flow, if the channel surface is asphalt.

P12.5 If the channel of P12.4 (with concrete surface) is divided into two equal sections by a central barrier of concrete, estimate the percentage change in flow.

P12.6 A trapezoidal channel with bed width of 2.0 m and side slopes 1.7 (H) : 1 (v) slopes at 1 in 500. Determine the flow rate for the normal depth of 80 cm. The channel surface is made of brick work.

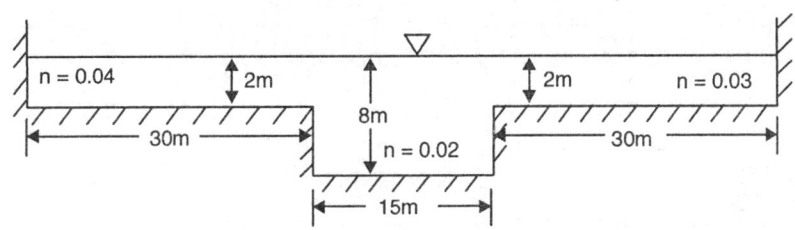

Fig. P.12.6

P12.7 Estimate the normal depth in the channel of P12.6 when the flow rate is 8 m³/s.

P12.8 For the channel section shown in Fig. P12.8, estimate the discharge if the channel slope is 0.0002.

P12.9 A rectangular channel has $B = 3$ m and $h = 1$ m. What is the diameter of a semicircular channel that will have the same discharge as in the rectangular channel. Assume that n and S are the same in the two cases. Compare the two wetted perimeters.

P12.10 What are the dimensions of the most efficient rectangular channel section to carry 5 m³/s at a slope of 1 in 900. The surface of the channel is of concrete.

P12.11 What is the most efficient depth for a brick channel of trapezoidal section with sides sloping at 45° to carry 3 m³/s. The bed slope is 0.0009.

P12.12 An irrigation canal is to be designed to carry water at 250 m³/s on a bed slope of 0.0004. Compute the required dimensions of the best trapezoidal channel with side slopes of 45° in clean earth. Estimate the reduction in excavation that would be possible if the canal were lined with concrete.

P12.13 A 3.0 m wide rectangular channel carries a discharge at a depth of 1.2 m. If the Manning's n and the bed slope remain unchanged, determine the diameter of a semi-circular channel that will have the same discharge. Compare the two wetted perimeters.

REFERENCES

(1) Asawa, GL: *"Irrigation and Water Resources Engineering"*, New Age International (P) Ltd., Publishers, New Delhi, 2005.

(2) Chow, VT: *"Open-Channel Hydraulics"*, McGraw-Hill Book Co., New York, 1959.

(3) Cowan, WL: *"Estimating hydraulic roughness coefficients"*, Agricultural Engineering, Vol. 37, No. 7, pp 473-475, July 1956 [Chow (2)].

(4) Einstein, HA and Barbarossa, NL: *"River channel roughness"*, Trans. ASCE, 1952 [Chow (2)].

(5) French, RH: *"Open-Channel Hydraulics"*, McGraw-Hill Book Co., Singapore, 1994.

(6) Henderson, FM: *"Open Channel Flow"*, The Macmillan Company, New York, 1966.

(7) Horton, RE: *"Separate roughness coefficients for channel bottom and sides"*, Engineering News Record, Vol. 111, 1933 [Chow (2)].

(8) Keulegan, GH: *"Laws of turbulent flow in open channels"*, Research Paper RP 1151, Journal of Research, U.S. National Bureau of Standards, Vol. 21, p. 707-741, Dec. 1938 [Chow (2)].

(9) Lane, E: *"Design of stable channels"*, Trans. ASCE, 1955 [Ranga Raju (11)].

(10) Meyer-Peter, E and Müller, R: *"Formulas for bed load transport"*, Proceedings of the 3rd meeting of IAHR, Stockholm, pp. 39-64, 1948 [Asawa (1)].

(11) Ranga Raju, KG: *"Flow through Open Channels"*, Tata McGraw-Hill Publishing Co. Ltd., New Delhi, 1993.

(12) White, F.M.: *"Fluid Mechanics"*, McGraw-Hill Companies, New York, 2003.

CHAPTER 13

Critical Flow in Channels

13.1 SPECIFIC ENERGY

Specific energy is defined as the energy per unit weight of the flowing fluid at any section of a channel measured with respect to the channel bed at the section. Since the flow in a channel is usually turbulent, the specific energy E, Eq. (11.5.10), can be expressed as (with $\alpha = 1$),

$$E = h + \frac{U^2}{2g} \qquad \qquad ...(13.1.1)$$

This means that the specific energy at any section in a channel flow is the total head at the section with respect to a datum that coincides with the channel bed. It is a convenient quantity to work with while studying open channel flow and was introduced by Bakhmeteff in 1913(1).

For uniform flow in an open channel, the energy grade line slopes downward remaining parallel to the channel bed indicating steady decrease in the flow energy. The specific energy, however, remains

unchanged along the channel since neither h nor $U^2/2g$ (and, hence, $\left[h + \left(U^2/2g \right) \right]$ changes in uni-

form flow. In non-uniform open channel flow too the energy grade line slopes downward, but, its slope is not equal to the slope of the channel bed. However, the specific energy may either increase or decrease depending upon the slope of the channel bed (or elevation of the channel bed), discharge, depth of flow, and channel characteristics.

Equation (13.1.1) can also be written as

$$E = h + \frac{Q^2}{2gA^2} \qquad \qquad ...(13.1.2)$$

Therefore, for a given channel section and discharge Q, the specific energy in the channel section is a function of the depth of flow h only and the resulting plot between E and h, Fig. 13.1.1, is known as *specific energy diagram or specific energy curve*. The curve has two limbs AC (asymptotic to the abscissa, i.e., E-axis) and BC (asymptotic to the line OD, i.e., $E = h$ line). For a channel of large slope, however, the line OD would have an inclination different than 45°. At any point P on this curve, the abscissa and ordinate represent, respectively, the specific energy $E(= h + (Q^2/2gA^2))$ and the depth of flow h. Further, for a specified specific energy there are two possible depths, corresponding to the solution of Eq. (13.1.2), that are known as *alternate depths*. That is, the low stage depth h_1 is the alternate depth of the high stage depth h_2, and vice versa.

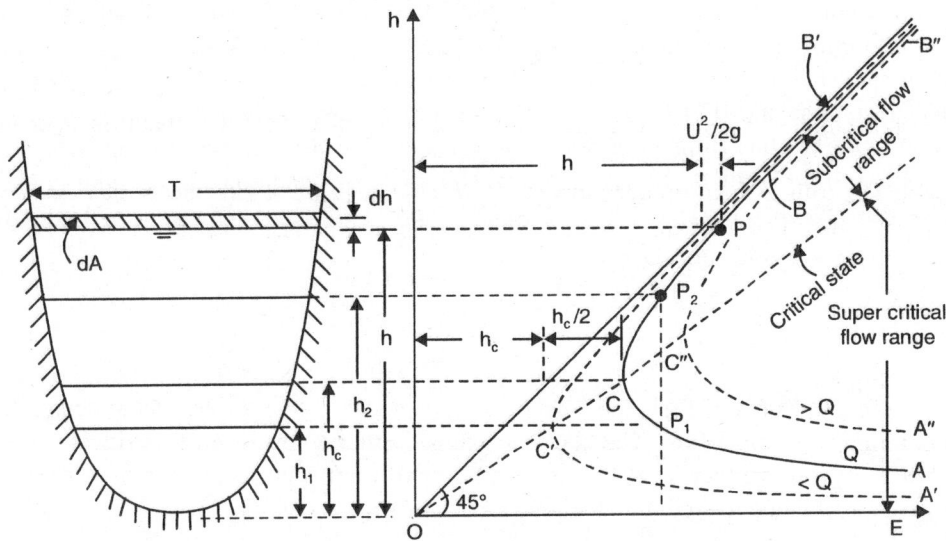

Fig. 13.1.1. Specific energy diagram for a channel of negligible slope

13.2 CRITERION FOR CRITICAL FLOW

From the specific energy curve, Fig. 13.1.1, one observes that at point C, the specific energy is minimum and the two alternate depths h_1 and h_2 merge into one depth, say h_c. One can obtain the condition for minimum specific energy by differentiating Eq. (13.1.2) with respect to h and equating dE/dh to zero. Thus,

$$\frac{dE}{dh} = 1 - \frac{Q^2}{gA^3}\frac{dA}{dh} \qquad \text{...(13.2.1)}$$

The discharge Q is, obviously, constant. Since the differential water area dA near the free surface, Fig. 12.5.1, is equal to Tdh, Eq. (13.2.1) becomes

$$\frac{dE}{dh} = 1 - \frac{Q^2 T}{gA^3} \qquad \text{...(13.2.2)}$$

$$\frac{dE}{dh} = 1 - \frac{U^2}{gD} \qquad \text{...(13.2.3)}$$

as $Q = UA$ and hydraulic depth, $D = A/T$. For specific energy to be minimum, $dE/dh = 0$. Therefore, The condition for flow to be critical in any channel of arbitrary cross-sectional shape is

$$1 - \frac{Q^2 T}{gA^3} = 1 - \frac{U^2}{gD} = 0$$

Hence,

$$\frac{Q^2 T}{g A^3} = 1 \qquad \qquad ...(13.2.4)$$

Also,

$$\frac{U}{\sqrt{gD}} = 1 \qquad \qquad ...(13.2.5)$$

That is, Froude number $\left(= \dfrac{U}{\sqrt{gD}} \right)$ is equal to 1. It may be recalled, Art. 11.4, that flow in open channel was classified as critical when Froude number, $U/\sqrt{gD}$ equals unity. This means that the *critical flow* in a channel is that state of flow at which the specific energy is minimum for a given discharge. Equation (13.2.5) further yields

$$\frac{U^2}{2g} = \frac{D}{2} \qquad \qquad ...(13.2.5)$$

Hence, at the critical state of flow, the velocity head is equal to half the hydraulic depth. This is yet another criterion for critical flow. The above-mentioned criteria developed for critical flow are valid provided that the flow is either parallel or gradually varied, energy correction factor α is unity, and the channel is of small slope $(\theta \approx 0)$. In the absence of these conditions, Eq. (13.2.5) gets modified to

$$\alpha \frac{U^2}{2g} = \frac{D \cos \theta}{2} \qquad \qquad ...(13.2.6)$$

and the Froude number may be defined as (2)

$$F_r = \frac{U}{\sqrt{\dfrac{g}{\alpha} D \cos \theta}} \qquad \qquad ...(13.2.7)$$

The depth of flow h_c at minimum specific energy, i.e., when the flow is in critical state, is called the *critical depth* that depends on Q for a given channel section as given by Eq. (13.2.4). Similarly, the location at which the state of critical flow exists in a channel is known as the *critical section*.

It should be noted that the concepts of specific energy and critical depth are essential in analyzing gradually varied flow and in determining control sections in open channel flows. Another application of critical flow concepts is in the measurement of discharge in open channels.

For specific energy E less than the minimum specific energy, no solution exists for Eq (13.1.2) for given Q and, hence, no flow is possible. However, for $E > E_{min}$ two feasible solutions exist : (i) large depth, i.e., $h > h_c$ and $U < U_c$ (i.e., the velocity of flow at critical condition) and, hence, $F_r < 1$ and the flow is *subcritical*, and (ii) small depth, i.e., $h < h_c$ and $U > U_c$ and, hence, $F_r > 1$ and the flow is *supercritical*. From Fig. 13.1.1, one observes that the specific energy curve is almost vertical at the critical state of flow. Therefore, when the flow is critical, a minor change in E due to, say, variations in channel cross-section, roughness, slope or deposits of sediment, causes large change in depth of flow. Hence, a flow in a channel at or near critical state is unstable. Therefore, channel designers should avoid long runs of near-critical flow by altering the shape or slope.

For critical flow in a rectangular channel, Eq. (13.2.4) yields

$$1 - \frac{Q^2 T}{gA^3} = 1 - \frac{Q^2 B}{gB^3 h_c^3} = 1 - \frac{q^2}{gh_c^3} = 0$$

Hence,

$$h_c = \left(\frac{q^2}{g}\right)^{1/3} = \left(\frac{Q^2}{gB^2}\right)^{1/3}$$ ...(13.2.8)

in which, $q = Q/B = Uh$, i.e., the discharge per unit width of rectangular channel. The corresponding minimum specific energy is

$$E_{min} = h_c + \frac{U^2}{2g} = h_c + \frac{q^2}{2gh_c^2} = h_c + \frac{gh_c^3}{2gh_c^2} = h_c + \frac{h_c}{2}$$

Thus,

$$E_{min} = \frac{3}{2} h_c$$ ...(13.2.9)

Further, Eq. (13.1.2) can be written as

$$Q = A\sqrt{2g(E - h)}$$ ...(13.2.10)

For any given channel section and specified specific energy E, graphical plot between Q and h, as obtained from Eq. (13.2.10), would be like the one shown in Fig. 13.2.1. From Eq. (13.2.10), it is

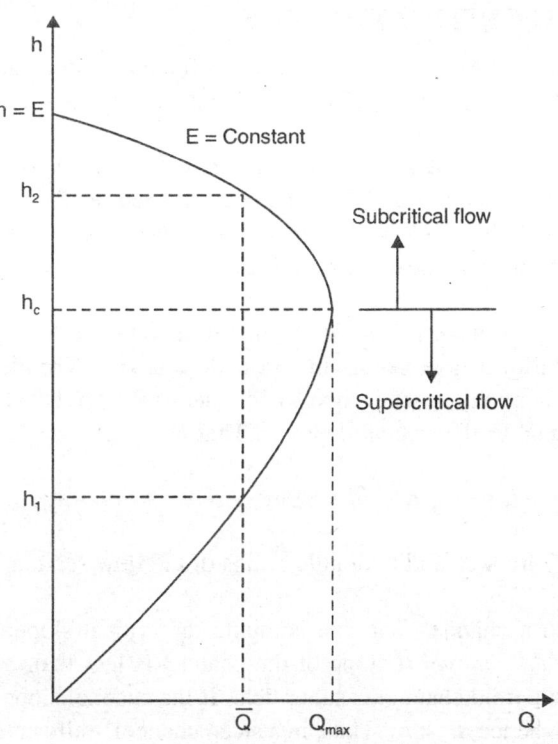

Fig. 13.2.1. Depth-discharge diagram

obvious that as $h \to 0$, $Q \to 0$ and also as $h \to E$, the term velocity head tends to zero and, therefore, $Q \to 0$. Hence, for given specific energy in a channel section, the discharge would be maximum at a depth of flow h such that $0 < h < E$.

For Q to be maximum, for given E, dQ/dh must be zero. Therefore,

$$\frac{dQ}{dh} = \sqrt{2g} \left[\sqrt{E-h}\, \frac{dA}{dh} - \frac{A}{2\sqrt{E-h}} \right] = 0 \qquad \qquad ...(13.2.11)$$

Since, $\dfrac{dA}{dh} = T$, Eq. (13.2.11) yields

$$2(E-h)T = A \qquad \qquad ...(13.2.12)$$

Combining Eqs. (13.1.2) and (13.2.12), one gets

$$\frac{Q^2 T}{gA^3} = 1$$

which condition is the same as the condition for minimum specific energy, Eq. (13.2.4). Therefore, at the critical state of flow, the discharge is maximum for given specific energy.

Figure 13.2.1 indicates that for given specific energy E, there are two possible depths, h_1 and h_2 for any given value of discharge except when the flow is critical when the discharge is maximum for the specified value of E.

13.3 NORMAL AND CRITICAL SLOPES

For given discharge Q and the Manning's roughness coefficient n, one can determine the slope of a prismatic channel for uniform flow with specified normal depth h_n. This slope is specifically termed the *normal slope S_n*.

Equation (13.2.4) indicates that for a given discharge, the depth of critical flow in a channel depends on the geometric elements A and T of the channel section. Therefore, the critical depth in a prismatic channel of uniform slope will be the same everywhere. Hence, the critical flow in a prismatic channel would be uniform flow and the corresponding slope that sustains a given discharge at a uniform and critical depth is called the *critical slope S_c*.

By varying the slope of a given prismatic channel to a certain value, it is possible to change the normal depth such that uniform flow occurs in a critical state (i.e., Froude number equals unity) for given discharge and the Manning's n. The slope so obtained is the critical slope S_c and the corresponding normal depth h_n is equal to the critical depth h_c. That is

$$Q = \frac{1}{n} A_c R_c^{2/3} S_c^{1/2} \qquad \qquad ...(13.3.1)$$

with $h = h_c$, and A_c and R_c are area and hydraulic radius of the flow section when depth of flow equals h_c.

For given Q and n in a channel, one can compute the critical slope S_c from Eq. (13.3.1). The channel is classified as *mild channel* if slope of the channel is less than S_c and, hence, $h_n > h_c$ and, therefore, uniform flow in a mild channel is subcritical. If the channel slope exceeds S_c, the channel is termed *steep channel* and, hence, $h_n < h_c$. Thus, in a steep channel, uniform flow is supercritical. When the slope of the channel equals S_c, $h_n = h_c$, i.e., uniform flow is in critical state.

From the Manning's equation, the discharge per unit width of a wide channel q is proportional to $h_n^{5/3}$. Also, $q \propto h_c^{3/2}$. Therefore, the classification of slope depends only slightly on the discharge and depends primarily on the magnitude of the slope itself and the roughness of the channel boundary.

13.4 SPECIFIC FORCE

While applying momentum principle, Eq. (11.5.14), for a short horizontal reach of a prismatic channel, one can ignore the external force of friction and the weight component of water for the reach under consideration. If the momentum correction factor β is also assumed to be unity, Eq. (11.5.14) gives

$$\rho Q(U_2 - U_1) = P_1 - P_2 \qquad \text{...(13.4.1)}$$

The hydrostatic forces P_1 and P_2 may be expressed as,

$$P_1 = \rho g \, \overline{z}_1 \, A_1 \qquad \text{...(13.4.2)}$$

and

$$P_2 = \rho g \, \overline{z}_2 \, A_2 \qquad \text{...(13.4.3)}$$

Here, $\overline{z}$ represents the distance of the centroid of the flow section A below the free surface. Equation (13.4.1), therefore, yields

or

$$\rho Q \left(\frac{Q}{A_2} - \frac{Q}{A_1} \right) = \rho g \, \overline{z}_1 \, A_1 - \rho g \overline{z}_2 A_2$$

or

$$\frac{Q^2}{gA_1} + \overline{z}_1 A_1 = \frac{Q^2}{gA_2} + \overline{z}_2 \, A_2 \qquad \text{...(13.4.4)}$$

Both sides of Eq. (13.4.4) are similar and, hence, may be expressed as, say,

$$F = \frac{Q^2}{gA} + \overline{z} \, A \qquad \text{...(13.4.5)}$$

Here, the term Q^2/gA is the momentum of the flow (*i.e.*, $\rho Q U$ or $\rho Q^2/A$) passing through a flow section of a channel per unit time per unit weight of the flowing fluid. The second term on the right side of Eq. (13.4.5), $\overline{z}A$ represents the hydrostatic force per unit weight of water. Thus, F is essentially sum of the two forces per unit weight of water and, therefore, called the *specific force*. Equation (13.4.4) can, accordingly, be written as

$$F_1 = F_2 \qquad \text{...(13.4.6)}$$

meaning that the specific forces at the end sections of a channel reach are equal provided that the channel is horizontal (or the bed slope of the channel is very small) and the external forces acting on the flowing fluid between the end sections can be ignored.

Equation (13.4.5) indicates that, like specific energy, specific force, for a given discharge and channel section, depends on the depth of flow in the channel. On plotting the depth of flow h (on ordinate) against F (on abscissa), for a given discharge in a channel section, one obtains the *specific force curve*, Fig. 13.4.1. This curve, like the specific-energy curve, too has two limbs one of which is asymptotic to the F-axis and other rises upward and extends indefinitely. For a given value of F, there would be two possible depths that are known as *conjugate* or *sequent depths*. At point C, the specific

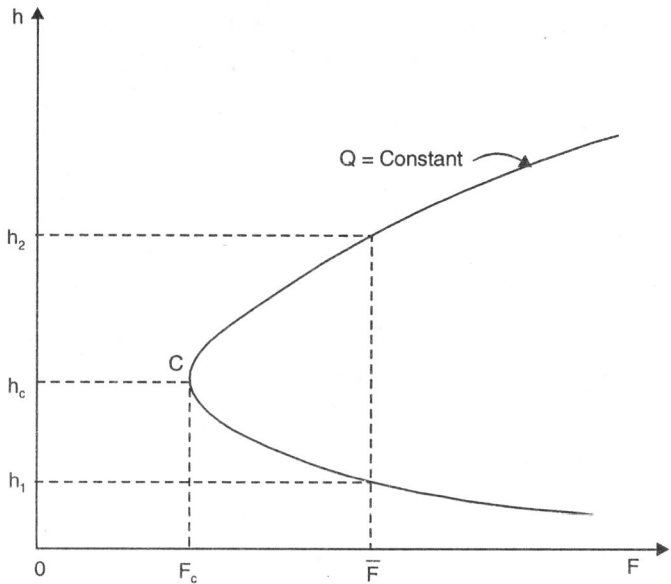

Fig. 13.4.1. Specific force diagram

force is minimum and the two depths merge into a single one. The condition for minimum specific force can be obtained by equating dF/dh, obtainable from Eq. (13.4.5) for given value of Q, to zero. Thus,

$$\frac{dF}{dh} = -\frac{Q^2}{gA^2}\frac{dA}{dh} + \frac{d(\bar{z}A)}{dh} = 0 \qquad \qquad ...(13.4.7)$$

For a change dh in the depth of flow, Fig. 13.1.1, the corresponding change in area is $dA(=Tdh)$ and that in the static moment of the area of the flow section about the free surface $(=\bar{z}A)$ is $d(\bar{z}A)$ that is equal to $[\{A(\bar{z}+dh)+T(dh)^2/2\}-\bar{z}A]$. Thus, $d(\bar{z}A)\cong Adh$ assuming that $(dh)^2 \approx 0$. Equation (13.4.7), therefore, gives

$$-\frac{Q^2T}{gA^2}+A=0$$

or
$$\frac{Q^2T}{gA^3}=1$$

which condition is the same as the condition for minimum specific energy, Eq. (13.2.4). Therefore, the depth at minimum specific force is, again, the critical depth. Hence, yet another definition of critical flow is that at the critical state of flow, the specific force is minimum for a given discharge.

Equation (13.4.5) may also be written as,

$$Q=\sqrt{gA(F-\bar{z}A)} \qquad \qquad ...(13.4.8)$$

and one may obtain the condition for the discharge to be maximum by equating dQ/dh (for a specified value of F) to zero. Hence,

$$\frac{dQ}{dh} = \sqrt{g}\left[(\sqrt{F - \overline{z}A})\frac{1}{2\sqrt{A}}\frac{dA}{dh} + \sqrt{A}\ \frac{1}{2\sqrt{F - \overline{z}A}}\left\{ -\frac{d}{dh}(A\overline{z}) \right\} \right] = 0 \qquad \text{...(13.4.9)}$$

Since $\dfrac{dA}{dh} = T$ and $\dfrac{d}{dh}(A\overline{z}) = A$, Eq. (13.4.9) yields

$$\sqrt{g}\left[(\sqrt{F - \overline{z}A})\frac{T}{2\sqrt{A}} - \frac{A^{3/2}}{2\sqrt{F - \overline{z}A}} \right] = 0$$

or
$$T(F - \overline{z}A) = A^2$$

Substituting the value of $F - \overline{z}A$ from Eq. (13.4.5),

$$\frac{Q^2 T}{gA} = A^2$$

or
$$\frac{Q^2 T}{gA^3} = 1$$

which condition is the same as the condition for minimum specific energy, Eq. (13.2.4). Hence, at the critical state of flow, the discharge is maximum for a given specific force. Recapitulating the discussions on the concepts of critical flow, one may list the following conditions that are fulfilled when the flow is in critical state:

(i) Froude number is unity and, hence, the velocity head is half the hydraulic depth.
(ii) Specific energy is minimum for a given discharge.
(iii) Discharge is maximum for a given specific energy.
(iv) Specific force is minimum for a given discharge.
(v) Discharge is maximum for a given specific force.

13.5 CONTROL SECTION

The flow condition in any given channel are decided primarily by the control mechanisms of the channel system. The notion of a control, therefore, means the establishment of definite stage-discharge (or depth-discharge) relationship and is of considerable importance in the study of open channel flow. A channel section that provides a definite depth-discharge relationship for the given channel is, therefore, known as *control section*. For example, sluice gate forms a control section as there is definite relationship between the opening of the gate and discharge for given upstream depth. Again, for any given channel section, Eq. (13.2.4) provides a definite relationship between the depth of flow and discharge. Equation (13.2.4) establishes that the stage-discharge relationship is theoretically unaffected by the channel roughness and other factors. Since Eq. (13.2.4) corresponds to the critical flow, one may conclude that a critical flow section too is a control section. Weir and spillways are other examples of control section.

The resistance equation for a uniform flow in a channel too provides definite relationship between the depth of flow and discharge in the channel. As such, one may think uniform flow to be a control. However, the uniform flow is not associated with a particular location or feature in the channel, but, exists in a long uniform channel in the absence of any other control.

It has been seen that the value of the Froude number $(U / \sqrt{gD})$ is unity for critical flow. In open channel flow, $\sqrt{gD}$ represents the velocity of propagation of a small disturbance (i.e., gravity wave) occurring at the free surface of a still channel. Hence, Froude number, besides being the ratio of inertial and gravity forces, is also the ratio of flow velocity and the velocity of a gravity wave. If Froude number is less than unity (i.e., subcritical flow), the gravity wave would be able to travel in both upstream and downstream directions as $\sqrt{gD} > U$. But, in supercritical flow (i.e., $U > \sqrt{gD}$), the wave can move only in the downstream direction. Ripples, formed on the free surface by dropping a pebble, would propagate in both upstream and downstream directions, if the flow is subcritical. In supercritical flow, however, the ripples propagate only in the downstream direction. One may, therefore, conclude that subcritical flow is controlled by a downstream control whereas superciritical flow can only be controlled from upstream.

13.6 COMPUTATIONS FOR CRITICAL DEPTH

The concepts of specific energy and critical depth are useful in analyzing gradually varied flow and also in determining control section (i.e., the section at which control of flow is exercised) that controls the flow in a channel in such a way that it restricts the transmission of the effect of changes in flow condition in either the upstream direction or the downstream direction depending upon the state of flow (i.e., subcritical or supercritical) in the channel. When the flow is in critical state, there exists a definite stage-discharge relationship, Eq. (13.2.4), that is independent of the channel roughness and other uncontrolled conditions. Therefore, a critical flow section can be used as a flow measuring section besides the section being a control section. Hence, concept of critical flow is used also for measurement of flow in channels.

For relatively simple problems, such as computation of the critical discharge in a channel for known depth of flow, one can obtain the solution directly by using Eq. (13.2.4). The critical depth for rectangular channels can be easily determined using Eq. (13.2.8). Using Eq. (13.2.4), one can obtain expression for critical depth in a triangular channel with side slope $z(H):1(V)$ as follows:

$$\frac{Q^2 T}{g A^3} = \frac{Q^2 (2z h_c)}{g(z h_c^2)^3} = \frac{2Q^2}{g z^2 h_c^5} = 1$$

$\therefore$
$$h_c = \left[\frac{2Q^2}{g z^2} \right]^{1/5} \qquad\qquad ...(13.6.1)$$

For trapezoidal channel with side slopes $z\ (H):1(V)$, Eq. (13.2.4) reduces to

$$\frac{Q^2 (B + 2z h_c)}{g[B h_c + z h_c^2]^3} = \frac{Q^2 (B + 2z h_c)}{g[B + z h_c]^3 h_c^3} = 1 \qquad\qquad ...(13.6.2)$$

The value of h_c for trapezoidal channel section can, therefore, be obtained only by trial and error procedure. One would obtain similar implicit relation for critical depth in circular channels. The computational procedure for critical depth in trapezoidal and circular channels can, however, be made simpler by using section factor curves.

13.6.1 Section Factor curves for critical flow computations

Equation (13.2.4) can also be written in the following form:

$$\frac{Q}{\sqrt{g}} = A\sqrt{A/T} = A\sqrt{D} = Z \qquad \qquad ...(13.6.3)$$

in which Z is the *section factor for critical flow computations*. Since Z equals $A\sqrt{D}$, Z depends on channel shape and the depth of flow. Therefore, there can be only one depth (i.e., critical depth) for maintaining the given discharge at critical state in a channel. Likewise, for the given depth, there can be only one discharge to maintain critical flow that makes the depth critical for the given channel. Equation (13.6.3) also states that at the critical state of flow, the section factor Z equals $Q/\sqrt{g}$ or $Q/\sqrt{g/\alpha}$ (when energy correction factor α cannot be approximated to unity).

For computations and analysis of critical flow in an open channel, Eq. (13.6.3) is very useful. For the given discharge, one can compute Z (i.e., the section factor for critical flow) as equal to $Q/\sqrt{g}$ and, thereafter, obtain the critical depth h_c corresponding to the value of Z. When the depth (and, hence, the section factor Z too) is given, the required discharge to maintain critical flow can simply be calculated using Eq. (13.6.3) or even Eq. (13.2.4). For trapezoidal and circular channels, however, it is advantageous to have dimensionless curves for the section factor Z as this would simplify the calculations particularly for the computation of the critical depth. These curves are shown in Fig. 13.6.1. The curves for trapezoidal channels of different side slopes (z (H):1(V)) can be merged into a single curve (5). Equation (13.6.2) yields,

$$\frac{Q^2}{g} = \frac{(B + zh_c)^3 h_c^3}{(B + 2zh_c)} \qquad \qquad ...(13.6.4)$$

or

$$\frac{z^3 Q^2}{g B^5} = \frac{(1 + zh_c/B)^3 (zh_c/B)^3}{(1 + 2zh_c/B)} \qquad \qquad ...(13.6.5)$$

Combining Eqs. (13.6.3) and (13.6.5), one obtains

$$\frac{z^3 Z^2}{B^5} = \frac{[(1 + zh_c/B)(zh_c/B)]^3}{[1 + 2zh_c/B]} \qquad \qquad ...(13.6.6)$$

or

$$\frac{z^{1.5} Z}{B^{2.5}} = \frac{[(1 + zh_c/B)(zh_c/B)]^{1.5}}{[1 + 2zh_c/B]^{0.5}} \qquad \qquad ...(13.6.7)$$

One can now obtain the values of $z^{1.5}Z/B^{2.5}$ for different values of zh_c/B and plot these values as shown in Fig. 13.6.2. The use of these curves for estimation of critical depth for given discharge in a channel requires computation of Z ($= Q/\sqrt{g}$) and the ordinate $z^{1.5}Z/B^{2.5}$ (for trapezoidal channel) or $Z/D_0^{2.5}$ (for circular channel) and read the corresponding value of zh_c/B (for trapezoidal channel) or h_c/D_0 (for circular channel) and thus obtain h_c.

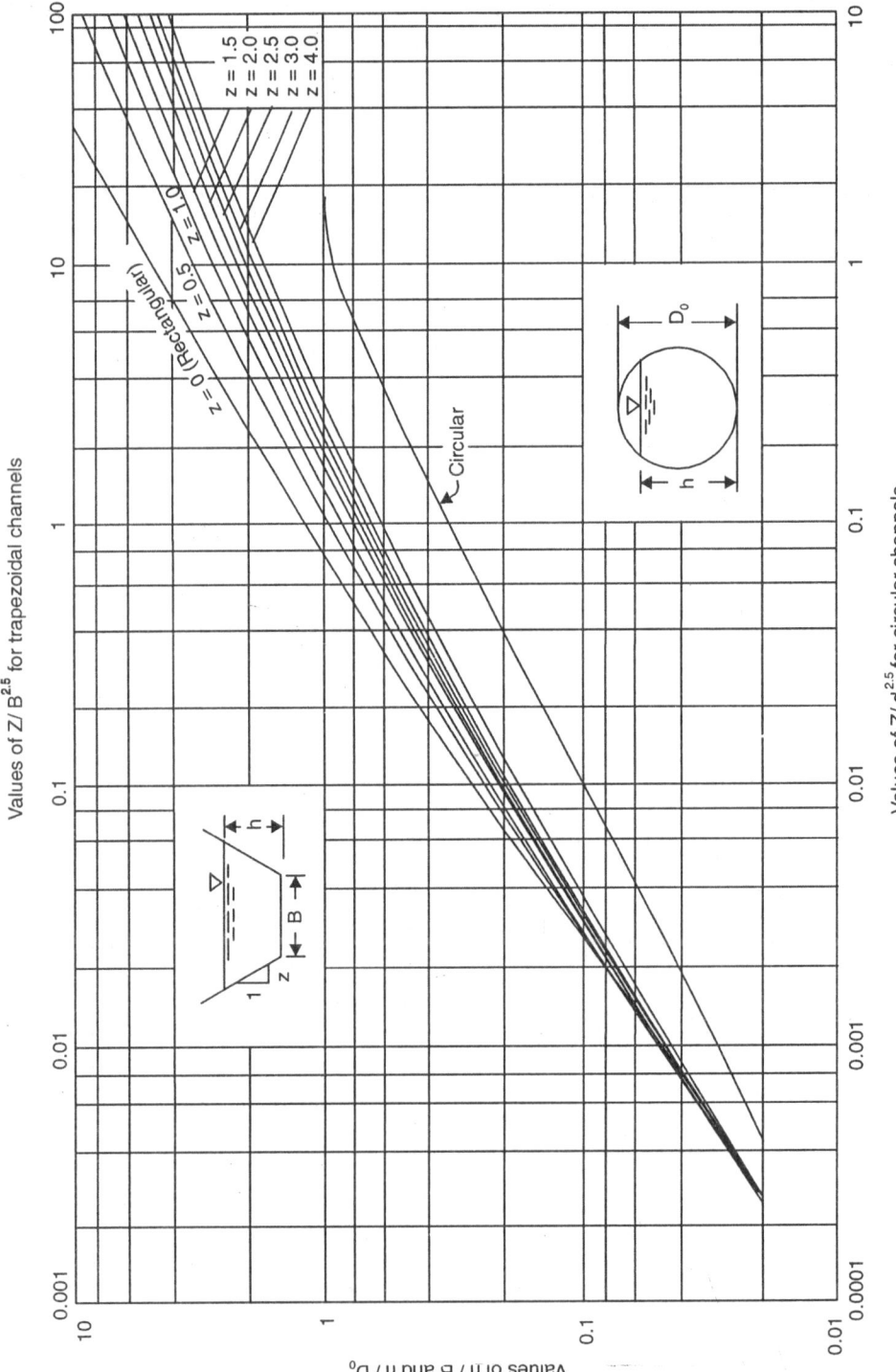

Fig. 13.6.1. Dimensionless curves for estimation of critical depth in tropezoidal and circular channels (2)

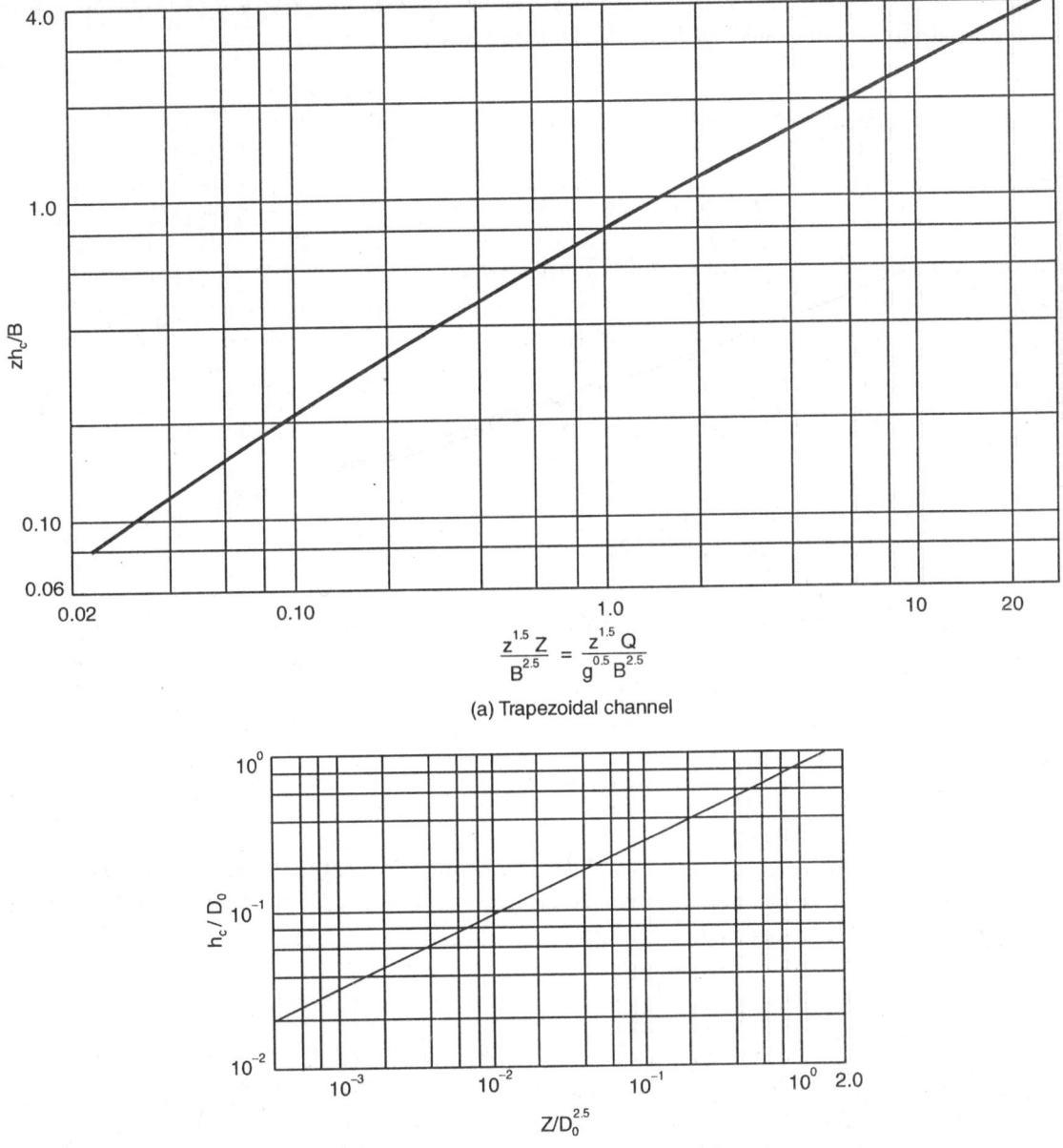

$$\frac{z^{1.5}\,Z}{B^{2.5}} = \frac{z^{1.5}}{g^{0.5}}\frac{Q}{B^{2.5}}$$

(a) Trapezoidal channel

$$Z/D_0^{2.5}$$

(b) Circular channel

Fig. 13.6.2. Dimensionless section factor for critical depth computations (5)

13.7 HYDRAULIC EXPONENT FOR CRITICAL FLOW COMPUTATIONS

Similar to that for uniform flow computations, it would be advantageous, for critical flow computations, to obtain a relation between the section factor Z and the depth of flow h by assuming that

$$Z^2 = c_2\,h^M \qquad\qquad ...(13.7.1)$$

where, c_2 is a suitable constant and M is the *hydraulic exponent for critical flow computations*. By taking logarithms on both sides of Eq. (13.7.1), one gets

$$2 \ln Z = M \ln h + \ln c_2 \qquad \qquad ...(13.7.2)$$

which, when differentiated with respect to h, yields

$$\frac{d(\ln Z)}{dh} = \frac{M}{2h} \qquad \qquad ...(13.7.3)$$

Since

$$Z = A\sqrt{A/T} = A^{3/2}/\sqrt{T}$$

∴

$$\ln Z = \frac{3}{2} \ln A - \frac{1}{2} \ln T$$

Hence,

$$\frac{d(\ln Z)}{dh} = \frac{3}{2A}\frac{dA}{dh} - \frac{1}{2T}\frac{dT}{dh}$$

or

$$\frac{d(\ln Z)}{dh} = \frac{3}{2}\frac{T}{A} - \frac{1}{2T}\frac{dT}{dh} \qquad \qquad ...(13.7.4)$$

as $\dfrac{dA}{dh} = T$ as explained earlier. Comparing Eqs. (13.7.3) and (13.7.4), one gets

$$\frac{M}{2h} = \frac{3}{2}\frac{T}{A} - \frac{1}{2T}\frac{dT}{dh}$$

∴

$$M = \frac{3Th}{A} - \frac{h}{T}\frac{dT}{dh} \qquad \qquad ...(13.7.5)$$

Equation (13.7.5) is a general equation for the hydraulic exponent M that depends on the depth of flow and the channel section. For example, the expressions for A and T for a trapezoidal channel with base width B and side slopes z (H): 1(V) are

$$A = (B + zh)h$$

and

$$T = B + 2zh$$

∴

$$\frac{dT}{dh} = 2z$$

Hence,

$$M = \frac{3(B + 2zh)h}{(B + zh)h} - \frac{2zh}{(B + 2zh)}$$

$$= \frac{3(1 + 2zh/B)}{(1 + zh/B)} - \frac{2zh/B}{(1 + 2zh/B)}$$

∴

$$M = \frac{3(1 + 2zh/B)^2 - (2zh/B)(1 + zh/B)}{(1 + zh/B)(1 + 2zh/B)} \qquad \qquad ...(13.7.6)$$

Equation (13.7.6) is shown plotted in Fig. 13.7.1. However, for rectangular channels (i.e., $z = 0$), Eq. (13.7.6) yields a constant value of M equal to 3 for all values of h. Values of M for a circular channel, obtained from Eq. (13.7.5), are listed in Table 13.7.1 which reveals that the value of M varies within a narrow range (3.85 – 4.0) for values of h/D_o less than about 0.6 and increases rapidly when h/D_o exceeds 0.6. This is due to the reason that as h increases beyond $0.5\,D_o$, T decreases and Z becomes large and, hence, the critical discharge, given by Eq. (13.6.3), too becomes large. This means that in a circular conduit (and, likewise, in other types of closed conduits with gradually closing crown) it is practically impossible to maintain critical flow when the free surface is close to the top of the section.

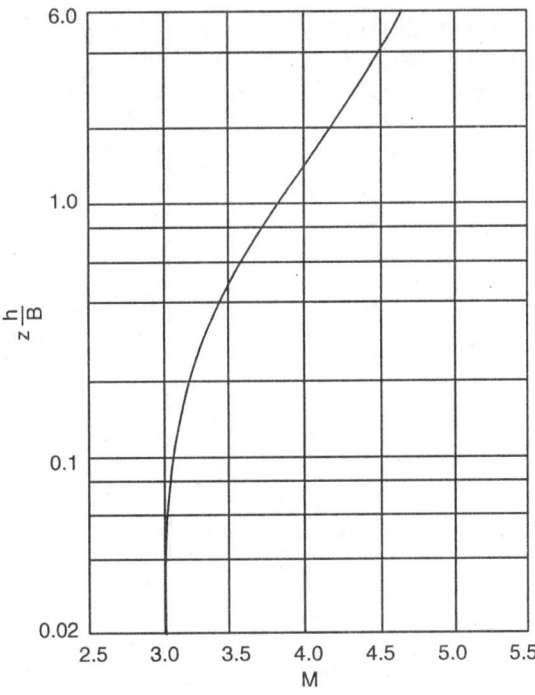

Fig. 13.7.1. Relation between M and zh/B for trapezoidal channel (5)

Table 13.7.1: Values of M for circular channels (5)

h/D_o	0.04	0.1	0.2	0.4	0.5	0.6	0.8	0.87
M	4.0	3.97	3.92	3.84	3.80	3.85	4.35	5.5

For channel sections other than the rectangular, trapezoidal, and circular shapes, one can, similarly, obtain the values of M provided that dT/dh is determinable. For most of the channel shapes (including the irregular ones that cannot be described by standard geometrical shapes and excluding the channels with abrupt changes in their cross-sectional form and the closed conduits with gradually closing top), a logarithmic plot of Z as ordinate against the depth h as abscissa, in accordance with Eq. (13.7.2), can be approximated as straight line. Using the coordinates of any two points on the straight line, one can compute the value of M from the following equation (2):

$$M = 2 \frac{\log(Z_1 / Z_2)}{\log(h_1 / h_2)} = 2 \times \text{slope of the line} \qquad ...(13.7.7)$$

For the curved portion of the plot, the value of M for a given depth would be equal to twice the slope of the tangent to the curve at that depth.

The hydraulic exponent M would prove very useful for computation of gradually varied flow.

13.8 APPLICATIONS OF CRITICAL FLOW CONCEPTS

A major application of the concepts of critical flow or specific energy is in the prediction of changes in the depth of flow in response to changes in the channel width and/or the channel bed elevation. This requires use of Eq. (13.2.4) and/or the use of a specific energy diagram, Fig. 13.1.1, or a depth-discharge curve, Fig. 13.2.1.

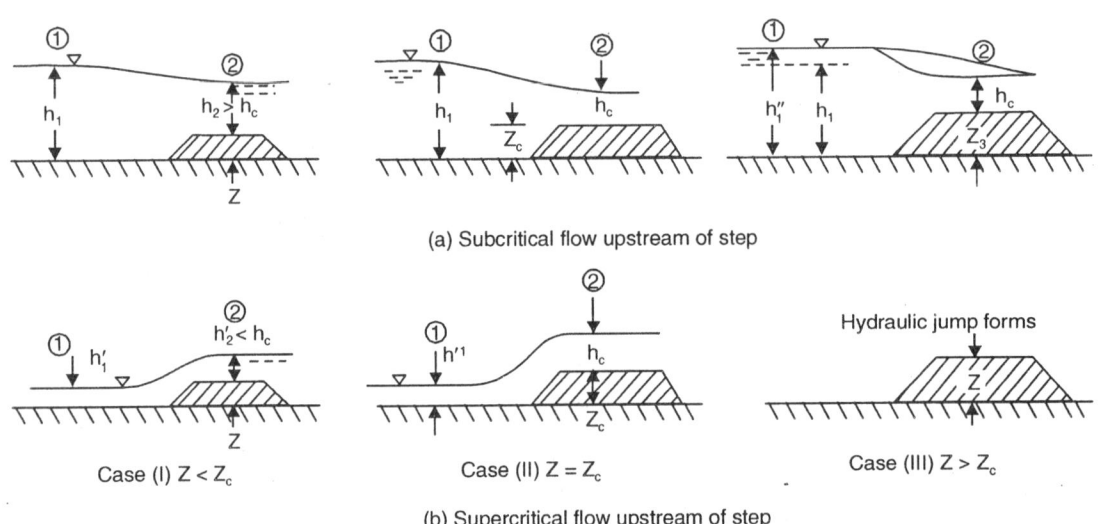

(a) Subcritical flow upstream of step

(b) Supercritical flow upstream of step

Fig. 13.8.1. Flow over a hump in channel bed

13.8.1 Change in Bed Elevation

Consider a horizontal rectangular channel section, Fig. 13.8.1, of width B that conveys a steady flow Q. Therefore, the unit discharge (i.e., the discharge per unit width of channel) is q (=Q/B). At a given location along the channel, let there be a smooth step (i.e., hump) of height z, Fig. 13.8.1. The prediction of the depth of flow over the step can be made by using the specific energy diagram shown in Fig. 13.8.2. Let the flow upstream of the step is represented by the point A (i.e., the flow is subcritical). Since the width of channel is constant, q remains the same over the step too and, hence, the point representing the flow over the step must also lie on the same specific energy diagram. Using Bernoulli's equation between sections 1 (upstream of the step) and 2 (over the step) and assuming that the energy (or head) loss between these two sections is negligible, one obtains

$$h_1 + \frac{U_1^2}{2g} = z + h_2 + \frac{U_2^2}{2g} \qquad ...(13.8.1)$$

or
$$E_1 = z + E_2$$

∴
$$E_2 = E_1 - z \qquad\qquad ...(13.8.2)$$

Having determined E_2, the corresponding depth of flow h_2 can be determined mathematically (Eq. 13.1.2) or using specific energy diagram, Fig. 13.8.2. One would notice that Eq. (13.1.2) would yield two feasible values of the depth h_2 besides one non-feasible value. Likewise, Fig. 13.8.2 yields two values of the depth of flow corresponding to points B and B' for the specific energy E_2. Since q is constant, the value of the "flow point" B or B' has to move only along the specific energy curve for $q(4)$. This means that the "flow point" B has to jump across the space separating the points B and B' to reach B'. This, is obviously, not possible. For the "flow point" to move from B to C and, then, to B', the increase in elevation of the channel bed must be greater than z and equal to z_c so that $E_2 = E_1 - z_c = E_c$ and the "flow point" reaches C along the specific energy curve. This increase in elevation must be followed by a suitable drop in the channel bed elevation. In the absence of such a channel bed step (or hump), the flow conditions over the hump would correspond to the point B only. If the height of the step is z_c, the two points B and B' merge into a single point C corresponding to the minimum specific energy and, hence, the flow over the hump is in critical state. If the step height z exceeds z_c so that $E_2 (= E_1 - z)$ is less than E_c, then, it is obvious that the solution is not feasible with prescribed values of q, E_1, and z. The physical explanation to this non-feasibility is that the flow has been choked because of the obstruction. The flow will, therefore, back up upstream of the hump so that the new value of the specific energy E_1'' upstream of the step is such that

$$E_1'' - z = E_c$$

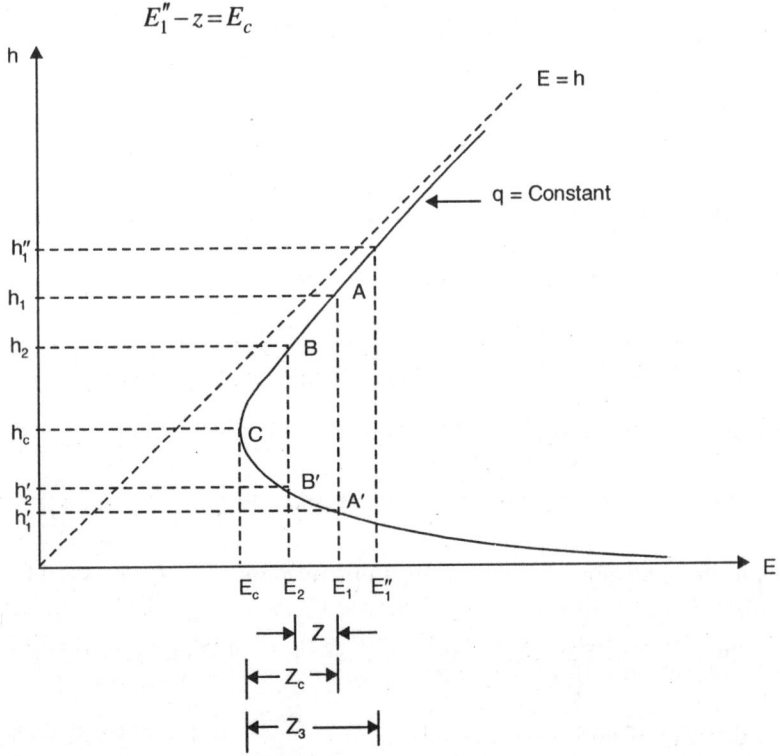

Fig. 13.8.2. Use of specific energy diagram for flow over a hump

Accordingly, the depth of flow upstream of the step (or hump) corresponds to the specific energy $E_1''(= E_c + z)$ and is h_1''.

If the approaching flow is supercritcal, the solution can be obtained in a similar way. Similarly, one can obtain solution if there is small drop in the channel bed. Similarly, one can analyze the flow conditions due to change in the elevation of a channel of other cross-sectional shapes.

Alternatively, for the case of flow with no head loss in a channel of rectangular section and constant width but varying bed level, the total energy (head) H and the discharge per unit width q are constant. Hence,

$$H = h + z + \frac{q^2}{2gh^2} = E + z = \text{constant} \qquad \text{...(13.8.3)}$$

$\therefore$
$$\frac{dE}{dx} + \frac{dz}{dx} = 0$$

or
$$\frac{dh}{dx}\frac{dE}{dh} + \frac{dz}{dx} = 0$$

Substituting the value of $\dfrac{dE}{dh}$ from Eq. (13.2.3), one obtains

$$\frac{dh}{dx}\left(1 - \frac{U^2}{gD}\right) + \frac{dz}{dx} = 0$$

or
$$\frac{dh}{dx}\left(1 - F_r^2\right) + \frac{dz}{dx} = 0 \qquad \text{...(13.8.4)}$$

Equation (13.8.4) indicates that if there is an upward step (i.e., $\dfrac{dz}{dx}$ is positive), then the term $\dfrac{dh}{dx}(1 - F_r^2)$ must be negative. Therefore, for subcritical flow (i.e., $F_r < 1$), $\dfrac{dh}{dx}$ must be negative, i.e., the depth will decrease over the step. Similarly, for supercritical flow (i.e., $F_r > 1$), $\dfrac{dh}{dx}$ must be positive, i.e., the depth over the step will increase. Similarly, one can obtain corresponding conclusions for a downward step (i.e., $\dfrac{dz}{dx} < 0$), in which case for subcritical flow $\left[\text{i.e., } F_r < 1 \text{ and } \left(1 - F_r^2 > 0\right)\right]$, $\dfrac{dh}{dx} > 0$, i.e., the depth of flow increases in the direction of flow. For supercritical flow $\left[\text{i.e., } F_r > 1 \text{ and } \left(1 - F_r^2 < 0\right)\right]$, however, $\dfrac{dh}{dx} < 0$, i.e., the depth of flow decreases in the direction of flow, if there is a downward step.

However, for the case of horizontal channel bed (i.e., $\dfrac{dz}{dx} = 0$), Eq. (13.8.4) yields

$$\frac{dh}{dx}\left(1 - F_r^2\right) = 0$$

Therefore, there can be three possible solutions:

(i) $\dfrac{dz}{dx} = 0$, $\dfrac{dh}{dx} = 0$; $F_r \neq 1$ i.e., uniform flow

(ii) $\dfrac{dz}{dx} = 0$, $\dfrac{dh}{dx} = 0$; $F_r = 1$ i.e., uniform critical flow

(iii) $\dfrac{dz}{dx} = 0$, $\dfrac{dh}{dx} \neq 0$; $F_r = 1$ i.e., non-uniform critical flow

To visualize a flow situation for which $\dfrac{dz}{dx} = 0$ and $\dfrac{dh}{dx} \neq 0,$ consider the flow of water being released from a lake (or spillway or broad-crested weir) over a short (but not sharp-edged to result in non-hydrostatic pressure distribution) crest (with gentle or no curvature so that vertical accelerations due to flow curvature are not large), Fig. 13.8.3, so that it can fall freely downstream (4). A short distance downstream of the crest, there may be either a free overfall or a slope so steep that the bed resistance imposes no effective restraint on the flow. In such a

situation, at the crest A, $\dfrac{dz}{dx} = 0$, but, $\dfrac{dh}{dx} \neq 0$ since the flow is accelerating. Hence, the Froude number must be unity and the flow, therefore, critical. Even if the pressure distribution is non-hydrostatic or the vertical accelerations are large, as for the case of a free overfall, the flow would still be close to critical condition. Experiments have indicated that the flow depth right at the brink of an overfall is approximately $(5/7)h_c$ and the depth of flow equals h_c at a distance of about $3h_c$ to $4h_c$ upstream from the overfall (6).

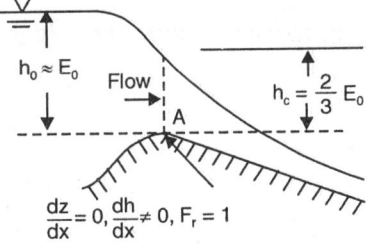

Fig. 13.8.3. Free outflow from a lake

13.8.2 Reduction in Channel Width

Consider, once again, a horizontal rectangular channel whose width B_1 is gradually reduced to B_2, Fig. 13.8.4, so that the energy loss due to the transition can be considered negligible. Since the bed elevation has not changed and there is no energy loss, the specific energy remains constant. However, the discharge per unit width q has changed. Hence, solution to such problems can be obtained by using q–h relation for constant E, Fig. 13.8.5. Let the depth of the approaching subcritical flow at section 1 be h_1 for the discharge $q_1 (= Q/B_1)$. In the contracted section, the unit discharge is $q_2 (= Q/B_2)$ which is greater than q_1 and the depth of flow h_2 at the contracted section can be determined by moving along the unit discharge – depth curve, Fig. 13.8.5. For the flow depth to be h_2' (i.e., in supercritical state), the "flow point" has to first reach h_c and, then, h_2'. This requires contraction of width to B_c followed by expansion so that the width of channel is B_2. In the absence of such a transition, the depth of flow would be h_2 only. Similarly, if the approaching flow were supercritical with the depth of flow being h_1', the depth of flow in the contracted section would be h_2'. However, if the contracted width is B_c so that the resulting unit discharge q_c is maximum for the given specific energy, then the depth of flow in the contracted section would be critical depth for both supercritical and subcritical approach flows.

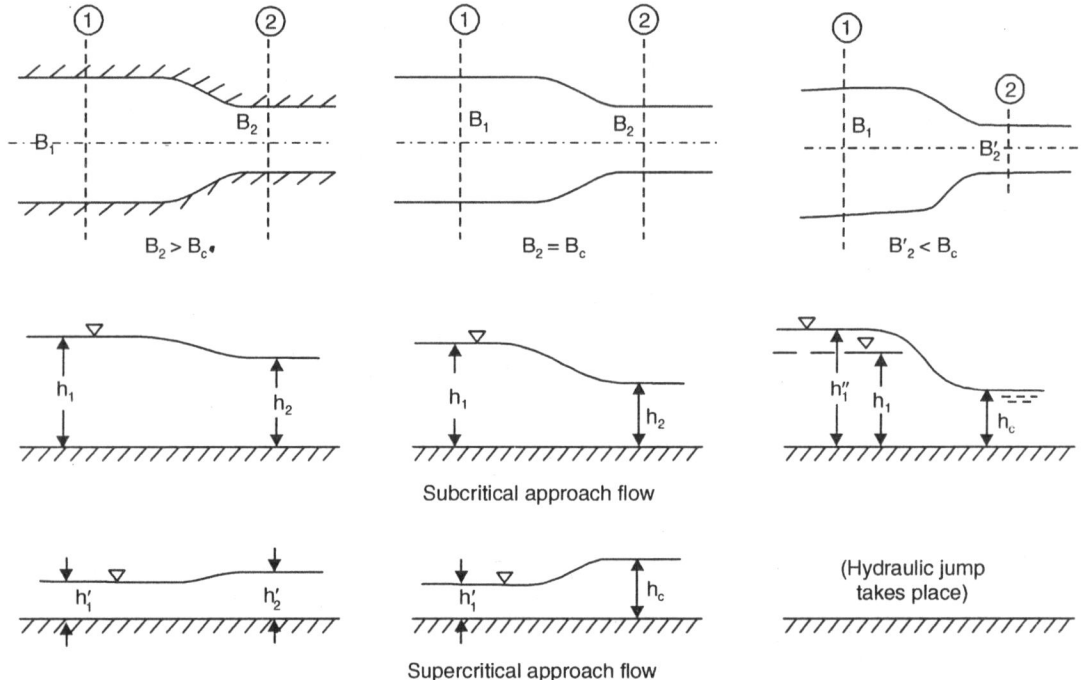

Fig. 13.8.4. Flow through a contraction

If the width of a given contracted channel is less than B'_c say B'_2, so that $q'_2(= Q/B'_2)$ exceeds $q_c(= Q/B_c)$, Fig. 13.8.5 shows that the flow cannot pass without modifying the upstream condition such that the specific energy at the upstream condition equals the specific energy at the contracted section with flow at the contracted section being in critical state corresponding to the unit discharge q'_2. The value of specific energy at the contracted section shall be 1.5 times the critical depth corresponding to the unit discharge q'_2. Thus, the modified specific energy E'_1 would be

$$E'_1 = \frac{3}{2}\left[\left(\frac{Q}{B'_2}\right)^2 \frac{1}{g}\right]^{1/3} \qquad\qquad ...(13.8.5)$$

and the depth of flow would be changed to h''_1, Fig. 13.8.5. Once again, one can, similarly, analyze the flow conditions due to change in the width of a channel of other cross-sectional shapes.

If one assumes a horizontal channel bed (i.e., $\dfrac{dz}{dx} = 0$) and a variable width B, one can differentiate Eq. (13.8.3) with respect to x to obtain

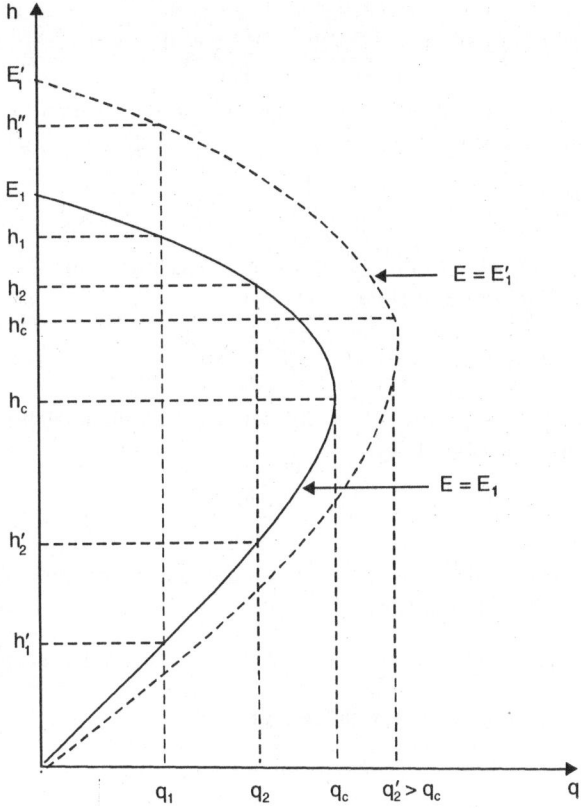

Fig. 13.8.5. Unit discharge-depth diagram for a rectangular channel

$$\frac{dh}{dx} - \frac{q^2}{gh^3}\frac{dh}{dx} + \frac{q}{gh^2}\frac{dq}{dx} = 0 \qquad \qquad ...(13.8.6)$$

and since $qB = Q = $ constant

$$B\frac{dq}{dx} + q\frac{dB}{dx} = 0 \qquad \qquad ...(13.8.7)$$

Substituting the value of $\dfrac{dq}{dx}$ in Eq. (13.8.6), one gets

$$\frac{dh}{dx}(1 - F_r^2) - \frac{q^2}{gh^2 B}\frac{dB}{dx} = 0$$

or

$$\frac{dh}{dx}(1 - F_r^2) - F_r^2\frac{h}{B}\frac{dB}{dx} = 0 \qquad \qquad ...(13.8.8)$$

In accordance with Eq. (13.8.8), there can be the following four flow situations corresponding to either subcritical or supercritical flow in either expanding or contracting channels (3):

(i) $\dfrac{dB}{dx} > 0$ and $F_r < 1$, i.e. $\left(1 - F_r^2\right) > 0$, $\therefore$ $\dfrac{dh}{dx} > 0$

This corresponds to subcritical flow in an expanding channel for which the depth of flow increases in the direction of flow.

(ii) $\dfrac{dB}{dx} > 0$ and $F_r > 1$, i.e. $\left(1 - F_r^2\right) < 0$, $\therefore$ $\dfrac{dh}{dx} < 0$

This corresponds to supercritical flow in an expanding channel for which the depth of flow decreases in the direction of flow.

(iii) $\dfrac{dB}{dx} < 0$ and $F_r < 1$, i.e. $\left(1 - F_r^2\right) > 0$, $\therefore$ $\dfrac{dh}{dx} < 0$

This corresponds to subcritical flow in a contracting channel for which the depth of flow decreases in the direction of flow.

(iv) $\dfrac{dB}{dx} < 0$ and $F_r > 1$, i.e. $\left(1 - F_r^2\right) < 0$, $\therefore$ $\dfrac{dh}{dx} > 0$

This corresponds to supercritical flow in a contracting channel for which the depth of flow increases in the direction of flow.

As in Art. 13.8.1, one can, again, conclude that critical flow occurs at the outflow from a lake (at which

location $\dfrac{dB}{dx} = 0$ and $\dfrac{dh}{dx} \neq 0$ as in real physical situations).

EXAMPLES

E13.1.   In a trapezoidal channel (width $B = 10$ m, side slope $z = 2$), a discharge of 90 m³/s flows under critical condition. Determine the critical depth of flow.

Solution: Substituting the value of Q, B, and z for the given channel in Eq. (13.6.2) one gets

$$\frac{(90)^2 (10 + 2(2)h_c)}{(9.8)[(10 + 2h_c)h_c]^3} = 1$$

$\therefore$ $[(10 + 2h_c)h_c]^3 = 826.53(10 + 4h_c)$

on solving this equation by trial, one obtains the desired critical depth of flow, $h_c = 1.78$ m.

E13.2.   Water is released through a sluice gate opening in a rectangular channel that is 1.5 m wide. The resulting depth and velocity of flow are, respectively, 0.6 m and 4.5 m/s. Determine whether the flow is supercritical or subcritical and also obtain

(a) critical depth for the resulting specific energy
(b) critical depth for the given discharge
(c) the alternate depth for the given flow depth

Solution: Froude number for the given flow condition,

$$F_r = \frac{U}{\sqrt{gD}} = \frac{4.5}{\sqrt{9.8 \times 0.6}} = 1.86$$

Hence, the flow is supercritical.
The specific energy of the flow is

$$E = h + \frac{U^2}{2g} = 0.6 + \frac{(4.5)^2}{2 \times 9.8} = 1.633 \, \text{m}$$

For this specific energy, the critical depth is given by Eq. (13.2.9), i.e.,

$$h_c = \frac{2}{3} E$$

$$= 1.089 \, \text{m}$$

The critical depth for the given discharge is given by Eq. (13.2.8), i.e.,

$$h_c = \left[\frac{q^2}{g} \right]^{1/3} = \left[\frac{(4.5 \times 0.6)^2}{9.8} \right]^{1/3} = 0.906 \, \text{m}$$

The specific energy for the given flow condition is 1.633 m. Therefore, from Eq. (13.1.2),

$$1.633 = h + \frac{Q^2}{2gB^2 h^2}$$

or

$$1.633 = h + \frac{(4.5 \times 0.6 \times 1.5)^2}{2 \times 9.8 \times (1.5 \times h)^2}$$

or

$$1.633 = h + \frac{0.372}{h^2}$$

Obviously, one of the roots for this equation is $h = 0.6$ m. Other real root of this equation would give the desired alternate depth. By solving this equation by trial, one gets the desired alternate depth

$$h = 1.458 \, \text{m}$$

E13.3. Water flows in a wide rectangular channel of concrete ($n = 0.014$) laid at a slope of 1 in 900. Determine the depth of flow and the unit discharge for the flow to take place in critical condition.

Solution: For a wide channel, the Manning's equation, Eq. (12.3.5) can be written as

$$q = \frac{1}{n} h_c^{5/3} S^{1/2}$$

with depth of flow equal to the critical depth h_c. Combining this equation with Eq. (13.2.8), one gets

$$\frac{\left[\frac{1}{n} h_c^{5/3} S^{1/2} \right]^2}{g h_c^3} = 1$$

i.e., $$\frac{S}{gn^2}h_c^{1/3} = 1$$

or $$h_c = \left[\frac{gn^2}{S}\right]^3$$

$$= \left[\frac{9.8 \times (0.014)^2}{(1/900)}\right]^3$$

∴ $$h_c = 5.166 \text{ m}$$

From Eq. (13.2.8)

$$q = \sqrt{g\,h_c^3}$$

$$= \sqrt{9.8(5.166)^3}$$

$$= 36.76 \text{ m}^3/\text{s/m}$$

E13.4.   Uniform flow occurs at a depth of 1.5 m in a 3.0 m wide rectangular channel (n = 0.015) laid at a slope of 1 in 1000. Assuming no transition loss, calculate minimum height of hump on the channel bed resulting into critical flow over the hump, and the width of contracted channel that will result in critical flow in the contracted section without affecting the depth of flow upstream of the contracted section.

If the width of the contracted channel section is made 2.0 m, what would be the depth of flow in the contracted section and the corresponding depth of flow upstream of the contraction?

Solution: The discharge Q in the channel is obtained by the Manning's equation, Eq. (12.3.5) as

$$Q = \frac{1}{0.015}(3 \times 1.5)\left[\frac{3 \times 1.5}{3 + (2 \times 1.5)}\right]^{2/3}\left(\frac{1}{1000}\right)^{1/2}$$

$$= 7.83 \text{ m}^3/\text{s}$$

∴ $$U = \frac{7.83}{3 \times 1.5}$$

$$= 1.74 \text{ m/s}$$

Hence, at section 1 (Fig. 13.8.1)

$$E_1 = h_1 + \frac{U_1^2}{2g} = 1.5 + \frac{(1.74)^2}{2 \times 9.8} = 1.655 \text{ m}$$

Also, unit discharge in the channel is

$$q = \frac{Q}{B} = \frac{7.83}{3} = 2.61 \text{ m}$$

Hence, the critical depth is

$$h_c = \left(\frac{q^2}{g}\right)^{1/3} = \left[\frac{(2.61)^2}{9.8}\right]^{1/3} = 0.886 \, \text{m}$$

Hence, $\quad E_c = \frac{3}{2} h_c = \frac{3}{2}(0.886) = 1.329 \, \text{m}$

Applying energy equation between sections 1 and 2 of Fig. 13.8.1, or using Eq. (13.8.1)

$$1.655 = z + 1.329$$

$\therefore \qquad\qquad z = 0.326 \, \text{m}$

Applying energy equation between sections 1 and 2 of Fig. 13.8.4, one gets

$$1.655 = E_2 = E_c = 1.5 h_c = 1.5 \left[\frac{(Q/B_2)^2}{g}\right]^{1/3}$$

i.e., $\qquad 1.655 = 1.5 \left[\frac{(7.83)^2}{9.8 \, B_2^2}\right]^{1/3}$

or $\qquad\qquad B_2 = 2.158 \, \text{m}$

Since the width of the contracted channel section is less than 2.158 m that is required for the flow to be critical at the section, the flow at the section would be critical and the upstream flow conditions would get modified.

For the contracted section,

$$q = \frac{7.83}{2.0} = 3.915 \, \text{m}^3/\text{s/m}$$

Hence, the depth of flow at the contracted section is

$$h_c = \left[\frac{(3.915)^2}{9.8}\right]^{1/3} = 1.16 \, \text{m}$$

$\therefore \qquad E_1' = E_2 = E_c = \frac{3}{2}(1.16) = 1.74 \, \text{m}$

Hence, $\qquad h_1' + \dfrac{Q^2}{2g \, B^2 h_1'^2} = 1.74 \, \text{m}$

or $\qquad h_1' + \dfrac{(7.83)^2}{2 \times 9.8 \times (3)^2 h_1'^2} = 1.74$

or $\qquad h_1' + \dfrac{0.3476}{h_1'^2} = 1.74$

By trial, $\qquad h_1' = 1.607 \, \text{m}$

PROBLEMS

P13.1 Compute critical depth of flow in a 6 m wide rectangular channel carrying a discharge of 30 m³/s. Also, determine the critical slope for this channel, if the Manning's *n* for the channel is 0.014.

P13.2 Estimate the value of the maximum discharge that can be carried by a 2.5 m wide rectangular channel when specific energy for the flow is 2.2 m.

P13.3 Compute the critical depth of flow for a discharge of 1.0 m³/s in a
   (a) rectangular channel of 1.5 m width,
   (b) trapezoidal channel of 1.0 m bed width and side slopes 1:1,
   (c) circular channel of 1.0 m diameter.

P13.4 A 3.0 m wide rectangular channel carries a discharge of 0.6 m³/s at a depth of 270 mm. A smooth hump of height 60 mm is placed on the channel bed. Determine the local change in the flow depth caused by the hump.

P13.5 Uniform flow occurs in a trapezoidal channel (bed width = 2.0 m, side slope = 2H:1V, the Manning's *n* = 0.015, and bed slope = 1.8 × 10⁻³). The depth of flow in the channel is 1.2 m. Determine the discharge flowing in the channel and the corresponding critical depth. Also, find the alternate and conjugate depths corresponding to the given depth of flow.

P13.6 Determine the maximum rise in the bed level of a 2.0 m wide rectangular channel (carrying a discharge of 3.0 m³/s at a depth of flow equal to 1.0 m) that can be allowed so as not to modify the upstream flow conditions. Neglect losses.

P13.7 A 5.0 m wide rectangular channel carries a discharge of 2.5 m³/s at a depth of 2.0 m. The channel is contracted to a width of (i) 4.75 m, and (ii) 3.0 m. Determine the depth of flow at the contracted section and the upstream flow condition, if modified. Neglect losses.

P13.8 A 11.0 m wide rectangular channel carries a discharge of 10 m³/s at a depth of 1.5 m. If the width is reduced to 8.0 m at a specified section, what change in bed elevation at the specified section will maintain the water surface elevation as before. Neglect losses.

REFERENCES

(1) Bakhmeteff, BA: *"Hydraulics of Open Channels"*, McGraw-Hill, New York, 1932 [White, F.M., Fluid Mechanics, McGraw-Hill, New York, 2003].

(2) Chow, VT: *"Open-Channel Hydraulics"*, McGraw-Hill Book Co., New York, 1959.

(3) French, RH: *"Open-Channel Hydraulics"*, McGraw-Hill Book Co., Singapore, 1994.

(4) Henderson, FM: *"Open-Channel Flow"*, The Macmillan Co., New York, 1966.

(5) Ranga Raju, KG: *"Flow through Open Channels"*, Tata McGraw-Hill Publishing Company Limited, New Delhi, 1993.

(6) Rouse, H: *"Discharge characteristics of the free overall"*, Civil Engineering, Vol. 6, No. 7, April 1936 [Chow(2)].

CHAPTER 14

Steady Non-uniform Flow in Channels

14.1 GRADUALLY VARIED FLOW

A control structure (such as fall) in a channel transforms uniform flow into a non-uniform (or varied) flow in which the depth of flow varies along the length of the channel. If the variation of the depth is gradual, the flow is termed gradually varied flow (GVF) and the streamlines are, therefore, practically parallel. A hydraulic (or river) engineer is, many a times, required to estimate with acceptable accuracy the gradual variation of depth of flow along a channel when flow in the channel is non-uniform. For example, the effect of a proposed dam on the upstream water levels in a river can only be ascertained by knowing as to what distance upstream does the dam produce a depth that is significantly different from the uniform flow depth. This chapter deals with gradually varied flow that is steady, i.e., the flow characteristics remain constant for the duration under consideration.

14.2 BASIC EQUATION FOR GRADUALLY VARIED FLOW

Consider an elementary length dx of an open channel reach, Fig. 14.2.1, that has gradually varied flow. Using Eq. (11.5.9), the total head above the datum at any section is

$$H = z + d\cos\theta + \alpha\frac{U^2}{2g} \qquad \qquad ...(14.2.1)$$

If one assumes θ and α to be constant for the channel reach under consideration and the channel bed to be the x-axis, then differentiation of Eq. (14.2.1) with respect to length x (measured along the x-axis) of the water surface profile yields

$$\frac{dH}{dx} = \frac{dz}{dx} + \cos\theta\frac{dd}{dx} + \alpha\frac{d}{dx}\left(\frac{U^2}{2g}\right) \qquad \qquad ...(14.2.2)$$

Here, dH/dx and dz/dx represent, respectively, the slope of the total energy line, i.e., $-S_f$ and slope of the channel bed, i.e., $-S_0$. Thus,

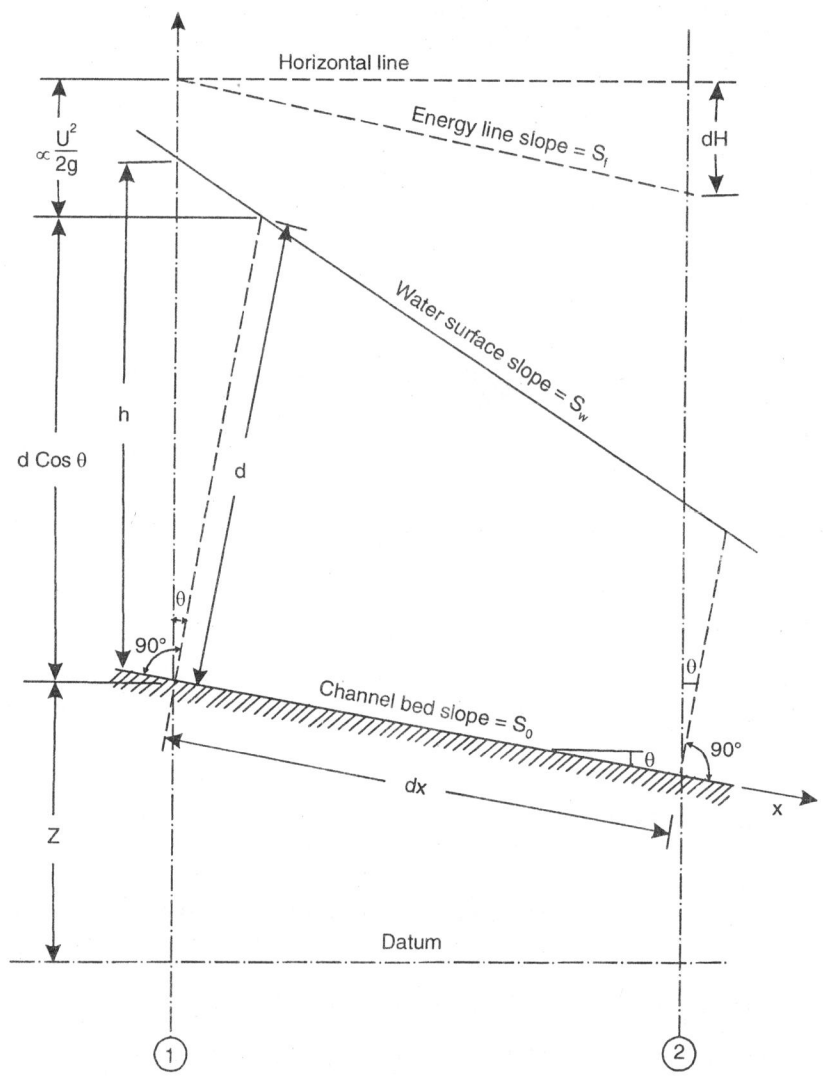

Fig. 14.2.1. Definition sketch for derivation of the GVF equation

$$\frac{dd}{dx} = \frac{S_0 - S_f}{\cos\theta + \alpha\left[d(U^2/2g)/dd\right]} \qquad \text{...(14.2.3)}$$

Equation (14.2.3) is the general differential equation for GVF, usually referred as the dynamic equation of GVF or, simply, the gradually varied flow equation. The term *dd/dx* in Eq. (14.2.3) represents the slope of the water surface with respect to the channel bed.

For small values of θ, $\cos\theta \approx 1$ and $d \approx h$ (i.e., the depth of flow section is approximately equal to the depth of flow *h* that is measured vertically from the channel bed). One may further assume the kinetic energy correction factor α to be constant and equal to unity as the flow in open channels is generally turbulent and the velocity distribution is logarithmic. Therefore, Eq.(14.2.3) reduces to

$$\frac{dh}{dx} = \frac{S_0 - S_f}{1 + \left[d(U^2/2g)/dh \right]} \qquad \qquad ...(14.2.4)$$

Further, the value of the term $d\left(U^2/2g\right)/dh$, representing the change in the velocity head, can be obtained as follows:

$$\frac{d}{dh}\left(\frac{U^2}{2g}\right) = \frac{Q^2}{2g}\frac{d}{dh}\left(\frac{1}{A^2}\right) = -\frac{Q^2}{gA^3}\frac{dA}{dh} = -\frac{Q^2 T}{gA^3}$$

Thus, Eq. (14.2.4) can be written as

$$\frac{dh}{dx} = \frac{S_0 - S_f}{1 - \dfrac{Q^2 T}{gA^3}} \qquad \qquad ...(14.2.5)$$

Since $\dfrac{Q^2 T}{gA^3} = \dfrac{U^2}{gD} = F_r^2$, Eq. (14.2.5) can also be written as

$$\frac{dh}{dx} = \frac{S_0 - S_f}{1 - F_r^2} \qquad \qquad ...(14.2.6)$$

The section factor for critical flow computation Z has been defined (Art. 11.3) as $A\sqrt{A/T}$ or $\sqrt{A^3/T}$.

Therefore, $Q^2 T/\left(gA^3\right)$ equals $Q^2/\left(gZ^2\right)$. If the discharge Q were to flow in a critical state (depth of flow being h_c) then Eq. (13.2.4) yields

$$\frac{Q^2}{g} = Z_c^2$$

Here, Z_c is the value of Z (i.e., $\sqrt{A^3/T}$) corresponding to the critical depth of flow h_c. Therefore, $\dfrac{Q^2}{gZ^2}$ can be written as Z_c^2/Z^2 with Z corresponding to the non-uniform flow depth h. Therefore, Eq. (14.2.5) can also be written as

$$\frac{dh}{dx} = \frac{S_0 - S_f}{1 - \dfrac{Z_c^2}{Z^2}} \qquad \qquad ...(14.2.7)$$

Assuming that the head loss for the non-uniform flow in the reach under consideration equals the head loss for uniform flow in the reach with the hydraulic radius and average velocity the same as that for the non-uniform flow, the slope of the energy line S_f is

$$S_f = \frac{n^2 U^2}{R^{4/3}} = \frac{n^2 Q^2}{A^2 R^{4/3}} \qquad \qquad ...(14.2.8)$$

if one uses the Manning's formula, Eq. (12.3.4). If the Chezy's formula, Eq. (12.3.2), is used, then,

$$S_f = \frac{U^2}{C^2 R} = \frac{Q^2}{C^2 A^2 R} \qquad \qquad ...(14.2.9)$$

Alternatively, one can express S_f in terms of the conveyance $K \left(= \frac{1}{n} A R^{2/3} \right)$ as

$$S_f = \frac{Q^2}{K^2} \qquad \qquad ...(14.2.10)$$

If the discharge Q were to flow in the channel under uniform flow condition so that the depth of flow is the normal depth and the slope of the energy line equals the bed slope of the channel and thus,

$$S_0 = \frac{Q^2}{K_n^2} \qquad \qquad ...(14.2.11)$$

Here, K_n is the conveyance for uniform flow depth h_n (i.e., the normal depth) whereas the conveyance K corresponds to the depth of the non-uniform flow (i.e., h) under consideration. Equations (14.2.10) and (14.2.11) yield

$$\frac{S_f}{S_0} = \frac{K_n^2}{K^2} \qquad \qquad ...(14.2.12)$$

Combining Eqs. (14.2.7) and (14.2.12), one obtains

$$\frac{dh}{dx} = S_0 \frac{1 - \left(K_n / K \right)^2}{1 - \left(Z_c / Z \right)^2} \qquad \qquad ...(14.2.13)$$

which is yet another form of the GVF equation.

It should be noted that Eqs. (14.2.5), (14.2.6), (14.2.7), and (14.2.13) are simplified forms of the general differential equation for GVF, Eq. (14.2.3), and have been obtained by making one or more of the following assumptions:

1. The head loss for the non-uniform flow in the reach under consideration equals the head loss for uniform flow in the reach with the hydraulic radius and average velocity the same as that for the non-uniform flow.
2. The channel bed slope is small and, hence, the depth of flow is the same whether it is measured vertically or perpendicular to the bed. Also, the pressure correction coefficient is unity and there is no air entrainment.
3. The velocity distribution in the channel is approximately uniform and fixed so that the kinetic energy correction factor is unity.
4. The resistance coefficient is invariant with the depth of flow and is the same for the reach under consideration.
5. The channel is prismatic.

14.3 GRADUALLY VARIED FLOW PROFILES

Different forms of the gradually varied flow equations, Eqs. (14.2.3), (14.2.5), (14.2.6), (14.2.7) and (14.2.13) enable one to compute the longitudinal water surface slope of the flow in a channel with respect to the channel bed. Hence, these equations can be used to describe the characteristics of profiles of the water surface of the flow often termed as the gradually varied flow (GVF) profile.

When a prismatic channel meets another prismatic channel having different slope there is a possibility of occurrence of critical flow in the vicinity of the junction. Consider Eq. (14.2.6) for a special case when $S_o = S_f$. Thus, either $dh/dx = 0$ (meaning the flow is uniform flow) or $F_r = 1$ (meaning that the flow is critical flow). To appreciate the physical meaning of the flow being critical, consider a long channel of two prismatic sections such that the upstream channel is of mild slope and the downstream channel is of steep slope both meeting at the junction point O, Fig. 14.3.1. In the mild channel far upstream of the junction, the flow will be subcritical ($h_n > h_c$) and in the steep channel far downstream of the junction, the flow will be supercritical ($h_n < h_c$). Therefore, the flow passes through critical state in the vicinity of the junction.

In the transition region upstream of the junction O, the depth of flow h is less than h_n and, therefore, the velocity is greater than that in uniform flow in the upstream channel. Therefore, $S_f\left(=\dfrac{V^2}{C^2 R}\right)$ is greater than S_0 in the transition region upstream of O. Similarly, S_f is less than S_0 in the transition region downstream of O. Assuming the junction to be a short curve (instead of a point) joining the two long channels, there must be some point on this curve where S_f equals S_o. Since the flow is non-uniform $\left(\text{i.e., } \dfrac{dh}{dx} \neq 0\right)$ in this region, the Froude number must be unity and the flow there must be critical.

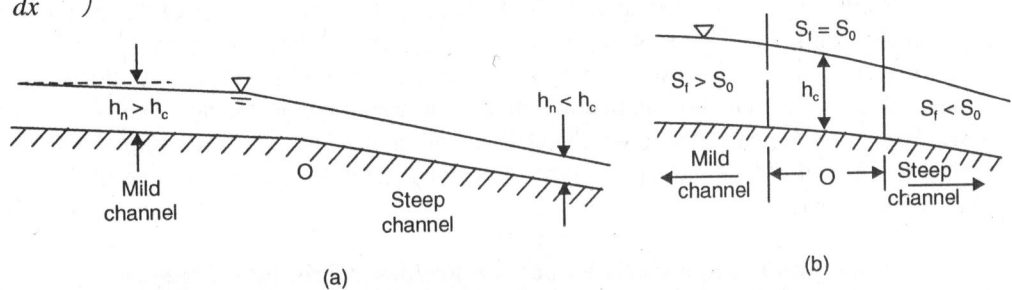

Fig. 14.3.1. Critical flow due to change of bed slope in a channel

The occurrence of critical flow at the outflow from a lake (Art. 13.8) into a steep channel (or free overfall) or through a constriction in width or at the upstream end of a steep slope preceded by a mild slope point to the physical concept of the mechanism of release of water. That is, the water that has previously been restrained, either by the boundaries (as in a lake) or the roughness of a channel bed (as in mild slope channel), is suddenly released into a region where restraint is either non-existent (as in free overfall) or is so small (as in steep channel) that the flow can no longer be forced into the subcritical condition. The flow is always critical at or in the vicinity of such a location of release and the critical flow section is a control where the depth-discharge relation is fixed.

14.4 CHARACTERISTICS OF GRADUALLY VARIED FLOW PROFILES

The flow profile shown in Fig. 14.3.1 is only one of a large number of GVF profiles that may conceivably occur when a control interferes with uniform flow, or when there is a transition from one state of uniform flow to another. The dynamic equation of GVF, Eq. (14.2.3) or one of its simpler forms, provides the slope (with respect to the channel bed) of the water surface of the channel flow. Hence, GVF equation can be used to describe the characteristics of GVF profiles or, simply, flow profiles.

From the Chezy's equation, Eq. (12.3.2) and the definition of the Froude number, one gets

$$S_f = \frac{U^2}{C^2 R} = \frac{Q^2}{C^2 A^2 R} = \frac{Q^2 P}{C^2 A^3} \qquad \qquad ...(14.4.1)$$

and

$$F_r^2 = \frac{U^2}{gD} = \frac{Q^2}{gA^2 D} = \frac{Q^2 T}{gA^3} \qquad \qquad ...(14.4.2)$$

Therefore, for a specified value of Q in a given channel, both S_f and F_r are functions of the depth h and vary in much the same way with depth h, as both (S_f and F_r) show strong inverse dependence on A. By definition, $S_f = S_o$ when $h = h_n$ (i.e., the depth of flow h equals the normal depth h_n). This means that

$$S_f > S_0 \text{ when } h < h_n \text{ and } S_f < S_0 \text{ when } h > h_n \qquad \qquad ...(14.4.3)$$

also

$$F_r > 1 \text{ when } h < h_c \text{ and } F_r < 1 \text{ when } h > h_c \qquad \qquad ...(14.4.4)$$

Here, h_c is the critical depth.

For the given discharge in a specified channel, the normal-depth and critical-depth lines divide the space (in vertical direction) in the channel into three zones as follows:

Zone 1 : The space above the upper line.
Zone 2 : The space between the two lines.
Zone 3 : The space below the lower line.

The GVF profiles (Fig. 14.4.1) are classified according to the nature of slope of the channel and the zone in which the profile lies. The profiles are named using an alphabet corresponding to the channel slope (like M for mild slope, S for steep slope, C for critical slope, H for horizontal slope, and A for adverse slope) followed by a number indicating the zone in which the profile lies. Therefore, using the inequalities, Eqs. (14.4.3) and (14.4.4) and the GVF equation, Eq. (14.2.3) or its simpler versions, one can establish the sign of *dh/dx* (and, hence, type of profile) and type of flow for various GVF profiles as given in Table 14.4.1.

Table 14.4.1: Characteristics of GVF Profiles in Prismatic Channels

Name of profile	Relation among h, h_n and h_c	S_0 and S_f	F_r	dh/dx	Type of curve	Type of flow
M1	$h > h_n > h_c$	$S_0 > S_f$	< 1	+ve	Backwater	Subcritical
M2	$h_n > h > h_c$	$S_0 < S_f$	< 1	-ve	Drawdown	Subcritical
M3	$h_n > h_c > h$	$S_0 < S_f$	> 1	+ve	Backwater	Supercritical
S1	$h > h_c > h_n$	$S_0 > S_f$	< 1	+ve	Backwater	Subcritical
S2	$h_c > h > h_n$	$S_0 > S_f$	> 1	-ve	Drawdown	Supercritical

(Contd.)

Name of profile	Relation among h, h_n and h_c	S_0 and S_f	F_r	dh/dx	Type of curve	Type of flow
S3	$h_c > h_n > h$	$S_0 < S_f$	> 1	+ve	Backwater	Supercritical
C1	$h > h_c (= h_n)$	$S_0 > S_f$	< 1	+ve	Backwater	Subcritical
C2	$h = h_c = h_n$	$S_0 = S_f$	1	zero	Parallel to channel bed	Uniform and critical
C3	$h_c (= h_n) > h$	$S_0 < S_f$	> 1	+ve	Backwater	Supercritical
H1	$h > h_n (= \infty) > h_c$		*does not exist*			
H2	$h_n(=\infty) > h > h_c$	$S_0 < S_f$	< 1	-ve	Drawdown	Subcritical
H3	$h_n(=\infty) > h_c > h$	$S_0 < S_f$	< 1	+ve	Backwater	Supercritical
A1	$h > h_n^* > h_c$		*does not exist*			
A2	$h > h_c$	$S_0 < 0$	< 1	-ve	Drawdown	Subcritical
A3	$h < h_c$	$S_0 < 0$	> 1	+ve	Drawdown	Supercritical

h_n^* *is an assumed positive value*

Having decided the type of curve and flow condition for various GVF profiles, one needs to determine the behaviour of the profiles at the boundaries of each zone. For example, when $h \to \infty$ (i.e., large depth of flow), Eq. (14.2.13) yields $\dfrac{dh}{dx} = S_0$ and, hence, GVF profile becomes asymptotic to a horizontal line. Similarly, when $h \to h_n$, $S_0 \to S_f$ and, therefore $\dfrac{dh}{dx} = 0$, that is, GVF profile is parallel to the channel bed and this signifies uniform flow. When $h \to 0$, both S_f and F_r tend to infinity and dh/dx tends to some positive finite value. This result is, however, of no practical importance as flow ceases to exist when h is zero.

Likewise, when $h \to h_c$, $\dfrac{dh}{dx} \to \infty$, that is, the GVF profile will be perpendicular to the channel bed in crossing the critical-depth line. However, under no circumstances, the water surface can be at right angles to the channel bed. If the depth of flow is changed abruptly from low stage (i.e., supercritical flow) to high stage (i.e., subcritical flow), a hydraulic jump forms resulting into a discontinuity in the profile. If the depth changes from high to low stage, then hydraulic drop occurs. It should be further noted that at or near the critical-depth line, the flow profile has too large a curvature and, therefore, violates the parallel flow assumption used in deriving GVF equation. Therefore, GVF equation becomes inapplicable for describing GVF profiles in the vicinity of critical section.

The shapes of various GVF profiles (characteristics of which are given in Table 14.4.1) are shown in Fig. 14.4.1. Since these profiles near the channel bed and critical-depth line cannot be accurately described by the GVF equation, they have been shown as dotted lines in those regions.

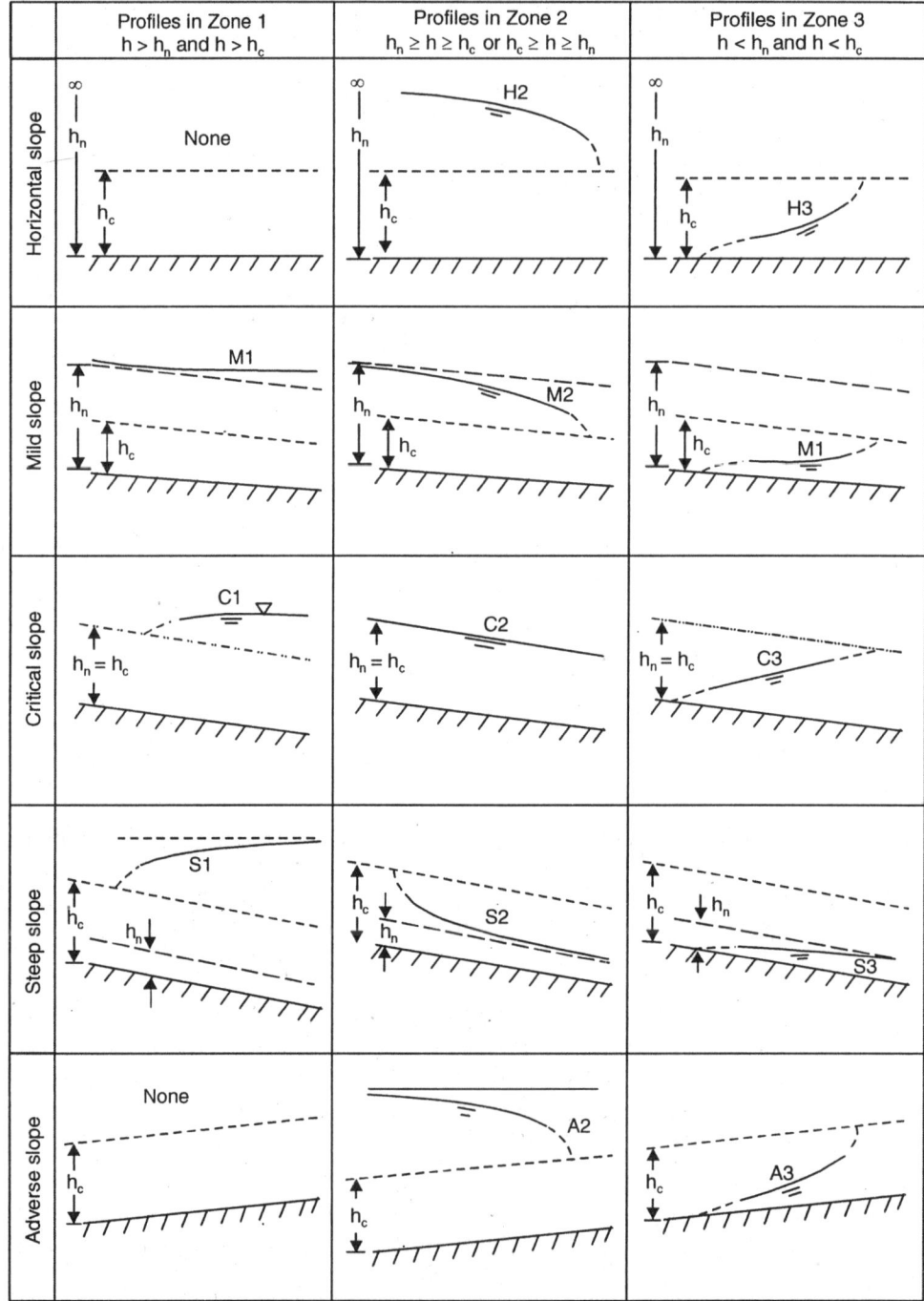

Fig. 14.4.1. Flow profiles of gradually varied flow (5)

14.5 EXAMPLES OF GVF PROFILES

14.5.1 M-Profiles ($S_0 < S_c$ and $h_n > h_c$)

M-profiles would occur in a channel having $S_o < S_c$, i.e., $h_n > h_c$ and the channel is mild. An M1 profile ($h > h_n > h_c$) occurs when the downstream end of a mild channel ends in a reservoir having depth greater than the normal depth h_n of the flow in the channel. At the upstream end of the profile, $h \to h_n$ and, hence, $dh/dx = 0$ and, therefore, the profile is asymptotic to the normal-depth line. Likewise, at the downstream end of the profile, $h \to \infty$ and, hence, $dh/dx = S_0$ and, therefore, the profile is asymptotic to the horizontal reservoir water surface. Common examples of M1 profile are the profile upstream of a dam in a natural river having mild slope, Fig. 14.5.1(a), and the profile in a canal (having mild slope) joining two reservoirs, Fig. 14.5.1(b).

An M2 profile would result when the downstream end of a mild channel meets a reservoir or steep channel in which the water surface is at a level lower than the normal depth. As in case of M1 profile, the upstream end of the M2 profile too is tangential to the normal–depth line. If the water surface in the reservoir (or the downstream channel) is lower than the critical-depth line, the flow profile tends to attain a large slope (almost tangential to a vertical line at a depth equal to critical as per GVF theory). This results in the creation of a hydraulic drop with critical depth at the junction, Fig. 14.5.1 (c). However, if the downstream water surface is higher than the critical-depth line, then only part of the profile (above the water surface in the reservoir) will form, Fig. 14.5.1 (d).

An M3 profile would occur when a supercritical flow enters a mild channel. In practice, the upstream end of M3 profile (meaning zero depth and infinite velocity as per theory) can never exist. An M3 profile would occur downstream of a sluice gate, Fig. 14.5.1 (e), or downstream of the junction of an upstream steep channel with downstream mild channel, Fig. 14.5.1 (f).

14.5.2 S-Profiles ($S_o > S_c$ and $h_n < h_c$)

S1 profile, a backwater curve, would result when a steep channel either meets a reservoir upstream of a dam, Fig. 14.5.1 (g), or meets a pool of high elevation, Fig. 14.5.1 (h). Such a profile has a jump at its upstream end and is asymptotic to horizontal pool/reservoir water surface at the downstream end.

S2 profile, a drawdown curve, forms on the downstream side of a section at which the channel expands, Fig. 14.5.1 (i), and also on the steep-slope side of a junction between steep (or mild) to steeper channel, Fig. 14.5.1 (j). It is like a transition between hydraulic drop and uniform flow and is usually a very short curve. The curve would, obviously, be asymptotic to the normal-depth line at its down-stream end.

S3 profile, a backwater curve, would form when a steep channel meets a milder steep channel, Fig. 14.5.1 (k). Another example of S3 profile is when a sluice gate issues a supercritical flow in a steep channel, Fig. 14.5.1 (l).

14.5.3 C-Profiles ($S_o = S_c$ and $h_n = h_c$)

Both C1 and C3 profiles are backwater (or rising) curves while C2 profile, representing the uniform critical flow, is parallel to the channel bed. C1 profile would form when a channel with bed slope equal to S_c meets a mild channel. The profile forms upstream of the junction, Fig. 14.5.1 (m). C3 profile forms downstream of a sluice gate issuing a supercritical flow in a critical slope channel, Fig. 14.5.1 (n).

14.5.4 H-Profiles ($S_o = 0$ and $h_n = \infty$)

Since h_n is infinite, there cannot form H1 profile. The H2 and H3 profiles are the limiting cases of M2 and M3 profiles when channel slope becomes horizontal, Figs. 14.5.1 (o and p).

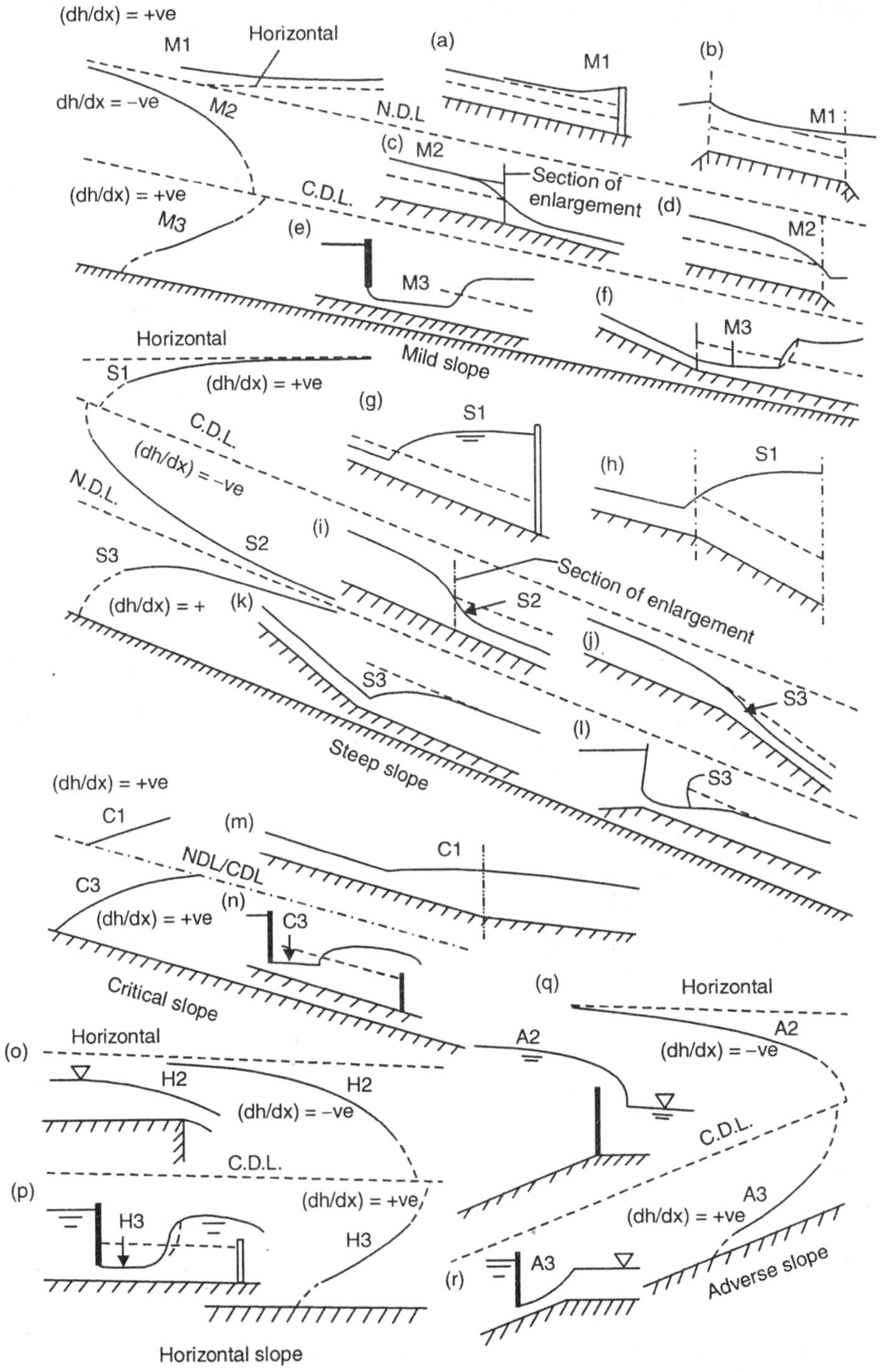

Fig. 14.5.1. Various kinds of gradually varied flow profiles (5)

14.5.5 A-Profiles ($S_o < 0$)

Since h_n is imaginary, A1 profiles would be impossible to form. A2 and A3 profiles, Figs. 14.5.1 (q and r), occur rather infrequently and are similar to H2 and H3 profiles.

14.6 COMPOSITE GVF PROFILES

Channels may have a single prismatic section. But more often, there may be a pair of connected prismatic channels of the same cross-section but with different slopes. Some typical GVF profiles in such channels are shown in Fig. 14.6.1. Similarly, a long prismatic channel may have several changes

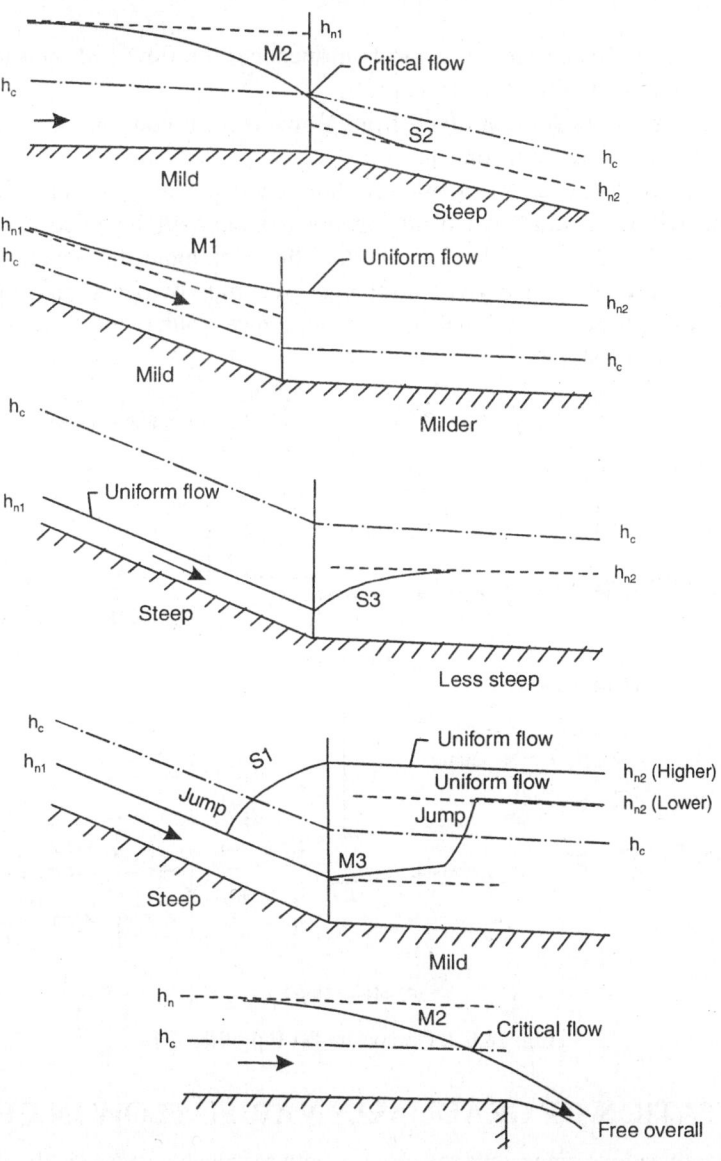

Fig. 14.6.1. Gradually varied flow profiles due to change in bed slope

in slopes or some artificial control sections such as a sluice gate. Two such simple cases are shown in Fig. 14.6.2 along with the resulting GVF profiles. For all these self-explanatory profiles of Figs. 14.6.1 and 14.6.2, and other similar profiles, the following rules are generally applicable (7,8):

1. The sign of *dh/dx* can be easily determined from GVF equation.
2. Water surface approaches the normal-depth line asymptotically.
3. Water surface meets the critical-depth line at a fairly large finite angle. The profile in the vicinity of the critical depth section cannot be predicted accurately by the GVF equation.
4. If a hydraulic jump forms, its location is decided by the relative steepness of the slopes of the two channels that meet, provided that roughness and shape of the two channels remain the same.
5. It is necessary to first locate the controls influencing the flow and then to trace the profiles upstream and downstream of these controls.
6. Subcritical flow is always controlled from a downstream control while supercritical flow is controlled from an upstream control.
7. The term normal depth has no meaning for horizontal and adverse slope channels.
8. For the channels with gradually closing top, the maximum discharge occurs not at full depth, h_o, but at about $0.938\ h_o$ (Art. 12.4.2.1). Around this maximum discharge, there would be two possible normal depths for a given discharge. Accordingly, one should expect an additional zone 4 between these two normal depths and, hence, four types of flow profiles would be possible for a given slope.

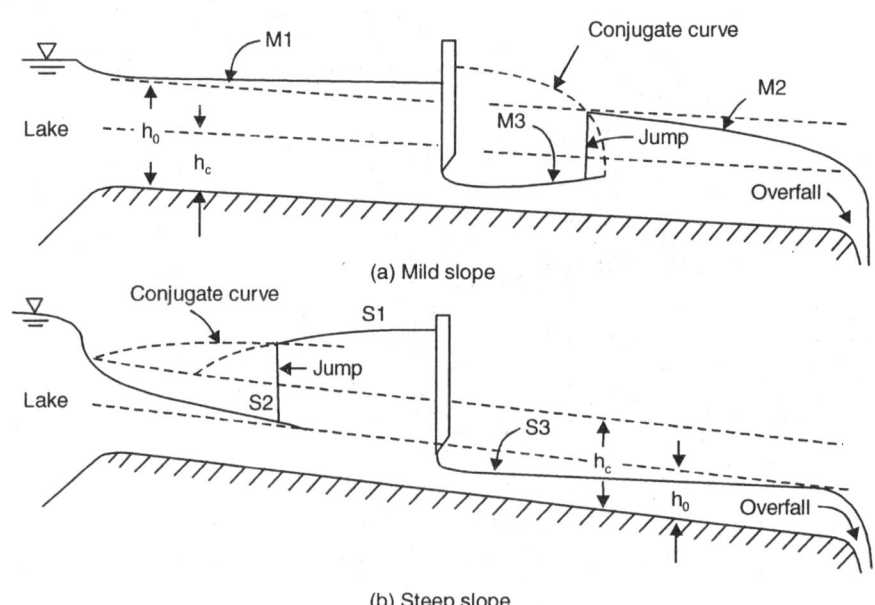

(a) Mild slope

(b) Steep slope

Fig. 14.6.2. Composite GVF profiles (8)

14.7 COMPUTATION OF GRADUALLY VARIED FLOW IN CHANNELS

In many practical applications, one needs to compute with reasonable accuracy the gradual variation of depth of flow along a channel when the flow is non-uniform. For example, when a dam is proposed to be constructed across a stream, one would like to know as to what distance upstream does the dam produce a depth that is substantially different from uniform flow depth in the absence of the dam. For this purpose, one needs to compute the GVF profile and this can be done only by solving the GVF equation in one of its several forms. The methods of computation of GVF profiles can be broadly classified into the following categories:

A. Graphical integration method
B. Numerical integration method
C. Direct-integration method
D. Step method
E. Standard step method
F. Computer programs

14.7.1 Graphical integration method

Consider any two channel sections at distances x_1 and x_2 from a chosen reference (O in Fig. 14.7.1 (a)), at which the depths of flow are h_1 and h_2 respectively. The distance between these two sections along the channel bed can be obtained as,

$$L = x_2 - x_1 = \int_{x_1}^{x_2} dx = \int_{h_1}^{h_2} \frac{dx}{dh} dh$$

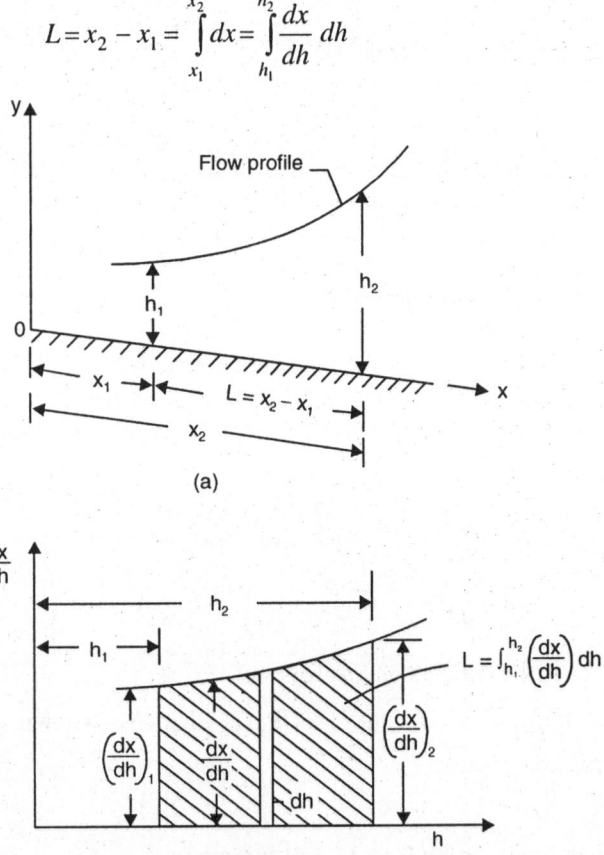

(a)

(b)

Fig. 14.7.1. Graphical integration method

The value of dx/dh is the reciprocal of the right side of the GVF equation, say, Eq. (14.2.13) so that

$$L = \int_{h_1}^{h_2} \frac{\left(1 - Z_c / Z\right)^2}{S_0 \left(1 - K_n / K\right)^2} dh \qquad \qquad ...(14.7.1)$$

Equation (14.7.1) can be graphically integrated, for a given channel and its discharge by plotting the

value of $\dfrac{dx}{dh} \left(= \left(1 - Z_c / Z\right)^2 / \left[S_0 \left(1 - K_n / K\right)^2\right]\right)$ on y-axis for various values of h plotted on x-axis,

Fig. 14.7.1(b). It is obvious that the value of L is the shaded area formed by the curve, the x-axis and the ordinates of dx/dh at $h = h_1$ and $h = h_2$. By measuring this area, the length L can be determined. The area can also be determined by computing the ordinates dx/dh for different values of h and, then, calculating the areas between the adjacent ordinates. Summing these areas, one can obtain the desired length L. This simple and straight forward method is applicable to both prismatic as well as non-prismatic channels.

14.7.2 Numerical Integration Method

On substituting the value of S_f from Eq. (14.2.8) in the GVF equation, Eq. (14.2.5) one gets

$$\frac{dh}{dx} = \frac{S_0 - \left(Q^2 n^2 / A^2 R^{4/3}\right)}{1 - \left(Q^2 T / g A^3\right)} \qquad \qquad ...(14.7.2)$$

For a given channel, A, T, and R depend on h and, hence, Eq. (14.7.2) is a non-linear differential equation in h that can be solved numerically as follows (10):

Describing the flow profile as $h = h(x)$

$$h_{i+1} = h_i + \Delta h = h_i + \frac{dh}{dx} \Delta x \qquad \qquad ...(14.7.3)$$

Here, i represents the channel section at which the depth of flow h_i is known and $(i+1)^{th}$ section is Δx distance from the ith section. The depth of flow h_{i+1} at $(i+1)^{th}$ section is to be determined. Assuming that the distance between i^{th} and $(i+1)^{th}$ sections is small, one can assume that dh/dx varies linearly over the distance Δx and, therefore,

$$h_{i+1} = h_i + \frac{1}{2}\left(\dot{h}_{i+1} + \dot{h}_i\right)\Delta x \qquad \qquad ...(14.7.4)$$

in which $\qquad \qquad \dot{h}_i = \left(dh / dx\right)_i$

The GVF profile $h(x)$ can now be obtained as follows:

1. Determine $\dot{h}_i$ $\left(i.e., dh/dx\right)$ for the control section at which h_i is known, using Eq. (14.7.2) and the known parameters at the section.
2. Assume $\dot{h}_{i+1} = \dot{h}_i$
3. Determine the value of h_{i+1} using Eq. (14.7.4) for known Δx.
4. Compute a new value of $\dot{h}_{i+1}$ using Eq. (14.7.2) and determine the next value of h_{i+1} using Eq. (14.7.4).

5. Compare the last value of h_{i+1} (obtained in step 4) with the value of h_{i+1} obtained in the preceding step (or iteration). If the two values match reasonably well, the depth of flow at distance Δx has been obtained. Otherwise, repeat steps 4 and 5 using the last value of h_{i+1}.

14.7.3 Direct Integration Method

It is impossible to express GVF equation explicitly in terms of h for all types of channel section so as to be able to integrate directly and exactly. Some simplifying assumptions are often made so as to make the equation amenable to mathematical integration.

Chow (4) made the GVF equation amenable to mathematical integration by expressing conveyance K and section factor for critical flow computations Z in terms of depth of flow using Eqs. (12.5.1) and (13.7.1) as follows:

$$K^2 = c_1 h^N$$

$$K_n^2 = c_1 h_n^N$$

$$Z^2 = c_2 h^M$$

and

$$Z_c^2 = c_2 h_c^M$$

Accordingly, GVF equation, Eq. (14.2.13) becomes

$$\frac{dh}{dx} = S_o \frac{1 - \left(h_n/h\right)^N}{1 - \left(h_c/h\right)^M} \qquad \qquad ...(14.7.5)$$

Let
$$u = h/h_n$$

so that
$$dh = h_n\, du$$

Equation (14.7.5), therefore, becomes

$$h_n \frac{du}{dx} = S_0 \frac{1 - \left(1/u\right)^N}{1 - \left(h_c/h_n\right)^M \left(1/u\right)^M}$$

$$= S_0 \frac{1 - u^N}{\left(h_c/h_n\right)^M - u^M}\left(1/u^{N-M}\right)$$

$\therefore$
$$dx = \frac{h_n}{S_0}\left[\frac{\left(u^{N-M}\right)\left[\left(h_c/h_n\right)^M - u^N + 1 - 1\right]}{1 - u^N}\right]du$$

or
$$dx = \frac{h_n}{S_0}\left[1 - \frac{1}{1 - u^N} + \left(\frac{h_c}{h_n}\right)^M \frac{u^{N-M}}{1 - u^N}\right]du \qquad \qquad ...(14.7.6)$$

One may assume the hydraulic exponents N and M to be reasonably constant within the range of limits of integration as the change in depth of a gradually varied flow is generally small. However, if one expects the exponents to be significantly dependent on the depth within the given reach, one should subdivide the reach and integrate assuming the hydraulic exponents in each subdivided reach to be constant. On integrating Eq. (14.7.6), one gets

$$x = \frac{h_n}{S_0}\left[u - \int_0^u \frac{du}{1-u^N} + \left(\frac{h_c}{h_n}\right)^M \int_0^u \frac{u^{N-M}}{1-u^N}du \right] + \text{const.} \qquad ...(14.7.7)$$

The integral $\int_0^u \frac{du}{1-u^N}$ is designated by F (u,N) and termed varied flow function. Thus,

$$F(u,N) = \int_0^u \frac{du}{1-u^N}$$

If one assumes $v = u^{N/J}$ and $J = N/(N-M+1)$ so that $u^N = v^J$

$$dv = \frac{N}{J}u^{\left(\frac{N}{J}-1\right)}du$$

$$= (N-M+1)u^{N-M}\,du$$

i.e., $$u^{N-M}\,du = \frac{1}{N-M+1}dv = \frac{J}{N}dv$$

$$\therefore \qquad \int_0^u \frac{u^{N-M}}{1-u^N}du = \frac{J}{N}\int_0^v \frac{dv}{1-v^J} = \frac{J}{N}F(v,J)$$

Here, $F(v, J)$ is the same function $F(u,N)$ with u and N replaced with v and J, respectively. Therefore, Eq. (14.7.7) can now be written as

$$x = \frac{h_n}{S_0}\left[u - F(u,N) + \left(\frac{h_c}{h_n}\right)^M \frac{J}{N}F(v,J) \right] + \text{const.} \qquad ...(14.7.8)$$

Thus, the length of flow profile between two consecutive sections 1 and 2 is equal to
$$L = x_2 - x_1$$

$$= \frac{h_n}{S_0}\left\{ (u_2 - u_1) - \left[F(u_2,N) - F(u_1,N)\right] + \left(\frac{h_c}{h_n}\right)^M \frac{J}{N}\left[F(v_2,J) - F(v_1,J)\right] \right\} \qquad ...(14.7.9)$$

Here, $u_1 = \dfrac{h_1}{h_n}$, $u_2 = \dfrac{h_2}{h_n}$, $v_1 = \left(\dfrac{h_1}{h_n}\right)^{N/J}$, and $u_2 = \left(\dfrac{h_2}{h_n}\right)^{N/J}$

Use of Eq. (14.7.9) requires the knowledge of the varied flow function F the values of which are given in Table 14.7.1. For computing a flow profile in a given channel, the channel is divided into subreaches and then the length of each subreach is computed using Eq. (14.7.9) and Table 14.7.1 as per the following procedure:

1. Compute the critical and normal depths, i.e., h_c and h_n for known discharge Q and channel slope S_0.
2. Compute the hydraulic exponents N and M for an estimated average depth of flow for the subreach under consideration. Determine $J = N/(N - M + 1)$.
3. Compute u_1, u_2, v_1, and v_2 at the two end sections of the subreach.
4. Obtain $F(u_1, N)$, $F(u_2, N)$, $F(v_1, J)$, and $F(v_2, J)$ from Table 14.7.1.
5. Compute the length of the subreach using Eq. (14.7.9).
6. Repeat steps 2-5 for the remaining subreaches to get the desired GVF profile.

For non-sustaining adverse slopes, the slope of the channel bed is negative and the equation corresponding to Eq. (14.7.8) would be (5)

$$x = -\frac{h_n}{S_0}\left[u - F(u,N)_{-S_0} - \left(\frac{h_c}{h_n}\right)^M \frac{J}{N} F(v,J)_{-S_0}\right] + \text{const.}$$

in which $F(u,N)_{-S_0} = \int_0^u \frac{u^{N-M}}{1+u^N} du$. The values of $F(u,N)_{-S_0}$ are given in Table 14.7.2.

Table 14.7.1: Values of the varied flow function, $F(u, N)$ for positive slopes (5)

N \ u	2.2	2.4	2.6	2.8	3.0	3.2	3.4	3.6	3.8	4.0
0.00	0.000	0.000	0.000	0.000	0.000	0.000	0.000	0.000	0.000	0.000
0.02	0.020	0.020	0.020	0.020	0.020	0.020	0.020	0.020	0.020	0.020
0.04	0.040	0.040	0.040	0.040	0.040	0.040	0.040	0.040	0.040	0.040
0.06	0.060	0.060	0.060	0.060	0.060	0.060	0.060	0.060	0.060	0.060
0.08	0.080	0.080	0.080	0.080	0.080	0.080	0.080	0.080	0.080	0.080
0.10	0. 100	0.100	0 .100	0.100	0.100	0.100	0.100	0.100	0.100	0.100
0.12	0.120	0.120	0.120	0.120	0.120	0.120	0.120	0.120	0.120	0.120
0.14	0.140	0.140	0.140	0.140	0.140	0.140	0.140	0.140	0.140	0.140
0.16	0. 161	0.161	0.160	0.160	0.160	0.160	0.160	0.160	0.160	0.160
0.18	0.181	0.181	0.181	0.180	0.180	1.180	0.180	0.180	0.180	0.180
0.20	0.202	0.201	0.201	0.201	0.200	0.200	0.200	0.200	0.200	0.200
0.22	0.223	0.222	0.221	0.221	0.221	0.220	0.220	0.220	0.220	0.220
0.24	0.244	0.243	0.242	0.241	0.241	0.241	0.240	0.240	0.240	0.240
0.26	0.265	0.263	0.262	0.262	0.261	0.261	0.261	0.260	0.260	0.260
0.28	0.286	0.284	0.283	0.282	0.282	0.281	0.281	0.281	0.280	0.280

(Contd.)

u \ N	2.2	2.4	2.6	2.8	3.0	3.2	3.4	3.6	3.8	4.0
0.30	0.307	0.305	0.304	0.303	0.302	0.302	0.301	0.301	0.301	0.300
0.32	0.329	0.326	0.325	0.324	0.323	0.322	0.322	0.321	0.321	0.321
0.34	0.351	0.348	0.346	0.344	0.343	0.343	0.342	0.342	0.341	0.341
0.36	0.372	0.369	0.367	0.366	0.364	0.363	0.363	0.362	0.362	0.361
0.38	0.395	0.392	0.389	0.387	0.385	0.384	0.383	0.383	0.382	0.382
0.40	0.418	0.414	0.411	0.408	0.407	0.405	0.404	0.403	0.403	0.402
0.42	0.442	0.437	0.433	0.430	0.428	0.426	0.425	0.424	0.423	0.423
0.44	0.465	0.460	0.456	0.452	0.450	0.448	0.446	0.445	0.444	0.443
0.46	0.489	0.483	0.479	0.475	0.472	0.470	0.468	0.466	0.465	0.464
0.48	0.514	0.507	0.502	0.497	0.494	0.492	0.489	0.488	0.486	0.485
0.50	0.539	0.531	0.525	0.521	0.517	0.514	0.511	0.509	0.508	0.506
0.52	0.565	0.557	0.550	0.544	0.540	0.536	0.534	0.531	0.529	0.528
0.54	0.592	0.582	0.574	0.568	0.563	0.559	0.556	0.554	0.551	0.550
0.56	0.619	0.608	0.599	0.593	0.587	0.583	0.579	0.576	0.574	0.572
0.58	0.648	0.635	0.626	0.618	0.612	0.607	0.603	0.599	0.596	0.594
0.60	0.676	0.663	0.653	0.644	0.637	0.631	0.627	0.623	0.620	0.617
0.61	0.691	0.678	0.667	0.657	0.650	0.644	0.639	0.635	0.631	0.628
0.62	0.706	0.692	0.680	0.671	0.663	0.657	0.651	0.647	0.643	0.640
0.63	0.722	0.707	0.694	0.684	0.676	0.669	0.664	0.659	0.655	0.652
0.64	0.738	0.722	0.709	0.698	0.690	0.683	0.677	0.672	0.667	0.664
0.65	0.754	0.737	0.724	0.712	0.703	0.696	0.689	0.684	0.680	0.676
0.66	0.771	0.753	0.738	0.727	0.717	0.709	0.703	0.697	0.692	0.688
0.67	0.787	0.769	0.754	0.742	0.731	0.723	0.716	0.710	0.705	0.701
0.68	0.804	0.785	0.769	0.757	0.746	0.737	0.729	0.723	0.718	0.713
0.69	0.822	0.804	0.785	0.772	0.761	0.751	0.743	0.737	0.731	0.726
0.70	0.840	0.810	0.802	0.787	0.776	0.763	0.757	0.750	0.744	0.739
0.71	0.858	0.836	0.819	0.804	0.791	0.781	0.772	0.764	0.758	0 752
0.72	0.878	0.855	0.836	0.820	0.807	0.796	0.786	0.779	0.772	0.766
0.73	0.898	0.871	0.854	0.837	0.823	0.811	0.802	0.793	0.786	0.780
0.74	0.918	0.892	0.868	0.854	0.840	0.827	0.817	0.808	0.800	0.794
0.75	0.940	0.913	0.890	0.872	0.857	0.844	0.833	0.823	0.815	0.808
0.76	0.961	0.933	0.909	0.890	0.874	0.861	0.849	0.839	0.830	0.823
0.77	0.985	0.954	0.930	0.909	0.892	0.878	0.866	0.855	0.846	0.838
0.78	1.007	0.976	0.950	0.929	0.911	0.896	0.883	0.872	0.862	0.854
0.79	1.031	0.998	0.971	0.949	0.930	0.914	0.901	0.889	0.879	0.870
0.80	1.056	1.022	0.994	0.970	0.950	0.934	0.919	0.907	0.896	0.887
0.81	1.083	1.046	1.017	0.992	0.971	0.954	0.938	0.925	0.914	0.904
0.82	1.110	1.072	1.041	1.015	0.993	0.974	0.958	0.945	0.932	0.922
0.83	1.139	1.099	1.067	1.039	1.016	0.996	0.979	0.965	0.952	0.940
0.84	1.171	1.129	1.094	1.064	1.040	1.019	1.001	0.985	0.972	0.960

(Contd.)

N \ u	2.2	2.4	2.6	2.8	3.0	3.2	3.4	3.6	3.8	4.0
0.85	1.201	1.157	1.121	1.091	1.165	1.043	1.024	1.007	0.993	0.980
0.86	1.238	1.192	1.153	1.119	1.092	1.068	1.048	1.031	1.015	1.002
0.87	1.272	1.223	1.182	1.149	1.120	1.095	1.074	1.055	1.039	1.025
0.88	1.314	1.262	1.228	1.181	1.151	1.124	1.101	1.081	1.064	1.049
0.89	1.357	1.302	1.255	1.216	1.183	1.155	1.131	1.110	1.091	1.075
0.90	1.401	1.343	1.294	1.253	1.218	1.189	1.163	1.140	1.120	1.103
0.91	1.452	1.389	1.338	1.294	1.257	1.225	1.197	1.173	1.152	1.133
0.92	1.505	1.438	1.351	1.340	1.300	1.266	1.236	1.210	1.187	1.166
0.93	1.564	1.493	1.435	1.391	1.348	1.311	1.279	1.251	1.226	1.204
0.94	1.645	1.568	1.504	1.449	1.403	1.363	1.328	1.297	1.270	1.246
0.950	1.737	1.652	1.582	1.518	1.467	1.423	1.385	1.352	1.322	1.296
0.960	1.833	1.741	1.665	1.601	1.545	1.497	1.454	1.417	1.385	1.355
0.970	1.969	1.866	1.780	1.707	1.644	1.590	1.543	1.501	1.464	1.431
0.975	2.055	1.945	1.853	1.773	1.707	1.649	1.598	1.554	1.514	1.479
0.980	2.164	2.045	1.946	1.855	1.783	1.720	1.666	1.617	1.575	1.536
0.985	2.294	2.165	2.056	1.959	1.880	1.812	1.752	1.699	1.652	1.610
0.990	2.477	2.333	2.212	2.106	2.017	1.940	1.873	1.814	1.761	1.714
0.995	2.792	2.621	2.478	2.355	2.250	2.159	2.079	2.008	1.945	1.889
0.999	3.523	3.292	3.097	2.931	2.788	2.663	2.554	2.457	2.370	2.293
1.000	∞	∞	∞	∞	∞	∞	∞	∞	∞	∞
1.001	3.317	2.931	2.640	2.399	2.184	2.008	1.856	1.725	1.610	1.508
1.005	2.587	2.266	2.022	1.818	1.649	1.506	1.384	1.279	1.188	1.107
1.010	2.273	1.977	1.757	1.572	1.419	1.291	1.182	1.089	1.007	0.936
1.015	2.090	1.807	1.602	1.428	1.286	1.166	1.065	0.978	0.902	0.836
1.020	1.961	1.711	1.493	1.327	1.191	1.078	0.982	0.900	0.828	0.766
1.03	1.779	1.531	1.340	1.1861	1.060	0.955	0.866	0.790	0.725	0.668
1.04	1.651	1.410	1.232	1.086	0.967	0.868	0.785	0.714	0.653	0.600
1.05	1.552	1.334	1.150	1.010	0.896	0.802	0.723	0.656	0.598	0.548
1.06	1.472	1.250	1.082	0.918	0.838	0.748	0.672	0.608	0.553	0.506
1.07	1.404	1.195	1.026	0.896	0.790	0.703	0.630	0.569	0.516	0.471
1.08	1.346	1.139	0.978	0.851	0.749	0.665	0.595	0.535	0.485	0.411
1.09	1.295	1.089	0.935	0.812	0.713	0.631	0.563	0.506	0.457	0.415
1.10	1.250	1.050	0.897	0.777	0.681	0.601	0.536	0.480	0.433	0.392
1.11	1.209	1.014	0.864	0.741	0.652	0.575	0.511	0.457	0.411	0.372
1.12	1.172	0.981	0.833	0.718	0.626	0.551	0.488	0.436	0.392	0.354
1.13	1.138	0.950	0.805	0.692	0.602	0.529	0.468	0.417	0.374	0.337
1.14	1.107	0.921	0.780	0.669	0.581	0.509	0.450	0.400	0.358	0.322
1.15	1.078	0.892	0.756	0.647	0.561	0.490	0.432	0.384	0.343	0.308
1.16	1.052	0.870	0.734	0.627	0.542	0.473	0.417	0.369	0.329	0.295
1.17	1.027	0.850	0.713	0.608	0.525	0.458	0.402	0.356	0.317	0.283
1.18	1.003	0.825	0.694	0.591	0.509	0.443	0.388	0.313	0.305	0.272
1.19	0.981	0.810	0.676	0.574	0.494	0.429	0.375	0.331	0.294	0.262
1.20	0.960	0.787	0.659	0.559	0.480	0.416	0.363	0.320	0.283	0.252
1.22	0.922	0.755	0.628	0.531	0.454	0.392	0.341	0.299	0.264	0.235

(Contd.)

N u	2.2	2.4	2.6	2.8	3.0	3.2	3.4	3.6	3.8	4.0
1.24	0.887	0.725	0.600	0.505	0.431	0.371	0.322	0.281	0.248	0.219
1.26	0.855	0.692	0.574	0.482	0.410	0.351	0.304	0.265	0.233	0.205
1.28	0.827	0.666	0.551	0.461	0.391	0.334	0.288	0.250	0.219	0.193
1.30	0.800	0.644	0.530	0.442	0.373	0.318	0.274	0.237	0.207	0.181
1.32	0.775	0.625	0.510	0.424	0.357	0.304	0.260	0.225	0.196	0.171
1.34	0.752	0.605	0.492	0.408	0.342	0.290	0.248	0.214	0.185	0.162
1.36	0.731	0.588	0.475	0.393	0.329	0.278	0.237	0.204	0.176	0.153
1.38	0.711	0.567	0.459	0.378	0.316	0.266	0.226	0.194	0.167	0.145
1.40	0.692	0.548	0.444	0.365	0.304	0.256	0.217	0. 185	0.159	0.138
1.42	0.674	0.533	0.431	0.353	0.293	0.246	0.208	0.177	0.152	0.131
1.44	0.658	0.517	0.417	0.341	0.282	0.236	0.199	0.169	0.145	0.125
1.46	0.642	0.505	0.405	0.330	0.273	0.227	0.191	0.162	0.139	0.119
1.48	0.627	0.493	0.394	0.320	0.263	0.219	0.184	0.156	0. 133	0.113
1.50	0.613	0.480	0.383	0.310	0.255	0.211	0.177	0.149	0.127	0.108
1.55	0.580	0.451	0.358	0.288	0.235	0.194	0.161	0.135	0.114	0.097
1.60	0.551	0.425	0.335	0.269	0.218	0.179	0.148	0.123	0.103	0.087
1.65	0.525	0.402	0.316	0.251	0.203	0.165	0.136	0.113	0.094	0.079
1.70	0.501	0.381	0.298	0.236	0.189	0.153	0.125	0.103	0.086	0.072
1.75	0.480	0.362	0.282	0.222	0.177	0.143	0.116	0.095	0.079	0.065
1.80	0.460	0.349	0.267	0.209	0.166	0.133	0.108	0.088	0.072	0.060
1.85	0.442	0.332	0.254	0.198	0.156	0.125	0.100	0.082	0.067	0.055
1.90	0.425	0.315	0.242	0.188	0.147	0.117	0.094	0.076	0.062	0.050
1.95	0.409	0.304	0.231	0.178	0.139	0.110	0.088	0.070	0.057	0.046
2.00	0.395	0.292	0.221	0.169	0.132	0.104	0.082	0.066	0.053	0.043
2.10	0.369	0.273	0.202	0.154	0.119	0.092	0.073	0.058	0.046	0.037
2.20	0.346	0.253	0.186	0.141	0.107	0.083	0.065	0.051	0.040	0.032
2.3	0.326	0.235	0.173	0.129	0.098	0.075	0.058	0.045	0.035	0.028
2.4	0.308	0.220	0.160	0.119	0.089	0.068	0.052	0.040	0.031	0.024
2.5	0.292	0.207	0.150	0.110	0.082	0.062	0.047	0.036	0.028	0.022
2.6	0.277	0.197	0.140	0.102	0.076	0.057	0.043	0.033	0.025	0.019
2.7	0.264	0.188	0.131	0.095	0.070	0.052	0.039	0.029	0.022	0.017
2.8	0.252	0.176	0.124	0.089	0.065	0.048	0.036	0.027	0.020	0.015
2.9	0.241	0.166	0.117	0.083	0.060	0.044	0.033	0.024	0.018	0.014
3.0	0.230	0.159	0.110	0.078	0.056	0.041	0.030	0.022	0.017	0.012
3.5	0.190	0.126	0.085	0.059	0.041	0.029	0.021	0.015	0.011	0.008
4.0	0.161	0.104	0.069	0.046	0.031	0.022	0.015	0.010	0.007	0.005
4.5	0.139	0.087	0.057	0.037	0.025	0.017	0.011	0.008	0.005	0.004
5.0	0.122	0.076	0.048	0.031	0.020	0.013	0.009	0.006	0.004	0.003
6.0	0.098	0.060	0.036	0.022	0.014	0.009	0.006	0.004	0.002	0.002
7.0	0.081	0.048	0.028	0.017	0.010	0.006	0.004	0.002	0.002	0.001
8.0	0.069	0.040	0.022	0.013	0.008	0.005	0.003	0.002	0.001	0.001
9.0	0.060	0.034	0.019	0.011	0.006	0.004	0.002	0.001	0.001	0.000
10.0	0.053	0.028	0.016	0.009	0.005	0.003	0.002	0.001	0.001	0.000
20.0	0.023	0.018	0.011	0.006	0.002	0.001	0.001	0.000	0.000	0.000

(Contd.)

Table 14.7.1: Values of the varied flow function $F(u, N)$ for positive slopes (5) (contd..)

u \ N	4.2	4.6	5.0	5.4	5.8	6.2	6.6	7.0	7.4	7.8
0.00	0.000	0.000	0.000	0.000	0.000	0.000	0.000	0.000	0.000	0.000
0.02	0.020	0.020	0.020	0.020	0.020	0.020	0.020	0.020	0.020	0.020
0.04	0.040	0.040	0.040	0.040	0.040	0.040	0·040	0.040	0.040	0.040
0.06	0.060	0.060	0.060	0.060	0.060	0.060	0.060	0.060	0.060	0.060
0.08	0.080	0.080	0.080	0.080	0.080	0.080	0.080	0.080	0.080	0.080
0.10	0.100	0.100	0.100	0.100	0.100	0.100	0.100	0.100	0.100	0.100
0.12	0.120	0.120	0.120	0.120	0.120	0.120	0.120	0.120	0.120	0.120
0.14	0.140	0.140	0.140	0.140	0.140	0.140	0.140	0.140	0.140	0.140
0.16	0.160	0.160	0.160	0.160	0.160	0.160	0.160	0.160	0.160	0.160
0.18	0.180	0.180	0.180	0.180	0.180	0.180	0.180	0.180	0.180	0.180
0.20	0.200	0.200	0.200	0.200	0.200	0.200	0.200	0.200	0.200	0.200
0.22	0.220	0.220	0.220	0.220	0.220	0.220	0.220	0.220	0.220	0.22()
0.24	0.240	0.240	0.240	0.240	0.240	0.240	0.240	0.240	0.240	0.240
0.26	0.260	0.260	0.260	0.260	0.260	0.260	0.260	0.260	0.260	0.260
0.28	0.280	0.280	0.280	0.280	0.280	0.280	0.280	0.280	0.280	0.280
0.30	0.300	0.300	0.300	0.300	0.300	0.300	0.300	0.300	0.300	0.300
0.32	0.321	0.320	0.320	0.320	0.320	0.320	0.320	0.320	0.320	0.320
0.34	0.341	0.340	0.340	0.340	0.340	0.340	0.340	0.340	0.340	0.340
0.36	0.361	0.361	0.360	0.360	0.360	0.360	0.360	0.360	0.360	0.360
0.38	0.381	0.381	0.381	0.380	0.380	0.380	0.380	0.380	0.380	0.380
0.40	0.402	0.401	0.401	0.400	0.400	0.400	0.400	0.400	0.400	0.400
0.42	0.422	0.421	0.421	0.421	0.420	0.420	0.420	0.420	0.420	0.420
0.44	0.443	0.442	0.441	0.441	0.441	0.441	0.440	0.440	0.440	0.440
0.46	0.463	0.462	0.462	0.461	0.461	0.461	0.460	0.460	0.460	0.460
0.48	0.484	0.483	0.482	0.481	0.481	0.481	0.480	0.480	0.480	0. 480
0.50	0.505	0.504	0.503	0.502	0.501	0.501	0.501	0.500	0.500	0.500
0.52	0.527	0.525	0.523	0.522	0.522	0.521	0.521	0.521	0.520	0.520
0.54	0.518	0.546	0.544	0.543	0.542	0.542	0.541	0.541	0.541	0.541
0.56	0.570	0.567	0.565	0.564	0.563	0.562	0.562	0.561	0.561	0.561
0.58	0.592	0.589	0.587	0.585	0.583	0.583	0.582	0.582	0.581	0.581
0.60	0.614	0.611	0.608	0.606	0.605	0.604	0.603	0.602	0.602	0.601
0.61	0.626	0.622	0.619	0.617	0.615	0.614	0.613	0.612	0.612	0.61 I
0.62	0.637	0.633	0.630	0.628	0.626	0.625	0.624	0.623	0.622	0.622
0.63	0.649	0.644	0.641	0.638	0.636	0.635	0.634	0.633	0.632	0.632
0.64	0.661	0.656	0.652	0.649	0.647	0.646	0.645	0.644	0.643	0.642
0.65	0.673	0.667	0.663	0.660	0.658	0.656	0.655	0.654	0.653	0.653
0.66	0.685	0.679	0.675	0.672	0.669	0.667	0.666	0.665	0.664	0.663
0.67	0.697	0.691	0.686	0.683	0.680	0.678	0.676	0.675	0.674	0.673
0.68	0.709	0.703	0.698	0.694	0.691	0.689	0.687	0.686	0.685	0.684
0.69	0.722	0.715	0.710	0.706	0.703	0.7 00	0.698	0.696	0.695	0.694

(Contd.)

N / u	4.2	4.6	5.0	5.4	5.8	6.2	6.6	7.0	7.4	7.8
0.70	0.735	0.727	0.722	0.717	0.714	0.712	0.710	0.708	0.706	0.705
0.71	0.748	0.740	0.734	0.729	0.726	0.723	0.721	0.719	0.717	0.716
0.72	0.761	0.752	0.746	0.741	0.737	0.734	0.732	0.730	0.728	0.727
0.73	0.774	0.765	0.759	0.753	0.749	0.746	0.743	0.741	0.739	0.737
0.74	0.788	0.779	0.771	0.766	0.761	0.757	0.754	0.752	0.750	0.745
0.75	0.802	0.792	0.784	0.778	0.773	0.769	0.706	0.763	0.761	0.759
0.76	0.817	0.806	0.798	0.791	0.782	0.778	0.786	0.775	0.773	0.771
0.77	0.831	0.820	0.811	0.804	0.798	0.794	0.790	0.787	0.784	0.782
0.78	0.847	0.834	0.825	0.817	0.811	0.806	0.802	0.799	0.796	0.791
0.79	0.862	0.849	0.839	0.831	0.824	0.819	0.815	0.811	0.808	0.805
0.80	0.878	0.865	0.854	0.845	0.838	0.832	0.828	0.823	0.820	0.818
0.81	0.895	0.881	0.869	0.860	0.852	0.846	0.841	0.836	0.833	0.830
0.82	0.913	0.897	0.885	0.875	0.866	0.860	0.854	0.850	0.846	0.842
0.83	0.931	0.914	0.901	0.890	0.881	0.874	0.868	0.863	0.859	0.855
0.84	0.949	0.932	0.918	0.906	0.897	0.889	0.882	0.877	0.872	0.868
0.85	0.969	0.950	0.935	0.923	0.912	0.905	0.898	0.891	0.887	0.88.
0.86	0.990	0.970	0.954	0.940	0.930	0.921	0.913	0.906	0.901	0.896
0.87	1.012	0.990	0.973	0.959	0.947	0.937	0.929	0.922	0.916	0.911
0.88	1.035	1.012	0.994	0.978	0.966	0.955	0.946	0.938	0.932	0.927
0.89	1.060	1.035	1.015	0.999	0.986	0.974	0.964	0.956	0.949	0.943
0.90	1.087	1.060	1.039	1.021	1.007	0.994	0.984	0.974	0.967	0.960
0.91	1.116	1.088	1.064	1.045	1.029	1.016	1.003	0.995	0.986	0.979
0.92	1.148	1.117	1.092	1.072	1.054	1.039	1.027	1.016	1.006	0.999
0.93	1.184	1.151	1.123	1.101	1.081	1.065	1.050	1.040	1.029	1.021
0.94	1.225	1.188	1.158	1.134	1.113	1.095	1.080	1.066	1.054	1.044
0.950	1.272	1.232	1.199	1.172	1.148	1.128	1.111	1.097	1.084	1.073
0.960	1.329	1.285	1.248	1.217	1.188	1.167	1.149	1.133	1.119	1.106
0.970	1.402	1.351	1.310	1.275	1.246	1.319	1.197	1.179	1.162	1.148
0.975	1.447	1.393	1.348	1.311	1.280	1.250	1.227	1.207	1.190	1.173
0.980	1.502	1.443	1.395	1.354	1.339	1.288	1.262	1.241	1.221	1.204
0.985	1.573	1.508	1.454	1.409	1.372	1.337	1.309	1.284	1.263	1.243
0.990	1.671	1.598	1.537	1.487	1.444	1.404	1.373	1.344	1.319	1.297
0.995	1.838	1.751	1.678	1.617	1.565	1.519	1.479	1.451	1.416	1.388
0.999	2.223	2.102	2.002	1.917	1.845	1.780	1.725	1.678	1.635	1.596
1.000	∞	∞	∞	∞	∞	∞	∞	∞	∞	∞
1.001	1.417	1.264	1.138	1.033	0.951	0.870	0.803	0.746	0.697	0.651
1.005	1.036	0.915	0.817	0.737	0.669	0.612	0.553	0.526	0.481	0.447
1.010	0.873	0.766	0.681	0.610	0.551	0.502	0.459	0.422	0.389	0.360
1.015	0.778	0.680	0.602	0.537	0.483	0.440	0.399	0.366	0.336	0.310
1.02	0.711	0.620	0.546	0.486	0.436	0.394	0.358	0.327	0.300	0.276
1.03	0.618	0.535	0.469	0.415	0.370	0.333	0.300	0.272	0.249	0.228
1.04	0.554	0.477	0.415	0.365	0.324	0.290	0.262	0.236	0.214	0.195
1.05	0.504	0.432	0.374	0.328	0.289	0.259	0.231	0.208	0.189	0.174

(Contd.)

N u	4.2	4.6	5.0	5.4	5.8	6.2	6.6	7.0	7.4	7.8
1.06	0.464	0.396	0.342	0.298	0.262	0.233	0.209	0.187	0.170	0.154
1.07	0.431	0.366	0.315	0.273	0.239	0.212	0.191	0.168	0.151	0.136
1.08	0.403	0.341	0.292	0.252	0.220	0.194	0.172	0.153	0.137	0.123
1.09	0.379	0.319	0.272	0.234	0.204	0.179	0.158	0.140	0.125	0.112
1.10	0.357	0.299	0.254	0.218	0.189	0.165	0.146	0.129	0.114	0.102
1.11	0.338	0.282	0.239	0.204	0.176	0.154	0.135	0.119	0.105	0.094
1.12	0.321	0.267	0.225	0.192	0.165	0.143	0.125	0.110	0.097	0.086
1.13	0.305	0.253	0.212	0.181	0.155	0.135	0.117	0.102	0.090	0.080
1.14	0.291	0.240	0.201	0.170	0.146	0.126	0.109	0.095	0.084	0.074
1.15	0.278	0.229	0.191	0.161	0.137	0.118	0.102	0.089	0.078	0.068
1.16	0.266	0.218	0.181	0.153	0.130	0.111	0.096	0.084	0.072	0.064
1.17	0.255	0.208	0.173	0.145	0.123	0.105	0.090	0.078	0.068	0.060
1.18	0.244	0.199	0.165	0.138	0.116	0.099	0.085	0.073	0.063	0.055
1.19	0.235	0.191	0.157	0.131	0.110	0.094	0.080	0.068	0.059	0.051
1.20	0.226	0.183	0.150	0.115	0.105	0.088	0.076	0.064	0.056	0.048
1.22	0.209	0. 168	0.138	0.114	0.095	0.080	0.068	0.057	0.049	0.042
1.24	0.195	0.156	0.127	0.104	0.086	0.072	0.060	0.051	0.044	0.038
1.26	0.182	0.145	0.117	0.095	0.079	0.065	0.055	0.046	0.039	0.033
1.28	0.170	0.135	0.108	0.088	0.072	0.060	0.050	0.041	0.035	0.030
1.30	0.160	0.126	0.100	0.081	0.066	0.054	0.045	0.037	0.031	0.026
1.32	0.150	0.118	0.093	0.075	0.061	0.050	0.041	0.034	0.028	0.024
1.34	0.142	0.110	0.087	0.069	0.056	0.045	0.037	0.030	0.025	0.021
1.36	0.134	0.103	0.081	0.064	0.052	0.042	0.034	0.028	0.023	0.019
1.38	0.127	0.097	0.076	0.060	0.048	0.038	0.032	0.026	0.021	0.017
1.40	0.120	0.092	0.071	0.056	0.044	0.036	0.028	0.023	0.019	0.016
1.42	0.114	0.087	0.067	0.052	0.041	0.033	0.026	0.021	0.017	0.014
1.44	0.108	0.082	0.063	0.049	0.038	0.030	0.024	0.019	0.016	0.013
1.46	0.103	0.077	0.059	0.046	0.036	0.028	0.022	0.018	0.014	0.012
1.48	0.098	0.073	0.056	0.043	0.033	0.026	0.021	0.017	0.013	0.010
1.50	0.093	0.069	0.053	0.040	0.031	0.024	0.020	0.015	0.012	0.009
1.55	0.083	0.061	0.046	0.035	0.026	0.020	0.016	0.012	0.010	0.008
1.60	0.074	0.054	0.040	0.030	0.023	0.017	0.013	0.010	0.008	0.006
1.65	0.067	0.048	0.035	0.026	0.019	0.014	0.011	0.008	0.006	0.005
1.70	0.060	0.043	0.031	0.023	0.016	0.012	0.009	0.007	0.005	0.004
1.75	0.054	0.038	0.027	0.020	0.014	0.010	0.008	0.006	0.004	0.003
1.80	0.049	0.034	0.024	0.017	0.012	0.009	0.007	0.005	0.004	0.003
1.85	0.045	0.031	0.022	0.015	0.011	0.008	0.006	0.004	0.003	0.002
1.90	0.041	0.028	0.020	0.014	0.010	0.007	0.005	0.004	0.003	0.002
1.95	0.038	0.026	0.018	0.012	0.008	0.006	0.004	0.003	0.002	0.002
2.00	0.035	0.023	0.016	0.011	0.007	0.005	0.004	0.003	0.002	0.001
2.10	0.030	0.019	0.013	0.009	0.006	0.004	0.003	0.002	0.001	0.001
2.20	0.025	0.016	0.011	0.007	0.005	0.004	0.002	0.001	0.001	0.001

(Contd.)

N \ u	4.2	4.6	5.0	5.4	5.8	6.2	6.6	7.0	7.4	7.8
2.3	0.022	0.014	0.009	0.006	0.004	0.003	0.002	0.001	0.001	0.001
2.4	0.019	0.012	0.008	0.005	0.003	0.002	0.001	0.001	0.001	0.001
2.5	0.017	0.010	0.006	0.004	0.003	0.002	0.001	0.001	0.000	0.000
2.6	0.015	0.009	0.005	0.003	0.002	0.001	0.001	0.001	0.000	0.000
2.7	0.013	0.008	0.005	0.003	0.002	0.001	0.001	0.000	0.000	0.000
2.8	0.012	0.007	0.004	0.002	0.001	0.001	0.001	0.000	0.000	0.000
2.9	0 010	0.006	0.004	0.002	0.001	0.001	0.000	0.000	0.000	0.000
3.0	0.009	0.005	0.003	0.002	0.001	0.001	0.000	0.000	0.000	0.000
3.5	0.006	0.003	0.002	0.001	0.001	0.000	0.000	0.000	0.000	0.000
4.0	0.004	0.002	0.001	0.000	0.000	0.000	0.000	0.000	0.000	0.000
4.5	0.003	0.001	0.001	0.000	0.000	0.000	0.000	0.000	0.000	0.000
5.0	1.002	0.001	0.000	0.000	0.000	0.000	0.000	0.000	0.000	0.000
6.0	0.001	0.000	0.000	0.000	0.000	0.000	0.000	0.000	0.000	0.000
7.0	0.001	0.000	0.000	0.000	0.000	0.000	0.000	0.000	0.000	0.000
8.0	0.000	0.000	0.000	0.000	0.000	0.000	0.000	0.000	0.000	0.000
10.0	0.000	0.000	0.000	0.000	0.000	0.000	0.000	0.000	0.000	0 000
20.0	0.000	0.000	0.000	0.000	0.000	0.000	0.000	o 000	0.000	0 000

N \ u	8.2	8.6	9.0	9.4	9.8
0.00	0.000	0.000	0.000	0.000	0.000
0.02	0.020	0.020	0.020	0.020	0.020
0.04	0.040	0.040	0.040	0.040	0.040
0.06	0.060	0.060	0.060	0.060	0.060
0.08	0.080	0.080	0.080	0.080	0.080
0.10	0.100	0.100	0.100	0.100	0.100
0.12	0.120	0.120	0.120	0.120	0.120
0.14	0.140	0.140	0.1-10	0.140	0.140
0.16	0.160	0.160	0.160	0.160	0.160
0.18	0.180	0.180	0.180	0.180	0.180
0.20	0.200	0.200	0.200	0.200	0.200
0.22	0.220	0.220	0.220	0.220	0.220
0.24	0.240	0.240	0.240	0.240	0.240
0.26	0.260	0.260	0.260	0.260	0.260
0.28	0.280	0.280	0.280	0.280	0.280
0.30	0.300	0.300	0.300	0.300	0.300
0.32	0.320	0.320	0.320	0.320	0.320
0.34	0.340	0.340	0.340	0.340	0.340
0.36	0.360	0.360	0.360	0.360	0.360
0.38	0.380	0.380	0.38()	0.380	0.380

(Contd.)

N u	8.2	8.6	9.0	9.4	9.8
0.40	0.400	0.400	0.400	0.400	0.400
0.42	0.420	0.420	0.420	0.420	0.420
0.44	0.440	0.440	0.440	0.440	0.440
0.46	0.460	0.460	0.460	0.460	0.460
0.48	0.480	0.480	0.480	0.480	0.480
0.50	0.500	0.500	0.500	0.500	0.500
0.52	0.520	0.520	0.520	0.520	0.520
0.54	0.540	0.540	0.540	0.540	0.540
0.56	0.561	0.560	0.560	0.560	0.560
0.58	0.581	0.581	0.580	0.580	0.580
0.60	0.601	0.601	0.601	0.600	0.600
0.61	0.611	0.611	0.611	0.611	0.610
0.62	0.621	0.621	0.621	0.621	0.621
0.63	0.632	0 631	0.631	0.631	0.631
0.64	0.642	0.641	0.641	0.641	0.641
0.65	0.652	0.652	0.651	0.651	0.651
0.66	0.662	0.662	0.662	0.661	0.661
0.67	0.673	0.672	0.672	0.672	0.671
0.68	0.683	0.683	0.682	0.682	0.681
0.69	0.694	0.693	0.692	0.692	0.692
0.70	0.704	0.704	0.703	0.702	0.702
0.71	0.715	0.714	0.713	0.713	0.712
0.72	0.726	0.725	0.724	0.723	0.723
0.73	0.736	0.735	0.731	0.734	0.733
0.74	0.747	0.746	0.745	0.744	0.744
0.75	0.758	0.757	0.756	0.755	0.754
0.76	0.769	0.768	0.767	0.766	0.765
0.77	0.780	0.779	0.778	0.777	0.776
0.78	0.792	0.790	0.789	0.788	0.787
0.79	0.804	0.802	0.800	0.799	0.798
0.80	0.815	0.813	0.811	0.810	0.809
0.81	0.827	0.825	0.823	0.822	0.820
0.82	0.839	0.837	0.835	0.833	0.831
0.83	0.852	0.849	0.847	0.845	0.844
0.84	0.865	0.862	0.860	0.858	0.856
0.85	0.878	0.875	0.873	0.870	0.868
0.86	0.892	0.889	0.886	0.883	0.881
0.87	0.907	0.903	0.900	0.897	0.894
0.88	0.921	0.918	0.914	0.911	0.908
0.89	0.937	0.933	0.929	0.925	0.922
0.90	0.954	0.949	0.944	0.940	0.937
0.91	0.972	0.967	0.961	0.957	0.953
0.92	0.991	0.986	0.980	0.975	0.970

(Contd.)

N u	8.2	8.6	9.0	9.4	9.8
0.93	1.012	1.006	0.999	0.994	0.989
0.94	1.036	1.029	1.022	1.016	1.010
0.950	1.062	1.055	1.047	1.040	1.033
0.960	1.097	1.085	1.074	1.063	1.053
0.970	1.136	1.124	1.112	1.100	1.087
0.975	1.157	1.147	1.134	1.122	1.108
0.980	1.187	1.175	1.160	1.150	1.132
0.985	1.224	1.210	1.196	1.183	1.165
0.990	1.275	1.260	1.243	1.223	1.208
0.995	1.363	1.342	1.320	1.302	1.280
0.999	1.560	1.530	1.500	1.476	1.447
1.000	∞	∞	∞	∞	∞
1.001	0.614	0.577	0.546	0.519	0.494
1.005	0.420	0.391	0.368	0.350	0.331
1.010	0.337	0.313	0.294	0.278	0.262
1.015	0.289	0.269	0.255	0.237	0.223
1.020	0.257	0.237	0.221	0.209	0.196
1.03	0.212	0.195	0.181	0.170	0.159
1.04	0.173	0.165	0.152	0.143	0.134
1.05	0.158	0.143	0.132	0.124	0.115
1.06	0.140	0.127	0.116	0.106	0.098
1.07	0.123	0.112	0.102	0.094	0.086
1.08	0.111	0.101	0.092	0.084	0.077
1.09	0.101	0.091	0.082	0.075	0.069
1.10	0.092	0.083	0.074	0.067	0.062
1.11	0.084	0.075	0.067	0.060	0.055
1.12	0.077	0.069	0.062	0.055	0.050
1.13	0.071	0.063	0.056	0.050	0.045
1.14	0.065	0.058	0.052	0.046	0.041
1.15	0.061	0.054	0.048	0.043	0.038
1.16	0.056	0.050	0.045	0.040	0.035
1.17	0.052	0.046	0.041	0.036	0.032
1.18	0.048	0.042	0.037	0.033	0.029
1.19	0.045	0.039	0.034	0.030	0.027
1.20	0.043	0.037	0.032	0.028	0.025
1.22	0.037	0.032	0 028	0.024	0.021
1.24	0.032	0.028	0.024	0.021	0.018
1.26	0.028	0.024	0.021	0.018	0.016
1.28	0.025	0.021	0.018	0.016	0.014
1.30	0.022	0.019	0.016	0.014	0.012
1.32	0.020	0.017	0.014	0.012	0.010
1.34	0.018	0.015	0.012	0.010	0.009
1.36	0.016	0.013	0.011	0.009	0.008

(Contd.)

N u	8.2	8.6	9.0	9.4	9.8
1.38	0.014	0.012	0.010	0.008	0.007
1.40	0.013	0.011	0.009	0.007	0.006
1.42	0.011	0.009	0.008	0.006	0.005
1.44	0.010	0.008	0.007	0.006	0.005
1.46	0.009	0.008	0.006	0.005	0.004
1.48	0.009	0.007	0.005	0.004	0.004
1.50	0.008	0.006	0.005	0.004	0.003
1.55	0.006	0.005	0.004	0.003	0.003
1.60	0.005	0.004	0.003	0.002	0.002
1.65	0.004	0.003	0.002	0.002	0.001
1.70	0.003	0.002	0.002	0.001	0.001
1.75	0.002	0.002	0.002	0.001	0.001
1.80	0.002	0.001	0.001	0.001	0.001
1.85	0.002	0.001	0.001	0.001	0.001
1.90	0.001	0.001	0.001	0.001	0.000
1.95	0.001	0.001	0.001	0.000	0.000
2.00	0.001	0.001	0.000	0.000	0.000
2.10	0.001	0.000	0.000	0.000	0.000
2.20	0.000	0.000	0.000	0.000	0.000
2.5	0.000	0.000	0.000	0.000	0.000
3.0	0.000	0.000	0.000	0.000	0.000
5.0	0.000	0.000	0.000	0.000	0.000
10.0	0.000	0.000	0.000	0.000	0.000
20.0	0.000	0.000	0.000	0.000	0.000

Table 14.7.2: Values of the varied flow function $F(u,N)_{-S_0}$ for negative slopes (5)

N u	2.0	2.2	2.4	2.6	2.8	3.0	3.2	3.4	3.6	3.8
0.00	0.000	0.000	0.000	0.000	0.000	0.000	0.000	0.000	0.000	0.000
0.02	0.020	0.020	0.020	0.020	0.020	0.020	0.020	0.020	0.020	0.020
0.04	0.040	0.040	0.040	0.040	0.040	0.040	0.040	0.040	0.040	0.040
0.06	0.060	0.060	0.060	0.060	0.060	0.060	0.060	0.060	0.060	0.060
0.08	0.080	0.080	0.080	0.080	0.080	0.080	0.080	0.080	0.080	0.080
0.10	0.099	0.100	0.100	0.100	0.100	0.100	0.100	0.100	0.100	0.100
0.12	0.119	0.119	0.120	0.120	0.120	0.120	0.120	0.120	0.120	0.120
0.14	0.139	0.139	0.140	0.140	0.140	0.140	0.140	0.140	0.140	0.140
0.16	0.158	0.159	0.159	0.160	0.160	0.160	0.160	0.160	0.160	0.160
0.18	0.178	0.179	0.179	0.180	0.180	0.180	0.180	0.180	0.180	0.180

(Contd.)

N \ u	2.0	2.2	2.4	2.6	2.8	3.0	3.2	3.4	3.6	3.8
0.20	0.197	0.198	0.199	0.199	0.200	0.200	0.200	0.200	0.200	0.200
0.22	0.216	0.217	0.218	0.219	0.219	0.220	0.220	0.220	0.220	0.220
0.24	0.234	0.236	0.237	0.238	0.239	0.240	0.240	0.240	0.240	0.210
0.26	0.253	0.255	0.256	0.257	0.258	0.259	0.259	0.260	0.260	0.260
0.28	0.272	0.274	0.275	0.276	0.277	0.278	0.278	0.279	0.280	0.280
0.30	0.291	0.293	0.294	0.295	0.296	0.297	0.298	0.298	0.299	0.299
0.32	0.308	0.311	0.313	0.314	0.316	0.317	0.318	0.318	0.319	0.319
0.34	0.326	0.329	0.331	0.333	0.335	0.337	0.338	0.338	0.339	0.339
0.36	0.344	0.347	0.350	0.352	0.354	0.356	0.357	0.357	0.358	0.358
0.38	0.362	0.355	0.368	0.371	0.373	0.374	0.375	0.376	0.377	0.377
0.40	0.380	0.384	0.387	0.390	0.392	0.393	0.394	0.395	0.396	0.396
0.42	0.397	0.401	0.405	0.407	0.409	0.411	0.412	0.413	0.414	0.414
0.44	0.414	0.419	0.423	0.426	0.429	0.430	0.432	0.433	0.434	0.435
0.46	0.431	0.437	0.440	0.444	0.447	0.449	0.451	0.452	0.453	0.454
0.48	0.447	0.453	0.458	0.461	0.464	0.467	0.469	0.471	0.472	0.473
0.50	0.463	0.470	0.475	0.479	0.482	0.485	0.487	0.489	0.491	0.492
0.52	0 479	0.485	0.491	0.494	0.499	0.502	0.505	0.507	0.509	0.511
0.54	0.494	0.501	0.507	0.512	0.516	0.520	0.522	0.525	0.527	0.529
0.56	0.509	0.517	0.523	0.528	0.533	0.537	0.540	0.543	0.545	0.517
0.58	0.524	0.533	0.539	0.545	0.550	0.554	0.558	0.561	0.563	0.567
0.60	0.540	0.548	0.555	0.561	0.566	0.571	0.575	0.578	0.581	0.583
0.61	0.547	0.556	0.563	0.569	0.575	0.579	0.583	0.587	0.589	0.592
0.62	0.554	0.563	0.571	0.578	0.583	0.578	0.591	0.595	0.598	0.600
0.63	0.562	0.571	0.579	0.585	0.590	0.595	0.599	0.603	0.607	0.609
0.64	0.569	0.579	0.586	0.592	0.598	0.602	0.607	0.611	0.615	0.618
0.65	0.576	0.585	0.592	0.599	0.606	0.610	0.615	0.619	0.623	0.620
0.66	0.583	0.593	0.600	0.607	0.613	0.618	0.622	0.626	0.630	0.634
0.67	0.590	0.599	0.607	0.614	0.621	0.626	0.631	0.635	0.639	0.613
0.68	0.597	0.607	0.615	0.622	0.628	0.634	0.639	0.643	0.647	0.651
0.69	0.603	0.613	0.621	0.629	0.635	0.641	0.646	0.651	0.655	0.659
0.70	0.610	0.620	0.629	0.637	0.644	0.649	0.654	0.659	0.663	0.667
0.71	0.617	0.621	0.636	0.644	0.651	0.657	0.661	0.666	0.671	0.674
0.72	0.624	0.634	0.643	0.651	0.658	0.664	0.669	0.674	0.679	0.682
0.73	0.630	0.641	0.650	0.659	0.665	0.672	0.677	0.682	0.687	0.691
0.74	0.637	0.648	0.657	0.665	0.672	0.679	0.684	0.689	0.694	0.698
0.75	0.543	0.655	0.664	0.671	0.679	0.686	0.691	0.696	0.701	0.705
0.76	0.649	0.661	0.670	0.679	0.687	0.693	0.699	0.704	0.709	0.713
0.77	0.656	0.667	0.677	0.685	0.693	0.700	0.705	0.711	0.715	0.719
0.78	0.662	0.673	0.683	0.692	0 700	0.707	0.713	0 713	0.723	0.727
0.79	0.668	0.680	0.689	0.698	0.705	0.713	0.719	0.724	0.729	0.733
0.80	0.674	0.685	0.695	0.703	0.712	0.720	0.726	0.732	0.737	0.741
0.81	0. 680	0.691	0.701	0.710	0.719	0.727	0.733	0.739	0.744	0.749
0.82	0.686	0.698	0.707	0.717	0.725	0.733	0.740	0.745	0.751	0.755

(Contd.)

N / u	2.0	2.2	2.4	2.6	2.8	3.0	3.2	3.4	3.6	3.8
0.83	0.092	0.703	0.713	0.722	0.731	0.740	0.746	0.752	0.757	0.762
0.84	0.698	0.709	0.719	0.729	0.737	0.746	0.752	0.758	0.764	0.769
0.85	0.704	0.715	0.725	0.735	0.744	0.752	0.759	0.765	0.770	0.775
0.86	0.710	0.721	0.731	0.741	0.750	0.758	0.765	0.771	0.777	0782
0.87	0.715	0.727	0.738	0.747	0.756	0.764	0.771	0.777	0.783	0.788
0.88	0.721	0.733	0.743	0.753	0.762	0.770	0.777	0.783	0.789	0.794
0.89	0 727	0.739	0.749	0.758	0.767	0.776	0.783	0.789	0.795	0.801
0.90	0.732	0.744	0.754	0.764	0.773	0.781	0.789	0.795	0.801	0.807
0.91	0.738	0.750	0.760	0.770	0.779	0.787	0.795	0.801	0.807	0.812
0.92	0.743	0.754	0.766	0.776	0.785	0.793	0.800	0.807	0.813	0.818
0.93	0.749	0.761	0.772	0.782	0.791	0.799	0.807	0.812	0.818	0.823
0.94	0.754	0.767	0.777	0.787	0.795	0.804	0.813	0.818	0.824	0. 829
0.950	0.759	0.772	0.783	0.793	0.801	0.809	0.819	0.823	0.829	0.835
0.960	0.764	0.777	0.788	0.798	0.807	0.815	0.824	0.829	0.835	0.841
0.970	0.770	0.782	0.793	0.803	0.812	0.820	0.826	0.834	0.840	0.846
0.975	0.772	0.785	0.796	0.805	0.814	0.822	0.828	0.836	0.843	0.848
0.980	0.775	0.787	0.798	0.808	0.818	0.825	0.830	0.839	0.845	0.851
0. 985	0.777	0.790	0.801	0.811	0.820	0.827	0.833	0.841	0.847	0.853
0.990	0.780	0.793	0.804	0.814	0.822	0.830	0.837	0.844	0.850	0.856
0.995	0.782	0.795	0.806	0.816	0.824	0.832	0.840	0.847	0.853	0.859
1 .000	0.785	0.797	0.808	0.818	0.826	0.834	0.842	0.849	0.856	0.862
1 .005	0.788	0.799	0.810	0.820	0.829	0.837	0.845	0.852	0.858	0.864
1.010	0.790	0.801	0.812	0.822	0.831	0.840	0.847	0.855	0.861	0.867
1.015	0.793	0.804	0.815	0.824	0.833	0.843	0.850	0.858	0.864	0.870
1.020	0.795	0.807	0.818	0.828	0.837	0.845	0.853	0.860	0.866	0.872
1.03	0.800	0.811	0.822	0.832	0.841	0.850	0.857	0.864	0.871	0.877
1.04	0.805	0.816	0.829	0.837	0.846	0.855	0.862	0.870	0.877	0.883
1.05	0.810	0.821	0.831	0.841	0.851	0.859	0.867	0.874	0.881	0.887
1.06	0.815	0.826	0.837	0.846	0.855	0.864	0.871	0.879	0.885	0.891
1.07	0.819	0.831	0.841	0.851	0.860	0.869	0.876	0.883	0.889	0.896
1.08	0.824	0.836	0.846	0.856	0.865	0.873	0.880	0.887	0.893	0.900
1.09	0.828	0.840	0.851	0.860	0.870	0.877	0.885	0.892	0.898	0.904
1.10	0.833	0.845	0.855	0.865	0.874	0.881	0.890	0.897	0.903	0.908
1.11	0.837	0.849	0.860	0.870	0.878	0.886	0.894	0.900	0.907	0.912
1.12	0.842	0.854	0.864	0.873	0.882	0.891	0.897	0.904	0.910	0.916
1.13	0.846	0.858	0.868	0.878	0.886	0.895	0.902	0.908	0.914	0.919
1.14	0.851	0.861	0.872	0.881	0.890	0.899	0.905	0.912	0.918	0.923
1.15	0.855	0.866	0.876	0.886	0.895	0.903	0.910	0.916	0.922	0.928
1.16	0.859	0.870	0.880	0.890	0.899	0.907	0.914	0.920	0.926	0.931
1.17	0.864	0.874	0.884	0.893	0.902	0.911	0.917	0.923	0.930	0.934
1.18	0.868	0.878	0.888	0.897	0.906	0.915	0.921	0.927	0.933	0.939
1.19	0.872	0.882	0.892	0.901	0.910	0.918	0.925	0.931	0.937	0.942
1.20	0.876	0.886	0.896	0.904	0.913	0.921	0.928	0.934	0.940	0.945

(Contd.)

N u	2.0	2.2	2.4	2.6	2.8	3.0	3.2	3.4	3.6	3.8
1.22	0.880	0.891	0.900	0.909	0.917	0.929	0.932	0.938	0.944	0.949
1.24	0.888	0.898	0.908	0.917	0.925	0.935	0.940	0.945	0.950	0.955
1.26	0.900	0.910	0.919	0.927	0.935	0.942	0.948	0.954	0.960	0.964
1.28	0.908	0.917	0.926	0.934	0.945	0.948	0.954	0.960	0.965	0.970
1.30	0.915	0.925	0.933	0.941	0.948	0.955	0.961	0.966	0.971	0.975
1.32	0.922	0.931	0.940	0.948	0.955	0.961	0.967	0.972	0.976	0.980
1.34	0.930	0.939	0.948	0.955	0.962	0.967	0.973	0.978	0.982	0.986
1.36	0.937	0.946	0.954	0.961	0.968	0.973	0.979	0.983	0.987	0.991
1.38	0.944	0.952	0.960	0.967	0.974	0.979	0.985	0.989	0.993	0.996
1.40	0.951	0.959	0.966	0.973	0.979	0.984	0.989	0.993	0.997	1.000
1.42	0.957	0.965	0.972	0.979	0.984	0.989	0.995	0.998	1.001	1.004
1.44	0.964	0.972	0.979	0.984	0.990	0.995	1.000	1.003	1.006	1.009
1.46	0.970	0.977	0.983	0.989	0.995	1.000	1.004	1.007	1.010	1.012
1.48	0.977	0.983	0.989	0.994	0.999	1.005	1.008	1.011	1.014	1.016
1.50	0.983	0.990	0.996	1.001	1.005	1.009	1.012	1.015	1.017	1.019
1.55	0.997	1.002	1.007	1.012	1.016	1.020	1.022	1.024	1.026	1.028
1.60	1.012	1.017	1.020	1.024	1.027	1.030	1.032	1.034	1.035	1.035
1.65	1.026	1.029	1.032	1.035	1.037	1.039	1.041	1.041	1.042	1.042
1.70	1.039	1.042	1.044	1.045	1.047	1.048	1.049	1.049	1.049	1.048
1.75	1.052	1.053	1.054	1.055	1.056	1.057	1.056	1.056	1.055	1.053
1.80	1.064	1.064	1.064	1.064	1.065	1.065	1.064	1.062	1.060	1.058
1.85	1.075	1.074	1.074	1.073	1.072	1.071	1.069	1.067	1.066	1.063
1.90	1.086	1.085	1.084	1.082	1.081	1.079	1.077	1.074	1.071	1.066
1.95	1.097	1.095	1.092	1.090	1.087	1.085	1.081	1.079	1.075	1.071
2.00	1.107	1.103	1.100	1.096	1.093	1.090	1.085	1.082	1.078	1.075
2.10	1.126	1.120	1.115	1.110	1.104	1.100	1.094	1.089	1.085	1.080
2.20	1.144	1.136	1.129	1.122	1.115	1.109	1.102	1.096	1.090	1.085
2.3	1.161	1.150	1.141	1.133	1.124	1.117	1.110	1.103	1.097	1.090
2.4	1.176	1.163	1.152	1.142	1.133	1.124	1.116	1.109	1.101	1.091
2.5	1.190	1.175	1.162	1.150	1.140	1.131	1.121	1.113	1.105	1.098
2.6	1.204	1.187	1.172	1.159	1.147	1.137	1.126	1.117	1.106	1.000
2.7	1.216	1.196	1.180	1.166	1.153	1.142	1.130	1.120	1.110	1.102
2.8	1.228	1.208	1.189	1.173	1.158	1.146	1.132	1.122	1.112	1.103
2.9	1.239	1.216	1.196	1.178	1.162	1.150	1.137	1.125	1.115	1.106
3.0	1.249	1.224	1.203	1.184	1.168	1.154	1.140	1.128	1.117	1.107
3.5	1.292	1.260	1.232	1.206	1.185	1.167	1.151	1.138	1.125	1.113
4.0	1.326	1.286	1.251	1.223	1.198	1.176	1.158	1.142	1.129	1.117
4.5	1.352	1.308	1.270	1.235	1.205	1.183	1.162	1.146	1.131	1.119
5.0	1.374	1.325	1.283	1.245	1.212	1.188	1.166	1.149	1.134	1.121
6.0	1.406	1.342	1.292	1.252	1.221	1.195	1.171	1.152	1.136	1.122
7.0	1.430	1.360	1.303	1.260	1.225	1.199	1.174	1.153	1.136	1.122
8.0	1.447	1.373	1.313	1.266	1.229	1.201	1.175	1.154	1.137	1.122
9.0	1.461	1.384	1.319	1.269	1.231	1.203	1.176	1.156	1.137	1.122
10.0	1.471	1.394	1.324	1.272	1.233	1203	1.176	1.156	1.137	1.122

Table 14.7.2: Values of the varied flow function $F(u, N)_{-S_0}$ for negative slopes (5) (contd..)

u \ N	4.0	4.2	4.5	5.0	5.5
0.00	0.000	0.000	0.000	0.000	0.000
0.02	0.020	0.020	0.020	0.020'	0.020
0.04	0.040	0.040	0.040	0.040	0.040
0.06	0.060	0.060	0.060	0.060	0.060
0.08	0.080	0.080	0.080	0.080	0.080
0.10	0.100	0.100	0.100	0.100	0.100
0.12	0.120	0.120	0.120	0.120	0.120
0.14	0.140	0.140	0.140	0.140	0.140
0.16	0.160	0.160	0.160	0.160	0.160
0.18	0.180	0.180	0.180	0.180	0.180
0.20	0.200	0.200	0.200	0.200	0.200
0.22	0.220	0.220	0.220	0.220	0.220
0.24	0.240	0.240	0.240	0.240	0.240
0.26	0.260	0.260	0.260	0.260	0.260
0.28	0.280	0.280	0.280	0.280	0.280
0.30	0.300	0.300	0.300	0.300	0.300
0.32	0.320	0.320	0.320	0.320	0.320
0.34	0.339	0.340	0.340	0.340	0.340
0.36	0.359	0.360	0.360	0.360	0.360
0.38	0.378	0.379	0.380	0.380	0.380
0.40	0.397	0.398	0.398	0.400	0.400
0.42	0.417	0.418	0.418	0.419	0.420
0.44	0.436	0.437	0.437	0.439	0.440
0.46	0.455	0.456	0.457	0.458	0.459
0.48	0.474	0.475	0.476	0.478	0.479
0.50	0.493	0.494	0.495	0.497	0.498
0.52	0.512	0.513	0.515	0.517	0.518
0.54	0.531	0.532	0.533	0.536	0.537
0.56	0.549	0.550	0.552	0.555	0.558
0.58	0.567	0.569	0.570	0.574	0.576
0.60	0.585	0.587	0.589	0.593	0.595
0.61	0.594	0.596	0.598	0.602	0.604
0.62	0.603	0.605	0.607	0.611	0.613
0.63	0.612	0.615	0.616	0.620	0.622
0.64	0.620	0.623	0.625	0.629	0.631
0.65	0.629	0.632	0.634	0.638	0.640
0.66	0.637	0.640	0.643	0.647	0.650
0.67	0.646	0.649	0.652	0.656	0.659
0.68	0.654	0.657	0.660	0.665	0.668
0.69	0.662	0.665	0.668	0.674	0.677

(Contd.)

N u	4.0	4.2	4.5	5.0	5.5
0.70	0.670	0.673	0.677	0.682	0.686
0.71	0.678	0.681	0.685	0.690	0.694
0.72	0.686	0.689	0.694	0.699	0.703
0.73	0.694	0.698	0.702	0.707	0.712
6.74	0.702	0.705	0.710	0.716	0.720
0.75	0.709	0.712	0.717	0.724	0.728
0.76	0.717	0.720	0.725	0.731	0.736
0.77	0.724	0.727	0.733	0.739	0.744
0.78	0.731	0.735	0.740	0.747	0.752
0.79	0.738	0.742	0.748	0.754	0.760
0.80	0.746	0.750	0.755	0.762	0.768
0.81	0.753	0.757	0.762	0.770	0.776
0.82	0.760	0.764	0.769	0.777	0.783
0.83	0.766	0.771	0.776	0.784	0.790
0.84	0.773	0.778	0.783	0.791	0.798
0.85	0.780	0.784	0.790	0.798	0.805
0.86	0.786	0.791	0.797	0.804	0.812
0.87	0.793	0.797	0.803	0.811	0.819
0.88	0.799	0.803	0.810	0.818	0.826
0.89	0.805	0.810	0.816	0.825	0.832
0.90	0.811	0.816	0.822	0.831	0.839
0.91	0.817	0.821	0.828	0.837	0.845
0.92	0.823	0.828	0.834	0.844	0.851
0.93	0.829	0.833	0.840	0.850	0.857
0.94	0.835	0.840	0.846	0.856	0.864
0.950	0.840	0.845	0.852	0.861	0.869
0.960	0.846	0.851	0.857	0.867	0.875
0.970	0.851	0.856	0.863	0.972	0.881
0.975	0.854	0.859	0.866	0.875	0.883
0.980	0.857	0.861	0.868	0.878	0.886
0.985	0.859	0.863	0.870	0.880	0.889
0.990	0.861	0.867	0.873	0.883	0.891
0.995	0.864	0.869	0.876	0.885	0.894
1.000	0.867	0.873	0.879	0.887	0.897
1.005	0.870	0.874	0.881	0.890	0.899
1.010	0.873	0.878	0.884	0.893	0.902
1.015	0.875	0.880	0.886	0.896	0.904
1.020	0.877	0.883	0.889	0.898	0.907
1.03	0.882	0.887	0.893	0.902	0.911
1.04	0.888	0.893	0.898	0.907	0.916
1.05	0.892	0.897	0.903	0.911	0.920
1.06	0.896	0.901	0.907	0.915	0.924
1.07	0.901	0.906	0.911	0.919	0.928

(Contd.)

N u	4.0	4.2	4.5	5.0	5.5
1.08	0.905	0.910	0.916	0.923	0.932
1.09	0.909	0.914	0.920	0.927	0.936
1.10	0.913	0.918	0.923	0.931	0.940
1.11	0.917	0.921	0.927	0.935	0.944
1.12	0.921	0.926	0.931	0.939	0.948
1.13	0.925	0.929	0.935	0.943	0.951
1.14	0.928	0.933	0.938	0.947	0.954
1.15	0.932	0.936	0.942	0.950	0.957
1.16	0.936	0.941	0.945	0.953	0.960
1.17	0.939	0.944	0.948	0.957	0.963
1.18	0.943	0.947	0.951	0.960	0.965
1.19	0.947	0.950	0.954	0.963	0.968
1.20	0.950	0.953	0.958	0.966	0.970
1.22	0.956	0.957	0.964	0.972	0.976
1.24	0.962	0.962	0.970	0.977	0.981
1.26	0.968	0.971	0.975	0.982	0.986
1.28	0.974	0.977	0.981	0.987	0.990
1.30	0.979	0.978	0.985	0.991	0.994
1.32	0.985	0.986	0.990	0.995	0.997
1.34	0.990	0.992	0.995	0.999	1.001
1.36	0.994	0.996	0.999	1.002	1.005
1.38	0.998	1.000	1.003	1.006	1.008
1.40	1.001	1.004	1.006	1.009	1.011
1.42	1.005	1.008	1.010	1.012	1.014
1.44	1.009	1.013	1.014	1.016	1.016
1.46	1.014	1.016	1.017	1.018	1.018
1.48	1.016	1.019	1.020	1.020	1.020
1.50	1.020	1.021	1.022	1.022	1.022
1.55	1.029	1.029	1.029	1.028	1.028
1.60	1.035	1.035	1.034	1.032	1.030
1.65	1.041	1.040	1.039	1.036	1.034
1.70	1.047	1.046	1.043	1.039	1.037
1.75	1.052	1.051	1.047	1.042	1.039
1.80	1.057	1.055	1.051	1.045	1.041
1.85	1.061	1.059	1.054	1.047	1.043
1.90	1.065	1.060	1.057	1.049	1.045
1.95	1.068	1.064	1.059	1.051	1.046
2.00	1.071	1.068	1.062	1.053	1.047
2.10	1.076	1.071	1.065	1.056	1.049
2.20	1.080	1.073	1.068	1.058	1.050
2.3	1.084	1.079	1.071	1.060	1.051
2.4	1.087	1.081	1.073	1.061	1.052
2.5	1.090	1.083	1.075	1.062	1.053

(Contd.)

N u	4.0	4.2	4.5	5.0	5.5
2.6	1.092	1.085	1.076	1.063	1.054
2.7	1.094	1.087	1.077	1.063	1.054
2.8	1.096	1.088	1.078	1.064	1.054
2.9	1.098	1.089	1.079	1.065	1.055
3.0	1.099	1.090	1.080	1.065	1.055
3.5	1.103	1.093	1.082	1.066	1.055
4.0	1.106	1.097	1.084	1.067	1.056
4.5	1.108	1.098	1.085	1.067	1.056
5.0	1.110	1.099	1.085	1.068	1.056
6.0	1.111	1.100	1.085	1.068	1.056
7.0	1.111	1.100	1.086	1.068	1.056
8.0	1.111	1.100	1.086	1.068	1.056
9.0	1.111	1.100	1.086	1.068	1.056
10.0	1.111	1.100	1.086	1.068	1.056

14.7.4 Direct Step Method

On equating the total heads at the two end sections 1 and 2 of a short channel reach, Fig. 14.7.2, one may write

$$S_0\,\Delta x + h_1 + \alpha_1\frac{U_1^2}{2g} = h_2 + \alpha_2\frac{U_2^2}{2g} + S_f\,\Delta x \qquad \ldots(14.7.10)$$

or

$$S_0\,\Delta x + E_1 = E_2 + S_f\,\Delta x$$

∴

$$\Delta x = \frac{E_2 - E_1}{S_0 - S_f} = \frac{\Delta E}{S_0 - S_f} \qquad \ldots(14.7.11)$$

Here, E is, obviously, the specific energy that equals $h + \alpha\dfrac{U^2}{2g}$. The friction slope S_f is computed at both sections using either the Manning's equation or Chezy's equation and their average value is used in Eq. (14.7.11) to obtain the distance Δx between sections 1 and 2. The step computation shall be carried from the control section (where the depth of flow and other hydraulic parameters are known) towards upstream if the flow is subcritical and towards downstream if the flow is supercritical. At the other end of the reach, the depth of flow h is assumed suitably, keeping in mind the nature of the water surface profile (i.e., backwater or drawdown), and the relevant hydraulic parameters are computed. Thereafter, the length of the reach Δx is determined. The next step is similarly computed for the next reach and so on. The direct step method is applicable to prismatic channels for which channel characteristics are invariant with distance along the channel.

14.7.5 Standard Step Method

The standard step method is applicable to both prismatic and nonprismatic channels. The nonprismatic channels are usually natural channels the characteristics of which are determined by carrying out a field survey. Here again, the computation is carried out step by step as in the direct step method. But, in this method, the distance between the sections and the depth of flow at one of the sections are known and the method is applied to determine the depth of flow at the other section. The method requires trial and error.

The application of energy equation between the two end sections 1 and 2 of a short channel reach, shown in Fig. 14.7.2, yields

$$z_1 + h_1 + \alpha_1 \frac{U_1^2}{2g} = z_2 + h_2 + \alpha_2 \frac{U_2^2}{2g} + h_f + h_e \qquad \qquad ...(14.7.12)$$

Here, h_f and h_e are the head loss due to friction and form loss (on account of variable cross-section of the channel, respectively.

If the flow is subcritical, the computations would proceed in the upstream direction from the control section 2 at which the depth of flow and other hydraulic parameters are known. For determining the depth of flow at section 1, located at a known distance Δx upstream, a suitable depth of flow is

assumed and the total energy head $\left(= z_1 + h_1 + \alpha_1 \dfrac{U_1^2}{2g} \right)$ is computed. For the assumed h_1 and known h_2,

compute h_f and h_e using

$$h_f = \frac{1}{2} \left(S_{f1} + S_{f2} \right) \Delta x \qquad \qquad ...(14.7.13)$$

and

$$h_e = k_L \left| \frac{U_1^2 - U_2^2}{2g} \right| \qquad \qquad ...(14.7.14)$$

Substituting these values in Eq. (14.7.12), the value of $z_1 + h_1 + \alpha_1 \dfrac{U_1^2}{2g}$ is computed and compared

with the corresponding value computed for the assumed h_1. If these two values match, the assumed value gives the desired depth h_1. Otherwise, another suitable value for h_1 is assumed and the procedure repeated. The procedure is repeated similarly for other sub-reaches.

14.7.6 Computer Programs

All the methods for computing GVF profiles discussed hitherto can be adapted for solution by a digital computer. However, general purpose computer programs have also been developed for the purpose of computing GVF profiles. HEC-2 is one such computer program (written in FORTRAN IV and developed by the Hydrologic Engineering Center (HEC) of the U.S. Army Corps of Engineers) that employs standard step method and, hence, is applicable to both prismatic and nonprismatic channels. The program takes into account the effects of various obstructions such as bridges, culverts, weirs etc. that may be present in the channel reach under investigation.

14.8 RAPIDLY VARIED FLOW

The streamlines in rapidly varied flow (RVF) have very large curvature. Many a times, an abrupt change in curvature breaks the flow profile and this results in a high state of turbulence causing considerable loss of energy. Hydraulic jump, an example of steady rapidly varied flow with broken profile, occurs when, in the same reach of a channel, the upstream control causes supercritical flow while the downstream control dictates subcritical flow. These two flows meet such that there is an abrupt change in the curvature of streamlines which breaks the flow profile resulting in a high state of turbulence and, thus, causing huge loss of energy.

The following characteristic features of RVF must be contrasted with those of GVF:

1. The pressure distribution is non-hydrostic due to huge curvature of flow.
2. The change in flow regime takes place over a relatively short reach of a channel. Therefore, boundary friction is relatively small and, in many cases, insignificant.
3. In RVF, velocity distribution coefficients, i.e., α and β are usually much greater than unity and cannot be accurately determined.
4. The separation zones, eddies, and rollers, occurring in RVF, make the flow pattern much more complex.

14.9 HYDRAULIC JUMP

Hydraulic jump, as explained in Art. 14.8, occurs when supercritical flow meets subcritical flow. Considerable turbulence and energy dissipation accompany a hydraulic jump that finds useful applications in many flow situations such as (1):

(i) The dissipation of energy of flow downstream of hydraulic structures such as dams, spillways, weirs, etc.
(ii) The reduction of net uplift pressures under hydraulic structures by raising the water depth on the apron of the structure.
(iii) The maintenance of high water levels in channels for water distribution purposes.
(iv) The mixing of chemicals for water purification or other purposes in chemical industries.

Newton's second law of motion, when applied to the control volume of the one-dimensional flow, Fig. 14.9.1, yields

$$P_1 - P_2 + W \sin\theta - F_f = \rho Q \left(\beta_2 u_2 - \beta_1 u_1 \right) \qquad \qquad ...(14.9.1)$$

where, P_1 and P_2 are the pressure forces at sections 1 and 2, W the weight of liquid between sections 1 and 2, F_f the component of unknown friction forces (along the direction of flow) acting between

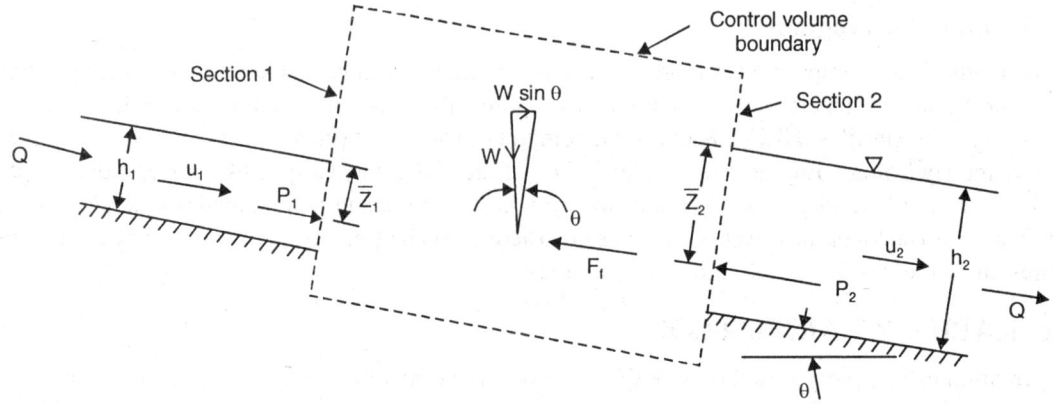

Fig. 14.9.1. Control volume for hydraulic jump

sections 1 and 2, θ the longitudinal slope of the channel, β_1 and β_2 the momentum correction coefficients at sections 1 and 2, and u_1 and u_2 are the average velocities at sections 1 and 2.

With the assumptions that θ is small (i.e., $\sin \theta \cong 0$), and $\beta_1 = \beta_2 = 1$, Eq. (14.9.1) becomes

$$\rho g \bar{z}_1 A_1 - \rho g \bar{z}_2 A_2 - F_f = \rho Q (u_2 - u_1) \qquad \ldots(14.9.2)$$

where, $\bar{z}_1$ and $\bar{z}_2$ are the distances to centroids of respective flow areas A_1 and A_2 from the free surface. Therefore,

$$P_1 = \rho g \bar{z}_1 A_1 \qquad \ldots(14.9.3)$$

and

$$P_2 = \rho g \bar{z}_2 A_2 \qquad \ldots(14.9.4)$$

Equation (14.9.2) can now be rewritten as

$$\frac{F_f}{\rho g} = \left(\frac{Q^2}{gA_1} + A_1 \bar{z}_1 \right) - \left(\frac{Q^2}{gA_2} + A_2 \bar{z}_2 \right) \qquad \ldots(14.9.5)$$

or

$$\frac{F_f}{\rho g} = M_1 - M_2 \qquad \ldots(14.9.6)$$

where,

$$M = \frac{Q^2}{gA} + A\bar{z} \qquad \ldots(14.9.7)$$

and M is called the *momentum function* or *specific momentum* or *force function* or *specific force*.

If the jump occurs in a horizontal channel and is not assisted by any other means, such as baffle blocks, then $F_f \cong 0$, and Eq. (14.9.6) becomes

$$M_1 = M_2$$

or

$$\frac{Q^2}{gA_1} + A_1 \bar{z}_1 = \frac{Q^2}{gA_2} + A_2 \bar{z}_2 \qquad \ldots(14.9.8)$$

14.9.1 Hydraulic Jump in Rectangular Channels

A hydraulic jump formed in a smooth, wide, and horizontal rectangular channel is termed classical hydraulic jump. For a rectangular channel of constant width B,

$$A_1 = Bh_1, \qquad A_2 = Bh_2, \qquad \bar{z}_1 = h_1/2 \qquad \text{and} \qquad \bar{z}_2 = h_2/2$$

Substituting these values in Eq. (14.9.8) one obtains

$$\frac{q^2 B^2}{gBh_1} + \frac{Bh_1^2}{2} = \frac{q^2 B^2}{gBh_2} + \frac{Bh_2^2}{2}$$

where, $q = Q/B$. On further simplification, one obtains

$$\frac{1}{2} h_1 h_2 (h_1 + h_2) = \frac{q^2}{g} \qquad \ldots(14.9.9)$$

Equation (14.9.9) has the following feasible solutions:

$$\frac{h_2}{h_1} = \frac{1}{2}\left(\sqrt{1+8F_1^2} - 1\right) \qquad \qquad ...(14.9.10)$$

and

$$\frac{h_1}{h_2} = \frac{1}{2}\left(\sqrt{1+8F_2^2} - 1\right) \qquad \qquad ...(14.9.11)$$

in which F_1 and F_2 are the Froude numbers at sections 1 and 2, respectively. Froude number F equals $u/\sqrt{gD}$ in which D is the hydraulic depth and equals A/T where T is the top width of flow section. Equation (14.9.10) and (14.9.11) are known as Belanger's equations.

14.9.2 Energy Loss in Hydraulic Jump in a Rectangular Channel

In a horizontal rectangular channel with the channel bed chosen as the datum (Fig. 14.9.1 with $\theta = 0$), the total energies (expressed as heads at sections 1 and 2) are equal to the specific energies E_1 and E_2 expressed as heads at sections 1 and 2, respectively, i.e.,

$$E_1 = h_1 + \frac{q^2}{2g\,h_1^2} \qquad \qquad ...(14.9.12)$$

$$E_2 = h_2 + \frac{q^2}{2g\,h_2^2} \qquad \qquad ...(14.9.13)$$

so that

$$\Delta E = E_1 - E_2$$

$$= h_1 - h_2 + \frac{q^2}{2g}\left[\frac{1}{h_1^2} - \frac{1}{h_2^2}\right]$$

$$= (h_1 - h_2) + \frac{1}{4}h_1\,h_2\,(h_1 + h_2)\left(\frac{h_2^2 - h_1^2}{h_1^2\,h_2^2}\right)$$

$$= \frac{4h_1^2 h_2 - 4h_2^2 h_1 + h_1 h_2^2 - h_1^3 + h_2^3 - h_1^2 h_2}{4h_1 h_2}$$

$$= \frac{h_2^3 - h_1^3 + 3h_1^2 h_2 - 3h_1 h_2^2}{4h_1\,h_2}$$

$\therefore$

$$\Delta E = \frac{(h_2 - h_1)^3}{4h_1 h_2} \qquad \qquad ...(14.9.14)$$

Also,

$$\frac{\Delta E}{E_1} = \frac{\left[h_1 + \left(u_1^2 / 2g\right)\right] - \left[h_2 + \left(u_2^2 / 2g\right)\right]}{\left[h_1 + \left(u_1^2 / 2g\right)\right]} \qquad \ldots(14.9.15)$$

$$= \frac{h_1\left[1 - \left(h_2 / h_1\right)\right] + \left(q^2 / 2g\, h_1^2\right)\left[1 - \left(h_1 / h_2\right)^2\right]}{\left(h_1 / 2\right)\left[2 + F_1^2\right]}$$

$$\therefore \qquad \frac{\Delta E}{E_1} = \frac{2 - 2\left(h_2 / h_1\right) + F_1^2\left[1 - \left(h_1 / h_2\right)^2\right]}{2 + F_1^2} \qquad \ldots(14.9.16)$$

Combining Eqs. (14.9.9), (14.9.12) and (14.9.14), one can obtain

$$\frac{\Delta E}{E_1} = \frac{8F_1^4 + 20F_1^2 - \left(8F_1^2 + 1\right)^{3/2} - 1}{8F_1^2\left(2 + F_1^2\right)} \qquad \ldots(14.9.17)$$

Hence, for a supercritical Froude number F_1 equal to 20, the energy loss ΔE is 0.86 E_1. This means that 86 per cent of the initial specific energy is dissipated due to hydraulic jump. Because of this energy dissipating capability, hydraulic jump is widely used as an energy dissipater at the foot of a spillway and for other hydraulic structures. In any hydraulic jump, the mean kinetic energy is first converted into turbulence and then dissipated through the action of viscosity. Equation (14.9.17) can also be rewritten as (5),

$$\frac{E_2}{E_1} = \frac{\left(8F_1^2 + 1\right)^{3/2} - 4F_1^2 + 1}{8F_1^2\left(2 + F_1^2\right)} \qquad \ldots(14.9.18)$$

The term E_2/E_1 is called the efficiency of the jump.

14.9.3 Length of Hydraulic Jump

The length of a hydraulic jump is defined as the distance from the front of the jump to a point of the surface of the flow immediately downstream of the roller associated with the jump. Although, a very important design parameter, the length of a hydraulic jump, L_j cannot be derived from theoretical considerations. Experimental studies have indicated that the length of a hydraulic jump is approximately equal to five times the height of the jump which is $(h_2 - h_1)$. Silvester (15) has shown that for horizontal rectangular channels, the ratio L_j/h_1 depends on the upstream supercritical Froude number. He obtained

$$\frac{L_j}{h_1} = 9.75\left(F_1 - 1\right)^{1.01} \qquad \ldots(14.9.19)$$

The length of a classical hydraulic jump L_j can also be estimated from Fig. 14.9.2 which shows the variation of L_j/h_2 with F_1.

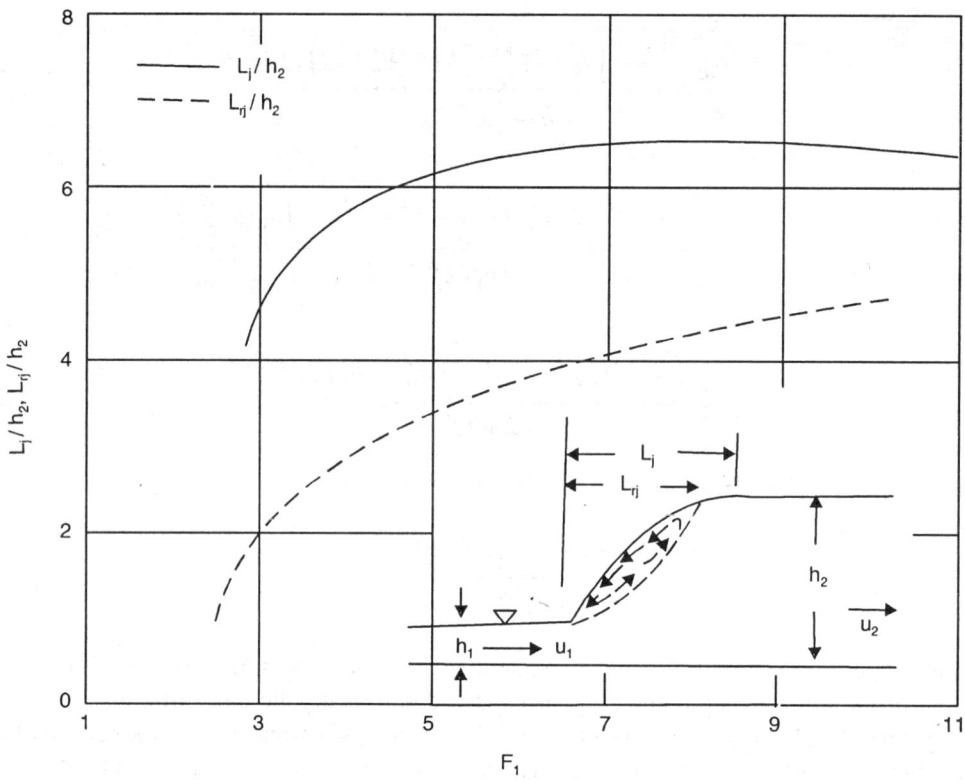

Fig. 14.9.2. Length characteristics of hydraulic jump (12)

14.9.4 Profile of Hydraulic Jump

The concrete aprons of the stilling basins of overflow structures located on permeable foundations are subjected to uplift pressures which are partly counterbalanced by the weight of water flowing on the apron. Therefore, in the hydraulic jump type stilling basins, determination of profile of the jump becomes necessary. Rajaratnam and Subramanya (13) have obtained an empirical relation, Fig. 14.9.3, between $h/[0.75\ (h_2 - h_1)]$ and $x/\bar{X}$. Here, $\bar{X}$ is the distance from the beginning of the jump to the section where the depth measured above the x-axis is $0.75\ (h_2 - h_1)$. The length $\bar{X}$ is empirically related to h_1 and F_1 as (13)

$$\frac{\bar{X}}{h_1} = 5.08\,F_1 - 7.82 \qquad\qquad\qquad ...(14.9.20)$$

14.9.5 Calculations for Hydraulic Jump in Horizontal Rectangular Channels

Equations (14.9.9), (14.9.12), (14.9.13), and (14.9.14) can be used to obtain direct solution of the parameters (i.e., E_2, h_2, ΔE, and E_1) for a hydraulic jump formed in a horizontal rectangular channel, if h_1 and q are known. These equations would also yield direct solution for E_1, h_1, ΔE, and E_2, if h_2 and q are known. One can also use Eq. (14.9.10) or Eq. (14.9.11) instead of Eq. (14.9.9) depending upon whether pre-jump or post-jump conditions are known.

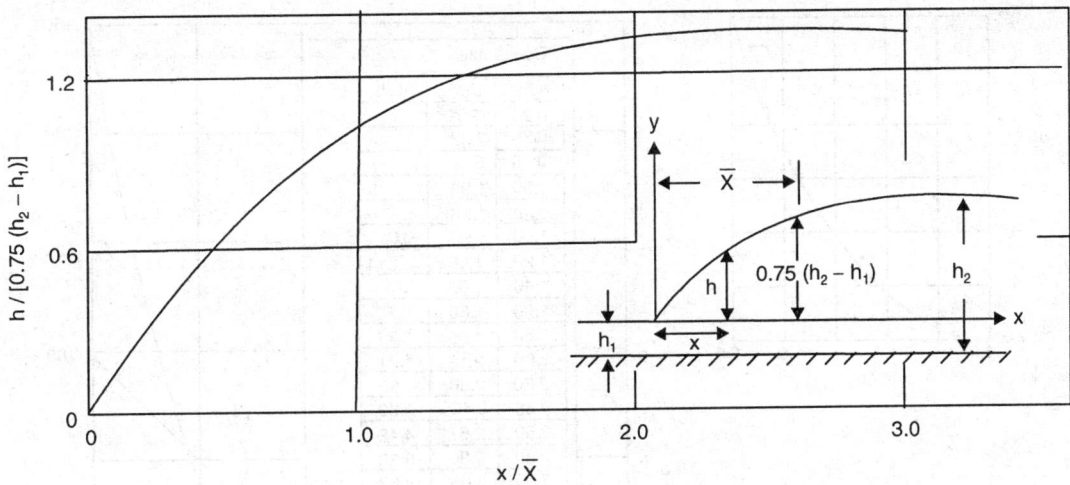

Fig. 14.9.3. Profile of hydraulic jump in a rectangular channel (13)

However, in an actual design problem, generally the discharge q and the levels of the upstream and downstream total energy lines are known. Thus, q and ΔE are known. Determination of the remaining four parameters of the jump from Eqs. (14.9.9), (14.9.12), (14.9.13), and (14.9.14) is rather difficult. This difficulty can be overcome by the use of critical depth h_c $(=(q^2/g)^{1/3})$ and defining

$$X = (h_1/h_c), \quad Y = (h_2/h_c), \quad Z = (\Delta E/h_c),$$

$$\xi = E_1/h_c, \text{ and } \eta = E_2/h_c$$

so that Eqs. (14.9.9), (14.9.12), (14.9.13), and (14.9.14) reduce to the following forms, respectively:

$$XY(X+Y) = 2 \qquad \qquad ...(14.9.21)$$

$$\xi = X + \frac{1}{2X^2} \qquad \qquad ...(14.9.22)$$

$$\eta = Y + \frac{1}{2Y^2} \qquad \qquad ...(14.9.23)$$

and
$$Z = \frac{(Y-X)^3}{4XY} \qquad \qquad ...(14.9.24)$$

Here, X can vary from 0 to 1 only. Using Eqs. (14.9.21) to (14.9.24), one can compute the sets of values of Y, ξ, η, and Z for different values of X and, thus, obtain the curves shown in Fig. 14.9.4. These curves are known as Crump's curves (6). The method to use these curves is as follows:

  (i) Calculate h_c from $h_c = (q^2/g)^{1/3}$.
 (ii) Compute Z, i.e., $\Delta E/h_c$.
(iii) Read h_2/h_c from the curve $\Delta E/h_c$ versus h_2/h_c.
 (iv) Read E_2/h_c from the curve E_2/h_c versus h_2/h_c.
  (v) Thus, $E_1 = \Delta E + E_2$.
 (vi) For known E_1/h_c, obtain h_1/h_c from the curve E_1/h_c versus h_1/h_c.

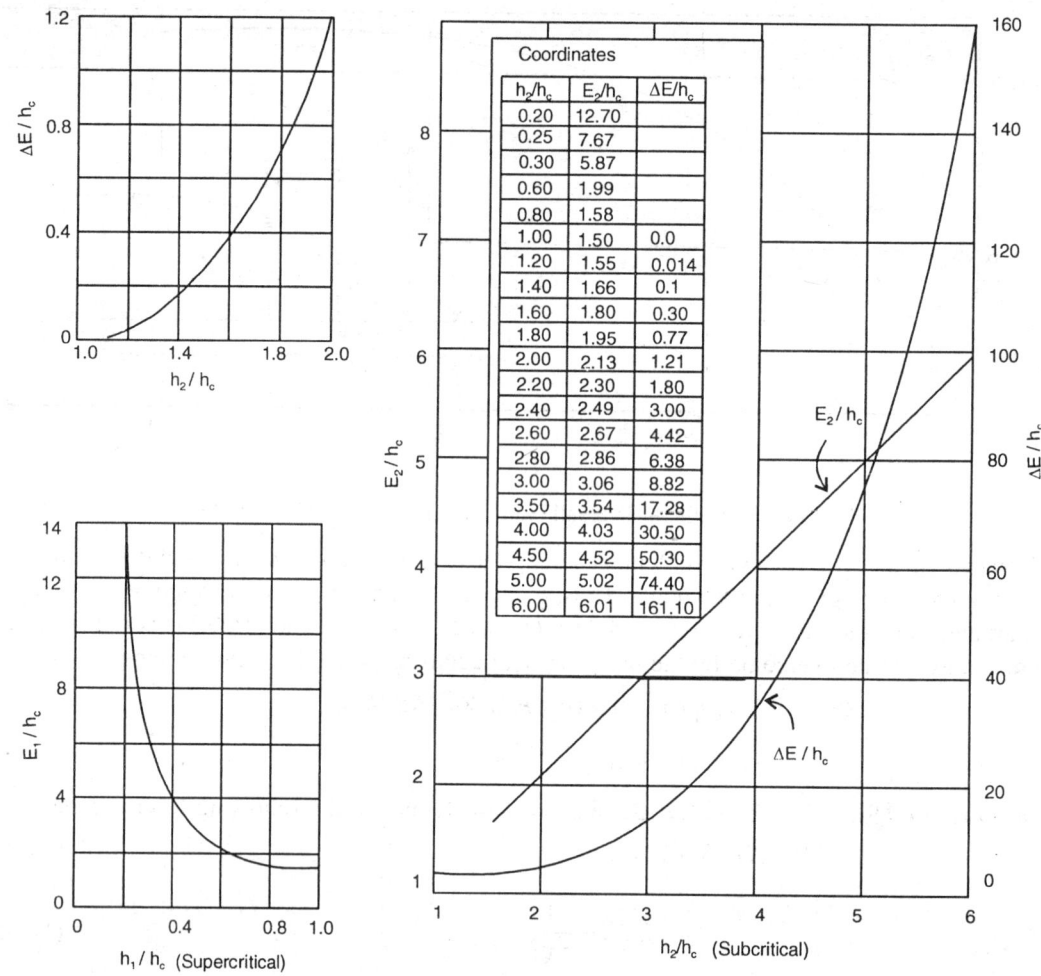

Fig. 14.9.4. Crump's curves (6)

Thus, for given q and ΔE, one can estimate h_2, E_2, E_1, and h_1 from the relationships for Crump's coefficients.

Combining Eqs. (14.9.21) and (14.9.24), one can obtain (16)

$$Z = \frac{-X^6 + 20X^3 + 8 - \left(X^4 + 8X\right)^{3/2}}{16X^2} \qquad \text{...(14.9.25)}$$

and also

$$Z = \frac{-Y^6 - 20Y^3 - 8 - \left(Y^4 + 8Y\right)^{3/2}}{16Y^2} \qquad \text{...(14.9.26)}$$

For different values of X and Y, the values of Z can be obtained and the curves X-Y and Y-Z prepared. Approximate equations for these curves were obtained as (16)

$$Y = 1 + 0.93556 Z^{0.368} \text{ for } Z < 1 \qquad \qquad ...(14.9.27)$$

and
$$Y = 1 + 0.93556 Z^{0.240} \text{ for } Z > 1 \qquad \qquad ...(14.9.28)$$

One of the two equations, viz., Eqs. (14.9.27) and (14.9.28), can be used for obtaining the value of Y for a specified value of Z. Equations (14.9.21), (14.9.22), and (14.9.23) can, then, be used for the determination of X, ξ, and η, respectively.

Equations (14.9.25) and (14.9.26), can be solved to prepare a table (Table 14.9.1) for computations of hydraulic jump elements in rectangular channels.

Table 14.9.1: Hydraulic jump elements in rectangular channels (16)

$Z = \dfrac{\Delta E}{h_c}$	X	Y	$\dfrac{E_1}{h_c}$	$\dfrac{E_2}{h_c}$	$\dfrac{Y}{X} = \dfrac{h_2}{h_1}$	$\dfrac{\Delta E}{E_1}$
0.01	0.839	1.180	1.549	1.539	1.406	0.006
0.10	0.681	1.407	1.760	1.660	2.067	0.057
0.50	0.516	1.728	2.396	1.896	3.351	0.209
1.00	0.436	1.936	3.069	2.069	2.442	0.326
1.50	0.389	2.082	3.700	2.200	4.356	0.406
2.00	0.356	2.199	4.303	2.303	6.179	0.465
2.50	0.331	2.298	4.893	2.393	6.940	0.511
3.00	0.311	2.384	5.472	2.472	7.659	0.548
4.00	0.281	2.531	6.609	2.609	9.002	0.605
4.50	0.269	2.594	7.169	2.669	9.638	0.628
5.00	0.259	2.654	6.725	2.725	10.254	0.647
6.00	0.241	2.761	8.826	2.826	11.439	0.680
7.00	0.227	2.856	9.917	2.917	12.572	0.706
8.00	0.215	2.942	11.000	3.000	13.663	0.727
10.00	0.197	3.094	13.146	3.146	15.744	0.761
15.00	0.165	3.396	18.439	3.439	20.527	0.814
20.00	0.145	3.640	23.878	3.878	25.079	0.846

14.9.6 Hydraulic Jump on Sloping Channels

Location of hydraulic jump on a horizontal floor with little or no friction varies considerably with a slight change in the depth or velocity of flow. But, on a sloping floor, location of hydraulic jump is relatively stable and can be closely predicted. However, energy dissipation in case of jumps on a sloping floor is less owing to the vertical component of velocity remaining intact.

Equation (14.9.1) is theoretically applicable to hydraulic jumps forming on sloping channels. But, the solution of the problem is difficult due to the following reasons:

(i) The length and shapes of the hydraulic jump are not well-defined and, hence, the term $W \sin \theta$ is poorly computed.

(ii) The specific weight of the liquid in the control volume can change considerably due to air entrainment.

(iii) The pressure terms cannot be determined accurately.

Figure 14.9.5 shows hydraulic jumps formed on sloping channels. In the experimental investigations of hydraulic jump on sloping channels, the end of the surface roller is taken as the end of the jump. This means that the length of roller (measured horizontally) is the length of the jump.

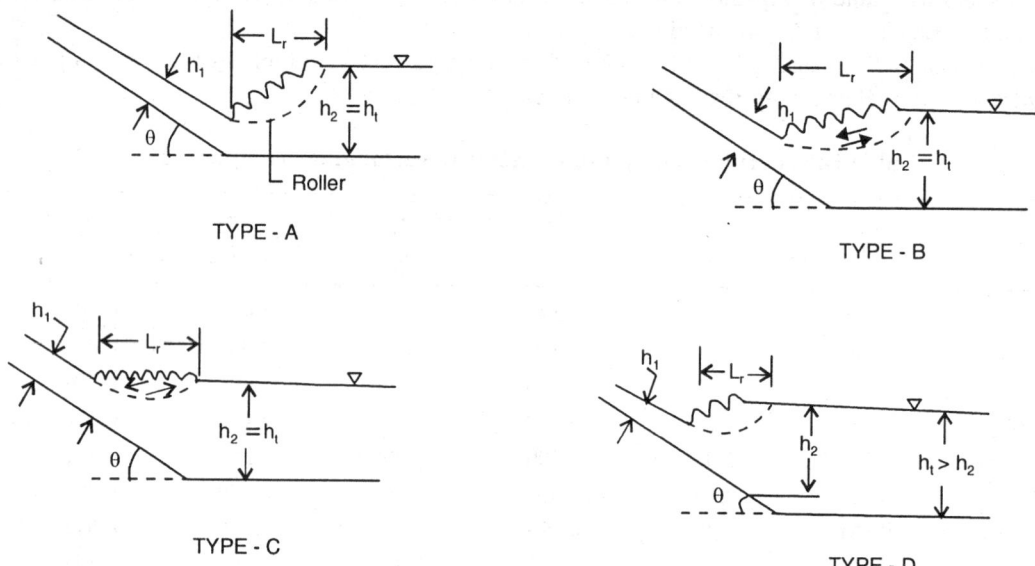

TYPE - A

TYPE - B

TYPE - C

TYPE - D

Fig. 14.9.5. Types of hydraulic jump in sloping channels

When a hydraulic jump begins at the end of the sloping apron, type A jump occurs and $h_2 = h_2^* = h_t$.

Here, h_2 is the subcritical sequent depth corresponding to h_1, h_t the tail-water depth, and h_2^* is the subcritical sequent depth h_2 given by Eq. (14.9.10). Type A jump is, obviously, governed by Eq. (14.9.10).

When the end of a hydraulic jump coincides with that of the sloping bed, type C jump occurs. For this case, Kindsvater (9) developed the following equation for the sequent depth h_2:

$$\frac{h_2}{h_1} = \frac{1}{2\cos\theta}\left[\sqrt{1 + 8F_1^2\left(\frac{\cos^2\theta}{1 - 2N\tan\theta}\right)} - 1\right] \qquad \text{...(14.9.29)}$$

in which θ is the longitudinal slope angle of the channel, and N is an empirical coefficient that depends on the length of the jump. Equation (14.9.29) can be rewritten as

$$\frac{h_2}{h_1'} = \frac{1}{2}\left[\sqrt{1 + 8G_1^2} - 1\right] \qquad \text{...(14.9.30)}$$

where $\qquad h_1' = h_1/\cos\theta \qquad \text{...(14.9.31)}$

and $\qquad G_1^2 = \frac{\cos^3\theta}{1 - 2N\tan\theta} \times F_1^2 \qquad \text{...(14.9.32)}$

Rajaratnam (12) gave the following simple expression:

$$\frac{\cos^3 \theta}{1-2N\tan\theta}=10^{0.054(\theta)} \qquad\qquad ...(14.9.33)$$

where, θ is in degrees.

If h_t is greater than the sequent depth h_2 required for type C jump, type D jump occurs completely on the sloping apron. Bradley and Peterka (3) found that Eqs. (14.9.30) to (14.9.33) valid for type C jump can be used for the type D jump also.

If h_t is less than that required for type C jump but greater than h_2^*, the toe of the jump is on the sloping bed, and the end of the jump on the horizontal bed. This jump is classed as type B jump. A graphical solution (Fig. 14.9.6) has been developed for this type of jump (12).

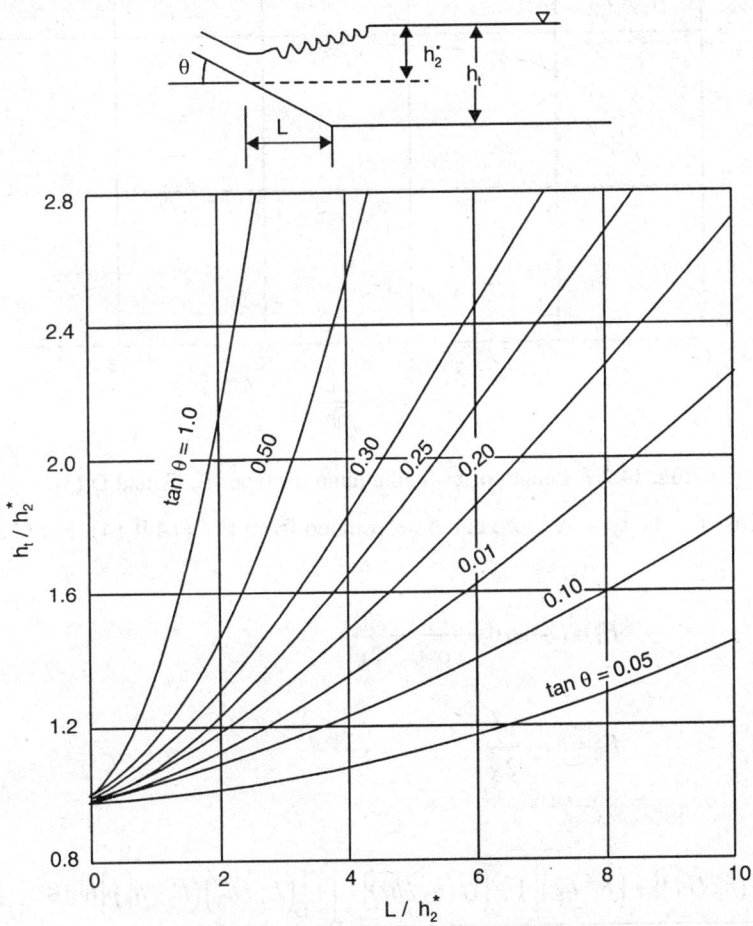

Fig. 14.9.6. Solution for B jump (12)

Bradley and Peterka (3) have developed plots for the estimation of the length of the type D jump (Fig. 14.9.7). These plots can also be used to estimate the lengths of the types B and C jumps.

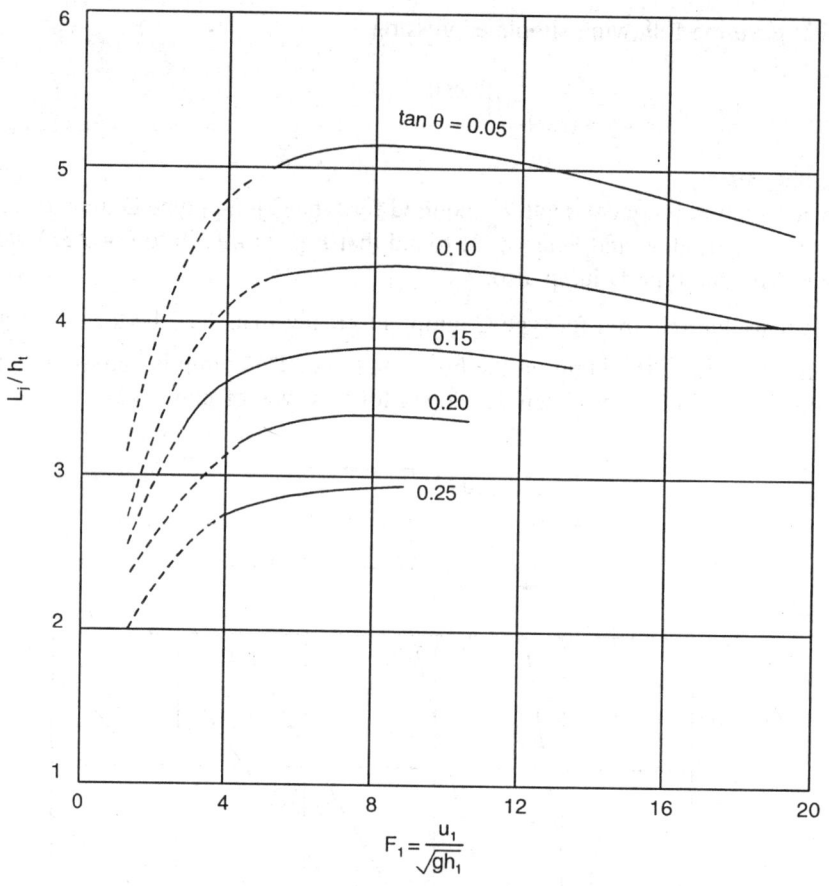

Fig. 14.9.7. Length of hydraulic jump for types B, C, and D (3)

The energy loss for the type A jump can be estimated from Eq. (14.9.14). For C and D jumps, one can write

$$E_1 = L_j \tan\theta + \frac{h_1}{\cos\theta} + \frac{u_1^2}{2g} \qquad \qquad ...(14.9.34)$$

and

$$E_2 = h_2 + \frac{u_2^2}{2g} \qquad \qquad ...(14.9.35)$$

Thus,

$$\frac{\Delta E}{E_1} = \frac{\left\{1-\left(h_2/h_1\right)\right\} + \left(F_1^2/2\right)\left[1-\left\{1/\left(h_2/h_1\right)\right\}^2\right] + \left[\left(L_j/h_2\right)\left(h_2/h_1\right)\right]\tan\theta}{1+\left(F_1^2/2\right)+\left(L_j/h_2\right)\left(h_2/h_1\right)\tan\theta} \qquad ...(14.9.36)$$

Here, the bed level at the end of the jump has been chosen as the datum and the potential energy term $h_1/\cos\theta$ has been approximated as h_1. Equation (14.9.36) should not be used if F_1 is less than 4 as in this range very little is known about L_j/h_2 that affects $\Delta E/E_1$. In order to solve a problem of hydraulic jump on a sloping channel, the first step is to determine the type of jump for the given slope, pre-jump supercritical depth, and tail-water condition. Figure 14.9.8 illustrates the procedure for the determination of the type of hydraulic jump.

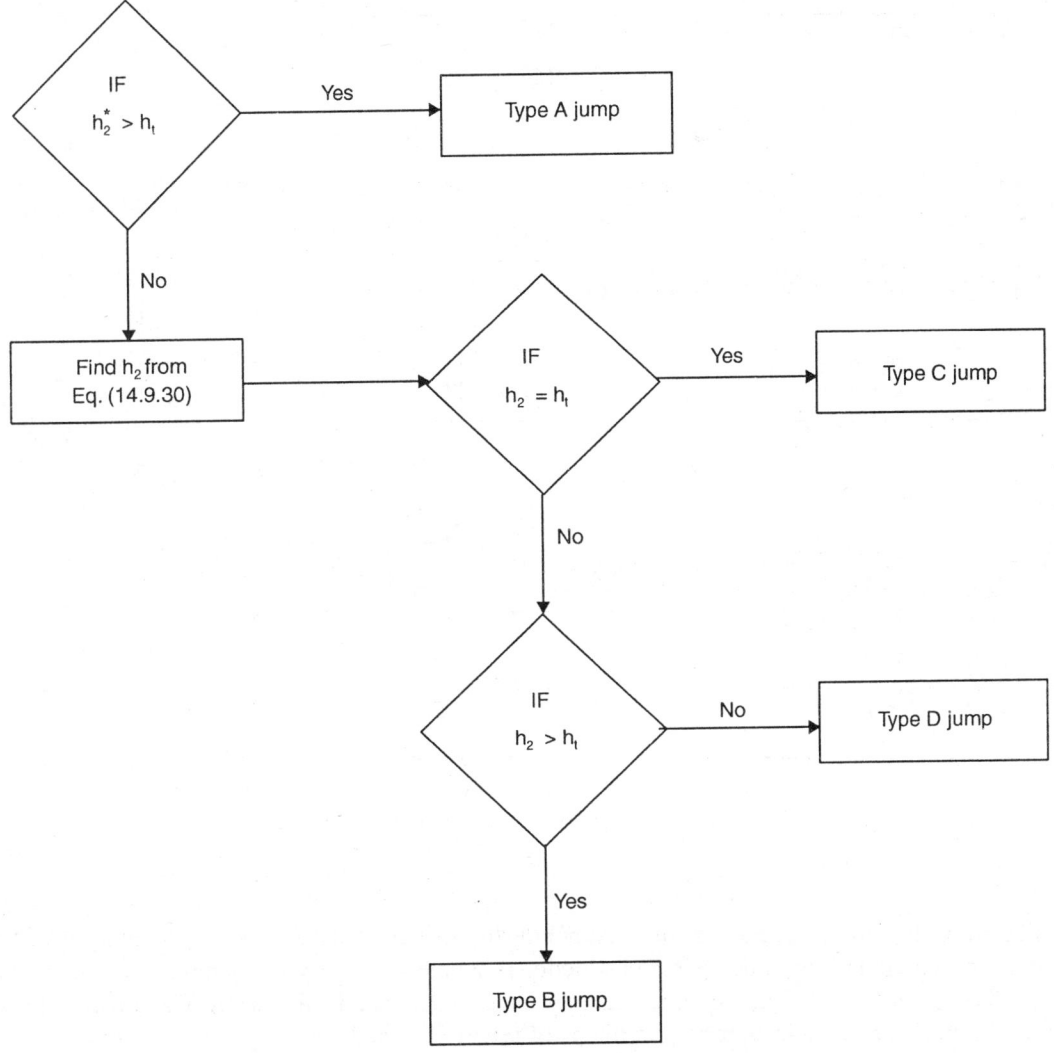

Fig. 14.9.8. Determination of type of hydraulic jump on sloping channels

14.9.7 Forced Hydraulic Jump

If the tail-water depth h_t is less than the required sequent depth h_2 corresponding to the pre-jump depth h_1, the jump is repelled downstream. However, by introducing devices such as baffle walls or baffle blocks and, thus, increasing the surface friction, the jump can be forced to form at the section it would have formed if sufficient tail-water depth $(=h_2)$ were available. Such a jump is called the forced hydraulic jump (Fig. 14.9.9).

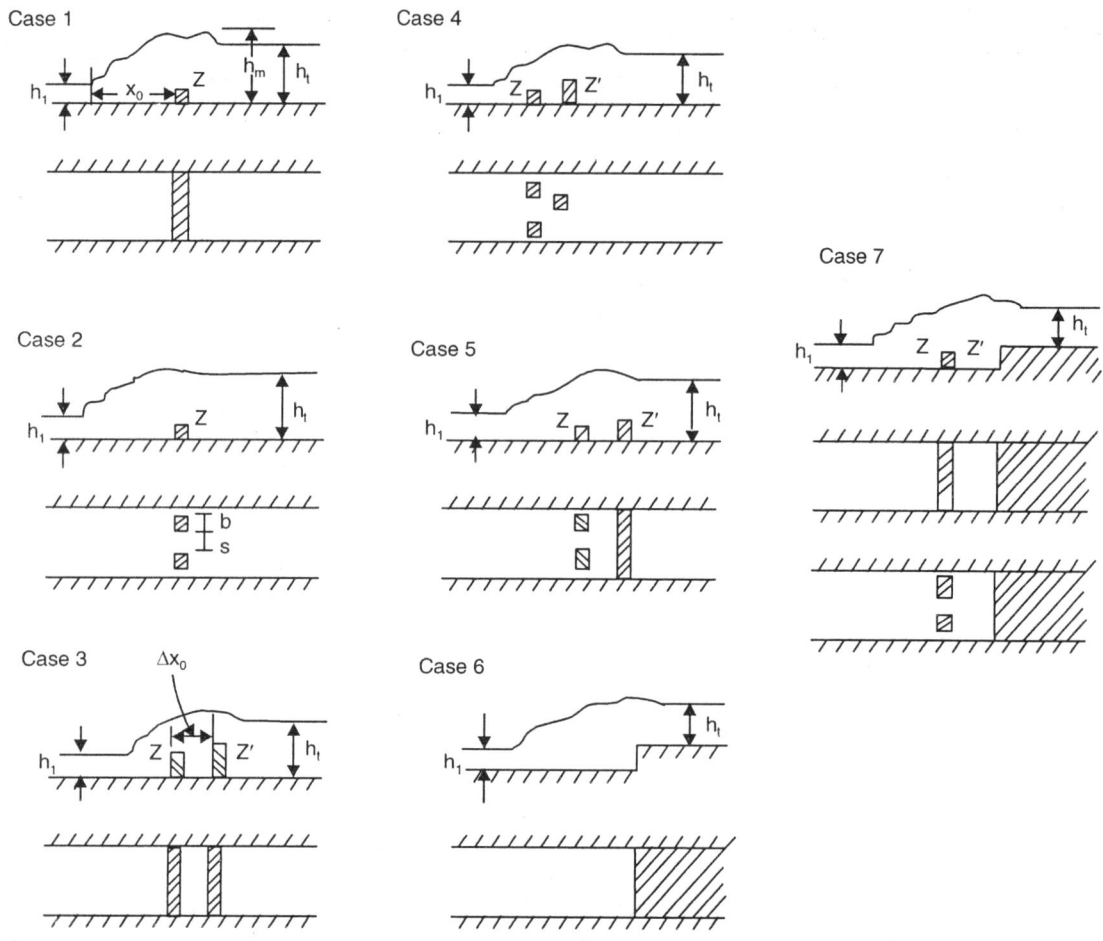

Fig. 14.9.9. Devices for forced hydraulic jump (11)

Figure 14.9.10 shows different types of forced hydraulic jump (11). For small Z and large x_0, type I jump is formed. This is similar to a normal jump. When Z is increased and x_0 is decreased, the baffle acts like an obstruction placed across the channel having free flow conditions and the jump is of type II*. With increase in tail-water depth, the obstruction is submerged and type II jump forms. On increasing Z and decreasing x_0 further, the jump becomes more violent and is of type III. The effect of further increase in Z and decrease in x_0 results in jumps of type IV, VI and VI*. The type VI* jump is a type VI jump with low tail-water depth. Between types IV and VI, there occurs an unstable transition phenomenon called type V (not shown in Fig. 14.9.10).

A large number of experimental studies have been conducted on the forced hydraulic jump as it forms the basic design element of the hydraulic jump type stilling basins. The simplest case of a forced hydraulic jump is the jump forced by a two-dimensional baffle, known as a baffle wall, of height Z kept at a distance x_0 from the toe (i.e., the beginning) of the jump (Fig. 14.9.9, case 1). Considering unit width of the channel, the momentum equation can be written as

$$\frac{1}{2}\rho g h_1^2 - \frac{1}{2}\rho g h_t^2 - F_b = \rho\left(\frac{q}{h_t} - \frac{q}{h_1}\right)$$

...(14.9.37)

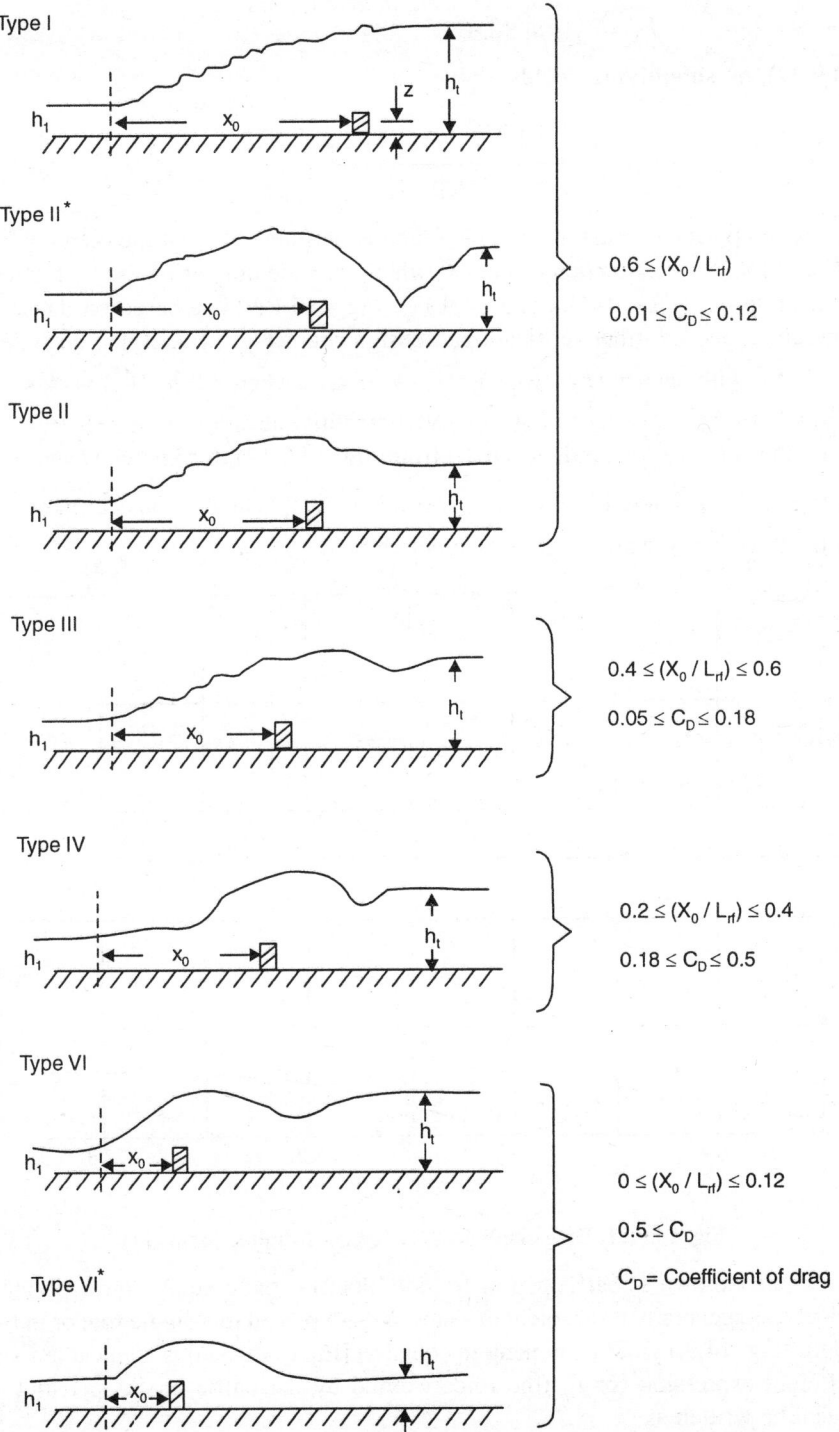

Fig. 14.9.10. Types of forced hydraulic jump (11)

where
$$F_b' = C_D \left(\rho u_1^2 / 2 \right) Z$$

Equation (14.9.37), on simplifying, yields

$$C_D = \frac{(\alpha - 1)\left[2F_1^2 - \alpha(1 + \alpha) \right]}{F_1^2 \, \alpha \beta}$$ ...(14.9.38)

where, $\alpha = h_t / h_1$ and $\beta = Z / h_1$. Rajaratnam (11) found experimentally that the drag coefficient C_D obtained from Eq. (14.9.38) is a function of only x_0 which is made dimensionless by the length of roller of the classical jump, L_{rj} (Figs. 14.9.2 and 14.9.11). Figure 14.9.11 is based on data from only one source and, therefore, needs further verification. A design chart (Fig. 14.9.12) has been developed (14), using Eq. (14.9.38), with $\psi (= h_t / h_2 = (h_t / h_1)(h_1 / h_2) = \alpha / \phi$ where $\phi \doteq h_2 / h_1$) versus F_1 for various values of βC_D. Choosing C_D equal to 0.4 (a very competitive design) or smaller (for a conservative design), obtain the value of $C_d \beta$ (and, hence, β) from Fig. 14.9.12), for known ψ and F_1. Now obtain x_0 / L_{rj} from Fig. 14.9.11. Using Fig. 14.9.2, L_{rj} (and, hence, x_0) can now be determined. From known β and h_1, the depth of baffle wall Z can also be determined.

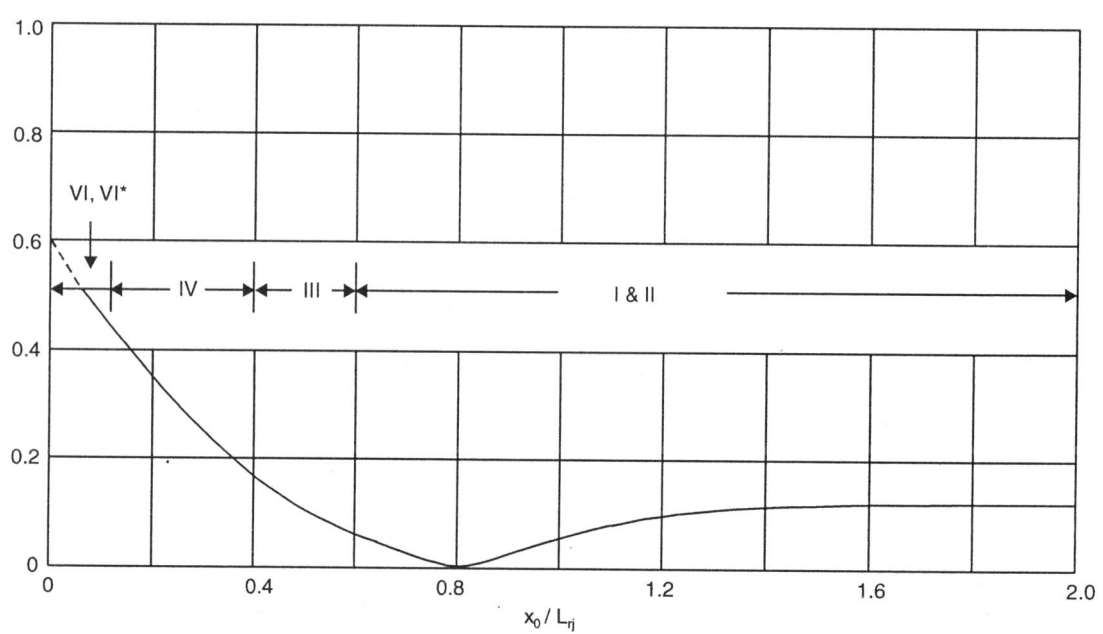

Fig. 14.9.11. Drag coefficient for forced hydraulic jump (11)

Baffle block (also known as baffle pier or friction block) is the case of a three-dimensional baffle wall. Baffle blocks are generally trapezoidal in shape and are placed in a single row or in two rows with staggered pattern (Fig. 14.9.13). The momentum equation [Eq. (14.9.37)] is applicable for this case too but with a different expression for F_b (the force exerted by the baffle blocks per unit width of the channel) that can be written as

$$\frac{F_b}{F_2} = f\left(\frac{x_o}{h_1}, \beta, F_1, \eta \right)$$ ...(14.9.39)

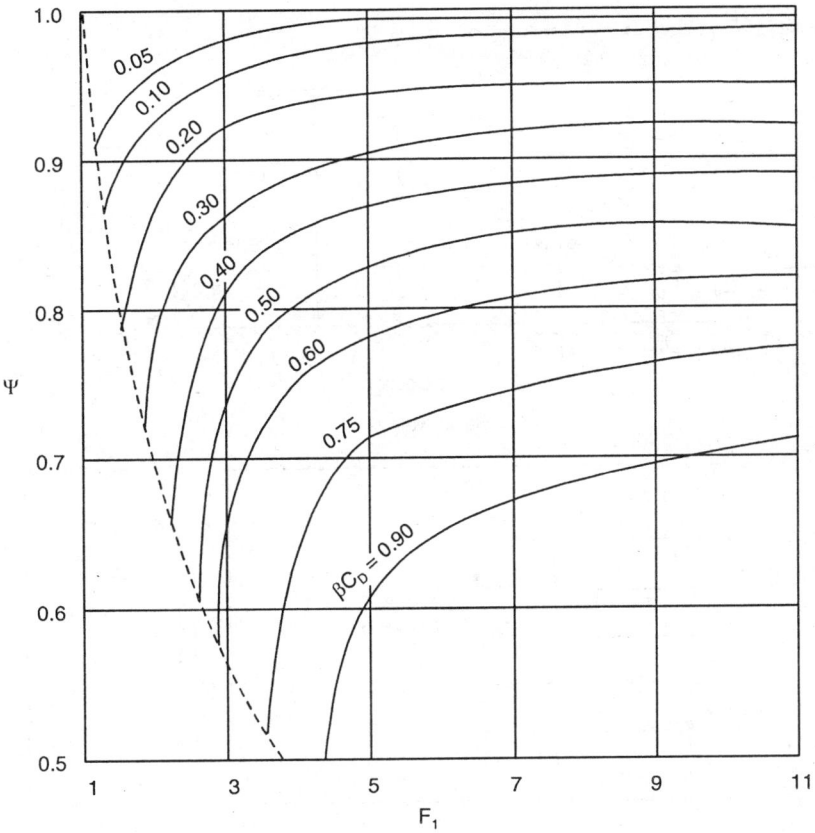

Fig. 14.9.12. Design chart for baffle wall (14)

Here $F_2 = \frac{1}{2}\rho g h_2^2$ and η (i.e., the blockage ratio) $= W/(W+S)$ (Fig. 14.9.13). Based on the analysis of data of Basco and Adams (2), Ranga Raju et al. (14) found that F_1 is unimportant in Eq. (14.9.39) and $\psi_1\psi_2 F_b/F_2$ is uniquely related to x_0/h_1 as shown in Fig. 14.9.14. ψ_1 and ψ_2 are empirical correction factors and are functions of Z/h_1 and η, respectively, as shown in Figs. 14.9.15 and 14.9.16. These data also indicated no change in the value of F_b when baffle blocks were placed in two rows for the range of r/Z from 2.5 to 5.0. Here, r is the spacing between the two rows of blocks.

14.9.8 Location of Hydraulic Jump on Glacis

It is generally assumed that the jump forms at the junction of the glacis with the horizontal floor and, therefore, the relations for classical hydraulic jump are used for determining the location of the jump on glacis. For known discharge intensity q, and the difference in the levels of the upstream and downstream levels of total energy line, ΔE, one can determine the downstream specific energy, E_2, using a suitable method or Blench's curves (Fig. 14.9.17). The location of the jump on a glacis can, then, be obtained by finding the intersection of the glacis with the horizontal plane E_2 below the downstream level of the total energy line.

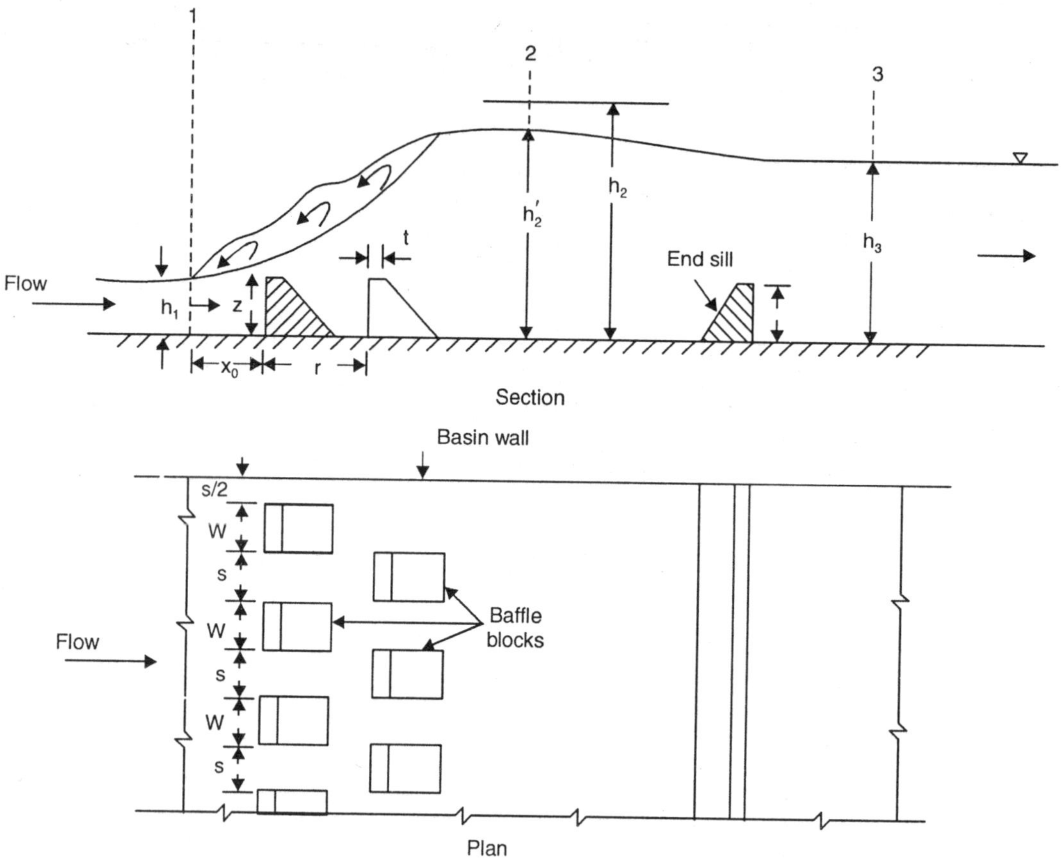

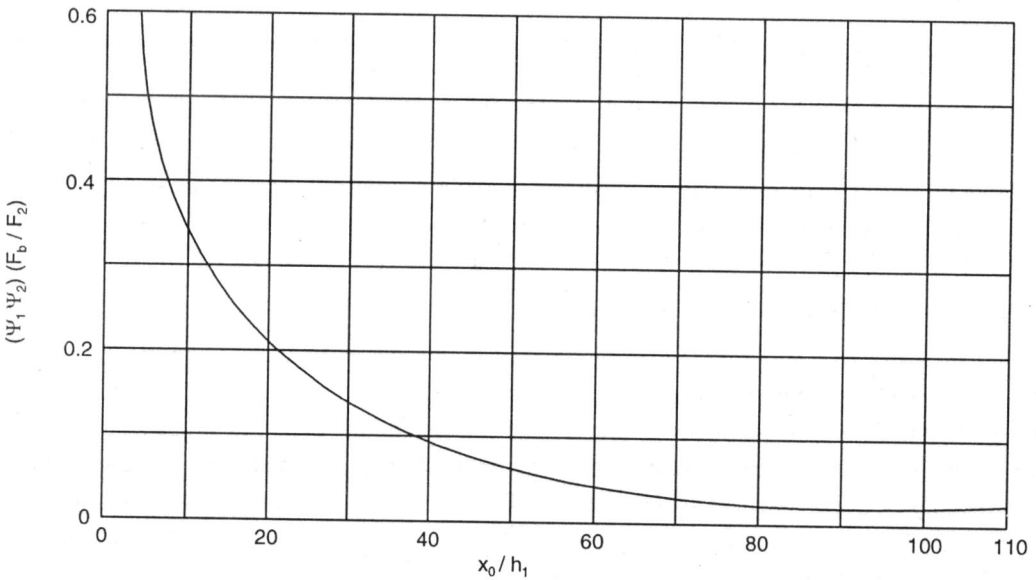

Fig. 14.9.13. Trapezoidal baffle blocks

Fig. 14.9.14. Variation of $\psi_1 \psi_2 F_b / F_2$ with x_0/h_1 for trapezoidal blocks (14)

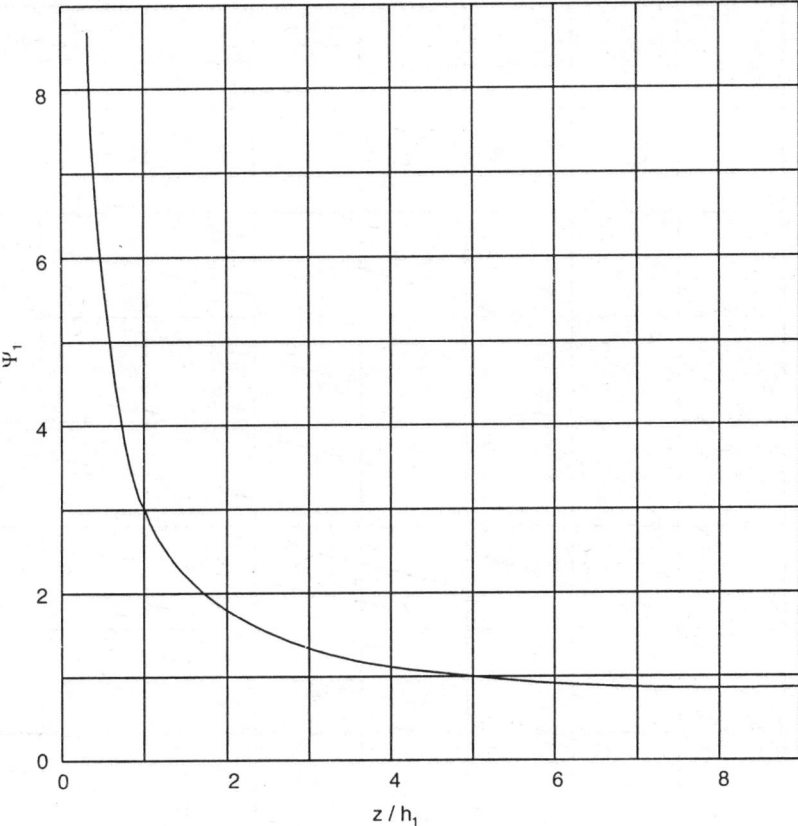

Fig. 14.9.15. Variation of ψ_1 with Z/h_1 (14)

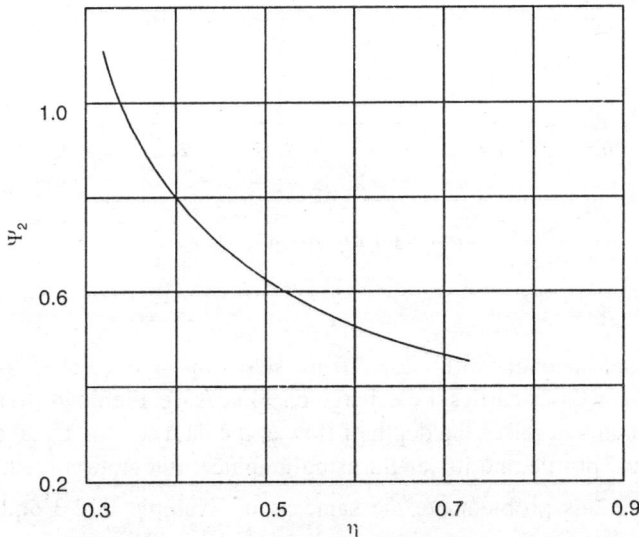

Fig. 14.9.16. Variation of ψ_2 with η (14)

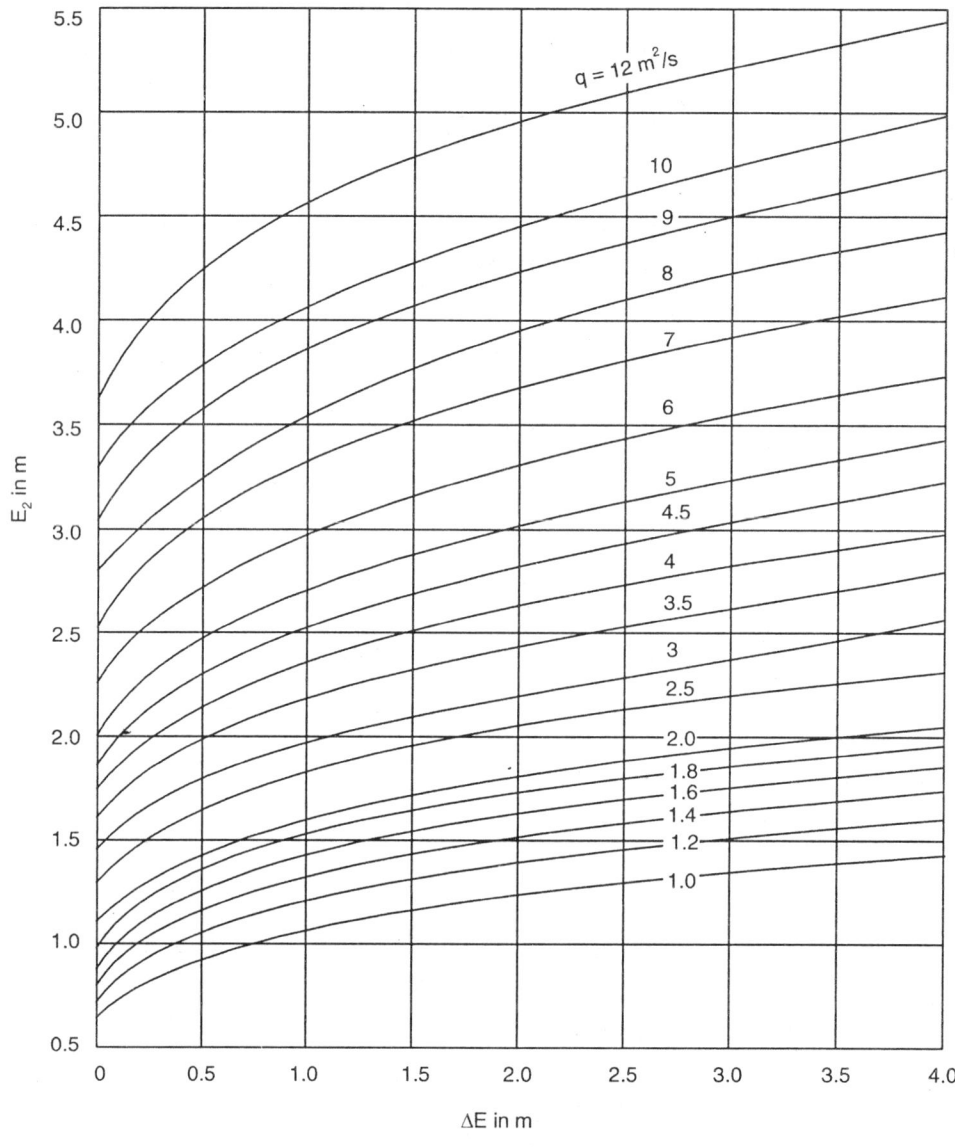

Fig. 14.9.17. Blench's curves

EXAMPLES

E14.1.   A trapezoidal channel (width $B = 10$ m, side slope $z = 2$, bed slope $= 0.0016$, and the Manning's $n = 0.02$) carries a discharge of 90 m³/s. At the downstream end of the channel is a small dam that raises the depth of flow at the dam section equal to 3.5 m. Compute the water surface profile and its length using graphical integration method.

Solution: The data for this problem are the same as in Example E12.3 and E13.1. Therefore, as obtained earlier for these problems,

$$h_n = 2.25 \text{ m}$$

and $h_c = 1.78$ m

Since $h_c < h_n$ ∴ $S_0 < S_c$ and, hence, the slope of the channel is mild. Also, since the depth of flow at the dam section is greater than the normal depth, the resulting water surface profile is a backwater profile (M1).

A backwater profile in a mild channel meets uniform depth line asymptotically. However, in practice, the upstream end of backwater profile is considered at that section of the channel where the depth of flow is 1% greater than the normal depth. Therefore, in this case, this depth of flow is 1.1 x 2.25 = 2.2725 m.

For the given problem, $Z_c = Q / \sqrt{g} = 90 / \sqrt{9.8} = 28.75$

$$K_n = Q / \sqrt{S_0} = 90 / \sqrt{0.0016} = 2250$$

For suitably selected values of h (between 3.5 m and 2.2725 m), required parameters (such as A, T, P, R, D, K and Z) are computed and then dx/dh is determined using Eq. (14.7.1). The computed values are tabulated in Table E14.1 prepared using 'Excel' spreadsheet).

Table E14.1: GRAPHICAL INTEGRATION METHOD

Depth	Area	Top width	Perimeter	Hyd. Radius	Hyd. Depth	Section Factors				
h	A	T	P	R	D	K	Z	dx/dh	Δx	X
3.5	59.50	24	25.65	2.32	2.48	5213	93.68	696		
3.4	57.12	23.6	25.21	2.27	2.42	4927	88.86	707	70	70
3.3	54.78	23.2	24.76	2.21	2.36	4651	84.18	721	71	141
3.2	52.48	22.8	24.31	2.16	2.30	4383	79.62	738	73	214
3.1	50.22	22.4	23.86	2.10	2.24	4124	75.20	760	75	289
3	48.00	22	23.42	2.05	2.18	3873	70.90	788	77	367
2.9	45.82	21.6	22.97	1.99	2.12	3630	66.74	826	81	447
2.8	43.68	21.2	22.52	1.94	2.06	3397	62.70	880	85	533
2.7	41.58	20.8	22.07	1.88	2.00	3171	58.79	958	92	625
2.6	39.52	20.4	21.63	1.83	1.94	2953	55.01	1083	102	727
2.5	37.50	20	21.18	1.77	1.88	2744	51.35	1309	120	846
2.4	35.52	19.6	20.73	1.71	1.81	2543	47.82	1839	157	1004
2.3	33.58	19.2	20.29	1.66	1.75	2349	44.41	4379	311	1314
2.2725	33.05	19.09	20.16	1.64	1.73	2298	43.49	8561	178	1492

The length of the backwater profile can be computed by either plotting the values of dx/dh (computed in Table E14.1) on y-axis for various values of h (between 3.5 m and 2.2725 m) on x-axis and, then, measuring the area between the curve, x-axis, and the ordinates at h = 3.5 m and 2.2725 m. Alternatively, the distance Δx between two sections having depth h_i and h_{i+1} can be obtained as

$$\Delta x = \frac{1}{2}\left[\left(\frac{dx}{dh}\right)_i + \left(\frac{dx}{dh}\right)_{i+1}\right]\left(h_i - h_{i+1}\right)$$

The values of Δx are shown in Col. 10 of Table E14.1. The cumulative sum of Δx is shown in Col. 11 of the same Table. The length of the backwater profile is, obviously, 1492 m.

E14.2. Using Chow's method of direct integration, compute the backwater profile for the trapezoidal channel of E14.1.

Solution: As for E14.1, $h_n = 2.25$ m, $h_c = 1.78$ m and the length of the backwater profile is the distance between sections at which the depths of flow are 3.5 m and 2.2725 m. Chow's equation, Eq. (14.7.9) also requires the knowledge of N, M, and J besides $u = h/h_n$ and $v = u^{N/J}$ and the corresponding values of the varied flow function.

For computation of hydraulic exponents N, Eq. (12.5.8), and M, Eq. (13.7.6), one can use the average of the depths of flow at the two ends of the backwater profiles, i.e., $(3.5+2.2725)/2 = 2.89$ m. Hence, h/B for use in Eqs. (12.5.8) and (13.7.6) is $2.89/10 = 0.289$. Accordingly,

$$N = \left(\frac{10}{3}\right)\frac{1+2(2)(0.289)}{1+2(0.289)} - \left(\frac{8}{3}\right)\frac{\sqrt{5}(0.289)}{1+2\sqrt{5}(0.289)}$$

$\therefore$ $\qquad$ $N = 3.8$

Similarly,

$$M = \frac{3\left[1+2(2)(0.289)\right]^2 - 2(2)(0.289)\left[1+(0.289)\right]}{\left[1+2(2)(0.289)\right]\left[1+2(0.289)\right]}$$

$\therefore$ $\qquad$ $M = 3.56$

The channel reach of the backwater profile has been divided into 04 subreaches. For these subreaches, computed values of u, v and the corresponding values of $F(u,N)$ and $F(v,J)$ as obtained from varied flow function table, Table 14.7.1, are listed in Table E14.2. The value of Δx is computed from (Eq. (14.7.9), i.e.,

$$\Delta x = \frac{h_n}{S_0}\left\{\left(u_1 - u_2\right) - \left[F\left(u_1,N\right) - F\left(u_2,N\right)\right] + \left(\frac{h_c}{h_n}\right)^M \left(\frac{J}{N}\right)\left[F\left(v_1,J\right) - \left(v_2,J\right)\right]\right\}$$

Table E14.2: Chow's integration method

h	u	v	$F(u,N)$	$F(v,J)$	Δx	L
3.5	1.556	1.730	0.112	0.180	0	0
3.2	1.422	1.547	0.151	0.235	216	216
2.9	1.289	1.370	0.213	0.326	229	445
2.6	1.156	1.197	0.336	0.484	282	727
2.2725	1.010	1.012	1.007	1.366	714	1441

Thus, the length of the backwater profile is 1441 m. Alternatively, the entire reach could be treated as single one and, then,

$$L = \frac{2.25}{0.0016}\left\{(1.556) - [0.112 - 1.007] + \left(\frac{1.78}{2.25}\right)^{3.56}\left(\frac{3.06}{3.8}\right)[0.180 - 1.3525]\right\}$$

$$= 1450 \text{ m}$$

E14.3. For the trapezoidal channel of E14.1, determine the depth of flow at a distance of 290 m upstream of the dam using numerical integration method.

Solution: At the control section (i.e., dam) the depth of flow h_i is 3.5 m. Using Eq. (14.7.2),

$$\dot{h}_i = 0.00144$$

Let the depth of flow at a section 290 m upstream of the dam be h_{i+1}. Assuming $\dot{h}_{i+1}$ to be the equal to $\dot{h}_i$, the first approximate value of h_{i+1} is given by Eq. (14.7.4), That is,

$$h_{i+1} = 3.5 + (0.00144)\,(-290) = 3.08 \text{ m}$$

Therefore, revised $\dot{h}_{i+1} = 0.00131$ and modified

$$h_{i+1} = 3.5 + \frac{1}{2}(0.00144 + 0.00131)(-290)$$

$$= 3.102$$

The computations are repeated as shown in Table E14.3 (prepared using 'Excel' spreadsheet). Thus, the required depth of flow works out to be 3.116 m.

Table E 14.3: Numerical Integration Method

Depth	Area	Top width	Perimeter	Hyd. Rad.		
h	A	T	P	R	dh/dx	h at 290 m
3.5	59.50	24	25.65	2.32	0.0014373	3.083
3.083	49.84	22.332	23.79	2.10	0.0013086	3.102
3.102	50.26	22.408	23.87	2.11	0.0013169	3.119
3.119	50.65	22.476	23.95	2.11	0.0013240	3.117
3.117	50.60	22.468	23.94	2.11	0.0013232	3.116
3.116	50.58	22.464	23.94	2.11	0.0013228	3.116

E14.4. Using the data of E14.1 and the direct step method obtain the distance of a section upstream of the dam such that the depth of flow at the section is 3.0 m.

Solution: For $h = 3.5$ m and other values of depth of flow (like, 3.4, 3.3, 3.2, 3.1 and 3.0 m), one can compute specific energy and also S_f (using the Manning's equation) as shown in the first six columns (Table E14.4).

Table E 14.4: Direct step method

Depth h	Area A	Velocity U	Sp. Energy E	Hyd. rad. R	Friction Slope Sf	Δx
3.5	59.50	1.513	3.617	0.42	0.00293	
						68.18
3.4	57.12	1.576	3.527	0.45	0.00291	
						68.46
3.3	54.78	1.643	3.438	0.48	0.00289	
						68.75
3.2	52.48	1.715	3.350	0.51	0.00287	
						67.98
3.1	50.22	1.792	3.264	0.55	0.00286	
						68
3	48.00	1.875	3.179	0.59	0.00284	

Total   341.37 m

Using Eq. (14.7.11) the values of Δx between adjacent sections can be determined. For example, the distance between the two sections where the depths of flow are 3.5 m and 3.4 m is

$$\Delta x = \frac{3.527 - 3.617}{0.0016 - \left[\frac{1}{2}(0.00293 + 0.00291)\right]}$$

$$= 68.18 \text{ m}$$

Likewise, the distances between adjacent sections of specified depths can be computed as shown in Table E14.4 (prepared using 'Excel' spreadsheet excepting the last column). Sum of these distances gives the desired result, i.e., 341.37 m.

One could also calculate the desired distance in one step in which case, using the relevant values from Table E14.4,

$$L = \frac{3.179 - 3.617}{0.0016 - \left[\frac{1}{2}(0.00293 + 0.00284)\right]}$$

Therefore, $L = 340.86$ m.

E14.5.   A discharge of 9.0 m³/s flows in a 6.0 m wide rectangular channel which is inclined at an angle of 3° with the horizontal. Determine the type of jump if $h_1 = 0.10$ m and $h_t = 2.6$ m.

Solution: $$F_1 = \frac{9.0/(6.0 \times 0.1)}{\sqrt{9.81 \times 0.10}} = 15.15$$

h_2^* (i.e., the sequent depth in a horizontal channel) can be calculated from

$$h_2^* = \frac{h_1}{2}\left[\sqrt{1+8F_1^2}-1\right] = \frac{0.1}{2}\left[\sqrt{1+8(15.15)^2}-1\right]$$

$$= 2.09 \text{ m}$$

Since $h_t > h_2^*$, the depth h_2 should be calculated from Eqs. (14.9.30)-(14.9.33).

$$G_1^2 = 10^{0.054(\theta)} F_1^2$$

$$= 10^{0.054(3)} \times (15.15)^2$$

$$= 333.29$$

$$\therefore \quad \frac{h_2}{(0.10/\cos 3^o)} = \frac{1}{2}\left[\sqrt{1+8G_1^2}-1\right]$$

or
$$h_2 = \frac{0.10}{2\cos(3^\circ)}\left[\sqrt{18(333.29)}-1\right] = 2.54 \text{ m}$$

since $h_2 < h_t$, the jump is classified as the type D jump. Using Fig. 14.9.7, the length of the jump, L_j, can be determined by obtaining L/h_t for tan 3° = 0.05 and F_1 = 15.15.

$$\frac{L_j}{h_t} = 4.9$$

$$L_j = 4.9 \times 2.6 = 12.74 \text{ m}$$

Now
$$\frac{h_t}{h_1} = \frac{2.6}{0.1} = 26$$

and
$$\frac{h_2}{h_1} = \frac{2.31}{0.1} = 23.1$$

Therefore, relative energy loss can be determined from Eq. (14.9.36) as follows:

$$\frac{\Delta E}{E_1} = \frac{(1-23.1)+\left[(15.15)^2/2\right]\left[1-(1/23.1)^2\right]+\tan(3^\circ)\left[4.9(26)\right]}{1+\left[(15.15)^2/2\right]+(4.9)(26)\tan(3^\circ)}$$

$$= 0.82$$

PROBLEMS

P14.1 A wide rectangular channel carrying 8 m³/s/m has a bed slope of 2.5 × 10⁻⁴ and the Manning's *n* equal to 0.013. If the channel ends in a free overfall, what type of surface profile would form? Assuming that the depth of flow at the overfall section of the channel is the critical depth, determine the distance upstream of the overfall where the depth of flow is 2.15 m using

    (a) graphical integration method   (c) direct integration method

    (b) numerical integration method   (d) step method

P14.2 A rectangular channel ($n = 0.014$) carries a unit discharge of 1.3 m³/s/m and ends in a free overall. Assuming that the depth of flow at the overfall section of the channel is the critical depth, determine the depth of flow at 300 m upstream of the overfall.

P14.3 A dam restricts the flow of a small river that can be approximated as a wide channel with bed slope and the Manning's n equal to 0.001 and 0.02, respectively. If the unit discharge in the river is 3 m³/s/m and the water depth immediately upstream of the dam is 5 m, determine the upstream distance to the location where the depth of flow is 3.0 m.

P14.4 The depth and velocity at the foot of a spillway are 0.34 m and 12.5 m/s respectively. The tail-water depth is 3.0 m. Assume the channel to be wide and the Manning's $n = 0.013$, determine the length of the horizontal concrete apron required to accommodate the hydraulic jump within the apron.

P14.5 A hydraulic jump occurs on a horizontal apron downstream of a wide spillway at a location where depth of flow is 0.8 m and speed is 24 m/s. Estimate the depth and flow speed downstream of the jump. Also, estimate the energy loss due to the jump.

P14.6 A hydraulic jump occurs in a 1 m wide rectangular channel. The flow rate is 7.0 m³/s and the pre-jump depth is 0.4 m. Determine the post-jump depth, length of the jump, and head loss due to the jump.

REFERENCES

(1) Asawa, GL: *"Irrigation and Water Resources Engineering"*, New Age International Publishers, New Delhi, 2005.

(2) Basco, DR and JR Adams: *"Drag forces on baffle blocks in hydraulic jumps"*, J. of Hydraulics Division, Proc. ASCE, Dec. 1971.

(3) Bradley, JN and AJ Peterka: *"The hydraulic design of stilling basins: hydraulic jumps on a horizontal apron (Basin I)"*, J. of Hydraulics Division, Proc. ASCE, Vol. 83, No. HY5, 1957, pp.1-24.

(4) Chow, VT: *"Integrating the equation of gradually varied flow"*, Proc. ASCE, Vol.81, Nov. 1955. [Chow, VT (5)]

(5) Chow, VT: *"Open-Channel Hydraulics"*, McGraw-Hill Book Co., New York, 1959.

(6) Crump, ES: *"A note on an approximate method of determining the position of a standing wave"*, Lahore, 1930 [Asawa, GL (1)].

(7) French, RH: *"Open-Channel Hydraulics"*, McGraw-Hill Book Co., Singapore, 1994.

(8) Henderson, FM: *"Open-Channel Flow"*, The Macmillan Company, New York, 1966.

(9) Kindsvater, CE: *"The hydraulic jump in sloping channels"*, Transactions of ASCE, Vol., 109, 1944.

(10) Prasad, R: *"Numerical method of computing flow profiles"*, Jour. of Hyd. Division, Proc. ASCE, Jan. 1970.

(11) Rajaratnam, N: *"The forced hydraulic jump'*, Water Power, Vol. 16, 1964.

(12) Rajaratnam, N: *"Hydraulic Jumps"*, Advances in Hydroscience, Vol. 4, Academic Press, New York, 1967.

(13) Rajaratnam, N and K Subramanaya: *"Profile of the hydraulic jump"*, J. of Hydraulics Division, Proc. ASCE, May 1968.

(14) Ranga Raju, KG, MK Mittal, MS Verma, and V Ganesan: *"Analysis of flow over baffle blocks and end Sills"*, J. of Hydraulic Research, Vol. 18, No.3, 1980.

(15) Silvester, R: *"Hydraulic jump in all shapes of horizontal channels,"* J. of Hydraulics Division, Proc. ASCE, Vol.90, No. HY1, Jan. 1965.

(16) Swamee, PK: *"Sequent depths in prismatic open channels"*, CBIP, J. of Irrigation and Power, Jan. 1970.

CHAPTER 15

Unsteady Non-uniform Flow in Channels

15.1 INTRODUCTION

Problems of the kind of the propagation of a flood wave (caused by excessive rainfall or failure of a dam) in a stream or movement of a tidal bore in estuaries involve variations of flow parameters with time as well as space and, hence, are few of the examples of many unsteady nonuniform flows in channels. To analyze such problems, one needs to use principles governing unsteady flow phenomenon.

15.2 CELERITY OF WAVES

Unsteady flow in a channel invariably implies some sort of wave motion. A *wave* (or surface wave for channel flows) is defined as a temporal variation in the water surface which is propagated through a fluid medium (3). Waves that are governed primarily by the compressibility of the fluid are termed *elastic waves* while those influenced by the surface tension are known as *capillary waves*. The waves in open channel flows are essentially governed by gravity (i.e., the fluid weight) and, hence, are known as *gravity waves*. Viscosity of the fluid opposes the deformation of the fluid besides dissipating the mechanical energy of the disturbance (causing the wave formation) through heat generation. *Tidal waves* in estuaries are caused by gravitational attraction of the moon. Waves (or gravity waves) in channels may be due to either natural phenomena such as due to floods or artificial factors such as sudden opening or closing of sluice gates or failure of a dam. It should be noted that a gravity wave can be propagated upstream in only subcritical flow. A gravity wave travels only downstream if the flow is supercritical.

Waves can also be classified as *oscillatory waves* (in which average mass transport is zero as in the case of sea waves) and *translatory waves* (in which there is net transport of fluid in the direction of wave travel as in the case of flood wave or moving. hydraulic jump). A *longitudinal wave* travels along the channel axis. A translatory wave having only a single peak, a rising limb, and a recession limb and that is followed and preceded by steady flow is termed a *solitary wave*. A *train of waves* (or wave train) is generated by several waves moving one after other.

A *positive wave* causes the water surface to be higher than that in the steady flow. If the water surface of the steady flow is lowered by the wave travel, the wave is termed *negative wave*.

Translatory wave motion is further classified as rapidly varying motion (e.g., surges and bores caused by sudden operation of gates or tidal changes etc.) or gradually varying motion (e.g., flood waves in rivers and lakes or the waves generated by gradual operation of gates etc.). In case of rapidly varying motion, the effect of channel friction is relatively insignificant compared to that of the acceleration due to large unsteadiness. However, for the gradually varying motion, the acceleration is negligible in comparison to the channel friction because of the longer length involved.

Waves in open channel flow are termed as *deep-water waves* when only the surface layers of the flow are disturbed by the wave motion. If the entire depth of flow is affected by the motion of a wave, the wave is termed *shallow-water wave*. If the ratio of the water depth to wave length exceeds about 0.5, the wave is considered as deep-water waves. For shallow-water waves, this ratio is less than 0.05 (4).

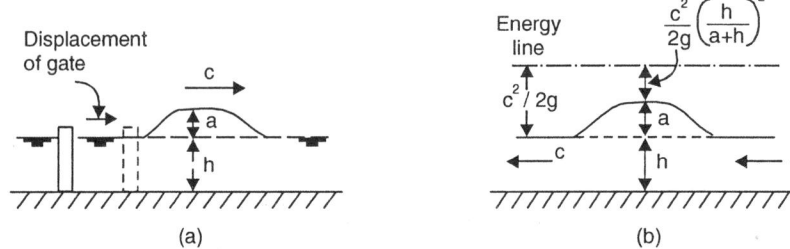

Fig. 15.2.1. Generation of a solitary wave (a) unsteady flow, (b) the flow that appears steady to an observer moving with the wave crest

Celerity of a wave is the speed of propagation of the wave relative to the fluid. Consider a solitary wave of height *a* above the normal water surface, Fig. 15.2.1, generated by a single movement of a paddle in a rectangular tank, travelling to the right with celerity *c*. Superimposing a velocity *c* in the direction opposite to that of wave travel makes the wave appear standstill while the flow moves to the left at a velocity equal to *c*. Neglecting viscous and surface tension effects and assuming the bed slope to be small and energy correction factor α to be unity at all sections, one can write the following energy equation between the normal flow section and the wave section (4):

$$h+\frac{c^2}{2g}=h+a+\frac{v^2}{2g}\qquad\qquad ...(15.2.1)$$

where v, the velocity at the wave section, is obtained from the continuity equation

$$v(h+a)=ch$$

$$\therefore\qquad\qquad v=c\left(\frac{h}{h+a}\right)$$

Hence, Eq. (15.2.1) reduces to

$$\frac{c^2}{2g}=a+\frac{c^2}{2g}\left(\frac{h}{h+a}\right)^2$$

or

$$a = \frac{c^2}{2g}\left[1 - \frac{h^2}{(h+a)^2}\right] = \frac{c^2}{2g}\left[\frac{2ah + a^2}{(h+a)^2}\right]$$

$\therefore$

$$c = \sqrt{\frac{2g(h+a)^2}{2h+a}} = \sqrt{\frac{2g\left(h^2 + a^2 + 2ah\right)}{2h+a}}$$

$$\approx \sqrt{\frac{2g\left(h^2 + 2ah\right)}{2h+a}} \qquad \text{if } a \ll h.$$

$$\approx \sqrt{gh}\ \sqrt{\frac{2h+4a}{2h+a}}$$

$$\approx \sqrt{gh}\ \sqrt{\left(1 + \frac{2a}{h}\right)\Big/\left(1 + \frac{a}{2h}\right)}$$

$$\approx \sqrt{gh}\ \sqrt{\left(1 + \frac{2a}{h}\right)\left(1 - \frac{a}{2h}\right)}$$

$\therefore$

$$c \approx \sqrt{gh}\ \sqrt{1 + \frac{3a}{2h}} \qquad \text{if } a \ll h \qquad\qquad \text{...(15.2.2)}$$

$\therefore$

$$c \approx \sqrt{gh} \qquad\qquad \text{if } a \ll h \qquad\qquad \text{...(15.2.3)}$$

Equations (15.2.2) and (15.2.3) are, respectively, known as the Saint-Venant celerity equation and Lagrange celerity equation in honour of their originators. For a non-rectangular channel, Eq. (15.2.3), reduces to

$$c = \sqrt{gD}$$

in which D is the hydraulic depth.

Based on the field and experimental data, a more suitable equation for the celerity of a solitary wave in a rectangular channel is (1)

$$c = \sqrt{g(h+a)} \qquad\qquad \text{...(15.2.4)}$$

A still more accurate equation for celerity of a solitary wave of small height in a rectangular channel is given as

$$c = \sqrt{\frac{g\lambda}{2\pi}(\tan h)\frac{2\pi h}{\lambda}} \qquad\qquad \text{...(15.2.5)}$$

in which λ is the wave length. This equation is generally known as the Airy celerity equation in honour of its originator. For deep-water waves (i.e., $h/\lambda > 0.5$ so that $(\tan h)\dfrac{2\pi h}{\lambda}$ tends to unity), Eq. (15.2.5) reduces to

$$c = \sqrt{\frac{g\lambda}{2\pi}} \qquad\qquad\qquad ...(15.2.6)$$

For shallow-water waves (i.e., $h/\lambda < 0.05$ and $(\tan h)\dfrac{2\pi h}{\lambda}$ tends to $2\pi h/\lambda$), Eq. (15.2.5) reduces to Eq. (15.2.3), i.e.,

$$c = \sqrt{gh}$$

15.3 UNSTEADY GRADUALLY VARIED FLOW

15.3.1 Continuity Equation

The continuity equation for unsteady flow in an open channel can be derived by considering the conservation of the mass between two channel sections that are infinitesimal distance apart, Fig. 15.3.1. Considering Q and h, respectively, to be the discharge and the depth of flow at the centre cc of the two sections at any time t, the net mass inflow into the infinitesimal fluid element between sections 1-1 and 2-2 (that are δx apart) during time δt is,

$$\rho\left\{\left(Q - \frac{\partial Q}{\partial x}\frac{\delta x}{2}\right) - \left(Q + \frac{\partial Q}{\partial x}\frac{\delta x}{2}\right)\right\}\delta t$$

$$= -\rho\frac{\partial Q}{\partial x}\delta x\,\delta t$$

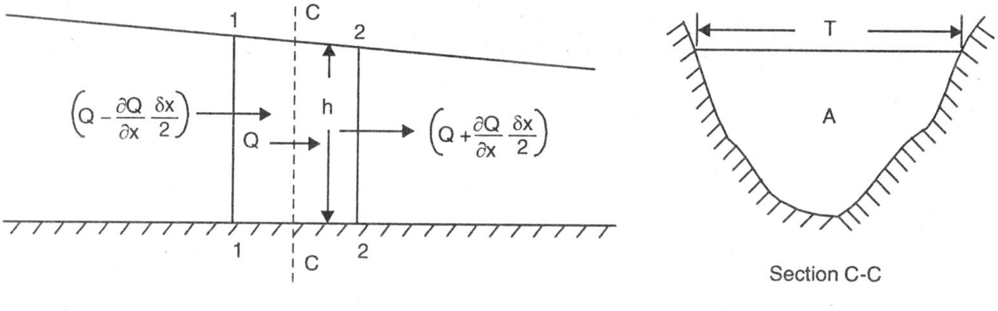

Fig. 15.3.1. Discharge flux through an element of a channel

The corresponding change in the mass of the fluid element in time δt is,

$$\frac{\partial}{\partial t}\left(\rho A\,\delta x\right)\delta t$$

in which A is the area of flow cross-section at cc at time t. According to the law of conservation of mass, the net mass inflow into the element must be equal to the corresponding change in the mass of the element. Hence, for incompressible fluid,

$$\rho \frac{\partial Q}{\partial x} \delta x \, \delta t + \frac{\partial}{\partial t}(\rho A \delta x) \delta t = 0$$

$\therefore$

$$\frac{\partial Q}{\partial x} + \frac{\partial A}{\partial t} = 0 \qquad \qquad ...(15.3.1)$$

In terms of the average velocity of flow U, Eq. (15.3.1) can be written as

$$\frac{\partial (UA)}{\partial x} + \frac{\partial A}{\partial t} = 0 \qquad \qquad ...(15.3.2)$$

or

$$A \frac{\partial U}{\partial x} + U \frac{\partial A}{\partial x} + \frac{\partial A}{\partial t} = 0 \qquad \qquad ...(15.3.3)$$

For a rectangular channel, $A = Bh$ and, hence, Eq.(15.3.3) reduces to

$$h \frac{\partial U}{\partial x} + U \frac{\partial h}{\partial x} + \frac{\partial h}{\partial t} = 0 \qquad \qquad ...(15.3.4)$$

If there is addition or withdrawal of water in the lateral direction at a rate of q_x per unit length, Eq.(15.3.1) gets modified to

$$\frac{\partial Q}{\partial x} + \frac{\partial A}{\partial t} = \pm q_x \qquad \qquad ...(15.3.5)$$

Obviously, plus sign for q_x is to be used when there is addition (as in side channel spillway) and negative sign for q_x is to be used for withdrawal (as in case of an open channel with bottom rack).

15.3.2 Energy Equation

Due to unsteadyness in a flow, an additional variable for the time element is to be considered along with other variables of the corresponding steady flow. This additional variable takes into account the change in velocity with time. This results into local acceleration which, in turn, produces a force and causes additional energy loss in the flow. This additional force for two-dimensional flow is, obviously, mass times acceleration and, hence, equals $(\rho A dx)\dfrac{\partial U}{\partial t}$ for the fluid between two sections dx apart and acts in the x-direction, Fig. 15.3.2. Assuming the slope of the channel to be small, the vertical component of the acceleration is negligible. The energy loss due to this force equals the work done by this force through a distance dx between the two channel sections and is, therefore, $(\rho A dx)\dfrac{\partial U}{\partial t}dx$. The energy loss per unit weight is, therefore, $\dfrac{1}{g}\dfrac{\partial U}{\partial t}dx$ and is, therefore, head loss due to local acceleration.

Using the energy principles between sections 1 and 2, one may write,

$$z + h + \alpha \frac{U^2}{2g} = (z + dz) + (h + dh) + \left[\alpha \frac{U^2}{2g} + d\left(\alpha \frac{U^2}{2g} \right) \right] + \frac{1}{g} \frac{\partial U}{\partial t} dx + S_f dx \qquad ...(15.3.6)$$

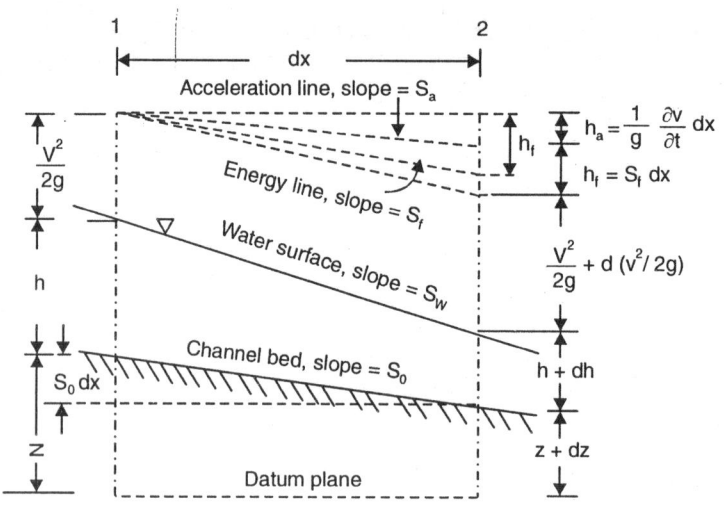

Fig. 15.3.2. Energy line in unsteady flow in a channel

Here, $S_f dx$ is the head loss due to friction. On simplification, Eq. (15.3.6) reduces to

$$d\left(z + h + \alpha \frac{U^2}{2g}\right) = -S_f dx - \frac{1}{g}\frac{\partial U}{\partial t}dx \qquad ...(15.3.7)$$

Therefore, the change in total head, $d\left(z + h + \alpha \dfrac{U^2}{2g}\right)$ in unsteady gradually varied flow depends on

loss due to friction, $S_f dx$ and the effect of acceleration, represented by $(1/g)(\partial U/\partial t)dx$, provided that other losses (such as form or transition losses) are negligible. On dividing Eq. (15.3.7) by dx, and using partial differentials, one gets

$$\frac{\partial(z + h)}{\partial x} + \frac{\partial}{\partial x}\left(\alpha \frac{U^2}{2g}\right) + S_f + \frac{1}{g}\frac{\partial U}{\partial t} = 0 \qquad ...(15.3.8)$$

or

$$\frac{\partial h}{\partial x} + \frac{\alpha U}{g}\frac{\partial U}{\partial x} + \frac{1}{g}\frac{\partial U}{\partial t} + \frac{\partial z}{\partial x} + S_f = 0 \qquad ...(15.3.9)$$

For prismatic channels, $-\partial z/\partial x = S_0$ and, hence, Eq. (15.3.9) reduces to

$$\frac{\partial h}{\partial x} + \frac{\alpha U}{g}\frac{\partial U}{\partial x} + \frac{1}{g}\frac{\partial U}{\partial t} = S_0 - S_f \qquad ...(15.3.10)$$

If the energy correction factor is further assumed unity, Eq. (15.3.10) reduces to

$$\frac{\partial h}{\partial x} + \frac{U}{g}\frac{\partial U}{\partial x} + \frac{1}{g}\frac{\partial U}{\partial t} = S_0 - S_f \qquad ...(15.3.11)$$

Equations (15.3.8 – 15.3.11) are known as Saint-Venant's dynamic equations for unsteady gradually varied flow. The friction slope in these equations can be estimated using the Manning's equation or the Chezy's equation or any other suitable uniform flow formula. The validity of these equations has been established by several field observations and laboratory experiments (1). However, the exact integration of the dynamic equation and continuity equation for the unsteady gradually varied flow is not possible. Solution of these equations is, therefore, obtained by methods based on simplifying assumptions.

15.4 METHOD OF CHARACTERISTICS

The method of characteristics is a semi-graphical method by which explicit solutions, if they exist, are readily obtained, and by which numerical solutions can be worked out in the more general cases where no explicit solutions are possible (2). This method reduces the problem of solving two partial differential equations into one of solving four ordinary differential equations.

Consider a simple case of unsteady gradually varied flow in a prismatic rectangular channel (i.e., constant width and constant bed slope). If c is the speed (i.e., celerity) of a long low wave in water of depth h, then h can be eliminated from Eqs. (15.3.4) and (15.3.11) by substituting for h from Eq. (15.2.3), i.e.,

$$c^2 = gh$$

so that c becomes the measure of the depth of flow h, and $d(c^2) = 2c\,dc = g\,dh$. On multiplying Eq. (15.3.11) by g, one gets

$$g\frac{\partial h}{\partial x} + U\frac{\partial U}{\partial x} + \frac{\partial U}{\partial t} = g\left(S_0 - S_f\right)$$

or

$$2c\frac{\partial c}{\partial x} + U\frac{\partial U}{\partial x} + \frac{\partial U}{\partial t} = g\left(S_0 - S_f\right) \qquad \qquad ...(15.4.1)$$

Similarly, on multiplying Eq. (15.3.4) by g, one gets

$$gh\frac{\partial U}{\partial x} + Ug\frac{\partial h}{\partial x} + g\frac{\partial h}{\partial t} = 0$$

$\therefore$

$$c^2\frac{\partial U}{\partial x} + 2cU\frac{\partial c}{\partial x} + 2c\frac{\partial c}{\partial t} = 0 \qquad (\text{as } gdh = 2cdc)$$

or

$$2U\frac{\partial c}{\partial x} + c\frac{\partial U}{\partial x} + 2\frac{\partial c}{\partial t} = 0 \qquad \qquad ...(15.4.2)$$

By adding Eqs. (15.4.1) and (15.4.2), one obtains

$$\frac{\partial U}{\partial t} + \left(U + c\right)\frac{\partial U}{\partial x} + 2\frac{\partial c}{\partial t} + 2\left(U + c\right)\frac{\partial c}{\partial x} = g\left(S_0 - S_f\right) \qquad ...(15.4.3)$$

Similarly, on substracting Eq. (15.4.2) from Eq. (15.4.1), one gets

$$\frac{\partial U}{\partial t} + \left(U - c\right)\frac{\partial U}{\partial x} - 2\frac{\partial c}{\partial t} - 2\left(U - c\right)\frac{\partial c}{\partial x} = g\left(S_0 - S_f\right) \qquad ...(15.4.4)$$

Equations (15.4.3) and (15.4.4) can be rewritten as follows:

$$\frac{\partial}{\partial t}(U+2c)+(U+c)\frac{\partial(U+2c)}{\partial x}=g\left(S_0-S_f\right) \qquad \text{...(15.4.5)}$$

and

$$\frac{\partial}{\partial t}(U-2c)+(U-c)\frac{\partial(U-2c)}{\partial x}=g\left(S_0-S_f\right) \qquad \text{...(15.4.6)}$$

If one considers a variable, h, dependent on x and t, i.e., $h(x, t)$, one can write the basic equation of partial differentiation as

$$dh=\frac{\partial h}{\partial t}dt+\frac{\partial h}{\partial x}dx$$

or

$$\frac{dh}{dt}=\frac{\partial h}{\partial t}+\frac{\partial h}{\partial x}\frac{dx}{dt} \qquad \text{...(15.4.7)}$$

Equation (15.4.7) gives rate of change of h if x and t are simultaneously varied in some specified manner, i.e., dx/dt. If the variables h, x, and t are assigned the usual meanings in open channel flow (i.e., depth of flow, distance along the flow direction, and time, respectively), one may interpret Eq. (15.4.7) in this manner: to an observer walking along a channel with any speed dx/dt, h would appear to vary with time at the rate given by Eq. (15.4.7). Therefore, the left side of Eq. (15.4.5) represents the rate of change of $(U+2c)$ from the viewpoint of an observer moving with velocity $(U+c)$. Similarly, the left side of Eq. (15.4.6) represents the rate of change of $(U-2c)$ from the viewpoint of an observer moving with velocity $(U-c)$. In terms of total derivative operator, Eqs. (15.4.5) and (15.4.6) can, alternatively, be written as

$$\frac{D_1(U+2c)}{D_1 t}=g(S_0-S_f) \qquad \text{...(15.4.8)}$$

$$\frac{D_2(U-2c)}{D_2 t}=g(S_0-S_f) \qquad \text{...(15.4.9)}$$

The importance of these discussions is that the paths of the two imaginary observers – one moving with velocity $(U+c)$ and other with velocity $(U-c)$ – can be traced on the x-t plane and a complete solution is obtained for any specified unsteady gradually varied flow condition. Explicit solution would, however, be obtained in the simplest cases. In other cases, numerical method would have to be used.

In many unsteady flow situations, the flow conditions change so rapidly that the acceleration term $\frac{1}{g}\frac{\partial U}{\partial t}$ in the Saint-Venant's equation is very large compared with S_o and S_f. An example is the release of water from a lock into a navigation canal. For such simple flow situation of horizontal channel bed (i.e., $S_0=0$) without frictional resistance (i.e., $S_f=0$), Eqs. (15.4.5) and (15.4.6) reduce to

$$\frac{\partial(U+2c)}{\partial t}+(U+c)\frac{\partial(U+2c)}{\partial x}=\frac{D_1(U+2c)}{D_1 t}=0 \qquad \text{...(15.4.10)}$$

with
$$\frac{dx}{dt} = U + c \qquad \qquad ...(15.4.11)$$

and
$$\frac{\partial(U - 2c)}{\partial t} + (U - c)\frac{\partial(U - 2c)}{\partial x} = \frac{D_2(U - 2c)}{D_2 t} \qquad \qquad ...(15.4.12)$$

with
$$\frac{dx}{dt} = U - c \qquad \qquad ...(15.4.13)$$

To an observer moving with velocity $(U \pm c)$, in accordance with Eqs. (15.4.10 – 15.4.13), the quantities $(U \pm 2c)$ would appear to remain constant as the total derivative of $(U + 2c)$ is zero vide Eqs. (15.4.10) and (15.4.12). The path of these observers would coincide with lines, on the x-t plane,

having $\frac{dx}{dt} = U \pm c$, Fig. 15.4.1. Thus, one would obtain two sets of lines that are called *characteristics*.

Along each line of the first set for which $\frac{dx}{dt} = U + c$, the inverse slope of the line is $(U+c)$, and $(U+2c)$,

known as positive Riemann invariant, is a constant. Similarly, along each member of the second set for

which $\frac{dx}{dt} = U - c$, the inverse slope of the line is $(U-c)$, and $(U-2c)$, known as negative Riemann

invariant, is a constant. The two sets of curves are, therefore, contours of $(U+2c)$ and $(U-2c)$. The first set is designated as C_1-family having positive characteristic direction while the second set is C_2-family having negative characteristic direction.

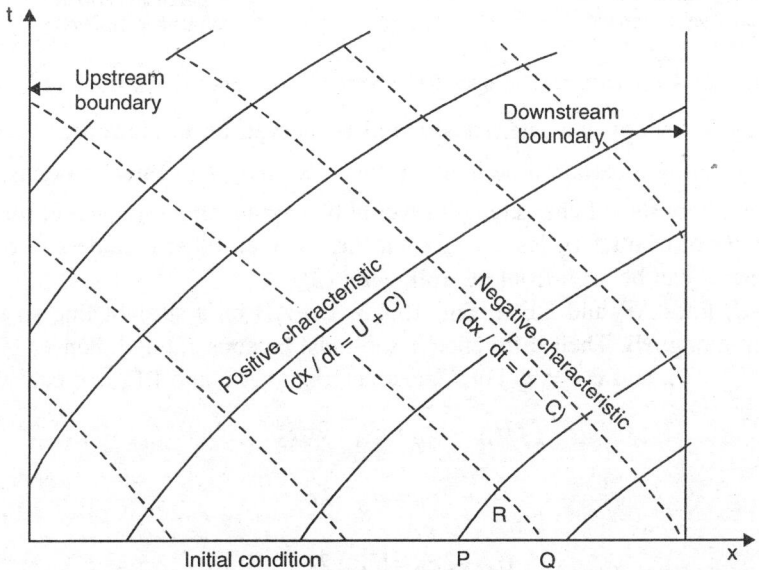

Fig. 15.4.1. Characteristic lines in the *x-t* plane (4)

For subcritical flow (in both positive or negative x-direction), $U < c$ and, therefore, C_1–lines would have positive slope but C_2- lines would have negative slope, Fig. 15.4.2. For supercritical flow $(U > c)$, however, both characteristics would have either positive slope (if the flow is in the positive x-direction) or negative slope (if the flow is in the negative x-direction), Fig. 15.4.2.

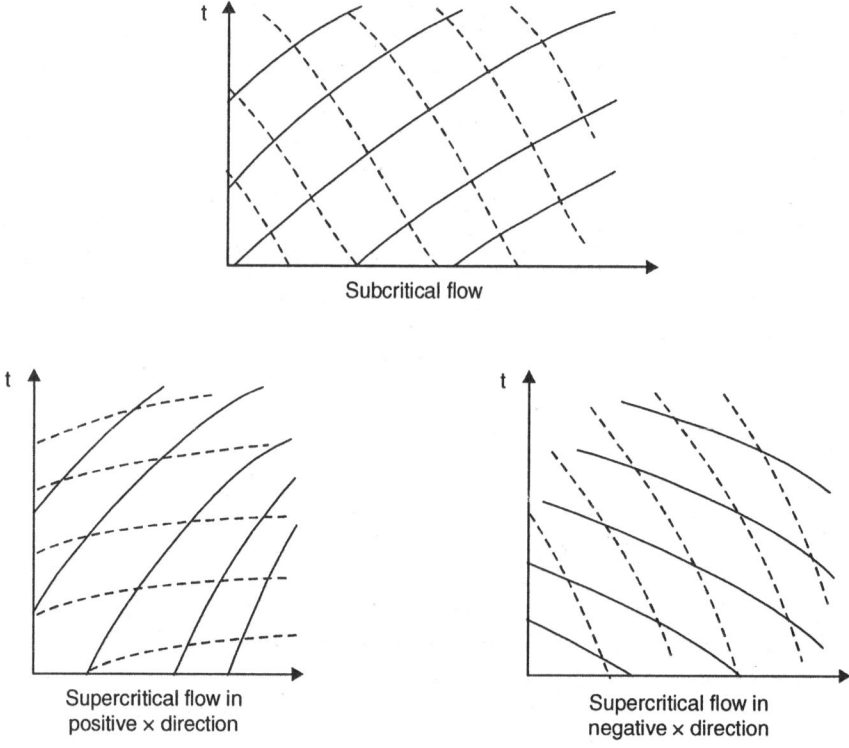

Subcritical flow

Supercritical flow in positive × direction

Supercritical flow in negative × direction

Fig. 15.4.2. Characteristic lines for subcritical and supercritical flows (4)

In general, the two sets of characteristics would be curved lines. However, when U and c are constant in a region during a certain time interval, then $dx/dt = (U \pm c)$ are also constant within this region and, therefore, both sets of characteristics would be parallel straight lines. Further, if any curve of the C_1 or C_2 family of characteristics is a straight line, then all other members of the same family are also straight line as can be seen from the following (2):

Consider two C_1-lines AB and DE in Fig. 15.4.3. Let AB be a straight line so that $(U+c)$ and $(U+2c)$ are constant along AB. Their difference, c too must be constant and, hence, U also. It, therefore, follows that $U_A = U_B$ and $c_A = c_B$. For C_2-characteristics AD and BE, one can write

$$U_D - 2c_D = U_A - 2c_A$$

$$U_E - 2c_E = U_B - 2c_B$$

$\therefore$
$$U_D - 2c_D = U_E - 2c_E \qquad ...(15.4.14)$$

Since DE is a C_1-characteristic,

$$U_D + 2c_D = U_E + 2c_E \qquad ...(15.4.15)$$

Obviously, Eqs. (15.4.14) and (15.4.15) can be satisfied only if $U_D = U_E$ and $c_D = c_E$ i.e., if AB is a straight line.

A simple problem of wave propagation (2) may be taken to illustrate the use of the method of characteristics. Consider a channel having uniform flow with $U = U_o$ and $c = c_o$. If one draws a straight line OF, of constant inverse slope equal to $U_o + c_o$, the line OF represents C_1-characteristic, Fig. 15.4.3. Since this characteristic is a straight line, it follows that all other members of the family too would be straight lines. At the upstream end (x = 0) of the channel, a disturbance (in the form of a prescribed variation of $U(t)$ and/or $c(t)$) is introduced at time t = 0. Therefore, line OF divides the undisturbed flow (i.e., zone of quiet) below OF from the disturbed flow above OF. If one can now calculate the values of U and c appropriate to every C_1-characteristic, one can obtain U and c at every point on the x-t plane and, thus, get the complete solution. For this, consider any point G on t-axis, Fig. 15.4.3. Obviously, the C_1-characteristic passing through this point G will have an inverse slope equal to

$$\frac{dx}{dt} = U(t) + c(t) \qquad \qquad ...(15.4.16)$$

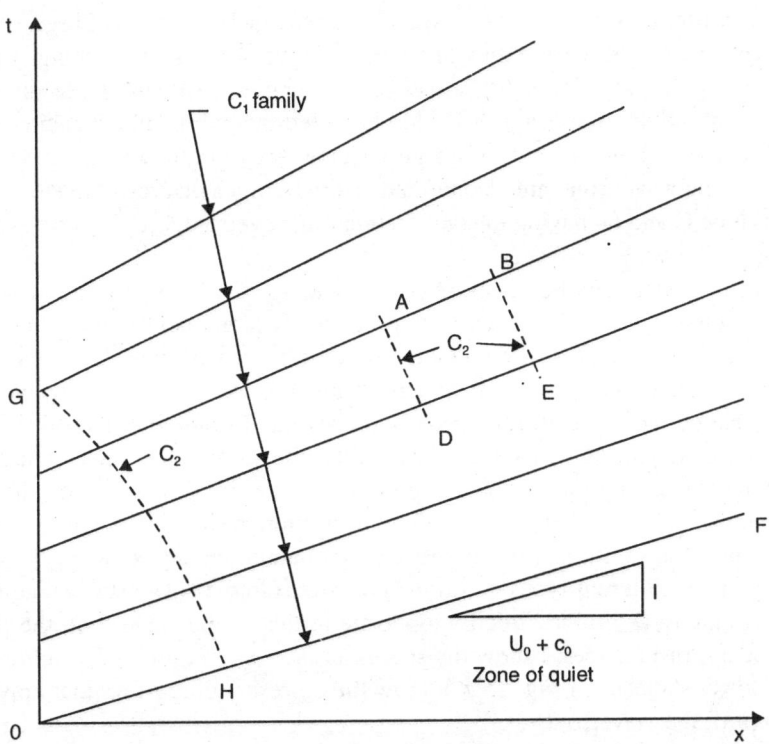

Fig. 15.4.3. Characteristic curves in the *x-t* plane (2)

If the curved line GH is assumed to be C_2-characteristic, then,

$$U(t) - 2c(t) = U_0 - 2c_0 \qquad \qquad ...(15.4.17)$$

Therefore, one may note that *U(t)* and *c(t)* are not independent and, hence, only one of these can be prescribed and the other, if required, can be determined from Eq. (15.4.17).

Combining Eqs. (15.4.16) and (15.4.17), one obtains

$$\frac{dx}{dt} = \frac{3}{2}U(t) - \frac{1}{2}U_0 + c_0 \qquad \qquad ...(15.4.18)$$

and also

$$\frac{dx}{dt} = 3c(t) + U_0 - 2c_0 \qquad \qquad ...(15.4.19)$$

Since $U(t)$ or $c(t)$ is known, dx/dt of C_1-characteristic passing through G (i.e., at any chosen time t after the disturbance was introduced) can be determined and, thereafter, the required information can be obtained.

15.5 UNSTEADY RAPIDLY VARIED FLOW

Surges (caused by sudden opening or closing of gates or blockage in a channel by, say, landslide) and bores (i.e., surges of tidal origin) are associated with an abrupt discontinuity at the water surface and, therefore, fall under the category of rapidly varied flow. A *surge* is a sort of moving wave front in a channel that causes an abrupt change in the depth of flow in the channel. Theoretically, there are four types of surge (or unsteady rapidly varied flow) as shown in Fig. 15.5.1. Surge of type A has an advancing wavefront moving downstream while surge of type B has an advancing wavefront moving upstream. Surge of type C has a retreating wavefront moving downstream while surge of type D has a retreating wavefront moving upstream. It should be noted that while an advancing wavefront has a stable wave profile, a retreating wavefront has un unstable wave profile (1). Surges of type A and B, having stable advancing wave front, are also termed positive surges that result in elevation of the water surface, surges of type C and D, having unstable retreating wavefront, are termed negative surges that result in lowering of the water surface.

The wavefront of a surge can be assumed to be made up of a large number of very small waves placed one over the other. The upper waves, having relatively larger depth, move faster in accordance with Eq. (15.2.3). The upper waves of an advancing wave front will, therefore, overtake the bottom waves in the forward direction of their movement. Thus, the waves have tendency to combine and eventually form a single large wavefront that is both steep and stable. On the other hand, the upper waves of a retreating wavefront will retreat faster than the lower ones and, hence, the wavefront becomes sloping and eventually flattens out. Due to the channel friction, however, the wave profile of an advancing wavefront actually changes. The effect of friction in short artificial smooth channels may not be noticeable, but it is significant in long reaches of natural channels having large friction.

Figure 15.5.1 shows different types of surges that are examples of unsteady rapidly varied flow. By superposing a velocity U_w in the direction opposite to that of the wavefront, the flow is rendered steady. Upper sketches of Fig. 15.5.1 show the surges as seen by an observer standing on the bank of the channel. The lower sketches of Fig. 15.5.1 show the corresponding flows that appear steady to an observer moving with the wavefront.

Considering surge type B, Fig. 15.5.1, occurring in a rectangular channel, one can write the continuity equation and momentum equation (neglecting boundary friction) as follows:

$$\left(U_1 + U_w\right)h_1 = \left(U_2 + U_w\right)h_2 \qquad \qquad ...(15.5.1)$$

and

$$\frac{1}{2}\rho g h_1^2 - \frac{1}{2}\rho g h_2^2 = \left[\rho h_1 \left(U_1 + U_w\right)\right]\left[\left(U_2 + U_w\right) - \left(U_1 + U_w\right)\right]$$

or

$$\frac{1}{2}\rho g h_1^2 - \frac{1}{2}\rho g h_2^2 = \left[\rho h_1 \left(U_1 + U_w\right)\right]\left(U_2 - U_1\right) \qquad \qquad ...(15.5.2)$$

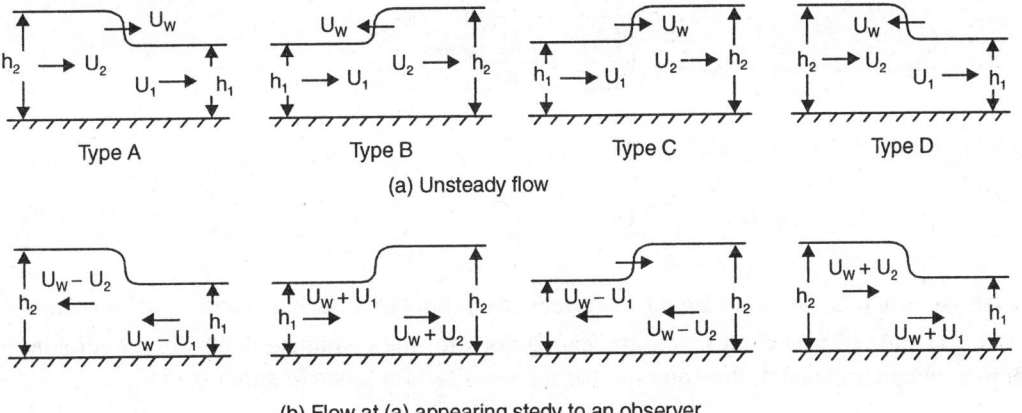

Type A   Type B   Type C   Type D

(a) Unsteady flow

(b) Flow at (a) appearing stedy to an observer
moving with the wave front

Fig. 15.5.1. Rapidly varied uniformly progressing flows (1)

The celerity of the surge c is the speed of propagation of the wavefront with respect to the fluid, i.e.,

$$c = U_1 + U_w \qquad\qquad ...(15.5.3)$$

From Eq. (15.5.1)

$$U_2 = U_1 \left(h_1 / h_2 \right) - U_w \left(1 - \frac{h_1}{h_2} \right)$$

Substituting the value of U_2 in Eq. (15.5.2), one gets

$$\frac{1}{2} g \left(h_1^2 - h_2^2 \right) = h_1 \left(U_1 + U_w \right) \left[\left\{ U_1 (h_1 / h_2) - U_w \left(1 - \frac{h_1}{h_2} \right) \right\} - U_1 \right]$$

$$= h_1 c \left[\frac{h_1}{h_2} (U_1 + U_w) - (U_1 + U_w) \right]$$

$$= h_1 c^2 \left[\frac{h_1}{h_2} - 1 \right]$$

$$= \frac{h_1}{h_2} c^2 \left(h_1 - h_2 \right)$$

Hence,

$$\frac{1}{2} g \left(h_1 + h_2 \right) = \frac{h_1}{h_2} c^2$$

$\therefore$

$$c^2 = \frac{1}{2} g h_1 \left(1 + \frac{h_2}{h_1} \right) \left(\frac{h_2}{h_1} \right)$$

or
$$c=\sqrt{gh_1}\left[\sqrt{\frac{1}{2}\frac{h_2}{h_1}\left(1+\frac{h_2}{h_1}\right)}\right] \qquad \text{...(15.5.4)}$$

Combining Eqs. (15.5.1) and (15.5.3), one gets

$$U_2=c\frac{h_1}{h_2}-U_w \qquad \text{...(15.5.5)}$$

For small (h_2-h_1), i.e., $h_1 \approx h_2$, Eq. (15.5.4) reduces to Eq. (15.2.3). For standing wave (i.e., $U_w = 0$), Eqs. (15.5.4) and (15.5.5) would yield the expressions for the conjugate depths in a hydraulic jump. For non-rectangular channel, the equation for the celerity of a positive surge is (5)

$$c=\sqrt{g\overline{z_1}}\left\{\frac{(\overline{z_2}A_2/\overline{z_1}A_1)-1}{1-(A_1/A_2)}\right\}^{1/2} \qquad \text{...(15.5.6)}$$

in which $\overline{z}$ is the distance of the centroid of the flow cross-section from the water surface.

EXAMPLES

E15.1.   A rectangular channel having uniform flow (depth = 1.5 m and velocity = 0.9 m/s) meets a large estuary having, initially, the same level as that in the channel. At some time, the estuary level falls at the rate of 0.3 m/hr for 3 hours. Neglecting the bed slope and resistance, determine how long it would take for the channel level to fall by 0.6 m at a section 1 km upstream of the junction of the channel and estuary? At this time, how far upstream will the channel level just be starting to fall?

Solution: Consider x to be positive upstream and origin at the junction of the channel and estuary. Hence,

$$U_0=-0.9 \text{ m/s}$$

and
$$c_0=\sqrt{gh_0} = \sqrt{9.80\times1.5}=3.834\,\text{m/s}$$

$\therefore$
$$U_0+c_0=-0.9+3.834=2.934\,\text{m/s}$$

Figure E15.1 shows C_1-line $\left(\dfrac{dx}{dt}=2.934\right)$. The objective is to find time t at which depth of flow would be 0.9 m (i.e., 0.6 m lower than initial depth of 1.5 m) at $x = 1000$ m. Let H represent the point (1000, t) at which $c(t)$ would be $\sqrt{9.8\times0.9}$ =2.97 m/s. The characteristic through H should originate at G ($x = 0$) at which point too $c = 2.97$ m/s and, hence, the depth $h = 0.9$ m. Therefore, time t at G would be 2 hours (as at this time the water level in the channel/estuary would fall by 0.6 m) or 7200 seconds.

Inverse slope of GH = $\dfrac{dx}{dt}=3c(t)+U_0-2c_0$

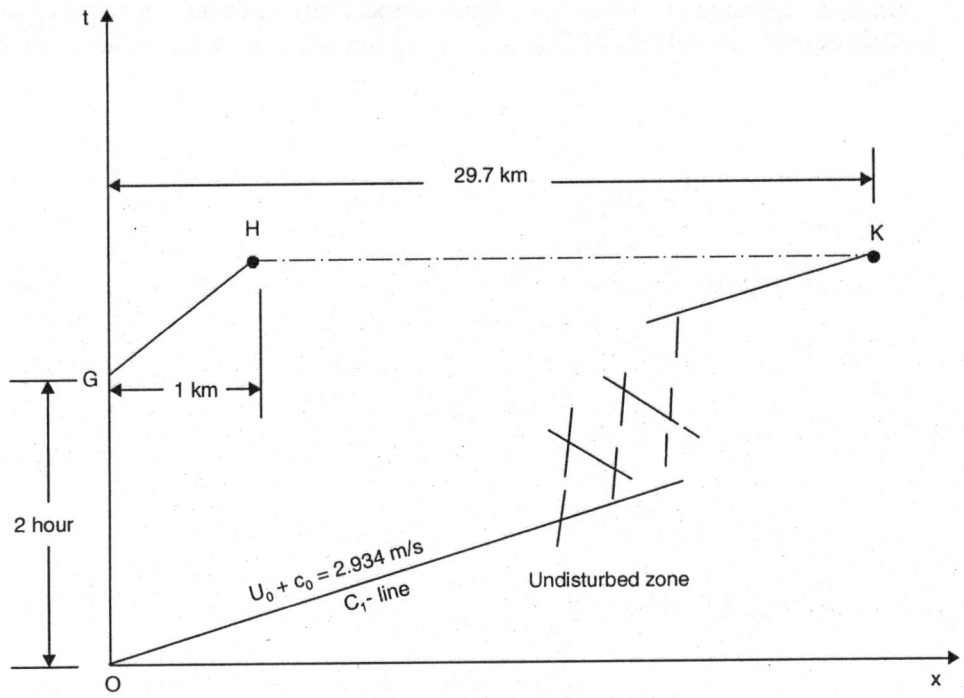

Fig. E 15.1

$$= 3\sqrt{9.8 \times 0.9} - 0.9 - 2(3.834)$$

$$= 0.342 \text{ m/s}$$

Hence, time interval between G and H $= \dfrac{1000}{0.342}$

$$= 2923.98 \text{ s}$$

$$= 2924 \text{ s (say)}$$

Therefore, total time elapsed since the water level started falling

$$= 7200 + 2924 = 10124 \text{ s}$$

$$= 2 \text{ hrs } 48 \text{ minutes } 44 \text{ s}$$

The x-t region below OK represents the undisturbed zone (i.e., zone of quiet) and the line OK represents the boundary of this zone. The distance to K from O (i.e., the junction) at time equal to 10124 s is $\left(\dfrac{dx}{dt}\right)_{OK}$ $(10124) = 3.834 \times 10124 = 29.70$ kms. This means that at $t = 10124$ s (since the water level started falling at the junction) the disturbance (i.e., falling of water level) would have reached upto only 29.70 kms. That is, at 29.70 km upstream of the junction, the water level would start falling at $t = 10124$ s.

E15.2.   The tidal surge in an estuary is observed to be moving upstream at a velocity of 5.0 m/s. The depth and velocity of flow under the steady uniform flow conditions prevailing prior to the arrival of the surge are 3.0 m and 1.0 m/s, respectively. Estimate the height of the surge. Also, estimate the velocity of flow in the upstream direction at any section after the surge has passed that section.

Solution: Referring to type B surge of Fig. 15.5.1,

$$h_1 = 3.0 \text{ m}$$

$$c = U_1 + U_w = 1.0 + 5.0 = 6.0 \text{ m/s}$$

Using Eq. (15.5.4), $c^2 = (gh_1)\left[\frac{1}{2}\left(\frac{h_2}{h_1}\right)\left(1 + \frac{h_2}{h_1}\right)\right]$

or $\qquad 36 = (9.8 \times 3.0)\left[0.5\left(\frac{h_2}{3.0}\right)\left(1 + \frac{h_2}{3.0}\right)\right]$

$\therefore \qquad h_2\left(1 + \frac{h_2}{3.0}\right) = 7.347$

or $\quad h_2^2 + 3h_2 - 22.041 = 0$

$\therefore \qquad h_2 = \dfrac{-3 \pm \sqrt{9 + 88.164}}{2}$

$\therefore \qquad h_2 = 3.429 \text{ m}$

$\therefore$ Height of the surge = 3.429 − 3.0 = 0.429 m

Further, from Eq. (15.5.5),

$$U_2 = c\frac{h_1}{h_2} - U_w$$

$$= (6)\frac{3.0}{3.429} - 5.0$$

$$= 5.249 - 5.0$$

i.e., $\qquad U_2 = 0.249 \text{ m/s}$

PROBLEMS

P15.1 Water flows in a frictionless horizontal rectangular channel with depth of flow equal to 2.4 m and velocity equal to 1.4 m/s. The channel joins a lake, the water level of which is, initially, the same as that of the channel. The water level of the lake suddenly starts falling resulting in an increase in the velocity at the junction of the channel with the lake at the rate of 0.5 m/s per hour for a period of three hours. How long would it take for the velocity in the channel at a distance of 1.5 km from the junction to increase to 2.4 m/s?

P15.2 If in P15.1, the depth of flow in the channel at the junction with the lake starts falling at the rate of 0.25 m per hour, determine how long would it take for the water level in the channel to fall by 0.4 m at a section 3 km from the junction. How far upstream of the junction has the water level started falling at this time?

P15.3 A tidal bore of 1.8 m height travels upstream in an estuary in which the depth and velocity of flow under steady uniform condition are 2.5 m and 0.8 m/s, respectively. Determine the velocity of the tidal bore and also the net unit discharge at a section in the estuary after the bore has passed the section.

P15.4 A tidal bore is observed travelling upstream with a speed of 5.0 m/s in an estuary having, prior to the arrival of the bore, depth and velocity equal to 3.0 m and 1.2 m/s. Estimate the height of the bore.

P15.5 Water flows at a velocity of 2.0 m/s and depth of 2.5 m in a rectangular channel the downstream gate of which is suddenly closed. Determine the height and speed of the surge.

REFERENCES

(1) Chow, VT: "*Open Channel Hydraulics*", McGraw-Hill Book Co., 1959.

(2) Henderson, FM: "*Open Channel Flow*", The Macmillan Company, New York, 1966.

(3) Mahmood, K and Yeyzevich, V: "*Unsteady Flow in Open Channels, Vols. I, II, and III*", Water Resources Publications, Fort Collins, Colorado, U.S.A., 1975 [Ranga Raju(4)].

(4) Ranga Raju, KG: "*Flow through Open Channels*", Tata McGraw-Hill Book Co., New Delhi. 1993.

(5) Rouse, H: "*Engineering Hydraulics*", John Wiley, New York, 1950 (Chapter-11) [Ranga Raju (4)].

CHAPTER 16

Controls and Flow Measurement in Channels

16.1 GENERAL

A channel control has been defined as any channel feature - natural or artificial - that fixes a relationship between depth of flow in its vicinity and the channel discharge. Weirs, overfalls, spillways, and sluices are typical examples of artificial controls while a rock outcrop in a river may act as natural control if it results in a unique depth-discharge relation at all stages of the river.

Local changes in flow cross-section cause variation from one uniform state to another uniform state. Such changes may be caused by channel bends, expansions, and contractions which are typical examples of channel transitions. The flow in the vicinity of a channel transition is necessarily nonuniform as is the flow in the vicinity of a channel control. This means that the channel control is a transition structure. But, a transition is not necessarily a control under all conditions. However, a common feature of transitions and control is that the flow in the vicinity of these is generally rapidly varied flow.

Accurate flow measurement is required for proper regulation, distribution, and charging of irrigation water and its importance cannot be overemphasized. There are several flow measuring devices that are used for flow measurement in irrigation systems. Weirs, flumes, and orifices are generally used for this purpose. Besides, there are some other indirect methods by which the discharge is computed using velocity measurements. In these methods, the channel section is divided into a suitable number of compartments and the mean velocity of flow for each of these compartments is measured by using devices such as current meter, surface floats, double floats, velocity rods, and so on. The discharge through any compartment is obtained by multiplying the mean velocity of flow in the compartment with the area of cross-section of the compartment. The sum of all compartmental discharges yields the channel discharge.

16.2 FREE OVERFALL

A vertical drop at the end of a channel is termed free overall, Fig. 16.2.1. The drop can be considered to be an example of a very steep slope and, hence, the lowermost streamline of the flow separates from the channel bed at the sharp corner resulting in the flow pattern shown in Fig. 16.2.1. A fall can also be considered as a special case of sharp-crested weir (dealt with in Art. 16.3) with $W = 0$. It was shown in Art. 14.3 that the depth of flow should be critical at the junction of a mild and steep slope. Even otherwise, the overfall section is the section of the minimum energy in the channel and, hence, theoretically, the depth at the free overall, usually called *end depth* or *brink depth* h_b, in a mild channel having subcritical flow should be the critical depth. Likewise, the depth of flow at the junction of a steep and steeper slope should, theoretically, be equal to the normal depth of flow on steep slope. That is, the brink depth at an overfall should be normal depth if the flow approaching the overfall is supercritical. However, in reality, the brink depth is different than the theoretically expected depth. This is due to the violation of the assumption of the hydrostatic pressure variation adopted in developing the relations for critical or normal depth. At the overfall, the pressure distribution is essentially non-hydrostatic with a mean pressure considerably less than hydrostatic owing to the pressure of strong vertical components of acceleration due to the curved streamlines and flow separation in the neighbourhood, Fig. 16.2.1. However, at some section A, some short distance upstream of the brink, the vertical acceleration would be small and the pressure distribution there would be hydrostatic (2).

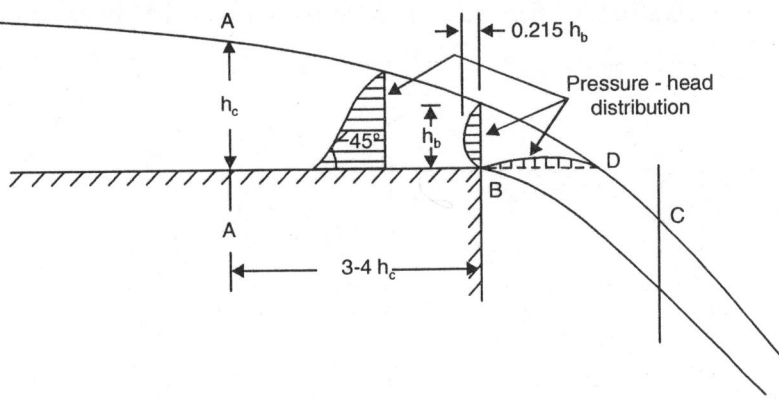

Fig. 16.2.1. Free overfall (2)

If slope of the channel upstream of the overfall is steep, the flow at A will be supercritical and, therefore, determined by the upstream conditions and the overfall section, is not the controlling section. If, on the other hand, the channel slope is mild or horizontal or adverse, the flow at A will be critical.

Experimental evidence indicates that the section A (with depth of flow equal to h_c) in mild and horizontal channels is about 3 to 4 times h_c upstream of the overfall. If one neglects the boundary friction between sections A and B that are only a short distance apart and assumes constant velocity over the flow cross-section, the application of momentum principle for the control volume between A and B yields

$$P_A - P_B = \rho q \left(U_B - U_A \right) \qquad \qquad ...(16.2.1)$$

The pressure force P_B can be assumed to be zero as the nappe (i.e., the water jet at B) is exposed to atmospheric pressure. As the depth of flow at A is the critical depth and pressure distribution there is hydrostatic, Eq. (16.2.1) reduces to

$$\frac{\rho g h_c^2}{2} = \rho q^2 \left(\frac{1}{h_b} - \frac{1}{h_c} \right)$$

or

$$\frac{gh_c^2}{2} = \left(gh_c^3\right)\left(\frac{h_c - h_b}{h_b h_c}\right)$$

$\therefore$

$$h_b = 2\left(h_c - h_B\right)$$

or

$$h_b = \frac{2}{3}h_c \qquad\qquad\qquad ...(16.2.2)$$

Experimental and semi-theoretical studies by Rouse (11) have indicated that the value of h_b/h_c for horizontal channels is about 0.715.

16.3 WEIRS

Weirs (Art. 5.7) are commonly used for measurement of discharge in an open channel. A *weir* is an obstruction placed across a channel over which flow occurs, Fig. 16.3.1. Weir is a control section and, therefore, has a unique depth-discharge relation and, hence, is a suitable discharge measuring device. Weirs can be broadly classified as sharp-crested (or thin-plate) weirs and broad-crested weirs.

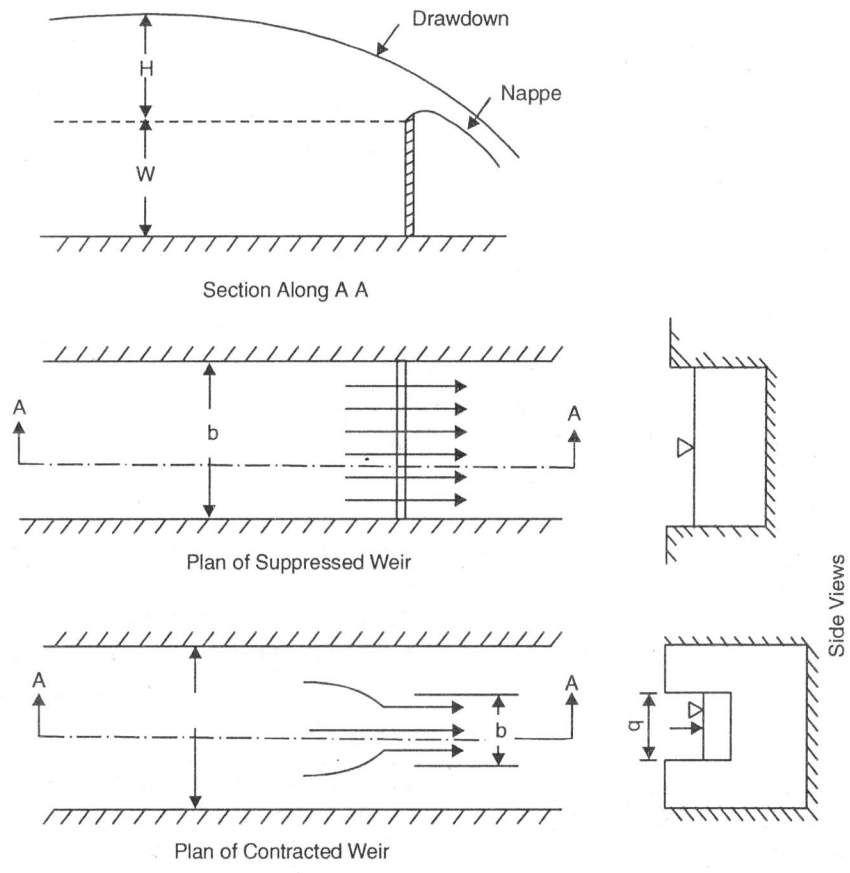

Fig. 16.3.1. Suppressed and contracted weirs (6)

16.3.1 Sharp-crested weirs

A sharp-crested weir normally consists of a vertical plate that is mounted at right angles to the flow and has a sharp-edged crest, Fig. 16.3.2. Theory of weirs forms the basis of spillway. Since the edge of the sharp-crested weir is sharp, opportunities for boundary layer development are limited to the vertical face of the weir.

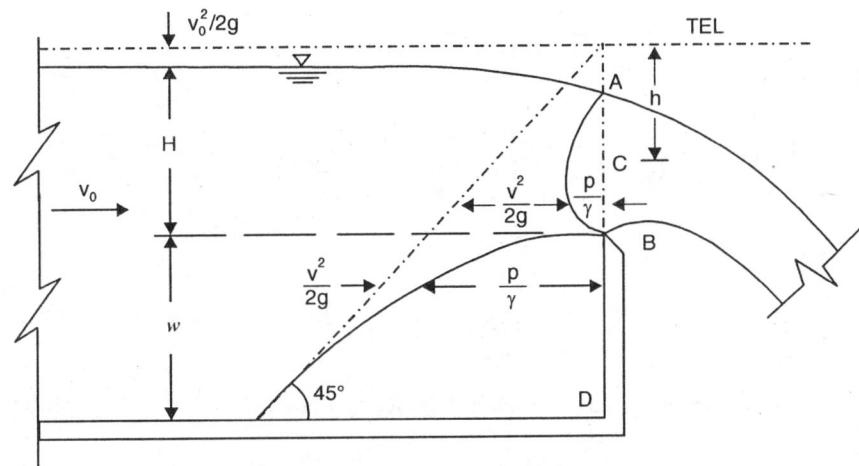

Fig. 16.3.2. Sharp-crested weir (2)

Consider a weir, Fig. 16.3.2, with its horizontal upper edge (i.e., crest) running the full width of a rectangular channel so that the flow is essentially two-dimensional with no lateral contraction. Since the contraction (and its effect) is suppressed by the sidewalls of the channel, this type of weir is the *suppressed weir* (2).

Assuming that the pressure is atmospheric across the whole flow section AB over the suppressed weir in rectangular channel of width b, Fig. 16.3.2, the velocity at any point (such as C) is equal to $\sqrt{2gh}$ and, accordingly, the discharge Q can be expressed as

$$Q = \int_{v_o^2/2g}^{H+\frac{v_o^2}{2g}} C_v \sqrt{2gh}\ bdh$$

or

$$Q = \frac{2}{3} b C_v \sqrt{2g} \left[\left(H + \frac{v_o^2}{2g} \right)^{3/2} - \left(\frac{v_o^2}{2g} \right)^{3/2} \right] \qquad \qquad ...(16.3.1)$$

Here, h is the depth to the point C measured from the total energy line while H is measured from the water surface at a location upstream of the weir and where the streamlines are parallel and the pressure distribution is hydrostatic. The theoretical velocity $\sqrt{2gh}$ has been multiplied by the coefficient of velocity C_v in order to obtain actual velocity.

Although, the lateral contraction effects have been suppressed, the flow contraction does occur in the vertical direction and this effect may be expressed by a contraction coefficient C_c. Equation (16.3.1) can, therefore, be written as

$$Q = \frac{2}{3} b \sqrt{2g}\ C_c C_v \left[\left(H + \frac{v_o^2}{2g} \right)^{3/2} - \left(\frac{v_o^2}{2g} \right)^{3/2} \right] \qquad ...(16.3.2)$$

This equation is usually written as

$$Q = \frac{2}{3} C_d\, b \sqrt{2g}\ H^{3/2} \qquad ...(16.3.3)$$

in which

$$C_d = C_c C_v \left[\left(1 + \frac{v_0^2}{2gH} \right)^{3/2} - \left(\frac{v_0^2}{2gH} \right)^{3/2} \right] \qquad ...(16.3.4)$$

One would expect both C_c and the ratio $\dfrac{v_0^2}{2gH}$ to depend on the boundary geometry and particularly the

ratio $\dfrac{H}{H+w}$ or, simply, H/w as the approach velocity v_0 depends on the cross-sectional area of the approaching flow. The experimental work has resulted into the following formula for C_d (10):

$$C_d = 0.611 + 0.075 \frac{H}{w} \qquad ...(16.3.5)$$

According to Eq. (16.3.5), the value of C_d equals 0.611 when w becomes very large. Combining Eqs. (16.3.3) and (16.3.5), one gets

$$Q = \frac{2}{3} \left(0.611 + 0.075 \frac{H}{w} \right) b \sqrt{2g}\ H^{3/2} \qquad ...(16.3.6)$$

Likewise, for a triangular sharp-crested weir, with notch angle equal to θ, one can obtain

$$Q = \frac{8}{15} C_d\ \sqrt{2g}\ \left(\tan(\theta/2) \right) H^{5/2} \qquad ...(16.3.7)$$

The pattern of flow over a thin-plate weir is very complex and cannot be analysed by theoretical analysis alone. This is primarily due to the non-hydrostatic pressure variation (on account of curvature of streamlines), turbulence and frictional effects, and the approach flow conditions. The effects of viscosity and surface tension also become significant at low heads. Therefore, the analytical relation (between the rate of flow and the head over the weir), derived after some simplifying assumptions, are suitably modified using experimental information. Following this approach, Ranga Raju and Asawa (6) modified Eqs. (16.3.7) and (16.3.6) and obtained the following discharge equations through experimental investigation:

For thin-plate triangular weir with notch angle θ,

$$Q = k_1 \frac{8}{15} C_d\ \sqrt{2g}\ \left(\tan(\theta/2) \right) H^{5/2} \qquad ...(16.3.8)$$

For a suppressed thin-plate rectangular weir,

$$Q = \left[\frac{2}{3} \left(0.611 + 0.075 \frac{H}{W} \right) b \sqrt{2g} \, H^{3/2} \right] k_1 \qquad \qquad ...(16.3.9)$$

where,

b	=	width of the weir,
C_d	=	coefficient of discharge for triangular weir (Fig. 16.3.3),
A	=	area of cross-section of approach flow,
k_1	=	correction factor to account for the effects of viscosity and surface tension (Fig. 16.3.4),
R_e	=	$g^{1/2} H^{3/2} / v$ (typical Reynolds number),

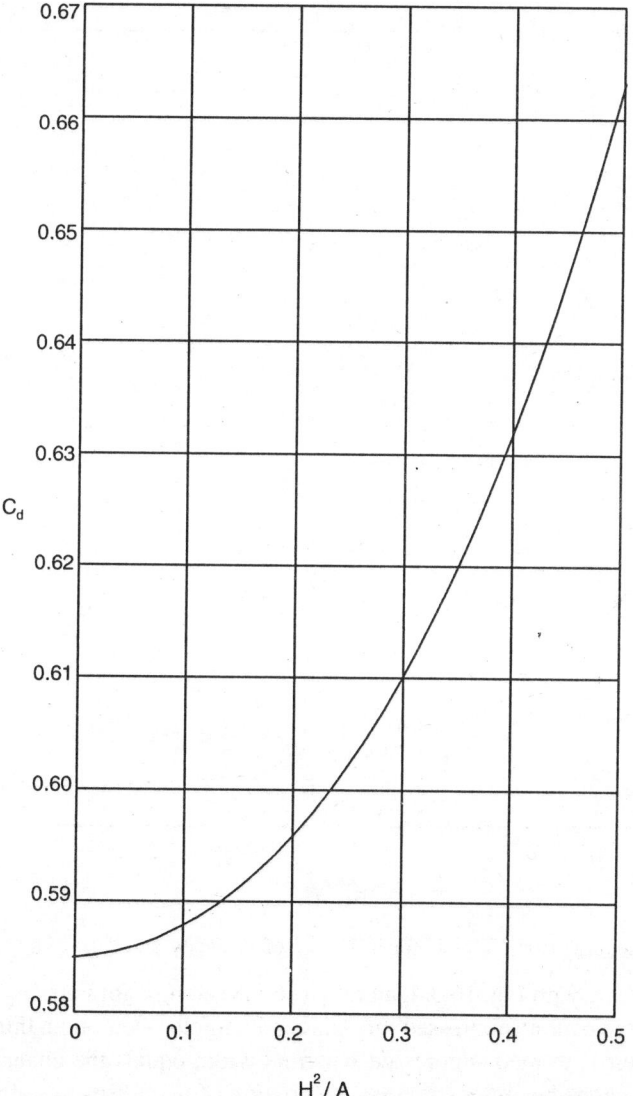

Fig. 16.3.3. Variation of coefficient of discharge for 90^0- triangular weir (6)

v = kinematic viscosity of the flowing liquid,

W_1 = $\rho g H^2 / \sigma$ (typical Weber number).

σ = surface tension of the flowing liquid,

ρ = mass density of the flowing liquid, and

g = acceleration due to gravity.

As noted from Fig. 16.3.4, that $k_1 = 1.0$ for $R_e^{0.2} W_1^{0.6}$ greater than 900. This limit corresponds to a head of 11.0 cm for water at 20°C. The mean line drawn in Fig. 16.3.4 can be used to find the value of k_1. The scatter of the experimental and field data (not shown in the figure) was generally less than 5 per cent implying maximum error of ±5 per cent in the prediction of discharge using Eqs. (16.3.8) and (16.3.9).

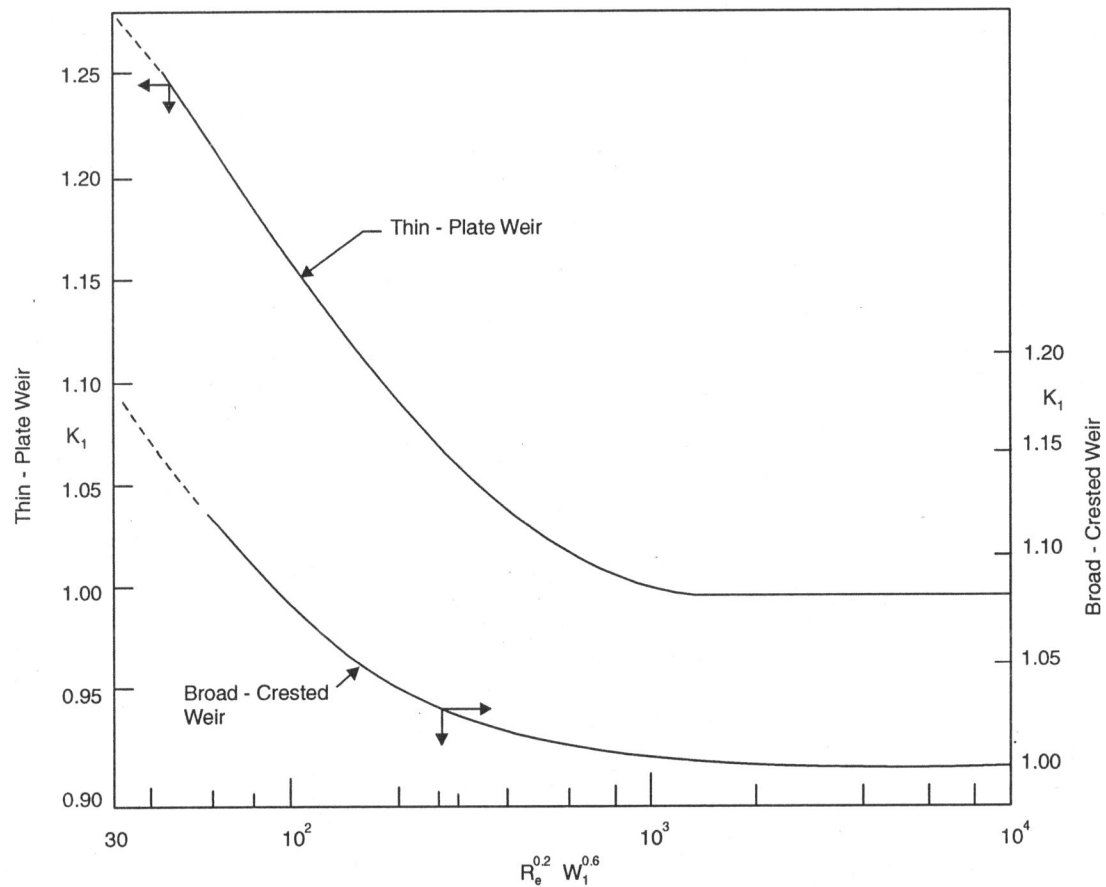

Fig. 16.3.4. Correction factor k_1 for influence of viscosity and surface tension (6,8)

Equation (16.3.9) along with Fig. 16.3.4, and Eq. (16.3.8) along with Figs. 16.3.3 and 16.3.4 enable computations of discharge over a suppressed thin-plate rectangular weir and a thin-plate 90°-triangular weir, respectively. A weir is termed suppressed when its width equals the channel width and in such cases the ventilation of nappe becomes essential. Ventilation of nappe ensures atmospheric pressure at the underside of nappe which is the requirement for the applicability of Eq. (16.3.9).

When the water level in the downstream channel rises above the crest level, the flow over the weir is submerged and the discharge for the same upstream head is reduced. The discharge under submerged conditios, Q_s is obtained from the following equations (12):

$$\frac{Q_s}{Q_f} = \left[1 - \left(\frac{H_2}{H_1} \right)^{5/2} \right]^{0.385} \quad \text{for triangular weirs} \qquad ...(16.3.10a)$$

$$\frac{Q_s}{Q_f} = \left[1 - \left(\frac{H_2}{H_1} \right)^{3/2} \right]^{0.385} \quad \text{for rectangular weirs} \qquad ...(16.3.10b)$$

16.3.2 Broad-Crested Weirs

Broad-crested weirs are generally used as diversion and metering structures in irrigation systems. The weir (Fig. 16.3.5) has a broad horizontal crest raised sufficiently above the bed so that the flow over the weir is in critical state and the cross-sectional area of the approaching flow is much larger than the cross-sectional area of flow over the top of the weir. The upstream edge of the weir is well-rounded to avoid undue eddy formation and consequent loss of energy. The derivation of the discharge equation for flow over a broad-crested weir is based on the concept of critical flow.

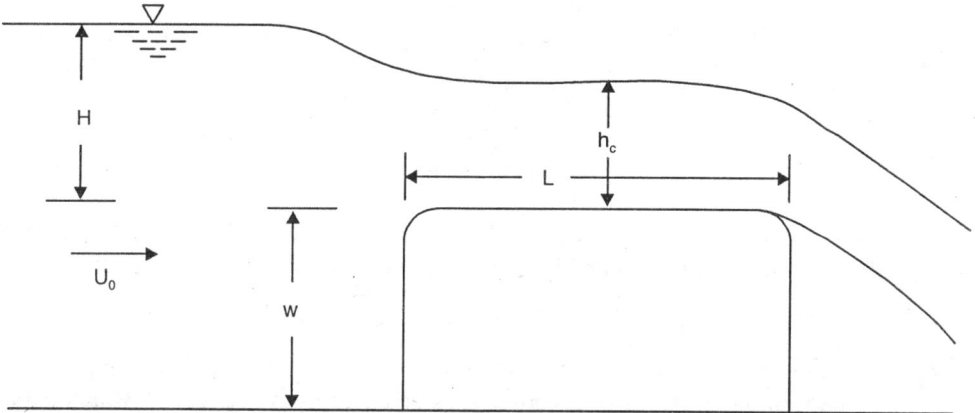

Fig. 16.3.5. Broad-crested weir

While discussing flow over a hump, Art. 13.8, it was noted that if the hump height is sufficiently large, the flow over the hump is in critical state. Therefore, a hump, under such condition, becomes a control section and there will be a unique relationship between the discharge in the channel and the depth of flow over the hump. Hence, such a hump provides a good means of flow measurement and is termed as broad-crested weir, Fig. 16.3.5. Considering a broad-crested weir of width b perpendicular to the flow, Fig. 16.3.5, in a rectangular channel of width b_1 and assuming (i) the flow to be in critical state on the weir crest, (ii) no energy loss between the section of the critical depth and the upstream section where the head H is measured, and (iii) $\alpha = 1.0$, one can write the specific energy equation as follows:

$$H + \frac{U_0^2}{2g} = \frac{3}{2} h_c \qquad ...(16.3.11)$$

As per the critical flow formula, Eq. (13.2.8),

$$Q = b\sqrt{g}\ h_c^{3/2} \qquad\qquad ...(16.3.12)$$

Since it is easier to measure H than h_c, the discharge can, alternatively, be expressed as

$$Q = Cb\sqrt{g}\ H^{3/2} \qquad\qquad ...(16.3.13)$$

$$\therefore \qquad U_0 = \frac{Q}{b_1(H+w)} = \frac{Cb\sqrt{g}\ H^{3/2}}{b_1(H+w)} \qquad\qquad ...(16.3.14)$$

Combining Eqs. (16.3.12) and (16.3.13), one gets

$$h_c = C^{3/2}\ H$$

Substituting the value of h_c and U_0 in Eq. (16.3.11), one obtains

$$H + \frac{C^2 H^3}{2(H+w)^2}\left(\frac{b}{b_1}\right)^2 = \frac{3}{2}H C^{2/3}$$

or

$$\frac{2}{3} + \frac{C^2 H^2 (b/b_1)^2}{3(H+w)^2} = C^{2/3}\left(\frac{b_1}{b}\right)^{2/3}$$

$$\therefore \qquad \frac{H}{H+w} = \frac{\sqrt{3}\left(C^{2/3} - \dfrac{2}{3}\right)^{1/2}}{C(b/b_1)} \qquad\qquad ...(16.3.15)$$

Ranga Raju and Asawa (8) further modified the discharge equation, Eq. (16.3.13), for a broad-crested weir with well-rounded upstream and downstream ends of the weir crest as follows:

$$Q = k_1 k_2 C b \sqrt{g}\ H^{3/2} \qquad\qquad ...(16.3.16)$$

Here, k_2 is the correction for the effect of curvature of flow over the weir crest (Fig. 16.3.6), and L is the length of the weir along the flow direction. C is obtained from Fig. 16.3.7 which is based on Eq. (16.3.15).

For suppressed broad-crested weirs $b = b_1$. Thus, one uses the curve for $b/b_1 = 1.0$ in Fig. 16.3.7 to find C. The value of k_1 is obtained from Fig. 16.3.4.

For broad-crested weir with sloping upstream and downstream faces, one can use Eq. (16.3.16) with different functional relations for k_1 and k_2 shown in Figs. 16.3.8 and 16.3.9, respectively.

For a submerged broad-crested weir, the discharge equation, Eq. (16.3.16) is modified to (7, 9)

$$Q = Cb\sqrt{g}\ H^{3/2} k_1 k_2 k_4 \qquad\qquad ...(16.3.17)$$

Assuming that C, k_1 and k_2 remain unaffected due to submergence, relationship for k_4 is as shown in Fig. 16.3.10. It should be noted that the discharge on broad-crested weirs remains unaffected up to submergence (H_2/H_1) as high as 75 per cent.

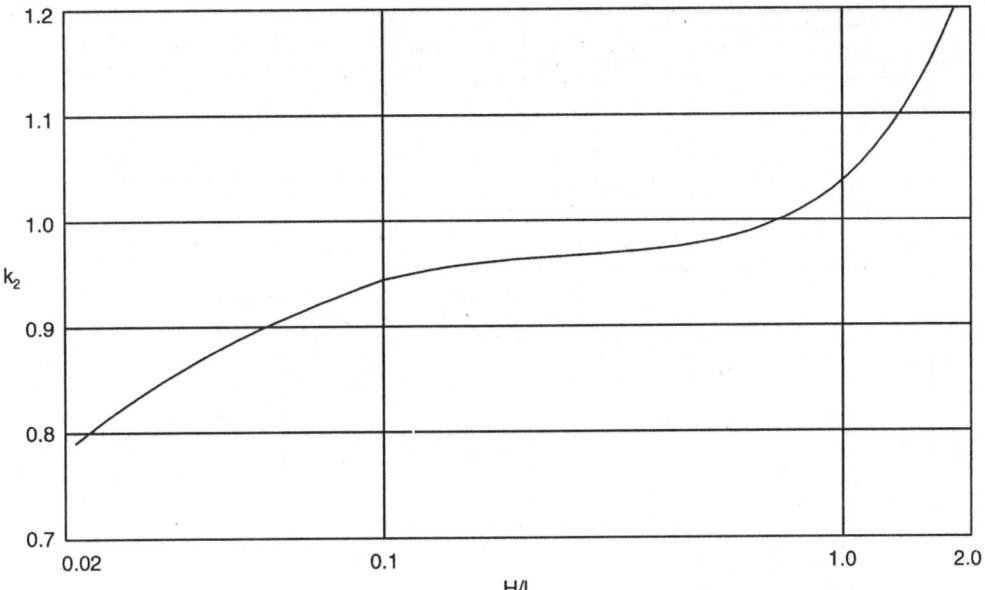

Fig. 16.3.6. Variation of k_2 with H/L for vertical-faced weir with rounded corner (8)

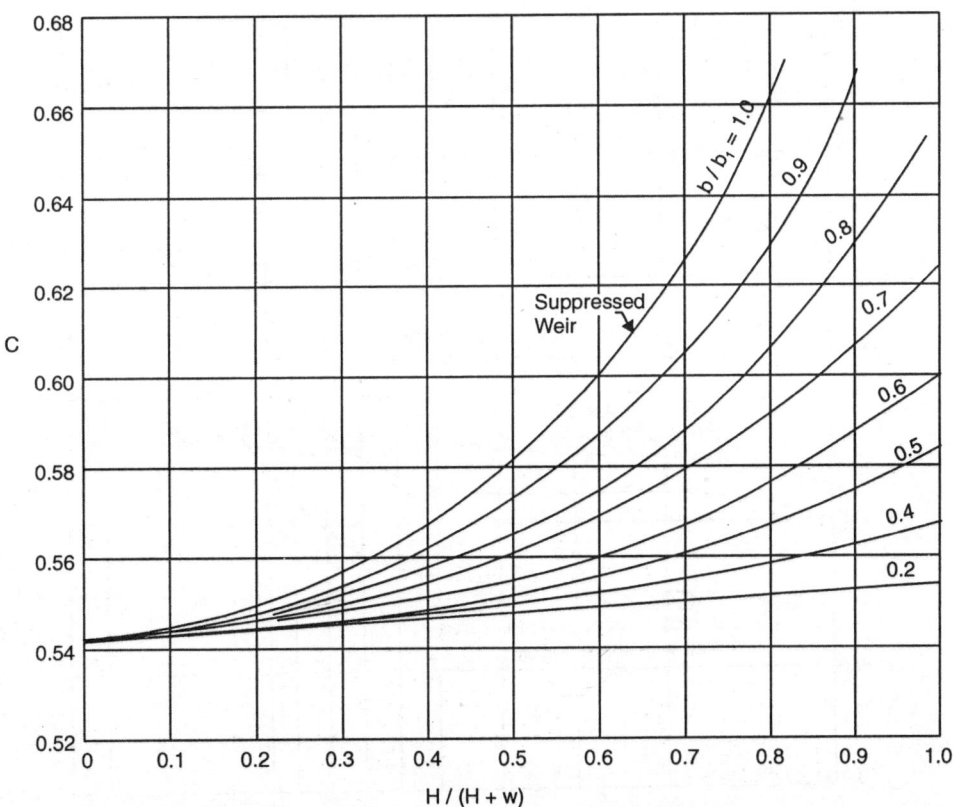

Fig. 16.3.7. Variation of C with $H/(H+w)$ and b/b_1 for broad-crested weirs (8)

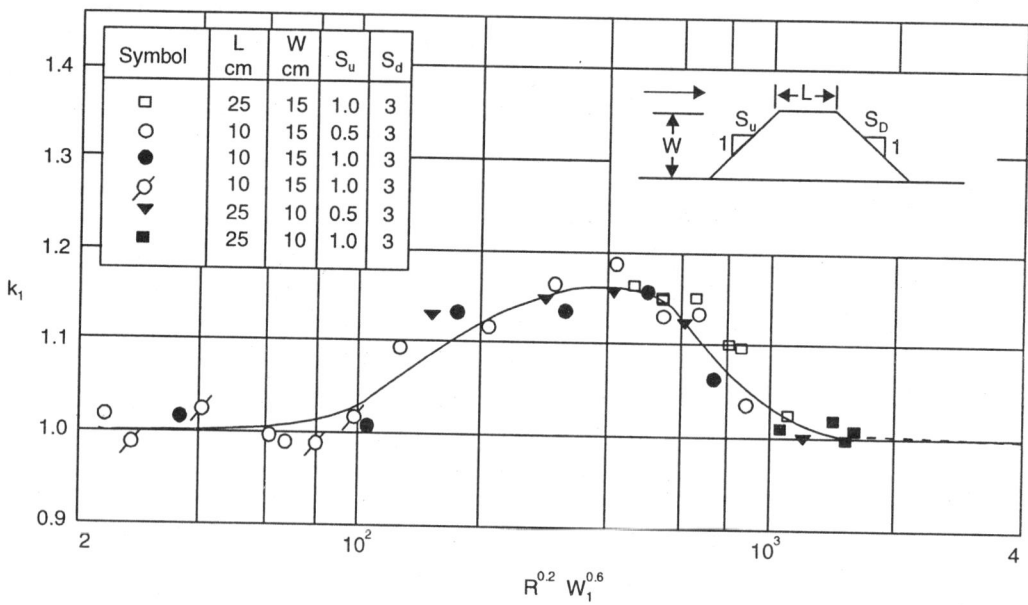

Fig. 16.3.8. Variation of k_1 for broad-crested weirs with sloping faces (7)

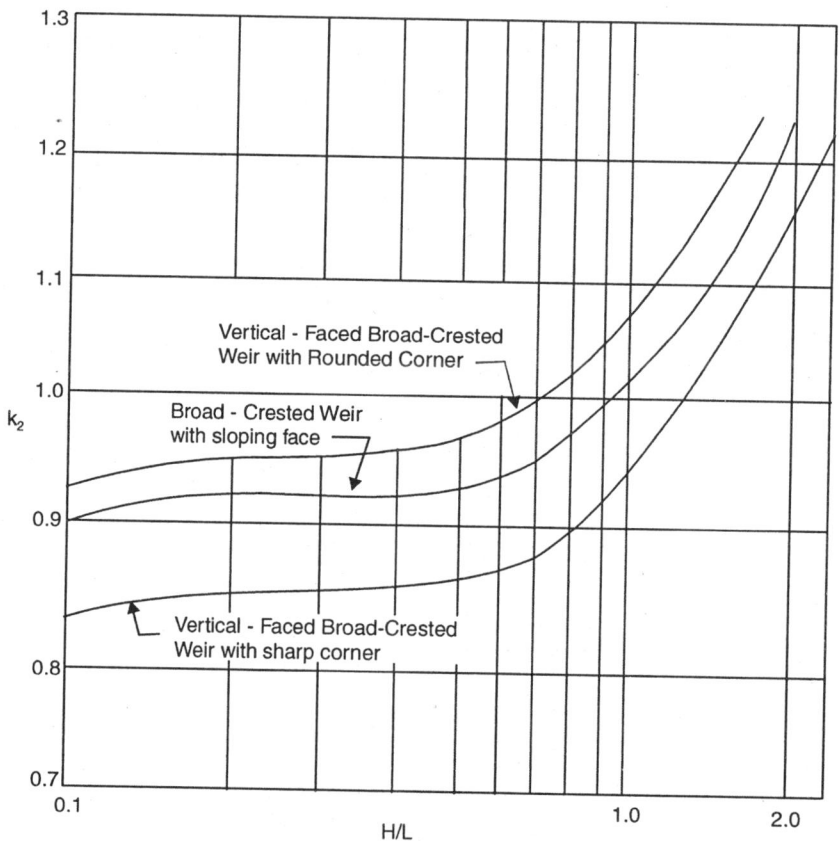

Fig. 16.3.9. Variation of k_2 with H/L for broad-crested weirs with sloping faces (7)

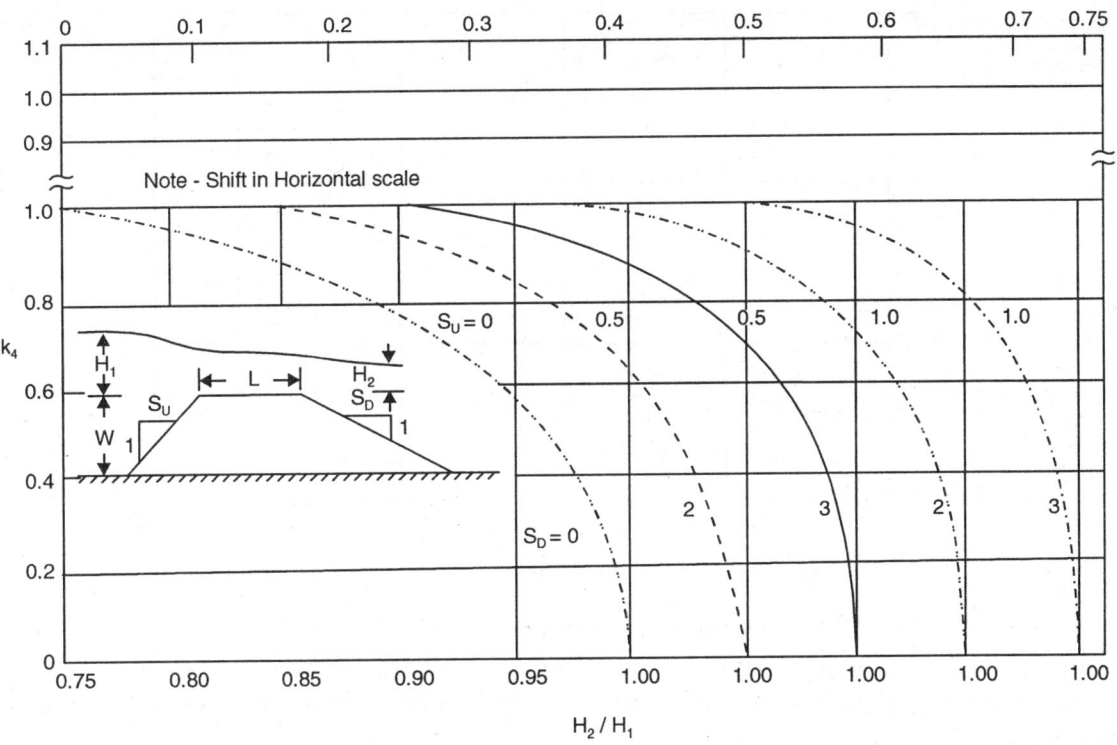

Fig. 16.3.10. Variation of k_4 with H_2/H_1 and weir geometry (7)

Weirs of widths smaller than that of the approach channel are termed contracted weirs. The above-mentioned relationships, Eq. (16.3.9) for sharp-crested rectangular weirs and Eq. (16.3.16) for broad-crested weir, require some modifications for contracted weirs. Ranga Raju and Asawa (8) suggested the following equation for the actual discharge over a contracted sharp-crested rectangular weir:

$$Q = k_1\, k_3\, \frac{2}{3}\Big[0.611 + C_1\big(H/W\big)\Big] b \sqrt{2g}\; H^{3/2} \qquad \text{...(16.3.18)}$$

in which k_3 is the correction factor for lateral contraction and C_1 is a function of b/b_1 as shown in Fig. 16.3.11. The value of k_3 should logically be a function of H/b. On analyzing the experimental data, the average value of k_3 was found to be 0.95 for H/b ranging from 0.1 to 1.0 (8).

Similarly, the actual discharge over a contracted broad-crested weir may be written as (8)

$$Q = k_1\, k_2\, k_3\, Cb \sqrt{g}\; H^{3/2} \qquad \text{...(16.3.19)}$$

in which k_3 is a correction factor for contraction effects and the value of C for various values of b/b_1, as obtained by solving Eq. (16.3.15), can be read from Fig. 16.3.7. Logically, k_3 should be a function of H/b. Based on the analysis of experimental data, the value of k_3 has been recommended as unity for H/b ranging from 0.1 to 1.0 (8).

The advantages of weirs for measurement of discharge in open channel are as follows:

(i) Simplicity and ease in construction,
(ii) Durability, and
(iii) Accuracy.

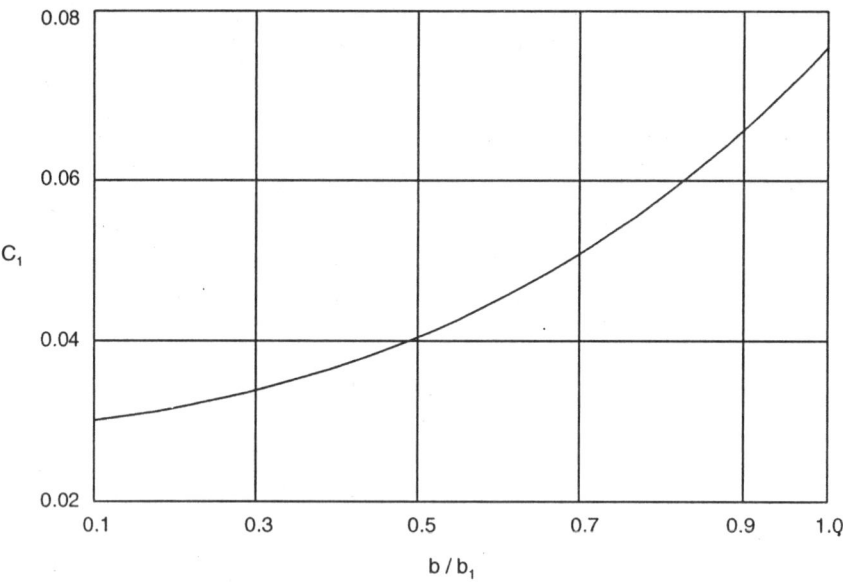

Fig. 16.3.11. Variation of C_1 with b/b_1 for contracted thin-plate weirs (8)

However, the main requirement of considerable fall of water surface makes their use in areas of level ground impracticable. Besides, deposition of sand, gravel, and silt upstream of the weir prevents accurate measurements.

16.4 FLUMES

A *flume* is a flow measuring device formed by a constriction in an open channel. The constriction can be either a narrowing of the channel or a narrowing in combination with a hump in the invert. By providing sufficient amount of constriction, it is possible to produce critical conditions there. When this happens, there exists a unique stage-discharge relationship independent of the downstream conditions. The use of critical-depth flumes for discharge measurement is based on this principle.

The main advantage of a critical-depth flume over a weir is in situations when material (sediment or sewage) is being transported by the flow. This material gets deposited upstream of the weir and affects the discharge relation and results in a foul-smelling site in case of sewage flow. The critical-depth flumes consisting only of horizontal contraction would easily carry the material through the flume. Critical-depth flumes can be grouped into two main categories, viz., long-throated flumes and short-throated flumes.

16.4.1 Long-Throated Flumes

The constriction of these flumes (Fig. 16.4.1) is sufficiently long (the length of the throat should be at least twice the maximum head of water that will occur upstream of the flume) so that it produces small curvatures in the water surface and the flow in the throat is virtually parallel to the invert of the flume.

This condition results in nearly hydrostatic pressure distribution at the control section (where critical depth occurs) which, in turn, allows analytical derivation of the stage-discharge relation. This gives the designer the freedom to vary the dimensions of the flume in order to meet specific requirements. Such flumes are usually of rectangular, trapezoidal, triangular or U-shaped cross-section. For a rectangular flume, the discharge of an ideal fluid is expressed as

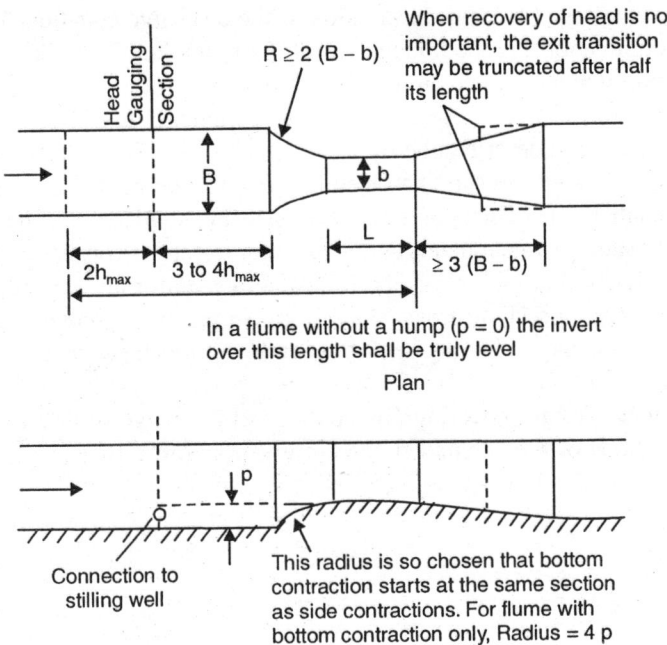

Fig. 16.4.1. Long-throated flume

$$Q = \left(\frac{2}{3}\right)^{3/2} \sqrt{g}\, b H^{3/2} \qquad\qquad ...(16.4.1)$$

Here, H represents the upstream energy and b is the typical width dimension for the particular cross-section shape of the flume. By introducing suitable coefficients, this equation can be generalized in the following form so that it applies to any cross-sectional shape (1):

$$Q = \left(\frac{2}{3}\right)^{3/2} \sqrt{g}\, C_v\, C_s\, C_d\, b h^{3/2} \qquad\qquad ...(16.4.2)$$

where C_v = coefficient to take into account the velocity head in the approach channels,

C_s = coefficient to take account of the cross-sectional shape of the flume,

C_d = coefficient for energy loss,

and h = depth of water, upstream of the flume, measured relative to the invert level of the throat (i.e., gauged head).

Short-Throated Flumes

In these flumes, the curvature of the water surface is large and the flow in the throat is not parallel to the invert of the flume. The principle of operation of these flumes is the same as that of long-throated flumes, viz., the creation of critical condition at the throat. However, non-hydrostatic pressure distribution (due to large curvature of flow) does not permit analytical derivation of the discharge equation. Further, energy loss also cannot be assessed. Therefore, it becomes necessary to rely on direct calibration either in the field or in the laboratory for the determination of the discharge equation. The designer does not have complete freedom in choosing the dimensions of the flume but has to select the closest standard design to meet his requirements. Such flumes, however, require lesser length and, hence, are more economical than long-throated flumes. One of the most commonly used short-throated flumes is the Parshall flume which has been described here.

Parshall (3, 4, and 5) designed a short-throated flume with a depressed bottom (Fig. 16.4.2) which is known as the Parshall flume. This was first developed in the 1920's in the USA and has given satisfactory service at water treatment plants and irrigation projects. It consists of a short parallel throat preceded by a uniformly converging section and followed by a uniformly expanding section. The floor is horizontal in the converging section, slopes downwards in the throat, and is inclined upwards in the expanding section. The control section at which the depth is critical, occurs near the downstream end of the contraction.

There are 22 standard designs covering a wide range of discharge from 0.1 litre per second to 93 m³/s. The main dimensions of the Parshall flume are given in Table 16.4.1 and Fig. 16.4.2. The discharge characteristics of these flumes are given in Table 16.4.2.

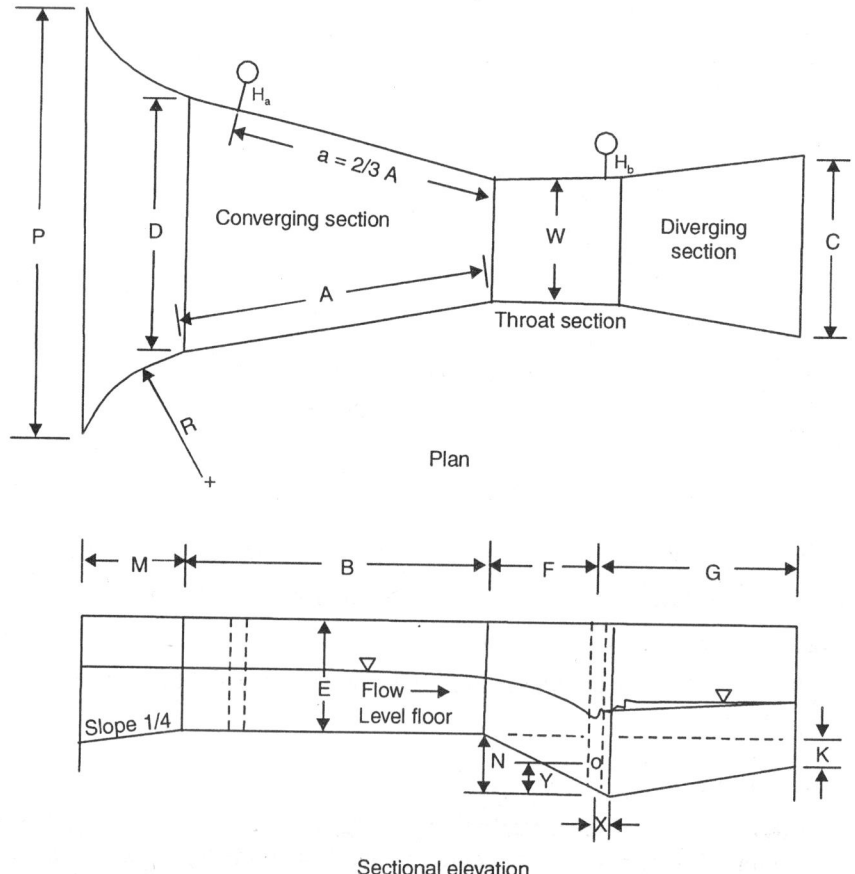

Fig. 16.4.2. Parshall flume

Table 16.4.1: Parshall Flume dimensions (mm) (1)

b	b (mm)	A	a	B	C	D	E	F	G	K	M	N	P	R	X	Y
1 in	25.4	363	242	356	93	167	229	76	203	19	-	29	-	-	8	13
2 in	50.8	414	276	406	135	214	254	114	254	22	-	43	-	-	16	25
3 in	76.2	467	311	457	178	259	457	152	305	25	-	57	-	-	25	38
6 in	152.4	621	414	610	394	397	610	305	610	76	305	114	902	406	51	76
9 in	228.6	879	587	864	381	575	762	305	457	76	305	114	1080	406	51	76
1 ft	304.8	1372	914	1343	610	845	914	610	914	76	381	229	1492	508	51	76
1.5ft	457.2	1448	965	1419	762	1026	914	610	914	76	381	229	1676	508	51	76
2 ft	609.6	1524	1016	1495	914	1206	914	610	914	76	381	229	1854	508	51	76
3 ft	914.4	1676	1118	1645	1219	1572	914	610	914	76	381	229	2222	508	51	76
4 ft	1219.2	1829	1219	1794	1524	1937	914	610	914	76	457	229	2411	610	51	76
5 ft	1524.0	1981	1321	1943	1829	2302	914	610	914	76	457	229	3080	610	51	76
6 ft	1828.8	2134	1422	2092	2134	2667	914	610	914	76	457	229	3442	610	51	76
7 ft	2133.6	2286	1524	2242	2438	3032	914	610	914	76	457	229	3810	610	51	76
8 ft	2438.4	2438	1626	2391	2743	3397	914	610	914	76	457	229	4172	610	51	76
10 ft	3048	-	1829	4267	3658	4756	1219	914	1829	152	-	343	-	-	305	229
12 ft	3658	-	2032	4877	4470	5607	1524	914	2438	152	-	343	-	-	305	229
15 ft	4572	-	2337	7620	5588	7620	1829	1219	3048	229	-	457	-	-	305	229
20 ft	6096	-	2845	7620	7315	9144	2134	1829	3658	305	-	686	-	-	305	229
25 ft	7620	-	3353	7620	8941	10668	2134	1829	3962	305	-	686	-	-	305	229
30 ft	9144	-	3861	7925	10566	12313	2134	1829	4267	305	-	686	-	-	305	229
40 ft	12192	-	4877	8230	13818	15481	2134	1829	4877	305	-	686	-	-	305	229
50 ft	15240	-	5893	8230	17272	18529	2134	1829	6096	305	-	686	-	-	305	229

Table 16.4.2: Discharge characteristics of parshal flumes $Q = kh^n$ (1)

Throat width b (cm)	Discharge range minimum (l/s)	Discharge range maximum (l/s)	k	n	Head range, m minimum	Head range, m maximum	Modular limit h_2/h_1
2.54	0.09	5.4	0.0604	1.55	0.015	0.21	0.50
5.08	0.18	13.2	0.1207	1.55	0.015	0.24	0.50
7.62	0.77	32.1	0.1771	1.55	0.03	0.33	0.50
15.29	1.50	111	0.3812	1.58	0.03	0.45	0.60
22.89	2.50	251	0.5354	1.53	0.03	0.61	0.60
30.48	3.32	457	0.6909	1.52	0.03	0.76	0.70
45.72	4.80	695	1.056	1.538	0.03	0.76	0.70
60.96	12.1	937	1.428	1.550	0.046	0.76	0.70
91.44	17.6	1427	2.184	1.566	0.046	0.76	0.70
121.92	35.8	1923	2.953	1.578	0.06	0.76	0.70
152.4	44.1	2424	3.732	1.587	0.06	0.76	0.70
182.88	74.1	2929	4.519	1.595	0.076	0.76	0.70
213.36	85.8	3438	5.312	1.601	0.076	0.76	0.70
243.84	97.2	3949	6.112	1.607	0.076	0.76	0.70
	m³/s	m³/s					
304.80	0.16	8.28	7.463	1.60	0.09	1.07	0.80
365.76	0.19	14.68	8.859	1.60	0.09	1.37	0.80
457.20	0.23	25.04	10.96	1.60	0.09	1.67	0.80
609.60	0.31	37.97	14.45	1.60	0.09	1.83	0.80
762.00	0.38	47.14	17.94	1.60	0.09	1.83	0.80
914.40	0.46	56.33	21.44	1.60	0.09	1.83	0.80
1219.20	0.60	74.70	28.43	1.60	0.09	1.83	0.80
1524.00	0.75	93.04	35.41	1.60	0.09	1.83	0.80

16.5 CURRENT METER

The current meter is a widely used mechanical device for the measurement of flow velocity and, hence, the discharge. It consists of a small wheel with cups at the periphery or propeller blades rotated by the force of flowing water, and a tail or fins to keep the instrument aligned in the direction of flow. The cup-type current meter has a vertical axis, and is a more rugged instrument which can be handled by relatively unskilled technicians. The propeller-type current meter has been used for relatively higher velocities (up to 6 to 9 m/s as against 3 to 5 m/s for the cup-type current meter). The small size of the propeller-type current meter is advantageous when the measurements have to be taken close to the wall. The properller-type current meter is less likely to be affected by floating weeds and debris.

For measurements, the current meter is mounted on a rod and moved vertically to measure the velocity at different points. The speed of rotation of cups or blades depends on the velocity of flow. The instrument has an automatic counter with which the number of rotations in a given duration is determined.

The current meter is calibrated by moving it with a known speed in still water and noting the number of revolutions per unit of time. During measurement, the current meter is held stationary in running water. Using the appropriate calibration (supplied by the manufacturer) the velocity can be predicted. By this method, one can obtain the velocity distribution and, hence, the discharge. Or, alternatively, one can measure the velocity at 0.2 h and 0.8 h (here, h is the depth of flow) below the free surface and the mean of the two values gives the average velocity of flow. Velocity at 0.6 h also approximates as the average velocity of flow.

16.6 OTHER METHODS

Mean velocities in open channels can, alternatively, be determined by measuring surface velocities using surface floats. The surface float is an easily visible object lighter than water, but sufficiently heavy not to be affected by wind. The surface velocity is measured by noting down the time the surface float takes in covering a specified distance which is generally not less than 30 metres and 15 metres for large and small channels, respectively. The surface velocity is multiplied by a suitable coefficient (less than unity) to get the average velocity of flow.

A double float consists of a surface float to which is attached a hollow metallic sphere heavier than water. Obviously, the observed velocity of the double float would be the mean of the surface velocity and the velocity at the level of the metallic sphere. By adjusting the metallic sphere at a depth nearly equal to 0.2 h above the bed, the observed velocity will be approximately equal to the mean velocity of flow.

Alternatively, velocity rods can be used for the measurement of average velocity of flow. Velocity rods are straight wooden rods or hollow tin tubes of 25 mm to 50 mm diameter and weighted down at the bottom so that these remain vertical and fully immersed except for a small portion at the top while moving in running water. These rods are either telescopic-type or are available in varying lengths so that they can be used for different depths of flow. As the rod floats vertically from the surface to very near the bed, its observed velocity equals the mean velocity of flow in that vertical plane.

EXAMPLE

E16.1.   Estimate the discharge of water over a suppressed rectangular sharp-crested weir of height 0.4 m in a channel of 1.5 m width. The head over the weir is 50 mm. Assume $v=10^{-6}$ m^2/s and $\sigma=0.075$ N/m.

Solution:
$$R_e = \frac{g^{1/2} H^{3/2}}{v} = \frac{(9.8)^{1/2}(0.05)^{3/2}}{10^{-6}} = 35\times 10^3$$

$$W_1 = \frac{\rho g H^2}{\sigma} = \frac{1000\times 9.8 \times (.05)^2}{0.075} = 326.67$$

$$R_e^{0.2}\, W_1^{0.6} = 261.38$$

From Fig. 16.3.4,
$$k_1 = 1.065$$

Therefore from Eq. (16.3.9),

$$Q = \frac{2}{3}\left(0.611 + 0.075\frac{50}{400}\right) \times 1.5 \times \sqrt{2 \times 9.8}\,(0.05)^{3/2} \times 1.065$$

$$= 0.0327 \text{ m}^3/\text{s}$$

PROBLEMS

P16.1 Estimate the discharge of water ($v = 9.8 \times 10^{-7}$ m^2/s, $\sigma = 0.075$ N/m) over a suppressed sharp-crested rectangular weir under a head of 58.5 mm. The height of the weir crest from the bottom of the channel is 0.25 m and width is 1.25 m.

P16.2 Develop the head-discharge curve for a 0.40 m high sharp-crested weir set to the full width of a 1.0 m wide channel up to a maximum discharge of 0.15 m^3/s, $v = 10^{-6}$ m^2/s, $\sigma = 0.075$ N/m.

P16.3 A weir 0.40 m wide is set to a height of 0.25 m in a 1.0 m wide channel. Find the discharge corresponding to a head of 0.20 m if $v = 10^{-6}$ m^2/s and $\sigma = 0.075$ N/m.

P16.4 The head and tail waters upstream and downstream of a suppressed rectangular sharp-crested weir are 0.40 m and 0.30 m respectively. The height of the weir crest is 0.30 m and the channel width is 0.75 m. Estimate the discharge if $v = 10^{-6}$ m^2/s and $\sigma = 0.075$ N/m.

P16.5 It is intended to measure the discharge of P16.1 using a 90° triangular weir set in the same channel. If the apex of the weir is at a height of 0.18 m from the channel bed, estimate the head over the weir.

REFERENCES

(1) Ackers, P, WR White, JA Perkins, and AJM Harrison: *"Weirs and Flumes for Flow Measurement"*, John Wiley and Sons, 1978.

(2) Henderson, FM: *"Open Channel flow"*, The Macmillan Co., New York, 1966.

(3) Parshall, RL: *"Discussion of Measurement of debris-laden stream flow with critical depth flumes,"* Trans. ASCE, Vol. 103, 1938.

(4) Parshall, RL: *"Measuring water in irrigation channels with Parshall flumes and small weirs,"* Soil Conservation Circular No. 843, US Department of Agriculture, May 1950.

(5) Parshall, RL: *"Parshall flumes of large size,"* Bulletin 426 A, Colorado Agricultural Experimental Station, Colorado State University, March 1953.

(6) Ranga Raju, KG and GL Asawa: *"Viscosity and surface tension effects on weir flow,"* J. of Hydraulics Division, Proc. ASCE, Oct. 1977.

(7) Ranga Raju, KG, GL Asawa, SK Gupta, and SJ Sahsrabudhe: *"Submerged broad-crested weirs,"* Proc. of IMEKO Symposium on Flow Measurement of Fluids, Groningen, The Netherlands, Sept. 1978.

(8) Ranga Raju, K.G. and GL Asawa: *"Comprehensive weir discharge formulae,"* Proc. of IMEKO Symposium on Flow Measurement and Control in Industry, Tokyo, Japan, Nov. 1979.

(9) Ranga Raju, KG, R. Srivastava and PD Porey: *"Scale effects in modelling of flow over broad-crested weirs"*, CBIP J. of Irrigation and Power, July 1992.

(10) Rehbok, T: Discussion on *"Precise Weir Measurements"* by E.W. Schoder and KB Turner, Trans. ASCE, Vol. 93, 1929 [Henderson (2)].

(11) Rouse, H: *"Discharge characteristics of the free overfall"*, Civil Engineering, Vol. 6, 1936 [Henderson (2)].

(12) Villemonte, JR: *"Submerged weir discharge studies"*, Engineering News Record, Dec. 1947 [Ranga Raju, KG, *"Flow through Open Channels"* Tata McGraw-Hill Publishing Co. Ltd., New Delhi, 1993].

CHAPTER 17

Flow in Alluvial Channels

17.1 GENERAL

Flow in natural river channels is relatively more complex than the flow in rigid boundary channels. The complexity arises due to the erodible bed and banks of the river channel. Depending upon the boundary flow conditions, the material of the river boundaries may get eroded and start moving with the flow or the material in motion may get deposited on the bed and banks. Therefore, the boundaries of the river channel are generally mobile and not rigid.

A change in discharge of water flowing in a rigid boundary channel will cause a change only in the depth of flow. But, in case of mobile (or loose) boundary channels, a change in discharge may vary cross-section, slopes, plan form of the channel, bed forms, and roughness coefficient. The theory of rigid boundary channels is, therefore, not applicable to loose boundary channels. The problem of mobile boundary channels is, obviously, more complicated.

The bed of a river channel usually consists of noncohesive sediment (i.e., silt, sand, and gravel). *Sediment* is defined as fragmental material that is transported by, or suspended in, or deposited by water or air, or accumulated in the beds by other natural agents. Sediment (also known as an alluvium) is, therefore, the loose and noncohesive material being carried or deposited by a river or channel flow. Ice, logs of wood, and organic materials flowing with water are excluded from the definition of sediment.

A channel (or river) flowing through sediment and transporting some of it along with the flowing water is called an *alluvial channel (or river)*. The analytical solution of alluvial channel problems is generally difficult, and experimental methods are, therefore, usually adopted for obtaining solutions of problems related to alluvial channels.

17.2 INCIPIENT MOTION OF SEDIMENT

Consider the case of flow of clear water in an open channel of a given slope with a movable bed of non-cohesive material. At low discharges, the bed material remains stationary and, hence, the channel can be considered as rigid. With the increase in discharge, a stage will come when the shear forces exerted by the flowing water on particles of the bed and banks of the mobile channel will just exceed the force opposing the movement of the particles. At this stage, a few particles on the bed move intermittently. This condition is called the *incipient motion* condition, or, simply, the *critical condition*.

455

An understanding of flow at the incipient motion condition is useful in fixing slope or depth for clear water flow in an alluvial channel. Knowledge of the incipient motion condition is also required for some methods of calculation of sediment load. Therefore, there is a need to understand the phenomenon that initiates motion of sediment particles.

The experimental data on incipient motion condition have been analysed by different investigations using one of the following three approaches (9):

  (i) Competent velocity approach,
 (ii) Lift force approach, and
(iii) Critical tractive force approach.

Competent velocity is the mean velocity of flow which just causes a particle to move. A relationship among the size of the bed material, its relative density, and the competent velocity is generally developed and used for this approach.

Investigators using the lift force approach assume that the incipient motion condition is established when the lift force exerted by the flow on a sediment particle just exceeds submerged weight of the particle.

The critical tractive force approach is based on the assumption that it is the drag (and not lift) force exerted by the flowing water on the channel bed that is responsible for the motion of the bed particles.

Of these three approaches, the critical tractive force approach is considered most logical and is very often used by hydraulic engineers. Hence, only this approach has been described here.

The *critical tractive (or shear) stress* is the average shear stress acting on the bed of a channel at which the sediment particles just begin to move. Shields (24) was the first investigator to give a semi-theoretical analysis of the problem of incipient motion. According to him, a particle begins to move when the fluid drag F_1 on the particle overcomes the particle resistance F_2. The fluid drag F_1 is given as

$$F_1 = k_1 \left[C_D d^2 \frac{1}{2} \rho u_d^2 \right]$$

and the particle resistance F_2 is expressed as

$$F_2 = k_2 \left[d^3 \left(\rho_s - \rho \right) g \right]$$

where,

$\quad\quad C_D$ = the drag coefficient,
$\quad\quad d$ = the size of the particle,
$\quad\quad \rho$ = the mass density of the flowing fluid,
$\quad\quad u_d$ = the velocity of flow at the top of the particle,
$\quad\quad \rho_s$ = the mass density of the particle,
$\quad\quad g$ = acceleration due to gravity,
$\quad\quad k_1$ = a factor dependent on the shape of the particle,
and $\quad k_2$ = a factor dependent on the shape of the particle and angle of internal friction.

Using the Karman-Prandtl equation for the velocity distribution, the velocity can be expressed as

$$\frac{u_d}{u_*} = f_1 \left(\frac{u_* d}{v} \right) = f_1 \left(R^* \right)$$

Here, ν is the kinematic viscosity of the flowing fluid, u_* the shear velocity equal to $\sqrt{\tau_0/\rho}$ and τ_0 is the shear stress acting on the boundary of the channel.
Similarly,

$$C_D = f\left(\frac{u_d d}{\nu}\right)$$

or

$$C_D = f_2\left(\frac{u_* d}{\nu}\right) = f_2\left(R^*\right)$$

Thus,

$$F_1 = k_1 f_2\left(R^*\right) d^2 \frac{1}{2}\rho u_*^2 \left[f_1\left(R^*\right)\right]^2$$

For the incipient motion condition, the two forces F_1 and F_2 will be equal. Hence,

$$k_1' f_2\left(R_c^*\right) d^2 \frac{1}{2}\rho u_{*c}^2 \left[f_1\left(R_c^*\right)\right]^2 = k_2'\left[d^3\left(\rho_s - \rho\right)g\right]$$

Here, the subscript c indicates the critical condition (or the incipient motion condition). The above equation can, alternatively, be rewritten as

$$\frac{\rho u_{*c}^2}{(\rho_s - \rho)gd} = \frac{2k'_2}{k'_1} f\left(R_c^*\right)$$

Alternatively,

$$\tau_c^* = f\left(R_c^*\right) \qquad\qquad ...(17.2.1)$$

where

$$\tau_c^* = \frac{\tau_c}{\Delta \rho_s\, gd}$$

$$\tau_c = \rho u_{*c}^2$$

and

$$\Delta \rho_s = \rho_s - \rho$$

Using the experimental data collected by different investigators, a unique relationship between τ_c^* and R_c^* was obtained by Shields (24) and is shown in Fig. 17.2.1. The curve shown in Fig. 17.2.1 is known as the Shields' curve for the incipient motion condition. The parameter $R_c^* = \left(\frac{u_{*c} d}{\nu}\right)$ is the ratio of the particle size d and ν/u_{*c}. Therefore, the parameter $\dfrac{\nu}{u_{*c}}$ is a measure of thickness of laminar sublayer, i.e., δ'. Hence, R_c^* can be taken as a measure of the roughness of the boundary surface. The boundary surface is rough at large values of R_c^* and, hence, τ_c^* attains a constant value of 0.06 and is independent of R_c^* at $R_c^* \geq 400$. This value of R_c^* (i.e., 400) is much higher than the value of 70 at which

the boundary becomes rough from the established criterion $\dfrac{d}{\delta'} > 6.0$. Also, the constant value of τ_c^*

equal to 0.06 is on the higher side.

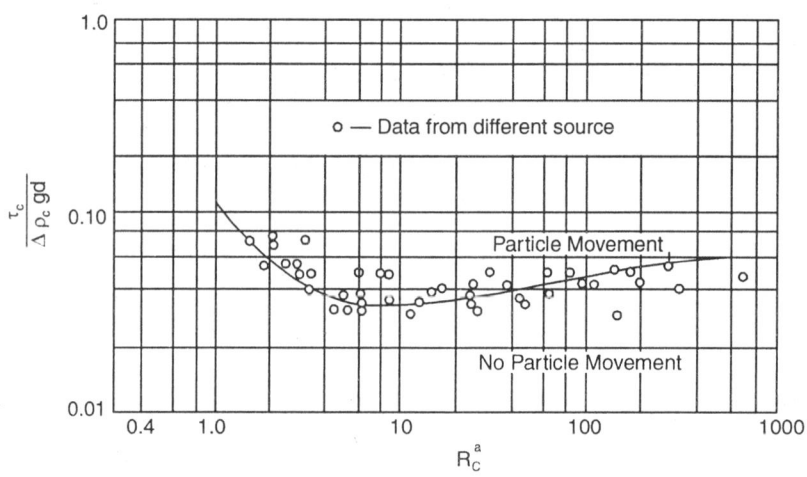

Fig. 17.2.1. Shields curve for incipient motion condition (24)

The use of the Shields curve for the estimation of τ_c requires trial procedure as τ_c appears in both τ_c^* and R_c^*. One can, alternatively, use the following equation of the Shields' curve for direct computation of τ_c (25):

$$\frac{\tau_c}{\Delta \rho_s g \left(\dfrac{\rho v^2}{\Delta \rho_s g}\right)^{1/3}} = 0.243 + \frac{0.06\, d_*^2}{\left(3600 + d_*^2\right)^{1/2}} \qquad \text{...(17.2.2)}$$

in which $\qquad d_* = \dfrac{d}{\left(\rho v^2 / \Delta \rho_s g\right)^{1/3}}$

For specific case of water (at 20°C) and the sediment of specific gravity 2.65, the above relation for τ_c simply becomes

$$\tau_c = 0.155 + \frac{0.409\, d^2}{\left(1 + 0.177\, d^2\right)^{1/2}} \qquad \text{...(17.2.3)}$$

in which τ_c is in N/m² and d is in mm. Equations (17.2) and (17.3) are expected to give the value of τ_c within about 5% of the value obtained from the Shields' curve (25).

Yalin and Karahan (26) developed a similar relationship (Fig. 17.2.2) between τ_c^* and R_c^* using a large amount of experimental data. It is noted that at higher values of $R_c^* (> 70)$ the constant value of τ_c^* is 0.045. This relation (Fig. 17.2.2) is considered better than the more commonly used Shields' relation (9).

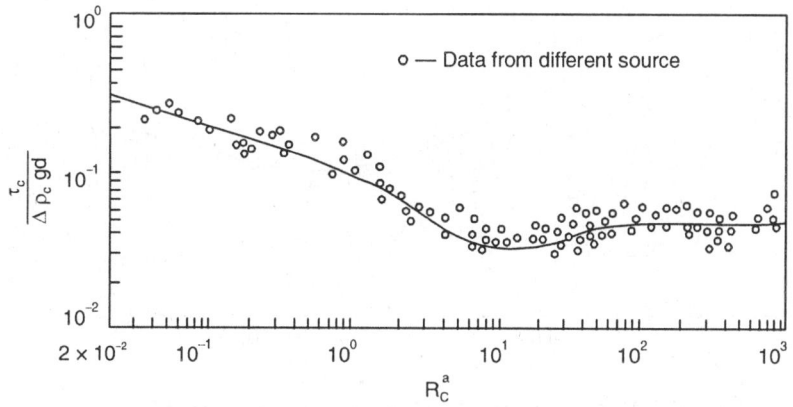

Fig. 17.2.2. Yalin and Karahan curve for incipient motion condition (26)

For given values of d, ρ_s, ρ, and v, the value of τ_c can be obtained from Fig. 17.2.1 or Fig. 17.2.2 by trial only as τ_c appears in both parameters τ_c^* and R_c^*. However, the ratio of R_c^* and $\sqrt{\tau_c^*}$ yields a parameter R_0^* which does not contain τ_c and is uniquely related to τ_c^*.

$$R_0^* = \frac{R_c^*}{\sqrt{\tau_c^*}} = \frac{u_{*c}d}{v}\left(\frac{\Delta\rho_s g d}{\tau_c}\right)^{1/2} = \left(\frac{\Delta\rho_s g d^3}{\rho v^2}\right)^{1/2} \qquad ...(17.2.4)$$

Since R_c^* is uniquely related to τ_c^* (Fig. 17.2.2), another relationship between R_0^* and τ_c^* can be obtained using Fig. 17.2.2 and Eq. 17.2.4.

The relationship between R_0^* and τ_c^*, based on Yalin and Karahan's curve, is shown in Fig. 17.2.3 and can be used to obtain direct solution for τ_c for given values of d, ρ_s, ρ and v.

17.3 REGIMES OF FLOW

If the average shear stress on the bed of an alluvial channel exceeds the critical shear stress, the bed particles start moving and, thus, disturb the plane bed condition. Depending upon the prevailing flow conditions and other relevant parameters, the bed and the water surface attain different forms. The features that form on the bed of an alluvial channel due to the flow of water are called '*bed forms*', or '*bed irregularities*' or '*sand waves*'. Garde and Albertson (6) introduced another term '*regimes of flow*' defined in the following manner:

'As the sediment characteristics, the flow characteristics and/or fluid characteristics are changed in an alluvial channel, the nature of the surface and water surface changes accordingly. These types of the bed and water surfaces are classified according to their characteristics and are called regimes of flow.'

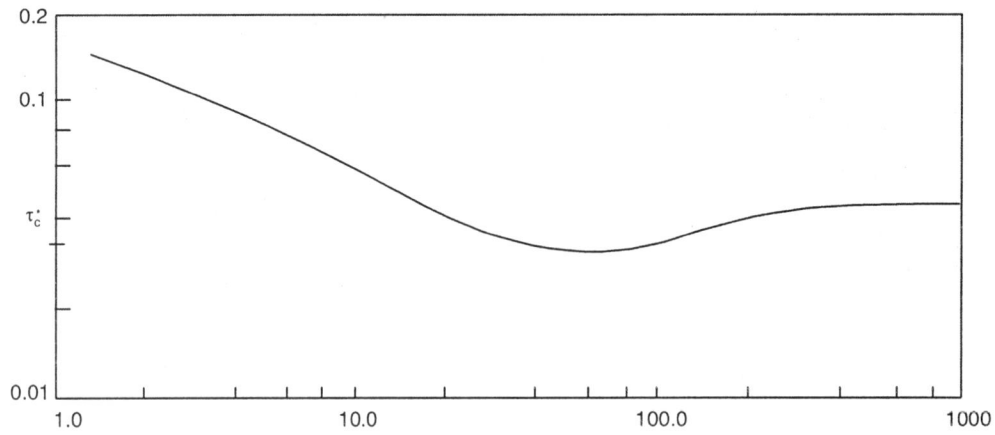

Fig. 17.2.3. Variation of R_0^* and τ_c^* based on Fig. 17.2.2

Regimes of flow will affect significantly the velocity distribution, resistance relations, and the transport of sediment. The regimes of flow may be divided into the following four categories:

    (i) Plane bed with no sediment movement,
   (ii) Ripples and dunes,
  (iii) Transition, and
  (iv) Antidunes.

17.3.1 Plane Bed with no Sediment Movement

If sediment and flow characteristics in a channel are such that the average shear stress on the bed is less than the critical shear stress, the sediment particles on the bed do not move. The bed remains plane and the channel boundary can, therefore, be treated as a rigid boundary. The water surface remains fairly smooth if the Froude number is low. Resistance offered to the flow is on account of the grain roughness only, and the Manning's equation may be used for prediction of the mean velocity of flow with the Manning's n obtained from the Strickler's equation, as discussed later in this chapter.

17.3.2 Ripples and Dunes

The sediment particles on the bed are set in motion when the average shear stress of the flow τ_0 exceeds the critical shear τ_c. As a result of this sediment motion, small triangular undulations known as ripples form on the bed [Fig. 17.3.1(a)]. Ripples would not occur for the sediment coarser than 0.6 mm. The length (between two adjacent troughs or crests) of the ripples is usually less than 0.4 m and their height (trough to crest) seldom exceeds 40 mm. The sediment motion is confined to the region near the bed and the sediment particles move either by sliding or taking a series of hops.

With increase in discharge (and, therefore, the average shear stress τ_0) the ripples grow into dunes [Fig. 17.3.1(b)]. Dunes, like ripples, are triangular undulations but of larger dimensions. These undulations are also unsymmetrical with a flat upstream face inclined at about 10-20° with the horizontal and steep downstream face whose angle of inclination with the horizontal is approximately equal to the angle of repose of the sediment material. Sometimes, ripples appear on the upstream face of a dune. The dunes in laboratory flumes may have length and height up to about 3 m and 0.4 m, respectively. But, in large rivers, the dunes may be several hundred metres long and up to about 15 m in height. The flow conditions correspond to the subcritical range and water surface falls over the crest of dunes (see Art. 13.8) and, hence, the water surface waves are out of phase with the bed waves. While most of the sediment particles move along the bed, some finer particles of the sediment may go and remain in suspension.

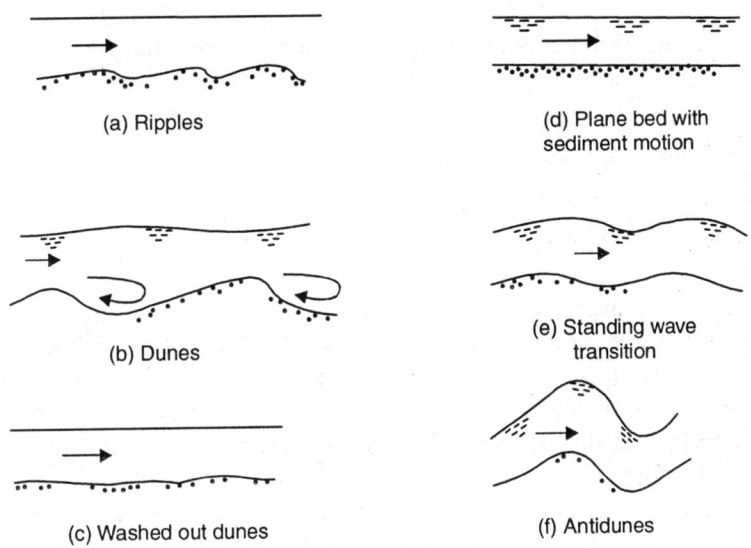

(a) Ripples

(d) Plane bed with sediment motion

(b) Dunes

(e) Standing wave transition

(c) Washed out dunes

(f) Antidunes

Fig. 17.3.1. Regimes of flow in alluvial channels

Ripples and dunes have many common features and, therefore, are generally dealt with together as one regime of flow. Both ripples and dunes move downstream slowly. Kondap and Garde (12) have obtained an approximate equation for the advance velocity of ripples and dunes, U_w, as follows:

$$\frac{U_w}{U} = 0.021 \left(\frac{U}{\sqrt{gh}}\right)^3 \qquad \qquad ...(17.3.1)$$

Here, U is the mean velocity of flow, and h is the average depth of flow. The average length (L) and height (H) of ripples and dunes can be estimated from the equations proposed by Ranga Raju and Soni (18):

$$\frac{H}{d}\left[\frac{U}{\sqrt{gR}}\right]\frac{U}{\sqrt{(\Delta\rho/\rho)gd}} = 6500\left(\tau'_*\right)^{8/3} \qquad \qquad ...(17.3.2)$$

$$\left[\frac{U}{\sqrt{(\Delta\rho/\rho)gd}}\right]^3 \frac{U}{\sqrt{gR}} \times \frac{L}{D} = 1.8\times10^8\left(\tau'_*\right)^{10/3} \qquad \qquad ...(17.3.3)$$

in which
$$\tau'_* = \frac{\rho R' S}{\Delta \rho_s d}$$
...(17.3.4)

and R is the hydraulic radius of the channel.

The value of R' (i.e., hydraulic radius corresponding to the grain roughness) is computed from the equation

$$U = \frac{1}{n_s} R^{2/3} S^{1/2}$$
...(17.3.5)

The value of n_s, (i.e., the Manning's roughness coefficient for the grains alone) is calculated from Strickler's equation,

$$n_s = \frac{d^{1/6}}{25.6}$$
...(17.3.6)

in which d is in metres.

17.3.3 Transition

With further increase in the discharge over the duned bed, the ripples and dunes are washed away, and only some very small undulations are left on the bed [Fig. 17.3.1(c)]. In some cases, however, the bed becomes plane but the sediment particles are in motion [Fig. 17.3.1(d)]. With slight increase in discharge, the bed and water surface attain the shape of a sinusoidal wave form. Such waves, known as standing waves [Fig. 17.3.1(e)], form and disappear and their size does not increase much. In this regime of transition, therefore, there is considerable variation in bed forms from washed-out dunes to plane bed with sediment motion and then to standing waves. The Froude number for this regime is relatively high. Besides the particles moving along the bed, large amount of sediment particles move in suspension. This regime is extremely unstable. The resistance to flow for this regime is relatively small as the bed is relatively plane.

17.3.4 Antidunes

When the discharge is further increased, the flow becomes supercritical (i.e., the Froude number is greater than unity), the standing waves (i.e., symmetrical bed and water surface waves) move upstream and break intermittently. However, the sediment particles keep on moving downstream only. Since the direction of movement of bed forms in this regime is opposite to that of the dunes, the regime is termed antidunes, Fig.17.3.1(f). The sediment transport rate is, obviously, very high. However, the resistance to flow is small compared to that of the ripple and dune regime. In canals and natural streams, antidunes would rarely occur.

17.3.5 Importance of Regimes of Flow

In rigid boundary channels, the resistance to flow is on account of the surface roughness (i.e., grain roughness) only except at very high Froude numbers when wave resistance may also be present. But, in alluvial channels, the total resistance to flow comprises the form resistance (due to bed forms) and grain resistance. In the ripple and dune regime, the form resistance may be an appreciable fraction of the total resistance. Because of the varying conditions of the bed of an alluvial channel, the form resistance is a highly varying quantity. Therefore, any meaningful resistance relation for alluvial channels should be regime-dependent. It is also evident that the stage-discharge relationship for an alluvial channel will also be affected by the regimes of flow.

The form resistance, which is on account of the difference in pressures on the upstream and downstream side of the undulations, acts normal to the surface of the undulations. As such, the form resistance is rather ineffective in affecting the transport of sediment. Only grain shear (i.e., the shear stress corresponding to grain resistance) affects the movement of sediment.

17.3.6 Prediction of Regimes of Flow

There are many methods for the prediction of regimes. The method proposed by Garde and Ranga Raju (7) has been described here.

The functional relationship for resistance to flow in alluvial channels is written, following the principles of dimensional analysis, as follows:

$$\frac{U}{\sqrt{(\Delta\rho_s/\rho)gd}} = f\left[\frac{R}{d}, \frac{S}{\Delta\rho_s/\rho}, \frac{g^{1/2}d^{3/2}}{v}\right] \qquad ...(17.3.7)$$

Here, S is the slope of the channel bed. Since resistance to flow and the regime of flow are closely related with each other, it was premised that the parameters on the right-hand side of Eq. (17.3.7) would predict the regime of flow. The third parameter (i.e., $g^{1/2}d^{3/2}/v$) was dropped from the analysis on the plea that the influence of viscosity in the formation of bed waves is rather small. The data from natural streams, canals, and laboratory flumes in which the regimes had also been observed, were used to prepare Fig. 17.3.2 on which lines demarcating the regimes of flow have been drawn. The data used in

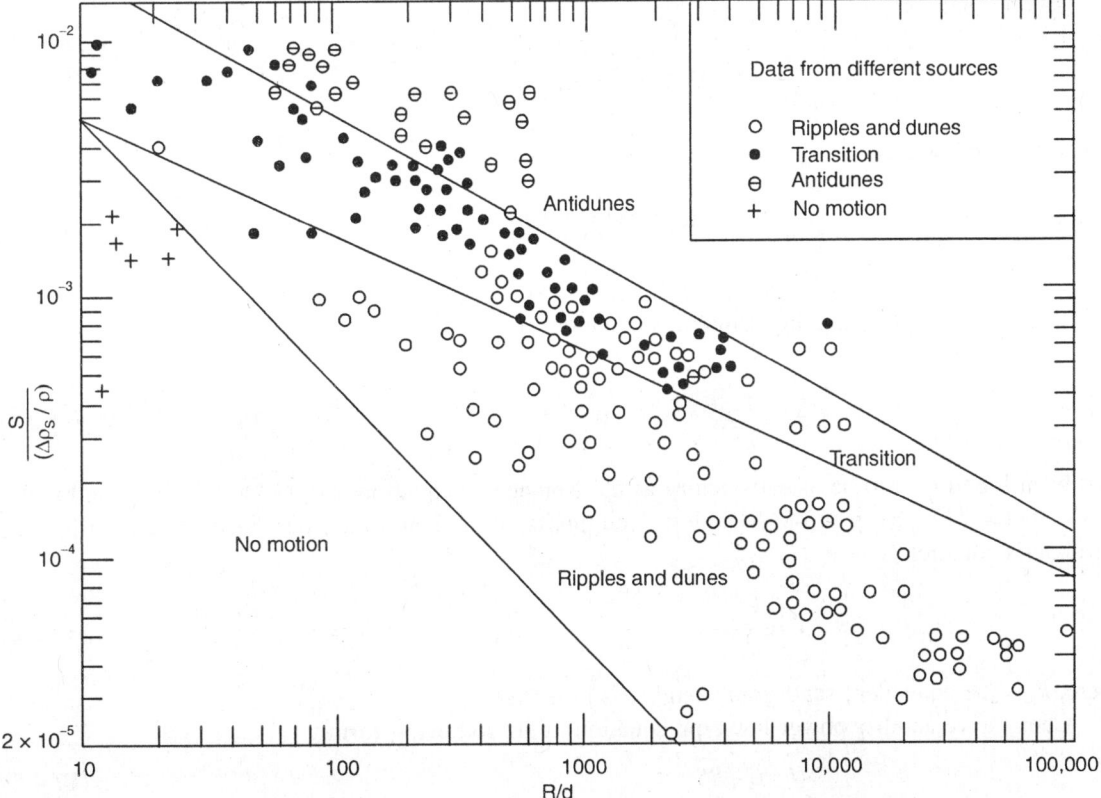

Fig. 17.3.2. Predictor for regimes of flow in alluvial channels (7)

developing Fig. 17.3.2 cover a wide range of depth of flow, slope, sediment size, and density of sediment.

It should be noted that the lines of 45° slope on Fig. 17.3.2 – such as the line demarcating 'no motion' and 'ripples and dunes' regimes – represent a line of constant value of $\tau_* \left(= \dfrac{\rho RS}{\Delta \rho_s d} \right)$. This means that different regimes of flow can be obtained at the same shear stress by varying suitably the individual values of R and S. Therefore, shear stress by itself cannot adequately define regimes of flow.

The method of using Fig. 17.3.2 for prediction of regimes of flow consists of simply calculating the parameters R/S and $S/(\Delta \rho_s / \rho)$ and then finding the region in which the corresponding point falls. One obvious advantage of this method is that it does not require knowledge of the mean velocity U and is, therefore, ideally suitable for prediction of regimes for resistance problems.

17.4 RESISTANCE TO FLOW IN ALLUVIAL CHANNELS

The resistance equation is a mathematical relation among the mean velocity of flow U, hydraulic radius R, and characteristics of the channel boundary. For steady and uniform flow in rigid boundary channels, the Keulegan's equations (logarithmic type) or power-law type equations (like the Chezy's and Manning's equations) are used. Keulegan (11) obtained the following logarithmic relations for rigid boundary channels:

For smooth boundaries,

$$\frac{U}{u_*} = 5.75 \log \left(\frac{u_* R}{v} \right) + 3.25 \qquad \qquad ...(17.4.1)$$

For rough boundaries,

$$\frac{U}{u_*} = 5.75 \log \left(R/k_s \right) + 6.25 \qquad \qquad ...(17.4.2)$$

For the range $5 < \dfrac{R}{k_s} < 700$, the Manning's equation

$$U = \frac{1}{n} R^{2/3} S^{1/2} \qquad \qquad ...(17.4.3)$$

has been found (11) to be as satisfactory as the Keulegan's equation [Eq. (17.4.2)] for rough boundaries. In Eq. (17.4.3), n is the Manning's roughness coefficient which can be calculated using the Strickler's equation

$$n = \frac{k_s^{1/6}}{25.6} \qquad \qquad ...(17.4.4)$$

Here, k_s is the equivalent sand grain roughness in metres.

Chezy gave another power-law type equation in the following form:

$$U = C \sqrt{RS} \qquad \qquad ...(17.4.5)$$

Comparing the Manning's and Chezy's equations, one obtains

$$\frac{U}{u_*} = \frac{C}{\sqrt{g}} = \frac{R^{1/6}}{n\sqrt{g}} = \left(\frac{R}{k_s}\right)^{1/6} \frac{25.6}{\sqrt{g}} \qquad \qquad ...(17.4.6)$$

So long as the average shear stress τ_0 on the boundary of an alluvial channel is less than the critical shear τ_c, the channel boundary can be considered rigid and any of the resistance equations valid for rigid boundary channels would yield results for the alluvial channel too. However, as soon as sediment movement starts, undulations develop on the bed, thereby increasing the boundary resistance. Besides, some energy is required to move the sediment. Further, the sediment particles in suspension also affect the resistance of alluvial streams. The suspended sediment particles dampen the turbulence or interfere with the production of turbulence near the bed where the concentration of these particles and the rate of turbulence production are maximum. It is, therefore, obvious that the problem of resistance in alluvial channels is very complex and the complexity further increases if one considers the effects of channel shape, non-uniformity of sediment size, discharge variation, and other factors on channel resistance. None of the resistance equations for alluvial channels developed so far takes all these factors into consideration.

The methods for computing resistance in alluvial channels can be grouped into two broad categories. The first category of methods deals with the overall resistance and uses either a logarithmic type relation or a power-law type relation for the mean velocity. The second category of methods separates the total resistance into grain resistance and form resistance (i.e., the resistance that develops on account of undulations on the channel bed). Both categories of methods generally deal with uniform steady flow.

17.4.1 Resistance Relationships based on Total Resistance Approach

The following equation, proposed by Lacey (13) on the basis of analysis of stable channel data from India, is the simplest relationship for alluvial channels:

$$U = 10.8 \, R^{2/3} \, S^{1/3} \qquad \qquad ...(17.4.7)$$

However, this equation is applicable only under regime conditions (see Art. 17.6.) and, hence, has only limited application.

Garde and Ranga Raju (8) analysed data from streams, canals, and laboratory flumes to obtain an empirical relation for prediction of mean velocity in an alluvial channel. The functional relation, Eq. (17.3.7), may be rewritten (8) as

$$\frac{U}{\sqrt{(\Delta \rho_s / \rho) g R}} = f\left[\frac{R}{d}, \frac{S}{\Delta \rho_s / \rho}, \frac{g^{1/2} d^{3/2}}{v}\right] \qquad \qquad ...(17.4.8)$$

By employing usual graphical techniques and using alluvial channel data of canals, rivers, and laboratory flumes, covering a large range of sediment size and depth of flow, a graphical relation (8, 19)

between $k_1 \dfrac{U}{\sqrt{(\Delta \rho_s / \rho) g R}}$ and $k_2 \left(\dfrac{R}{d}\right)^{1/3} \dfrac{S}{\Delta \rho_s / \rho}$, Fig. 17.4.1, was obtained for the prediction of the

mean velocity U. The coefficients k_1 and k_2 were related to the sediment size d by the graphical relations shown in Fig. 17.4.2. It may be noted that the dimensionless parameter $g^{1/2} d^{3/2} / v$ has been replaced by the sediment size alone on the plea that the viscosity of the liquid for a majority of the data used in the analysis did not change much (19). This method is expected to yield results with an accuracy of

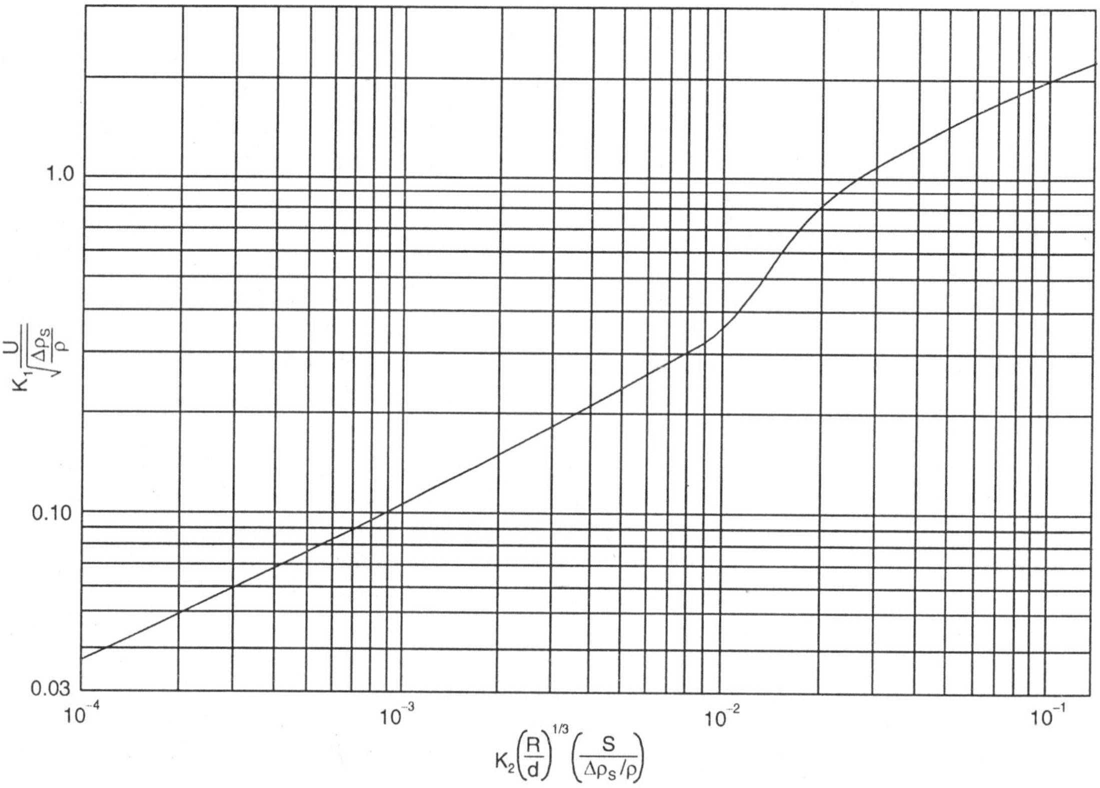

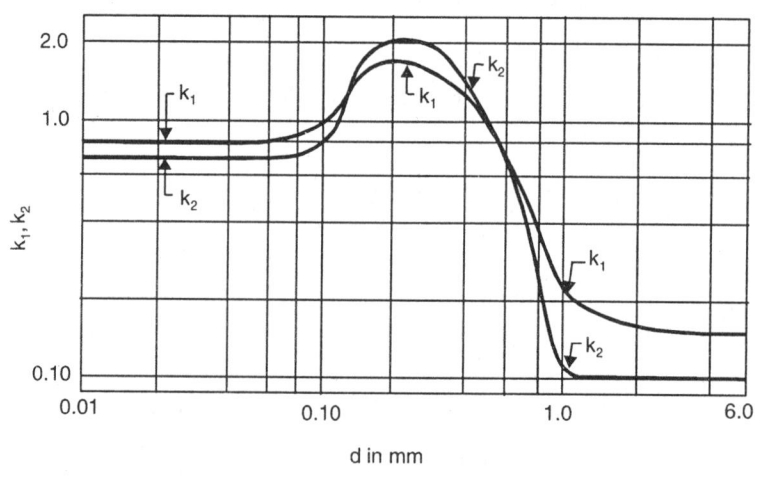

Fig. 17.4.1. Resistance relationship for alluvial channels (19)

Fig. 17.4.2. Variation of k_1 and k_2 with sediment size (19)

about 30 per cent (22). For given S, d, $\Delta\rho_s$, ρ, and the stage-hydraulic radius curve and stage-area curve of the channel cross-section, the stage-discharge curve for an alluvial channel can be computed as follows:

(i) Assume a stage and read the values of hydraulic radius and area of cross-section A from stage-hydraulic radius and stage-area curves, respectively.

(ii) Determine k_1 and k_2 for known value of d using Fig. 17.4.2.

(iii) Compute $k_2 (R/d)^{1/3} \dfrac{S}{\Delta\rho_s/\rho}$ and read the value of $k_1 \dfrac{U}{\sqrt{(\Delta\rho_s/\rho)gR}}$ from Fig. 17.4.1.

(iv) Calculate the value of the mean velocity U and, hence, the discharge.

(v) Repeat the above steps for other values of stages.

Finally, a graphical relation between stage and discharge can be prepared.

17.4.2 Resistance Relationship based on Division of Resistance

While dealing with open channel flows, hydraulic radius R of the flow cross-section is taken as the characteristic depth parameter. The use of this parameter requires that the roughness over the whole wetted perimeter is the same. Such a condition can be expected in a very wide channel with alluvial bed and banks. However, laboratory flumes with glass walls and sand bed would have different roughnesses on the bed and side walls. In such situations, therefore, the hydraulic radius of the bed R_b is used instead of R in the resistance relations. The hydraulic radius of the bed R_b can be calculated using Einstein's method (2) that assumes that the velocity is uniformly distributed over the whole cross-section. Assuming that the total area of cross-section of flow A can be divided into areas A_b and A_w corresponding to the bed and walls, respectively, one can write

$$A = A_w + A_b$$

For rectangular channels, one can, therefore, obtain

$$(B + 2h)R = 2hR_w + BR_b$$

$$\therefore \quad R_b = \left(1 + \frac{2h}{B}\right)R - \frac{2h}{B}R_w \qquad \qquad ...(17.4.9)$$

$$= (B + 2h)(R/B) - (2hR_w/B)$$

$$= (PR/B) - (2hR_w/B)$$

$$= (A/B) - (2hR_w/B)$$

$$= h - (2hR_w/B)$$

Using the Manning's equation for the walls, i.e.,

$$U = \frac{1}{n_w}R_w^{2/3} S^{1/2} \qquad \qquad ...(17.4.10)$$

one can compute the hydraulic radius of the wall R_w if the Manning's coefficient for the walls, n_w is known. Using Eq. (17.4.9), the hydraulic radius of the bed R_b can be computed.

Einstein and Barbarossa (3) developed a rational solution to the problem of resistance in alluvial channels by dividing the total bed resistance (or shear) τ_{0b} into resistance (or shear) due to sand grains τ'_{0b} and resistance (or shear) due to the bed forms τ''_{0b}, i.e.,

$$\tau_{0b} = \tau'_{0b} + \tau''_{0b} \qquad \qquad ...(17.4.11)$$

or
$$\rho g\, R_b\, S = \rho g\, R'_b S + \rho g\, R''_b\, S$$

i.e.,
$$R_b = R'_b + R''_b \qquad \qquad ...(17.4.12)$$

where, R'_b and R''_b are hydraulic radii of the bed corresponding to grain and form resistances (or roughnesses).

For a hydrodynamically rough plane boundary, the Manning's roughness coefficient for the grain roughness n_s, is given by the Strickler's equation, i.e.,

$$n_s = \frac{d_{65}^{1/6}}{24.0} \qquad \qquad ...(17.4.13)$$

Here, d_{65} (in metres) represents the sieve diameter through which 65 per cent of the sediment will pass through, i.e., 65 per cent of the sediment is finer than d_{65}. Therefore, the Manning's equation can be written as

$$U = \frac{1}{n_s} R_b^{2/3}\, S^{1/2}$$

$$U = \frac{24}{d_{65}^{1/6}} R_b'^{2/3}\, S^{1/2} \qquad \qquad ...(17.4.14)$$

Since
$$U'_* = \sqrt{\tau'_{0b}/\rho} = \sqrt{gR'_b S}$$

$$\frac{U}{U'_*} = \left(\frac{24}{d_{65}^{1/6}}\right) \frac{R_b'^{2/3}\, S^{1/2}}{\sqrt{gR'_b S}}$$

or
$$\frac{U}{U'_*} = 7.66 \left(\frac{R'_b}{d_{65}}\right)^{1/6} \qquad \qquad ...(17.4.15)$$

Einstein and Barbarossa (3) replaced this equation with the following logarithmic relation having theoretical support:

$$\frac{U}{U'_b} = 5.75 \log \frac{12.27\, R'_b}{d_{65}} \qquad \qquad ...(17.4.16)$$

Equation (17.4.16) is valid for a hydrodynamically rough boundary. A viscous correction factor x (which is dependent on d_{65}/δ', Table 17.4.1, Fig. 17.4.3) was introduced in this equation to make it applicable to boundaries consisting of finer material $\left(d_{65}/\delta' < 10\right)$. The modified equation is (3)

$$\frac{U}{U'_*} = 5.75 \log \frac{12.27\, R'_b\, x}{d_{65}} \qquad\qquad ...(17.4.17)$$

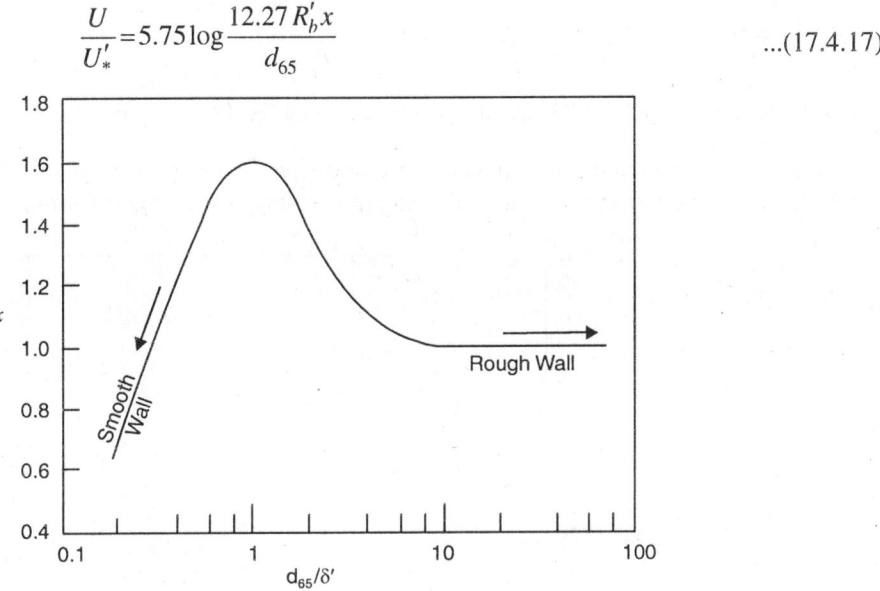

Fig. 17.4.3. Correction x in Eq. (17.4.17)

Table 17.4.1 Variation of x with d_{65}/δ' (3)

d_{65}/δ'	0.2	0.3	0.5	0.7	1.0	2.0	4.0	6.0	10
x	0.7	1.0	1.38	1.56	1.61	1.38	1.10	1.03	1.0

Einstein and Barbarossa (3) recommended that one of the equations, Eq. (17.4.15) or Eq. (17.4.16), may be used for practical problems. The resistance (or shear) due to bed forms τ'_{0b} is calculated by considering that there are N undulations of cross-sectional area a in a length of channel L with total wetted perimeter P. Total form drag F on these undulations is given by

$$F = C_D\, a\left(\frac{1}{2}\right)\rho U^2 N \qquad\qquad ...(17.4.18)$$

Here, C_D is the average drag coefficient of the undulations. Since this drag force acts on area LP, the average shear stress τ''_{0b} will be given as

$$\tau''_{0b} = \frac{F}{LP} = \frac{C_D a N}{LP}\,\rho\,\frac{U^2}{2}$$

$$\therefore \qquad \frac{\tau''_{0b}}{\rho} = U''^2_* = \frac{C_D\, aN}{LP}\,\frac{U^2}{2}$$

or

$$\frac{U}{U''_*} = \sqrt{2LP/\left(C_D\, aN\right)} \qquad\qquad ...(17.4.19)$$

Here, U_*'' is the shear velocity corresponding to bed undulations. According to Einstein and Barbarossa, the parameters on the right hand side of Eq. (17.4.19) would primarily depend on sediment transport rate which is a function of Einstein's parameter $\psi' = \Delta\rho_s d_{35} / \rho R_b' S$. Therefore, they obtained an empirical relation, Fig. 17.4.4, between $\dfrac{U}{U_*''}$ and ψ' using field data from natural streams. The relationship proposed by Einstein and Barbarossa can be used to compute mean velocity of flow for a given stage (i.e., depth of flow) of the river and also to prepare stage–discharge relationship.

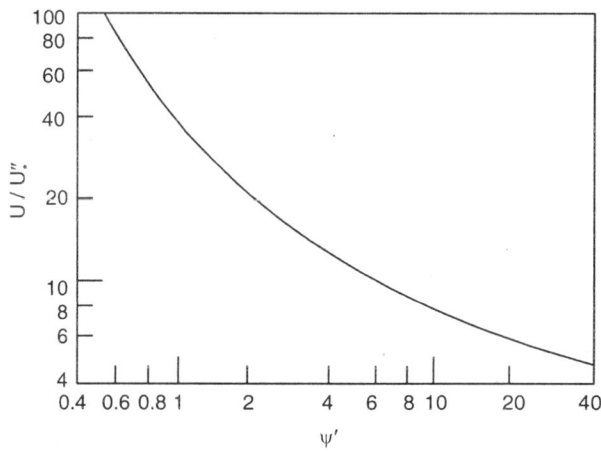

Fig. 17.4.4. Einstein and Barbarossa relation between U/U_*'' and ψ' (3)

The computation of mean velocity of flow for a given stage requires a trial procedure. From the known channel characteristics, the hydraulic radius R of the flow section can be determined for a given stage (or depth of flow) of the river for which the mean velocity of flow is to be predicted. For a wide alluvial river, this hydraulic radius R approximately equals R_b. A value of R_b' smaller than R_b is assumed and a trial value of the mean velocity U is calculated from Eq. (17.4.14) or Eq. (17.4.15) or Eq. (17.4.16). The value of $\dfrac{U}{U_*''}$ is read from Fig. 17.4.4 for ψ' corresponding to the assumed value of R_b'.

From known values of U (trial value) and U/U_*'', U_*'' and, hence, R_b'' can be computed. If the sum of R_b' and R_b'' equals R_b, the assumed value of R_b' and, hence, the corresponding mean velocity of flow U computed from Eq. (17.4.14) or Eq. (17.4.15) or Eq. (17.4.16) are okay. Otherwise, repeat the procedure for another trial value of R_b' till the sum of R_b' and R_b'' equals R_b. The computations can be carried out easily in a tabular form as illustrated in the example, E17.5.

For preparing a stage-discharge curve, one needs to obtain discharges corresponding to different stages of the river. If one neglects bank friction (*i.e.*, $R = R_b$), the procedure, requiring no trial, is as follows:

For an assumed value of R_b', the mean velocity of flow U is computed from Eq. (17.4.15) and U/U_*'' is read from Fig. 17.4.4 for ψ' corresponding to the assumed value of R_b'. From known values of U and U/U_*'', one can determine U_*'' and, hence, R_b''. The sum of R_b' and R_b'' gives R_b which equals R (if bank friction is neglected). Corresponding to this value of R, one can determine the stage and, hence, the area of flow cross-section A. The product of U and A gives the discharge, Q corresponding to the stage. Likewise, for another value of R_b', one can determine stage and the corresponding discharge.

17.5 TRANSPORT OF SEDIMENT

When the average shear stress τ_0 on the bed of an alluvial channel exceeds the critical shear stress τ_c, the sediment particles start moving in different manners depending on the flow condition, sediment size, fluid and sediment densities, and channel condition.

At relatively low shear stresses, the particles roll or slide along the bed. The particles remain in continuous contact with the bed and the movement is generally discontinuous. Sediment material transported in this manner is termed *contact load*.

On increasing the shear stress further, some sediment particles lose contact with the bed for some time, and 'hop' or 'bounce'. The sediment particles moving in this manner fall into the category of *saltation load*. This mode of transport is significant only in case of noncohesive materials of relatively high fall velocities such as sand in air and, to a lesser extent, gravel in water.

Since saltation load is insignificant in case of flow of water and also because it is difficult to distinguish between saltation load and contact load, the two are grouped together and termed *bed load*, and is defined as the sediment load transported on or near the bed.

With further increase in the shear stress, the particles may go in suspension and remain in suspension. Such sediment load is included in the *suspended load*. For sediment particles to move in suspension, $u_*/w_0 > 0.5$. Here, w_0 is the fall velocity for the sediment particle of given size.

The material for bed load as well as a part of the suspended load originates from the bed of the channel and, hence, both are grouped together and termed *bed-material load*.

Analysis of suspended load data from rivers and canals has shown that the suspended load comprises the sediment particles originating from the bed and the sediment particles that are not available in the bed. The former is the bed-material load in suspension and the latter is the product of erosion in the catchment and is, therefore, called *wash load*. The wash load, having entered the stream, is unlikely to deposit unless the velocity (or the shear stress) is reduced considerably or the concentration of such fine sediments is very high. The transport rate of wash load is related to the availability of fine material in the catchment and its erodibility and is, normally, independent of the hydraulic characteristics of the stream. As such, it is difficult to make an estimate of wash load.

When the bed-material load in suspension is added to the bed-material load moving as bed load, one gets the total bed-material load which may be major or minor fraction of the total load comprising bed-material load and wash load of the stream depending on the catchment characteristics.

Irrigation channels carrying silt-laden water and flowing through alluvial bed are designed to carry certain amounts of water as well as sediment discharges. This means that the total sediment load transport will affect the design of an alluvial channel. Similarly, problems related to reservoir sedimentation, aggradation, degradation, etc. can be solved only if the total sediment load being transported by river (or channel) is known. One obvious method of estimation of total load is to determine bed load, suspended load, and wash load individually and then add these together. The wash load is usually carried without being deposited and is also not easy to estimate. This load is, therefore, ignored while analyzing flow in alluvial channels.

It should, however, be noted that the available methods of computation of bed-material load are such that errors of the order of one magnitude are not uncommon. If the bed-material load is a large fraction of the total load, the foregoing likely error would considerably reduce the validity of the computations. This aspect of sediment load computations must always be kept in mind while evaluating the result of the computations.

17.5.1 Bed Load

The prediction of the bed load transport is not an easy task because it is interrelated with the resistance to flow that, in turn, is dependent on flow conditions. Nevertheless, several attempts have been made to propose methods - empirical as well as semi-theoretical - for the computation of bed load. The most commonly used empirical relation is given by Meyer-Peter and Müller (17). This relation is based on: (i) the division of total shear into grain shear and form shear, and (ii) the assumption that the bed load transport is a function of only the grain shear. This equation, written in dimensionless form, is as follows:

$$\left[\frac{n_s}{n}\right]^{3/2} \frac{\rho RS}{\Delta\rho_s d_a} = 0.047 + 0.25 \frac{q_B^{2/3}}{\rho_s^{2/3} g d_a \left(\frac{\Delta\rho_s}{\rho}\right)^{1/3}} \qquad ...(17.5.1)$$

which is rewritten as

$$\tau_*' = 0.047 + 0.25 \phi_B^{2/3} \qquad ...(17.5.2)$$

or

$$\phi_B = 8\left(\tau_*' - 0.047\right)^{3/2} \qquad ...(17.5.3)$$

where, ϕ_B is the bed load function and equals $\dfrac{q_B}{\rho_s g^{3/2} d_a^{3/2} \sqrt{\Delta\rho_s / \rho}}$,

τ_*' is the dimensionless grain shear and equals $\dfrac{\tau_0'}{\Delta\rho_s g d_a}$,

and τ_0' is the grain shear and equals $\left[\dfrac{n_s}{n}\right]^{3/2} \rho g RS.$

Here, q_B is the rate of bed load transport in weight per unit width, i.e., N/m/s and d_a is the arithmetic mean size of the sediment particles which generally varies between d_{50} and d_{60} (9).

From Eq. (17.5.3), it may be noted that the value of the dimensionless shear τ_*' at the incipient motion condition (i.e., when q_B and, hence, ϕ_B is zero) is 0.047. Therefore, $\left(\tau_*' - 0.047\right)$ can be interpreted as the effective shear stress causing bed load movement.

The layer in which the bed load moves is called the bed layer and its thickness is generally approximated as 2d.

A semi-theoretical analysis of the problem of the bed load transport was first attempted by Einstien (2) in 1942 when he did not consider the effect of bed forms on bed load transport. Later, he presented a modified solution (4) to the problem of bed load transport. Einstein's solution does not use the concept of critical tractive stress but, instead, is based on the premise that a sediment particle resting on the bed is set in motion when the instantaneous hydrodynamic lift force exceeds the submerged weight of the particle. Based on his semi-theoretical analysis, a curve, Figure 17.5.1, between the Einstein's bed load parameter

$$\phi_B\left(=\frac{q_B}{\rho_s g^{3/2} d^{3/2}\sqrt{\Delta\rho/\rho}}\right) \text{ and } \psi'\left(=\Delta\rho_s d/\rho R'S\right)$$

can be used to compute the bed load transport in case of uniform sediment. The coordinates of the curve of Fig. 17.5.1 are given in Table 17.5.1. The method involves computation of ψ' for given sediment characteristics and flow conditions and reading the corresponding value of ϕ_B from Fig. 17.5.1 to obtain the value of q_B.

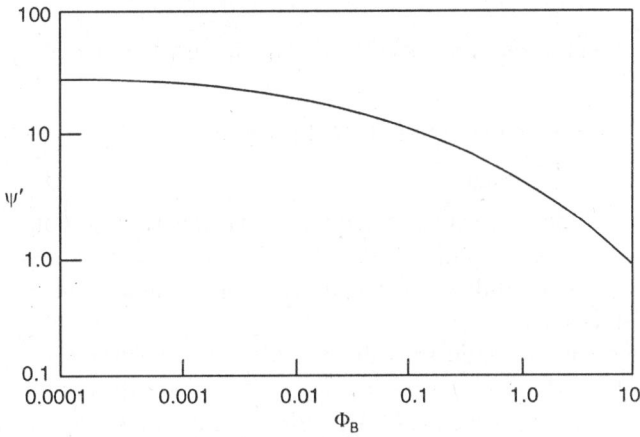

Fig. 17.5.1. Einstein's bed load transport relation (4)

Table 17.5.1: Relationship between ϕ_B and ψ' (4)

ψ'	27.0	24.0	22.4	18.4	16.4	11.5	9.5	5.5	4.08	1.4	0.70
ϕ_B	10^{-4}	5×10^{-4}	10^{-3}	5×10^{-3}	10^{-2}	5×10^{-2}	10^{-1}	5×10^{-1}	1.0	5.0	10.0

17.5.2 Suspended Load

At the advanced stage of bed load movement, the average shear stress is relatively high and finer particles may move into suspension. With the increase in the shear stress, coarser fractions of the bed material will also move into suspension. The particles in suspension move with a velocity almost equal to the flow velocity. It is also evident that the concentration of sediment particles will be maximum at or near the bed and that it would decrease as the distance from the bed increases. The concentration of suspended sediment is generally expressed as follows:

(i) *Volume concentration*: The ratio of absolute volume of solids and the volume of sediment-water mixture is termed the volume concentration and can be expressed as percentage by volume.

(ii) *Weight concentration*: The ratio of the dry weight of solids and the weight of sediment-water mixture is termed the weight concentration and is usually expressed in parts per million (ppm).

1% of volume concentration equals 10,000 ppm by volume.

Starting from the differential equation for the distribution of suspended material in the vertical and using an appropriate diffusion equation, Rouse (23) developed the following equation for sediment distribution (i.e., variation of sediment concentration along the vertical):

$$\frac{C}{C_a} = \left(\frac{h-y}{y} \times \frac{a}{h-a} \right)^{Z_0} \qquad ...(17.5.4)$$

where, C = the concentration at a distance y from the bed,

C_a = the reference concentration at $y = a$,

h = the depth of flow,

$Z_0 = \dfrac{w_0}{u_* k}$ and is the exponent in the sediment distribution equation,

w_0 = the fall velocity of the sediment particles,

and k = Karman's constant.

Rouse's equation, Eq. (17.5.4), assumes two-dimensional steady flow, constant fall velocity, and fixed Karman's constant. However, it is known that the fall velocity and Karman's constant vary with concentration and turbulence (9). Further, knowledge of some reference concentration C_a at $y = a$ is required for the use of Eq. (17.5.4).

Knowledge of the velocity distribution and the concentration variation (Fig. 17.5.2) would enable one to compute the rate of transport of suspended load q_s. Consider a strip of unit width and thickness dy at an elevation y. The volume of suspended load transported past this strip in a unit time is equal to $\dfrac{1}{100} Cudy$. Here, C is the volume concentration (expressed as percentage) at an elevation y where the velocity of flow is u. Thus,

$$q_s = \frac{\rho_s g}{100} \int_a^h Cudy \qquad ...(17.5.5)$$

where, q_s is the weight of the suspended load transported per unit width per unit time. Since the suspended sediment moves only on top of the bed layer, the lower limit of integration, a can be considered equal to the thickness of the bed layer, i.e., 2d.

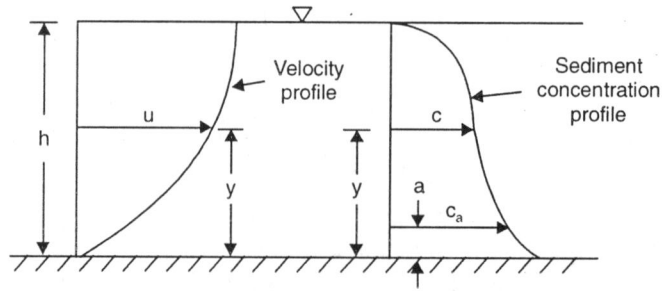

Fig. 17.5.2. Variation of velocity of flow and sediment concentration in the vertical

Instead of using the curves of the type shown in Fig. 17.5.2, one may use a suitable velocity distribution law and the sediment distribution equation, Eq. (17.5.4). For the estimation of the reference concentration C_a appearing in Eq. (17.5.4), Einstein (4) assumed that the average concentration of bed load in the bed layer equals the concentration of suspended load at $y = 2d$. This assumption is based on the fact that there will be continuity in the distribution of suspended load and bed load. Making use of suitable velocity distribution laws, the velocity of the bed layer was determined as $11.6u'_*$ and as such

the concentration in the bed layer was obtained as $\dfrac{(q_B/\rho_s g)}{(11.6u'_*)(2d)}$. Hence, the reference concentration C_a (in per cent) at $y = 2d$ is given as

$$C_a = C_{2d} = \frac{(q_B/\rho_s g)}{(23.2u'_*)(d)} \times 100 \qquad \qquad ...(17.5.6)$$

Equation (17.5.5) can now be integrated in a suitable manner.

17.5.3 Total Bed-Material Load

The total bed-material load can be determined by adding together the bed load and suspended load. There is, however, another category of methods too for the estimation of the total bed-material load. The supporters of these methods argue that the process of suspension is an advanced stage of tractive shear along the bed, and, therefore, the total load should be related to the shear parameter. One such method is proposed by Engelund and Hansen (5) who developed a relationship for the total bed-material load q_T (expressed as weight per unit width per unit time) by relating the sediment transport to the shear stress and friction factor f. The relationship is expressed as

$$f\phi_T = 0.4\,\tau_*^{5/2} \qquad \qquad ...(17.5.7)$$

where,

$$\phi_T = \frac{q_T}{\rho_s g^{3/2} d^{3/2} \sqrt{\Delta\rho_s/\rho}} \qquad \qquad ...(17.5.8)$$

and

$$f = \frac{8gRS}{U^2} \qquad \qquad ...(17.5.9)$$

The median size d_{50} is used for d in the above equation.

17.6 DESIGN OF ALLUVIAL CHANNELS

Surface water for either irrigation or power generation is generally conveyed from its source to the place of demand by means of canals or channels. These channels generally have alluvial boundaries and carry sediment-laden water. A hydraulic engineer is concerned with the design, construction, operation, maintenance, and improvement of such channels. Lane (14) gave the definition of stable channel as follows:

"A stable channel is an unlined earth channel: (a) which carries water, (b) the banks and bed of which are not scoured objectionably by moving water, and (c) in which objectionable deposits of sediment do not occur". This means that over a long period, the bed and banks of a stable channel remain unaltered even if minor deposition and scouring occur in the channel. Obviously, silting and scouring in a stable channel balance each other over a long period of time.

An irrigation channel can have either a rigid boundary or one consisting of alluvial material. These channels may have to carry either clear water or sediment-laden water. Accordingly, there can be four different types of problems related to the design of a stable channel. These are as follows:

(i) Rigid-boundary (i.e., non-erodible) channels carrying clear water,
(ii) Rigid-boundary channels carrying sediment-laden water,
(iii) Alluvial channels carrying clear water, and
(iv) Alluvial channels carrying sediment-laden water.

The design of a stable channel aims at obtaining the values of mean velocity, depth (or hydraulic radius), width and slope of the channel for known values of discharge Q, sediment discharge Q_T, sediment size d and the channel roughness characteristics without causing undue silting or scouring of the channel bed.

The design of rigid-boundary channels carrying clear water has been dealt with in Chapter-12. The design of rigid-boundary channels carrying sediment-laden water and the design of alluvial channels are being dealt with in this section.

17.6.1 Rigid Boundary Channels carrying Sediment-Laden Water

These channels are to be designed in such a way that the sediment in suspension does not settle on the channel boundary. The design is, therefore, based on the concept of minimum permissible velocity which will prevent both sedimentation as well as growth of vegetation. In general, velocities of 0.7 to 1.0 m/s will be adequate for this purpose if the sediment load is small. If the sediment concentration is large, Fig. 17.6.1 can be used to ensure that the sediment does not deposit (1). In Fig. 17.6.1, C_s is the concentration of sediment in ppm (by volume), f_b the friction factor of the channel bed, D_0 the central depth, T the top width and S_c equals $S/(\Delta\rho_s/\rho)$. If the designed channel section is not able to carry the specified sediment load, the slope S of the channel is increased.

17.6.2 Alluvial Channels carrying Clear Water

Compared to the design of rigid boundary channels, the design of stable alluvial channels is more complex. Alluvial channels carrying clear water should be designed so that the erodible material of the channel boundary is not scoured. The design of such channels is, therefore, based on the concept of tractive force.

Scour on a channel bed occurs when the tractive force on the bed exerted by the flow is sufficient to cause the movement of the bed particles. A sediment particle resting on the sloping side of a channel will move due to the resultant of the tractive force in the flow direction and the component of gravitational force that makes the particle roll or slide down the side slope. If the tractive force acting on the bed or the resultant of the tractive force and the component of the gravitational force both acting on the side slopes is larger than the force resisting the movement of the particle, erosion starts. Based on this principle, the following method of design of stable alluvial channels was proposed by Lane (15) and is known as Lane's method or the USBR method.

In uniform flow, the average tractive stress is given as

$$\tau_0 = \rho gRS \qquad \qquad ...(17.6.1)$$

The shear stress is not uniformly distributed over the channel perimeter. Figure 12.2.5 shows the variation of the maximum shear stresses acting on the side, τ_{sm} and the bed, τ_{bm} of a channel. It may be noted that for the trapezoidal channels, the maximum tractive shear on the side is approximately 0.76 times the tractive shear stress on the channel bed.

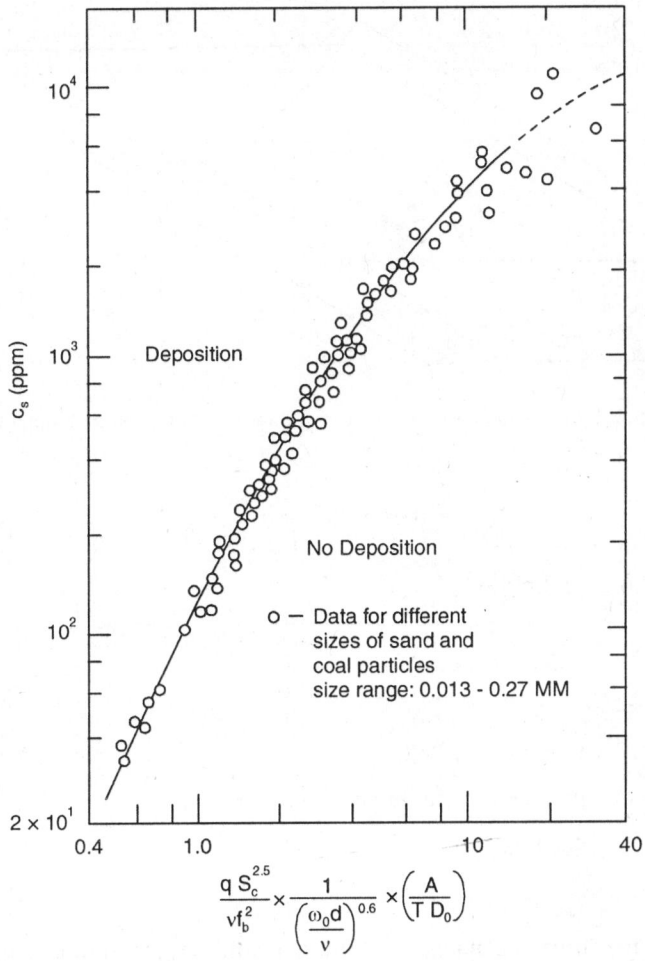

Fig. 17.6.1. Relation for limiting concentration of suspended sediment in channels (1)

For a particle resting on a level or mildly sloping bed, one can write the following expression for the incipient motion condition:

$$\tau_{bl} a = W_s \tan\theta \qquad \qquad ...(17.6.2)$$

where, W_s is the submerged weight of the particle and a, the effective area of the particle over which the tractive stress τ_{bl} is acting, and θ is the angle of repose for the particle. τ_{bl} is, obviously, the critical shear stress τ_c for the bed particles. Since the aim is to avoid the movement of the particles (15), τ_{bl} may be kept less than τ_c, say $0.9\ \tau_c$.

For any particle resting on either of the sloping sides of a channel (Fig. 17.6.2), the condition for the incipient motion is

$$\left(W_s \sin\alpha\right)^2 + \left(\tau_{sl} a\right)^2 = \left(W_s \cos\alpha \tan\theta\right)^2$$

$$\therefore \qquad W_s^2 \sin^2\alpha + \left(\tau_{sl} a\right)^2 = W_s^2 \cos^2\alpha \tan^2\theta$$

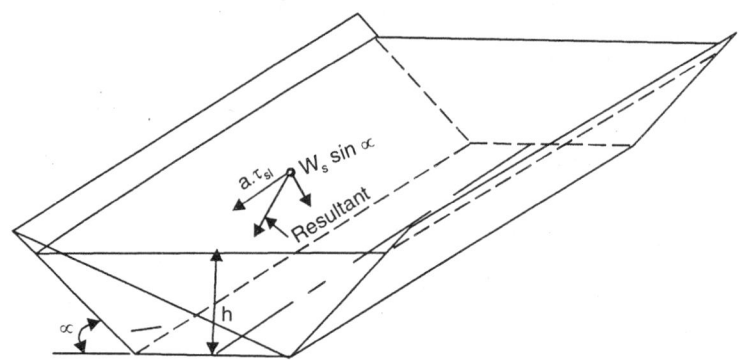

Fig. 17.6.2. Forces causing movement of a particle resting on a channel bank

or

$$\tau_{sl} = \frac{W_s}{a} \cos\alpha \tan\theta \sqrt{1 - \frac{\tan^2\alpha}{\tan^2\theta}} \qquad \qquad ...(17.6.3)$$

On combining Eqs. (17.6.2) and (17.6.3)

$$k = \frac{\tau_{sl}}{\tau_{bl}} = \cos\alpha \sqrt{1 - \frac{\tan^2\alpha}{\tan^2\theta}} \qquad \qquad ...(17.6.4)$$

i.e.,

$$\tau_{sl} = k\,\tau_{bl} = 0.9k\,\tau_c$$

For non-scouring condition, the design criteria becomes

$$\tau_{bm} \leq \tau_{bl}$$

$$\tau_{sm} \leq \tau_{sl}$$

Lane also observed that the curved channels, scour more readily. He, therefore, suggested some correction factors which should be multiplied with the critical value of tractive stress (15). The values of critical shear stress (and also τ_{bl}) for the bed and sides of curved channels are given in Table 17.6.1.

Table 17.6.1: Critical Shear Stress and the values of τ_{bl} for Curved Channels

Type of channel	Critical shear stress (τ_c)	τ_{bl}
Straight channels	τ_c	$0.900\,\tau_c$
Slightly curved channels	$0.90\tau_c$	$0.810\,\tau_c$
Moderately curved channels	$0.75\tau_c$	$0.675\,\tau_c$
Very curved channel	$0.60\tau_c$	$0.540\,\tau_c$

17.6.3 Alluvial Channels carrying Sediment-Laden Water

The cross-section of a stable alluvial channel would depend on the flow rate, sediment transport rate, and sediment size. There are two methods commonly used for the design of alluvial channels carrying sediment-laden water. The first is based on the 'regime' approach in which a set of empirical equations is used. These equations have been obtained by analyzing the data of stable field channels. A more logical method of design of stable alluvial channel should include the sediment load too.

17.6.3.1 Regime methods

Regime methods for the design of stable channels were first developed by the British engineers working for canal irrigation in India in the nineteenth century. At that time, the problem of sediment deposition was one of the major problems of channel design in India. To find a solution to this problem, some of the British engineers studied the behaviour of such stretches of the existing canals where the bed was in a state of stable equilibrium. The stable reaches had not required any sediment clearance for several years of the canal operation. Such channels were called *regime channels*. These channels generally carried a sediment load smaller than 500 ppm. Suitable relationships for the velocity of flow in regime channels were evolved. These relationships are now known as regime equations that find acceptance in other parts of the world as well. The regime relations do not account for the sediment load and, therefore, should be considered applicable when the sediment load is not large.

Kennedy's Method

Kennedy (10) collected data from 22 channels of Upper Bari Doab canal system in Punjab. His observations on this canal system led him to conclude that the sediment in a channel is kept in suspension solely by the vertical component of the eddies that are generated on the channel bed. In his opinion, the eddies generating on the sides of the channel had horizontal movement for greater part and, hence, did not have sediment supporting power. Therefore, the sediment supporting power of a channel is proportional to its width (and not wetted perimeter).

On plotting the observed data, Kennedy obtained the following relation, known as Kennedy's equation:

$$U_0 = 0.55 h^{0.64} \qquad \qquad ...(17.6.5)$$

Kennedy termed U_0 as the critical velocity[1] (in m/s) defined as the mean velocity which will not allow scouring or silting in a channel having depth of flow equal to h (in metres). This equation is, obviously, applicable to such channels that have the same type of sediment as was presented in the Upper Bari Doab canal system. On recognizing the effect of the sediment size on the critical velocity, Kennedy modified Eq. (17.6.5) to

$$U = 0.55 m h^{0.64} \qquad \qquad ...(17.6.6)$$

in which m is the critical velocity ratio and is equal to U/U_0. Here, the velocity U is the critical velocity for the relevant size of sediment, while U_0 is the critical velocity for the Upper Bari Doab sediment. Therefore, the value of m is unity for sediment of the size of Upper Bari Doab sediment. For sediment coarser than Upper Bari Doab sediment, m is greater than 1 while for sediment finer than Upper Bari Doab sediment, m is less than 1. Kennedy did not establish any other relationship for the slope of regime channels in terms of either the critical velocity or the depth of flow. He suggested the use of the Kutter's equation along with the Manning's roughness coefficient. The final results do not differ much if one uses the Manning's equation instead of the Kutter's equation. Thus, the Kennedy equation, Eq. (17.6.6), the continuity equation

---

[1] This critical velocity should be distinguished from the critical velocity of flow in open channels corresponding to Froude number equal to unity (Chapter-13).

$$Q = AU \qquad \qquad ...(17.6.7)$$

and the Manning's equation

$$U = \frac{1}{n} R^{2/3} S^{1/2} \qquad \qquad ...(17.6.8)$$

enable one to determine the unknowns B, h and U for given Q, n, m and the longitudinal slope S.

The longitudinal slope S is decided primarily on the basis of ground considerations. Such considerations, therefore, limit the range of slopes. However, within this range of slopes, one can obtain different combinations of B and h satisfying Eqs. (17.6.6) to (17.6.8). The resulting channel sections can vary from very narrow to very wide. While all these channel sections would be able to carry the given discharge, not all of them would behave satisfactorily. Table 17.6.2 gives values of recommended width-depth ratio, i.e., B/h for stable channels (22).

Table 17.6.2: Recommended values of B/h for stable channels (22)

Q, m^3/s	5.0	10.0	15.0	50.0	100.0	200.0	300.0
B/h	4.5	5.0	6.0	9.0	12.0	15.0	18.0

Several studies carried out on similar lines indicated that the constant C' and the exponent x in the Kennedy's equation, $U = C'mh^x$ are different for different canal systems. Table 17.6.3 gives the values of C' and x in the Kennedy's equation for some regions.

Table 17.6.3: Values of C' and x in the Kennedy's equation for different regions (22)

Region	C'	x
Egypt	0.25 to 0.31	0.64 to 0.73
Thailand	0.34	0.66
Rio Negro (Argentina)	0.66	0.44
Krishna River (India)	0.61	0.52
Chenab River (India)	0.62	0.57
Pennar River (India)	0.60	0.64
Shwebo (Burma)	0.60	0.57
Imperial Valley (USA)	0.64 to 1.20	0.61 to 0.64

The design steps based on Kennedy's method involves trial and are as follows:

(i) For known Q, n, m, and S, assume a trial value of h and obtain the critical velocity U from the Kennedy's equation, Eq. (17.6.6).

(ii) From the continuity equation, Eq. (17.6.7), one can calculate the area of cross-section A and, thus, know the value of B for the assumed value of h.

(iii) Using these values of B and h, compute mean velocity from the Manning's equation, Eq. (17.6.8). If this value of the mean velocity matches with the value of the critical velocity obtained earlier, the assumed value of h (step (i)) and the computed value of B (step (ii)) provide channel dimensions. If the two velocities do not match, assume another value of h and repeat steps (i) to (iii).

Ranga Raju and Misri (21) suggested a simplified procedure that does not require trial. The method is based on the observed final side slope of $1H : 2V$ attained by most of the alluvial channels. During construction, however, the side slopes of a channel are kept flatter than the angle of repose of the soil. But, after some time of canal operation, the side slopes become steeper due to the deposition of sediment. The final shape of the channel cross-section is approximately trapezoidal with side slopes $1H : 2V$. For this final cross-section of channel, one can write

$$A = Bh + 0.5h^2 = h^2(p + 0.5) \qquad \text{...(17.6.9)}$$

$$P = B + 2.236h = h(p + 2.236) \qquad \text{...(17.6.10)}$$

where,

$$p = B/h$$

Now

$$R = \frac{h(p + 0.5)}{p + 2.236} \qquad \text{...(17.6.11)}$$

and

$$U = \frac{Q}{h^2(p + 0.5)} \qquad \text{...(17.6.12)}$$

Substituting the values of U and R in the Manning's equation, Eq. (17.6.6), one obtains

$$S = \frac{Q^2 n^2 (p + 2.236)^{4/3}}{h^{16/3}(p + 0.5)^{10/3}} \qquad \text{...(17.6.13)}$$

Similarly, substituting the value of U in the Kennedy's equation, Eq. (17.6.6), one gets

$$h = \left[\frac{1.818Q}{(p + 0.5)m}\right]^{0.378} \qquad \text{...(17.6.14)}$$

On substituting the value of h from Eq. (17.6.14) in Eq. (17.6.13), one finally obtains

$$\frac{SQ^{0.02}}{n^2 m^2} = 0.299 \frac{(B/h + 2.236)^{1.333}}{(B/h + 0.5)^{1.313}} \qquad \text{...(17.6.15)}$$

Figure 17.6.4 shows the graphical form of Eq. (17.6.15).

For given Q, n, m, and a suitably selected value of S, compute $SQ^{0.02}/n^2 m^2$ and read the value of B/h from Fig. 17.6.3. From Table 17.6.2, check if this value of B/h is satisfactory for the given discharge. If the value of B/h needs modification, choose another slope. Having obtained B/h, calculate h from Eq. (17.6.14) and then calculate B.

Lindley's Method

Lindley (16) was the first to recognize that width, depth, and slope of a channel adjust themselves in an alluvial channel for a given set of conditions. He stated that when an artificial channel is used to

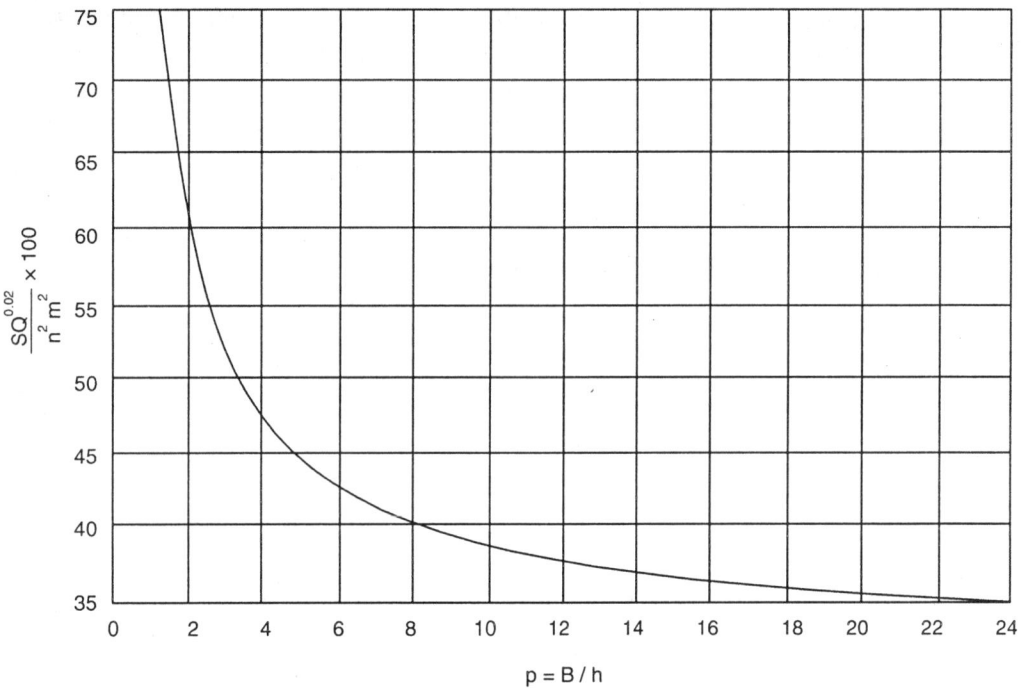

Fig. 17.6.3. Diagram for design of alluvial channels using the Kennedy's equation (21)

carry sediment-laden water, both the bed and banks either scour or silt and, thus, change depth, gradient, and width until a state of balance is attained at which condition the channel is said to be in regime.

The observed width, slope, and depth of the Lower Chenab canal system were analysed by Lindley (16) using $n = 0.025$ and side slopes as $0.5H : 1V$. He obtained the following equations:

$$U = 0.57 h^{0.57} \qquad \qquad ...(17.6.16)$$

$$U = 0.27 B^{0.355} \qquad \qquad ...(17.6.17)$$

From Eqs. (17.6.16) and (17.6.17), one can get

$$B = 7.8 h^{1.61} \qquad \qquad ...(17.6.18)$$

It should be noted that these equations do not include the effect of sediment size on the multiplying coefficient.

Lacey's Method

Lacey (13) stated that the width, depth, and slope of a regime channel to carry a given water discharge loaded with a given sediment discharge are all fixed by nature. According to him, the fundamental requirements for a channel to be in regime are as follows:

(i) The channel flows uniformly in incorherent alluvium. Incoherent alluvium is the loose granular material which can scour or deposit with the same ease. The material may range from very fine sand to gravel, pebbles, and boulders of small size.

(ii) The characteristics and the discharge of the sediment are constant.

(iii) The water discharge in the channel is constant.

The perfect 'regime' conditions rarely exist. The channels which have lateral restraint (because of rigid banks) or imposed slope are not considered as regime channels. For example, an artificial channel, excavated with width and longitudinal slope smaller than those required, will tend to widen its width and steepen its slope, if the banks and bed are of incoherent alluvium and non-rigid. In case of rigid banks, the width is not widened but the slope becomes steeper. Lacey termed this regime as the initial regime. A channel in initial regime is narrower than it would have been if the banks were not rigid. This channel is said to have attained working stability. If the continued flow of water overcomes the resistance to bank erosion so that the channel now has freedom to adjust its perimeter, slope, and depth in accordance with the discharge, the channel is likely to attain what Lacey termed the final regime.

The river bed material may not be active at low stages of the river, particularly, if the bed is composed of coarse sand and boulders. However, at higher stages, the bed material becomes active, i.e., it starts moving. As such, it is only during the high stages that the river may achieve regime conditions. This fact is utilized in solving problems related to floods in river channels.

Lacey also suggested that for a regime channel, the roughness coefficient as well as the critical velocity ratio should be dependent on sediment size alone. However, it is now well known that in a movable bed channel, the total roughness includes grain as well as form roughness. Likewise, the non-silting and non-scouring velocity (included in the critical velocity ratio) shall depend on the sediment load and size of the sediment.

Lacey felt that the sediment in an alluvial channel is kept in suspension by the vertical components of eddies generated at all points along the wetted perimeter. He, therefore, plotted the available data of regime channels to obtain a relationship between the regime velocity U (in m/s) and the hydraulic radius R (in metres). He, thus, found that $U \propto R^{1/2}$ and that the exponential power did not change with data. He, therefore, formulated

$$U = C'\sqrt{R} \qquad\qquad ...(17.6.19)$$

in which C' is a proportionality constant. Including a factor f_1 to account for the size and density of the sediment. Lacey finally obtained

$$U = \sqrt{\frac{2}{5} f_1 R} = 0.632\sqrt{f_1 R} \qquad\qquad ...(17.6.20)$$

The factor f_1 has been named as the *silt factor*. For natural sediment of relative density equal to 2.65, the silt factor f_1 can be obtained by Lacey's relation

$$f_1 = 1.76\sqrt{d} \qquad\qquad ...(17.6.21)$$

where d is the median size of sediment in millimeter.

On plotting Lindley's data and other data of regime channels, Lacey obtained

$$R^{1/2} S = C''$$

and

$$C' = 10.8\, C''^{1/3}$$

$$\therefore \qquad \frac{U}{R^{1/2}} = 10.8\left(R^{1/2} S\right)^{1/3}$$

which gives

$$U = 10.8\, R^{2/3} S^{1/3} \qquad\qquad ...(17.6.22)$$

Equation (17.6.22) is known as Lacey's regime equation and is of considerable use in evaluating flood discharges.

On the basis of the data of regime channels, Lacey also obtained

$$A f_1^2 = 140 U^5$$ ...(17.6.23)

i.e., $$Q f_1^2 = 140 U^6$$ ...(17.6.24)

On substituting the value of f_1 from Eq. (17.6.20), one obtains

$$Q \left(\frac{5U^2}{2R} \right)^2 = 140 U^6$$

or $$\frac{25}{4} \times \frac{Q}{R^2} = 140 U^2$$

or $$\frac{25}{4} Q \frac{P^2}{A^2} = 140 U^2$$

or $$P^2 = \frac{560}{25} Q$$

i.e., $$P = 4.733 \sqrt{Q} \cong 4.75 \sqrt{Q}$$ ...(17.6.25)

Equation (17.6.25) with multiplying constant modified to 4.75 has been verified by a large amount of data and is very useful for fixing clear waterways for structures, such as bridges on rivers. Again, substituting the value of f_1 in Eq. (17.6.23),

$$A \left(\frac{5U^2}{2R} \right)^2 = 140 U^5$$

$\therefore$ $$\frac{A}{R^2} = \frac{560}{25} U$$

or $$P = 22.4 RU$$

or $$4.75 \sqrt{Q} = 22.4 RU$$

or $$RU = 0.212 Q^{1/2}$$ ...(17.6.26)

For wide channels, RU equals the discharge per unit width. Hence,

$$q = 0.212 Q^{1/2}$$ ...(17.6.27)

Equation (17.6.27) relates the discharge per unit width of a regime channel with the total discharge flowing in the channel.

On substituting the value of U from Eq. (17.6.20) in Eq.(17.6.26), one gets

$$R\sqrt{\frac{2}{5}f_1 R} = 0.212 Q^{1/2}$$

$$\therefore \qquad R = 0.48\left(\frac{Q}{f_1}\right)^{1/3} \qquad \qquad ...(17.6.28)$$

For wide channels, the hydraulic radius is almost equal to the depth of flow. Equation (17.6.28), therefore, gives the depth of scour below high flood level. Hence, Eq. (17.6.28) can be utilized to estimate the depth of flow in a river during flood. This information forms the basis for the determination of the levels of foundations, vertical cutoffs, and lengths of launching aprons of a structure constructed across a river.

On combining Eqs. (17.6.27) and (17.6.28),

$$R = 1.35\left(\frac{q^2}{f_1}\right)^{1/3} \qquad \qquad ...(17.6.29)$$

Further, eliminating U from Eqs. (17.6.20) and (17.6.22), one can obtain a relationship for the slope of a regime channel. Thus,

$$\sqrt{\frac{2}{5}f_1 R} = 10.8 R^{2/3} S^{1/3}$$

$$\therefore \qquad S = 0.0002\frac{f_1^{3/2}}{R^{1/2}} \qquad \qquad ...(17.6.30)$$

On substituting the value of R from either Eq. (17.6.28) or Eq. (17.6.29), in Eq. (17.6.30), one obtains

$$S = 0.0003\frac{f_1^{5/3}}{Q^{1/6}} \qquad \qquad ...(17.6.31)$$

and $$S = 0.000178\frac{f_1^{5/3}}{q^{1/3}} \qquad \qquad ...(17.6.32)$$

Lacey's regime relations, Eq. (17.6.19) to (17.6.32), are applicable for regime channels and can be used suitably to design a regime channel for a given discharge and sediment size.

The following flow equation was also obtained by Lacey (13):

$$U = \frac{1}{N_a} R^{3/4} S^{1/2} \qquad \qquad ...(17.6.33)$$

where $$N_a = 0.0225 f_1^{1/4} \qquad \qquad ...(17.6.34)$$

Hence, the absolute roughness coefficient N_a depends only on the sediment size. On examining the data from many different channels, the value of N_a was, however, not found to be constant. Lacey introduced the concept of 'shock' to explain the variation in N_a. He opined that a non-regime channel requires a larger slope (i.e., large value of N_a) to overcome what he termed as 'shock resistance' or the resistance due to bed irregularities. The shock resistance can, therefore, be considered similar to the form resistance of the bed undulations. This concept of Lacey leads one to conclude that a regime channel is free from shock. It is, however, known that the geometry of bed undulations can change even for the same sediment size and, therefore, Lacey's contention that a regime channel is free from shock, is unacceptable (9).

Lacey's equations, commonly used for the design of alluvial channels, are summarized below:

$$f_1 = 1.76\sqrt{d} \qquad \qquad ...(17.6.21)$$

$$U = 10.8R^{2/3}S^{1/3} \qquad \qquad ...(17.6.22)$$

$$P = 4.75\sqrt{Q} \qquad \qquad ...(17.6.25)$$

$$R = 0.48\left(\frac{Q}{f_1}\right)^{1/3} \qquad \qquad ...(17.6.28)$$

$$S = 0.0003\frac{f_1^{5/3}}{Q^{1/6}} \qquad \qquad ...(17.6.31)$$

Comments on Regime Equations

The regime equations have been empirically developed using data of regime channels. These channels carried relatively less sediment load (approximately 500 ppm). Therefore, the equations can be expected to yield meaningful results only for the conditions in which the sediment load is of the same order as was being carried by the channels whose data have been used in developing the regime relations. The dimensions of a stable channel will be affected by the water discharge, the sediment characteristics, and the amount of sediment load in the channel. The regime relations take into account only the water discharge and the sediment size, and do not take into account the amount of sediment material being transported by the channels. Nevertheless, the regime equations do provide useful information which is very helpful in the design of unlined channels and other structures on alluvial rivers.

17.6.3.2 Method of design of alluvial channels including sediment load as a variable

Experiments have indicated that the stable width of a regime channel is practically independent of sediment load (20). Hence, one can use Lacey's equation, Eq. (17.6.25) for stable perimeter P even when the sediment load is varying. If the bank soil is cohesive, the perimeter may be kept smaller than that given by Eq. (17.6.25). However, the sediment load is known to affect the regime slope of a channel and, therefore, the sediment load should also be included in the design of stable alluvial channels.

If the sediment load is moving mainly in the form of bed load, one can estimate the unknowns R and S for given Q, q_B, d, and n using the Meyer-Peter and Müller's equation, Eq. (17.5.3), and the Manning's equation, Eq. (17.6.8).

If the suspended sediment load is also considerable, then one may use the total load equation instead of Meyer-Peter and Müller's equation. Combining Engelund and Hansen equation, Eq. (17.5.7), for sediment load,

$$f\phi_T = 0.4\tau_*^{5/2} \qquad \qquad ...(17.6.35)$$

with their resistance equation (5),

$$U = 10.97 d^{-3/4} h^{5/4} S^{9/8} \qquad \qquad ...(17.6.36)$$

a design chart (Fig. 17.6.4) was prepared. This chart can be used to estimate h and S, for known values of

$$\phi_T \left[= \frac{q_T / \rho_s g}{\sqrt{\dfrac{\Delta \rho_s}{\rho} g d^3}} \right] \text{ and } \frac{q}{\sqrt{\dfrac{\Delta \rho_s}{\rho} g d^3}}$$

For the width of stable channel, Engelund and Hansen (5) suggested the use of the empirical relation

$$B = \frac{0.786 Q^{0.525}}{d^{0.316}} \qquad \qquad ...(17.6.37)$$

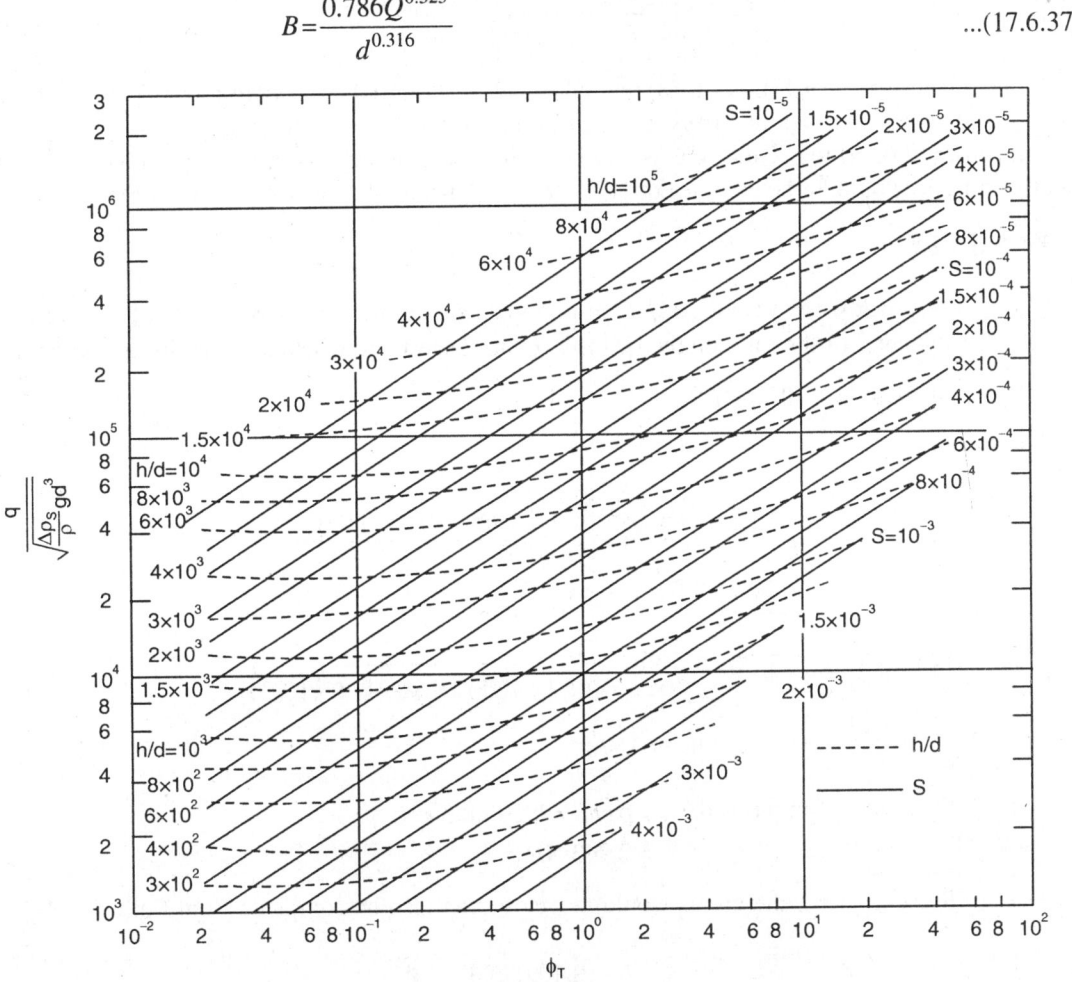

Fig. 17.6.4. Engelund and Hansen's chart for stable channel design

Equation (17.6.37) is valid for channels with sandy bed and banks (9). However, because of cohesive soils of India, the width of stable channels is usually kept smaller than that given by Eq. (17.6.37).

Therefore, one can adopt a value of B which is slightly less than or equal to P obtained from Lacey's equation, Eq. (17.6.25).

Equations (17.6.35) and (17.6.36) can be solved for h and S to yield the following explicit relations:

$$h=0.17\frac{q^{20/21}}{\left(g\phi_T\right)^{2/7} d^{3/7} \left(\Delta\rho_s/\rho\right)^{5/7}} \qquad\qquad ...(17.6.38)$$

$$S=8.4\frac{q^{8/9}d^{2/3}}{h^2} \qquad\qquad ...(17.6.39)$$

or

$$S=4.12\frac{\left(g\phi_T\right)^{4/7} d^{32/21} \left(\Delta\rho_s/\rho\right)^{10/7}}{q^{64/63}} \qquad\qquad ...(17.6.40)$$

Equations (17.6.37), (17.6.38), and (17.6.39) or (17.6.40) will yield direct solutions for B, h, and S, and avoid interpolation errors involved in the use of design chart (Fig. 17.6.4).

In Eqs. (17.6.36) to (17.6.40), U is expressed in m/s, Q in m³/s, q in m²/s, q_T in N/m/s, and h, d, and B in m. It should be noted that Eq. (17.6.38) is applicable for duned-bed conditions.

EXAMPLES

E17.1.   Water flows at a depth of 0.30 m in a wide stream having a slope of 1×10^{-3}. The median diameter of the sand on the bed is 1.0 mm. Determine whether the grains are stationary or moving $\left(v=10^{-6}\,\mathrm{m}^2/\mathrm{s}\right)$.

Solution:

$$R_0^* = \left(\frac{\Delta\rho_s g d^3}{\rho v^2}\right)^{1/2} = \left(\frac{1.65\times9.80\times\left(1\times10^{-3}\right)^3}{\left(10^{-6}\right)^2}\right)^{1/2} = 127.23$$

From Fig. 17.2.3, $\tau_c^* = 0.035 = \dfrac{\tau_c}{\Delta\rho_s g d} = \dfrac{\tau_c}{1650\times9.80\times\left(1.0\times10^{-3}\right)}$

$\therefore$ $\tau_c = 0.5665\,\mathrm{N/m}^2$

Shear stress on the bed, $\tau_0 = \rho\,ghS = 9800\times0.3\times10^{-3}$
$$= 2.943\ \mathrm{N/m}^2$$

Since $\tau_0 > \tau_c$, the grains would move. τ_c can also be computed from Eq. (17.2.3).

$$\tau_c = 0.155+\frac{0.409\left(1.0\right)^2}{\left[1+0.177\left(1.0\right)^2\right]^{1/2}}$$

$$= 0.532\ \mathrm{N/m}^2$$

E17.2. An irrigation canal has been designed to have $R = 2.5$ m and $S = 1.6 \times 10^{-4}$. The sediment on the bed has a median size of 0.30 mm. Find: (i) the bed condition that may be expected, (ii) the height and spacing of undulations, and (iii) the advance velocity of the undulations. Assume depth of flow and mean velocity of flow to be 2.8 m and 0.95 m/s, respectively.

Solution:

$$\frac{R}{d} = \frac{2.5}{0.3 \times 10^{-3}} = 8333.33$$

$$\frac{S}{\Delta \rho_s / \rho} = \frac{1.6 \times 10^{-4}}{1.65} = 9.7 \times 10^{-5}$$

From Fig. 17.3.2, the expected bed condition would correspond to 'ripples and dunes' regime.

From Eq. (17.3.6),

$$n_s = \frac{d^{1/6}}{25.6} = \frac{\left(0.3 \times 10^{-3}\right)^{1/6}}{25.6} = 0.01$$

and from the Manning's equation, Eq. (17.3.5),

$$U = \frac{1}{n_s} R'^{2/3} S^{1/2}$$

$$\therefore \quad R' = \left(\frac{U n_s}{S^{1/2}}\right)^{3/2} = \left[\frac{0.95 \times 0.01}{\left(1.6 \times 10^{-4}\right)^{1/2}}\right]^{3/2} = 0.651 \text{ m}$$

$$\tau'_* = \frac{\rho R' S}{\Delta \rho_s d} = \frac{0.651 \times 1.6 \times 10^{-4}}{1.65 \times 0.3 \times 10^{-3}} = 0.21$$

Using Eq. (17.3.2)

$$\left(\frac{H}{d}\right)\left(\frac{U}{\sqrt{gR}}\right)^3 \frac{U}{\sqrt{(\Delta \rho_s / \rho) gd}} = 6500 \left(\tau'_*\right)^{8/3}$$

$$\frac{H}{0.3 \times 10^{-3}} \left(\frac{0.95}{\sqrt{9.80 \times 2.5}}\right)^3 \left(\frac{0.95}{\sqrt{9.80 \times 1.65 \times 0.3 \times 10^{-3}}}\right) = 6500 (0.21)^{8/3}$$

$$\therefore \quad H = 0.316 \text{ m}$$

Similarly, from Eq. (17.3.3)

$$\left[\frac{U}{\sqrt{(\Delta \rho_s / \rho) gd}}\right]^3 \frac{U}{\sqrt{gR}} \times \frac{L}{d} = 1.8 \times 10^8 \left(\tau'_*\right)^{10/3}$$

i.e., $$\left[\frac{0.95}{\sqrt{9.80 \times 1.65 \times 0.3 \times 10^{-3}}}\right]^3 \frac{0.95}{\sqrt{9.80 \times 2.5}} \times \frac{L}{0.3 \times 10^{-3}} = 1.8 \times 10^8 (0.21)^{10/3}$$

$$\therefore \qquad L = 0.611\,\text{m}$$

$$\frac{U_w}{U} = 0.021\left[\frac{U}{\sqrt{gh}}\right]^3$$

$$\therefore \qquad U_w = 0.95 \times 0.021\left[\frac{0.95}{\sqrt{9.80 \times 2.8}}\right]^3 = 0.43\,\text{m/hr}$$

E17.3.   An alluvial stream $(d = 0.60\,\text{mm})$ has a bed slope of 3×10^{-4}. Find the mean velocity of flow when the hydraulic radius is 1.40 m.

Solution: From Fig. 17.4.2, $k_1 = 0.75$ and $k_2 = 0.70$.

$$k_2 (R/d)^{1/3} \frac{S}{\Delta\rho_s/\rho} = 0.70\left(\frac{1.40}{0.60 \times 10^{-3}}\right)^{1/3} \times \frac{3 \times 10^{-4}}{1.65} = 1.69 \times 10^{-3}$$

From Fig. 17.4.1,

$$k_1 \frac{U}{\sqrt{(\Delta\rho_s/\rho)gR}} = 0.135$$

or $$0.75 \times \frac{U}{\sqrt{1.65 \times 9.80 \times 1.4}} = 0.135$$

$$\therefore \qquad U = 0.86\,\text{m/s}.$$

E17.4.   A 0.40 m wide laboratory flume with glass walls $(n_w = 0.01)$ and mobile bed of 2.0 mm particles carries a discharge of 0.1 m³/s at a depth of 0.30 m. The bed slope is 3×10^{-3}. Determine whether the particles would move or not. Neglect viscous effects.

Solution: Hydraulic radius, R

$$= \left(\frac{0.833 \times 0.01}{\left(3 \times 10^{-3}\right)^{1/2}}\right)^{3/2}$$

$$= 0.0593\,\text{m}$$

Using Eq. (17.4.9),

$$R_b = \left(1 + \frac{2 \times 0.3}{0.4}\right)(0.12) - \frac{2 \times 0.3}{0.4} \times 0.0593$$

$$= 0.211 \text{ m}$$

∴ Bed shear, $\tau_b = \rho g R_b S$

$$= 9800 \times 0.211 \times 3 \times 10^{-3}$$

$$= 6.21 \text{ N/m}^2$$

On neglecting viscous effects and using Yalin and Karahan's curve,

$$\frac{\tau_c}{\Delta \rho_s g d} = 0.045$$

∴   Critical shear, $\tau_c = 1.65 \times 9800 \times 2 \times 10^{-3} \times 0.045$

$$= 1.457 \text{ N/m}^2$$

Since $\tau_b > \tau_c$, the particles would move.

E17.5.   Solve E17.3 using Einstein and Barbarossa method.

Solution: For given $d = 0.6$ mm and bed slope $S = 3 \times 10^{-4}$,

$$U_*' = \sqrt{g R_b' S} = \sqrt{9.80 \times R_b' \times 3 \times 10^{-4}} = 0.054 \sqrt{R_b'}$$

From Eq. (17.4.4),

$$U = 7.66 U_*' \left(R_b' / d\right)^{1/6} = 7.66 \times 0.054 \sqrt{R_b'} \left(\frac{R_b'}{0.6 \times 10^{-3}}\right)^{1/6}$$

∴ $U = 1.4243 R_b'^{2/3}$

$$\psi' = \frac{\Delta \rho_s d_{35}}{\rho R_b' S} = \frac{1.65 \times 0.6 \times 10^{-3}}{R_b' (3 \times 10^{-4})} = \frac{3.3}{R_b'}$$

From $U_*'' = \sqrt{g R_b'' S}$

$$R_b'' = \frac{\left(U_*''\right)^2}{gS} = \frac{\left(U_*''\right)^2}{9.80 \times 3 \times 10^{-4}} = 339.79 \left(U_*''\right)^2$$

The trial procedure for computation of mean velocity can now be carried out in a tabular form. It is assumed that the alluvial river is wide and, therefore,

$$R_b \cong R$$

Trial No.	R_b' (m)	U_*' (m/s)	U (m/s)	ψ'	U/U_*''	U_*'' (m/s)	R_b'' (m)	R_b (m)	Comments
1	1.2	0.059	1.6084	2.75	16.0	0.1005	3.432	4.632	higher than 1.4
2	0.5	0.038	0.8973	6.60	10.5	0.0855	2.484	2.984	higher than 1.4
3	0.2	0.024	0.4871	16.50	7.0	0.0696	1.646	1.846	higher than 1.4
4	0.1	0.017	0.3069	33.00	5.0	0.0614	1.281	1.381	close to 1.4
5	0.11	0.018	0.3270	30.00	5.2	0.0629	1.344	1.444	higher than 1.4
6	0.105	0.175	0.3170	31.43	5.1	0.0622	1.315	1.420	close to 1.4

Values of $R_b (\cong R)$ in row nos. 4 and 6 are reasonably close to the given value of 1.4 m. Thus, the velocity of flow is taken as the average of 0.3069 m and 0.3170 m, i.e., 0.312 m/s. The difference in the value of mean velocity obtained by Einstein and Barbarossa method compared with that obtained by Garde and Ranga Raju method (E17.3) should be noted.

E17.6. Determine the amount of bed load for the data given in E17.2.

Solution: From the solution of E17.2,

$$\tau_*' = 0.21$$

From Eq. (17.5.3)

$\therefore$
$$\phi_B = 8 \times (0.21 - 0.047)^{3/2} = 0.5265$$

i.e.,
$$\frac{q_B}{\rho_s g^{3/2} d^{3/2} \sqrt{\Delta \rho_s / \rho}} = 0.5265$$

$\therefore$
$$q_B = 0.5265 \times 2650 \times \left(9.80 \times 0.3 \times 10^{-3}\right)^{1.5} (1.65)^{1/2} \text{ N/m/s}$$

$$= 0.286 \text{ N/m/s}$$

E17.7. Determine the amount of bed load for the data given in E17.2 using Einstein's method.

Solution: From the solution of E17.2,

$$R' = 0.651 \text{ m}$$

$\therefore$
$$\psi' = \frac{\Delta \rho_s d_{35}}{\rho R' S}$$

$$= \frac{1.65 \times 0.3 \times 10^{-3}}{0.651 \times 1.6 \times 10^{-4}}$$

$$= 4.752$$

$\therefore$
$$\phi_B = 0.763 \text{ (from Fig. 17.5.1)}$$

$$\therefore \qquad q_B = 0.763 \times 2650 \times \left(9.80 \times 0.3 \times 10^{-3}\right)^{1.5} \left(1.65\right)^{1/2}$$

$$= 0.415 \text{ N/m/s}$$

E17.8. Prepare a table for the distribution of sediment concentration in the vertical for E17.2. Assume fall velocity of the particles as 0.01 m/s.

Solution: From the solution of E17.2 and E17.3,

$$q_B = 0.286 \, \text{N/m/s}$$

and $\qquad R' = 0.651$ m

Using Eq. (17.5.6)

$$C_a = \frac{(q_B / \rho_s g)}{23.2 u'_* \, d} \times 100$$

$$= \frac{0.286 \times 100}{\left(2650 \times 9.80\right) \times 23.2 \times \left(9.80 \times 0.651 \times 1.6 \times 10^{-4}\right)^{1/2} \left(0.3 \times 10^{-3}\right)}$$

$$= 5\%$$

Now

$$\frac{C}{C_\alpha} = \left[\frac{h-y}{y} \times \frac{a}{h-a}\right]^{Z_0}$$

$$a = 2d = 2 \times 0.3 \times 10^{-3} \text{ m}$$

$$Z_0 = \frac{w_0}{u_* k} = \frac{0.01}{\left(9.80 \times 2.5 \times 1.6 \times 10^{-4}\right)^{1/2} \times 0.4} = 0.4$$

$$\therefore \qquad \frac{C}{5.0} = \left[\frac{2.8-y}{y} \times \frac{2 \times 0.3 \times 10^{-3}}{2.8 - \left(2 \times 0.3 \times 10^{-3}\right)}\right]^{0.4}$$

$$\therefore \qquad C = 0.17 \left(\frac{2.8-y}{y}\right)^{0.4}$$

The variation of C with y can now be computed as shown in the following table:

y (m)	0.1	0.2	0.5	1.0	1.5	2.0	2.5	2.7	2.8
C (%)	0.635	0.474	0.313	0.215	0.161	0.118	0.0728	0.0455	0
C (ppm)	6350	4740	3130	2150	1610	1180	728	455	0

E17.9. Determine the total bed-material load transport rate for E17.2.

Solution: Using Eqs. (17.5.7) and (17.5.9),

$$f\,\phi_T = 0.4\,\tau_*^{5/2}$$

$$\therefore \qquad \frac{8gRS}{U^2}\phi_T = 0.4\left(\frac{\rho RS}{\Delta\rho_s d}\right)^{5/2}$$

$$\text{or} \quad \phi_T = 0.4\left[\frac{2.5\times1.6\times10^{-4}}{1.65\times0.3\times10^{-3}}\right]^{5/2}\times\left[\frac{0.95\times0.95}{8\times9.80\times2.5\times1.6\times10^{-4}}\right]$$

$$\therefore \qquad \frac{q_T}{\rho_s g^{3/2}d^{3/2}\sqrt{\Delta\rho_s/\rho}} = 6.75$$

$$\therefore \qquad q_T = 6.75\times2650\times\left(9.80\times0.3\times10^{-3}\right)^{3/2}(1.65)^{1/2}$$

$$= 3.67 \text{ N/m/s}$$

E17.10. A rectangular channel 5 m wide is to carry 2.5 m³/s on a slope of 1 in 2000 at a depth of 0.75 m. It is expected that fine silt of 0.04 mm size will enter the channel. What is the maximum concentration of this sediment that can be allowed into the channel without causing objectionable deposition? Assume that the fall velocity of the given sediment in water is 1.5 mm/s, kinematic viscosity of water is 10^{-6} m³/s, and specific gravity of the sediment is 2.65.

Solution: Since the wall and bed of the channel are of the same material, the friction factor for the channel bed f_b is given as

$$f_b = \frac{8gRS}{U^2}$$

Since
$$U = \frac{2.5}{5\times0.75} = 0.7143 \text{ m/s}$$

and
$$R = \frac{5\times0.75}{5+2\times0.75} = 0.5769 \text{ m}$$

$$\therefore \qquad f_b = \frac{8\times9.80\times0.5769\times(1/2000)}{(0.7143)^2}$$

$$= 0.0444$$

$$S_c = S/(\Delta\rho_s/\rho) = \frac{1}{2000\times1.65} = \frac{1}{3300}$$

Now

$$\left(\frac{qS_c^{2.5}}{vf_b^2}\right)\frac{1}{\left(w_0 d/v\right)^{0.6}}\left(\frac{A}{TD_0}\right)^2$$

$$=\frac{\left(2.5/5.0\right)\left(1/3300\right)^{2.5}}{\left(10^{-6}\right)\left(0.0444\right)^2\left(1.5\times10^{-3}\times0.4\times10^{-3}\right)\Big/\left(10^{-6}\right)^{0.6}}\left(\frac{5\times0.75}{5\times0.75}\right)^2=2.2$$

Therefore, from Fig. 17.6.1, maximum concentration for no deposition,

$$C_s = 500 \text{ ppm}$$

E17.11. Design a trapezoidal channel (side slopes 2H : 1V) to carry 25 m³/s of clear water with a slope equal to 10^{-4}. The channel bed and banks comprise gravel (angle of repose = 31°) of size 3.0 mm. The kinematic viscosity of water may be taken as 10^{-6} m²/s.

Solution: From Eq. (17.2.2),

$$R_0^* = \sqrt{\frac{\Delta\rho_s g d^3}{\rho v^2}} = \sqrt{\frac{(1.65)(9.80)(3.0\times10^{-3})^3}{(10^{-6})^2}}$$

∴ $$R_0^* = 661.1$$

From Fig. 17.2.3 $\tau_c^* = 0.045$

∴ $$\tau_c = (0.045)\Delta\rho_s g d = 0.045\times1650\times9.80\times(3.0\times10^{-3})$$

$$= 2.185 \text{ N/m}^2$$

Taking

$$\tau_{bl} = 0.9\tau_c$$

$$= 1.97 \text{ N/m}^2$$

∴ $$\tau_{bm} = 1.97 \text{ N/m}^2$$

Now

$$k = \cos\alpha\sqrt{1-\frac{\tan^2\alpha}{\tan^2\theta}} = \frac{2}{\sqrt{5}}\sqrt{1-\frac{(0.5)^2}{(0.6)^2}} = 0.494$$

∴ $$\tau_{sl} = 0.494\times\tau_{bl} = 0.494\times1.97 = 0.973 \text{ N/m}^2 = \tau_{sm}$$

Rest of the computation is by trial and can be carried out as follows:

Assume $B/h = 10$

Therefore, from Fig. 12.2.5,

$$\frac{\tau_{sm}}{\rho g h s} = 0.78$$

$$h=\frac{0.973}{9800\times10^{-4}\times0.78}=1.27\,\text{m}$$

Also from Fig. 12.2.5,

$$\frac{\tau_{bm}}{\rho ghs}=0.99$$

$$\therefore \qquad h=\frac{1.97}{9800\times10^{-4}\times0.99}=2.03\,\text{m}$$

Choosing the lesser of the two values of h

$$h=1.27\text{ m}$$

$$\therefore \qquad B=10h=12.7\text{ m}$$

$$A=Bh+2h^2=12.7\times1.27+2\times(1.27)^2=19.355\text{ m}^2$$

$$P=B+2\sqrt{5}h=12.7+\left(2\sqrt{5}\times1.27\right)=18.38\text{ m}$$

$$R=1.053\text{ m}$$

and

$$n=\frac{d^{1/6}}{25.6}=\frac{\left(3\times10^{-3}\right)^{1/6}}{25.6}=0.0148$$

$$\therefore Q=\frac{1}{n}AR^{2/3}S^{1/2}=\frac{1}{0.0148}\times19.355\times(1.053)^{2/3}\left(10^{-4}\right)^{1/2}$$

$$=13.5\text{ m}^3/\text{s}$$

Since this value of Q is less than the given value, another value of B/h, say, 20.0 is assumed. Using Fig. 12.2.5 and repeating the above steps, it will be seen that

$$h=1.27\text{ m}$$

$$\therefore \qquad B=25.4\text{ m}$$

$$\therefore \qquad A=35.484\text{ m}^2,\ P=31.08\text{ m},\ R=1.142\text{ m and }Q=26.195\text{ m}^3/\text{s}.$$

This value of Q is only slightly greater than the desired value 25.00 m³/s. Hence, $B=25.4$ m and $h=1.27$ m. The trial calculations can be done in a tabular form as shown below:

B/h	$\dfrac{\tau_{sm}}{\rho ghS}$	$\dfrac{\tau_{bm}}{\rho ghS}$	$h(m)$	$B(m)$	$A(m^2)$	$P(m)$	$R(m)$	$Q(m^3/s)$
10.0	0.78	0.99	1.27	12.7	19.355	18.38	1.053	13.500
20.0	0.78	0.99	1.27	25.4	35.484	32.08	1.142	26.195

E17.12.   Design a channel carrying a discharge of 30 m³/s with critical velocity ratio and the Manning's n equal to 1.0 and 0.0225, respectively. Assume that the bed slope is equal to 1 in 5000.

Solution: Kennedy's method:

Assume $h = 2.0$ m. From Kennedy's equation, Eq. (17.6.6),

$$U = 0.55\,mh^{0.64} = 0.55 \times 1 \times (2.0)^{0.64}$$

$$= 0.857 \text{ m/s}$$

$$\therefore \quad A = Q/U = \frac{30}{0.857} = 35.01\,\text{m}^2$$

For a trapezoidal channel with side slope $1H:2V$

$$Bh + \frac{h^2}{2} = B(2.0) + \frac{2 \times 2}{2} = 2B + 2 = 35.01$$

$$\therefore \quad B = 16.51\,\text{m}$$

$$\therefore \quad R = \frac{35.01}{16.51 + 2.0\sqrt{5}} = 1.67\,\text{m}$$

Therefore, from the Manning's equation, Eq. (17.6.8),

$$U = \frac{1}{n} R^{2/3} S^{1/2} = \frac{1}{0.0225}(1.67)^{2/3}\left(\frac{1}{5000}\right)^{1/2} = 0.885\,\text{m/s}$$

Since the velocities obtained from the Kennedy's equation and Manning's equation are appreciably different, assume $h = 2.25$ m and repeat the above steps.

$$U = 0.55 \times 1 \times (2.25)^{0.64} = 0.924\,\text{m/s}$$

$$A = \frac{30}{0.924} = 32.47\,\text{m}^2$$

$$\therefore \quad B(2.25) + (2.25)^2 = 32.47$$

$$\therefore \quad B = 13.31\,\text{m}$$

$$R = \frac{32.47}{13.31 \times \left(\sqrt{5} \times 2.25\right)} = 1.77 \text{ m}$$

$$U = \frac{1}{0.0255}(1.77)^{2/3}\left(\frac{1}{5000}\right)^{1/2}$$

$$= 0.92 \text{ m/s}$$

Since the two values of the velocities are matching, the depth of flow can be taken as equal to 2.25 m and the width of trapezoidal channel = 13.31 m.

Ranga Raju and Misri's method:

$$\frac{SQ^{0.02}}{n^2 m^2} = \frac{\frac{1}{5000}(30)^{0.02}}{(0.0225)^2 \times (1)^2} = 0.423$$

Hence, from Fig. 17.6.3,

$$p = \frac{B}{h} = 6.0$$

and

$$h = \left[\frac{1.818Q}{(p+0.5)m}\right]^{0.378}$$

$$= \left(\frac{1.818 \times 30}{(6.0+0.5) \times 1}\right)^{0.378} = 2.235\,\text{m}$$

$\therefore$ $\qquad B = 6.0h = 13.41\,\text{m}$

and

$$U = \frac{Q}{h^2(p+0.5)} = \frac{30}{(2.235)^2(6.0+0.5)}$$

$$= 0.924 \text{ m/s}$$

E17.13.   Design a stable channel for carrying a discharge of 30 m³/s using Lacey's method assuming silt factor equal to 1.0.

Solution: From Eq. (17.6.25),

$$P = 4.75\sqrt{Q} = 4.75\sqrt{30.0} = 26.02 \text{ m}$$

From Eq. (17.6.28),

$$R = 0.48(Q/f_1)^{1/3} = 0.48\left(\frac{30.0}{1.0}\right)^{1/3} = 1.49 \text{ m}$$

From Eq. (17.6.31),

$$S = 3 \times 10^{-4} f_1^{5/3} / Q^{1/6} = 3 \times 10^{-4}(1.0)^{5/3} / (30)^{1/6}$$

$$= 1.702 \times 10^{-4}$$

From Eq. (17.6.22),

$$U = 10.8 R^{2/3} S^{1/3} = 10.8(1.49)^{2/3}(1.702 \times 10^{-4})^{1/3}$$

$$= 0.781 \text{ m/s}$$

Assuming the final side slope of the channel as $0.5H : 1V$ (i.e., generally observed field value),

$$P = B + \sqrt{5}h = 26.02 \text{ m}$$

$\therefore$
$$B = 26.02 - 2.24h$$

and
$$A = Bh + \frac{h^2}{2} = PR = 26.02 \times 1.49 = 38.77 \text{ m}^2$$

$\therefore$
$$26.02h - 2.24h^2 + 0.5h^2 = 38.77$$

or
$$1.74h^2 - 26.02h + 38.77 = 0$$

$\therefore$
$$h = \frac{26.02 \pm \sqrt{(26.02)^2 - 4 \times 1.74 \times 38.77}}{2 \times 1.74}$$

$$= \frac{26.02 \pm 20.18}{3.48}$$

$$= 13.28 \text{ m and } 1.68 \text{ m}$$

The value of h equal to 13.28 m gives negative B and is, therefore, not acceptable. Hence, $h = 1.68$m, and

$$B = 26.02 - 2.24 \times 1.68$$

$$= 22.23 \text{ m}$$

E17.14. An irrigation channel is to be designed for a discharge of 50 m³/s adopting the available ground slope of 1.5×10^{-4}. The river bed material has a median size of 2.00 mm. Design the channel and recommend the size of coarser material to be excluded or ejected from the channel for its efficient functioning.

Solution: From Eqs. (17.6.21) and (17.6.31)

$$f_1 = 1.76\sqrt{d}$$

$$= 1.76\sqrt{2}$$

$$= 2.49$$

and
$$S = 0.0003 f_1^{5/3} / Q^{1/6}$$

$$= 0.0003(2.49)^{5/3} / (50)^{1/6}$$

$$= 7.15 \times 10^{-4}$$

The computed slope is much larger than the available ground slope of 1.5×10^{-4} which is to be adopted as the channel bed slope. Therefore, the median size of sediment, which the channel would be able to carry, can be determined by computing the new value of f_1 for $S = 1.5 \times 10^{-4}$ and the given discharge and then obtaining the value of d for this value of f_1 using Eq. (17.6.21). Thus,

$$1.5 \times 10^{-4} = 0.0003 \, f_l^{5/3} / (50)^{1/6}$$

$$\therefore \qquad f_l = 0.976$$

and

$$d = (f_l / 1.76)^2$$

$$= 0.30 \text{ mm}$$

Therefore, the material coarser than 0.30 mm will have to be removed for the efficient functioning of the channel. The hydraulic radius of this channel R is obtained from Eq. (17.6.28):

$$R = 0.48 \left(\frac{50}{0.976} \right)^{1/3}$$

$$\therefore \qquad R = 1.783 \text{ m}$$

Using Eq. (17.6.25)

$$P = 4.75 \sqrt{50}$$

$$= 33.59 \text{ m}$$

$$\therefore \qquad B + \sqrt{5}h = 33.59$$

or

$$B = 33.59 - 2.24 h$$

and

$$A = Bh + \frac{h^2}{2} = PR = 33.59 \times 1.783 = 59.89 \, \text{m}^2$$

or

$$33.59h - 2.24h^2 + 0.5h^2 = 59.89$$

or

$$1.74h^2 - 33.59h + 59.89 = 0$$

$$\therefore \qquad h = \frac{33.59 \pm \sqrt{(33.59)^2 - 4 \times 1.74 \times 59.83}}{2 \times 1.74}$$

$$= \frac{33.59 \pm 26.67}{3.84}$$

$$= 17.32 \text{ m or } 1.99 \text{ m}$$

Obviously, $h = 1.99$ m as the other root of h would result in too narrow a channel section

$$\therefore \qquad B = 33.29 - 2.24(1.99)$$

$$= 29.13 \text{ m}$$

E17.15.   Design a channel to carry a discharge of 30 m³/s with sediment load concentration of 50 ppm by weight. The average grain size of the bed material is 0.3 mm. Assume the cross-section of the channel as trapezoidal with side slopes $\frac{1}{2}H : 1V$.

Solution: Using (17.6.25)

$$P = 4.75\sqrt{Q} = 4.75\sqrt{30} = 26.02 \text{ m}$$

Choose B to be slightly less than P, say $B = 24.0$ m.

$$Q_T = 50 \times 10^{-6} \times 9810 \times 30 = 14.71 \,\text{N/s}$$

$$\therefore \qquad q_T = \frac{14.71}{24.0} = 0.613 \,\text{N/m/s}$$

and

$$q = \frac{30.0}{24.0} = 1.25 \,\text{m}^2/\text{s}$$

Now

$$\phi_T = \frac{0.613/(2.65 \times 9800)}{\sqrt{1.65 \times 9.80 \times \left(0.3 \times 10^{-3}\right)^3}}$$

$$= 1.128$$

and

$$\frac{q}{\sqrt{\left(\dfrac{\Delta \rho_s}{\rho}\right) gd^3}} = \frac{1.25}{\sqrt{1.65 \times 9.80 \times \left(0.3 \times 10^{-3}\right)^3}}$$

$$= 59793.2$$

From Fig. 17.6.4,

$$S = 1.1 \times 10^{-4}$$

and

$$\frac{h}{d} = 7.2 \times 10^3$$

$$\therefore \qquad h = 7.2 \times 10^3 \times \left(0.3 \times 10^{-3}\right) = 2.16 \text{ m}$$

Alternatively, using Eqs. (17.6.38) and (17.6.39)

$$h = \frac{0.17 \times (1.25)^{20/21}}{(9.81 \times 1.128)^{2/7} \left(0.3 \times 10^{-3}\right)^{3/7} (1.65)^{5/7}}$$

$$= 2.38 \text{ m}$$

$$S = \frac{8.4 \times (1.25)^{8/9} \times (0.3 \times 10^{-3})^{2/3}}{(2.38)^2}$$

$$= 8.1 \times 10^{-3}$$

E17.16. Solve Example 17.15 considering the entire sediment load as bed load.

Solution: $Q_B = 50 \times 10^{-6} \times 9800 \times 30 = 14.71 \text{ N/s}$

Silt factor, $f_1 = 1.76\sqrt{d} = 1.76\sqrt{0.3} = 0.964$

Since the value of the Manning's n is 0.0225 for $f_1 = 1.0$ one can take $n = 0.222$ for $f_1 = 0.964$. Using Eq. (17.6.25),

$$P = 4.75\sqrt{30.0} = 26.02 \text{ m}$$

Let $\qquad B = 24.0 \text{ m}$

$\therefore \qquad q_B = \dfrac{14.71}{24.0} = 0.613 \text{ N/m/s}$

Hence,

$$\phi_B = \frac{Q_B / (\rho_s g)}{\sqrt{\dfrac{\Delta \rho_s}{\rho} g d^3}} = \frac{0.613 / (2650 \times 9.80)}{\sqrt{1.65 \times 9.80 \times (0.3 \times 10^{-3})^3}}$$

$$= 1.128$$

and from Eq. (17.4.4) with k_s replaced by d

$$n_s = \frac{d^{1/6}}{25.6}$$

$$= \frac{(0.3 \times 10^{-3})^{1/6}}{25.6}$$

$$= 0.01$$

$\therefore \qquad \tau'_* = \left(\dfrac{n_s}{n}\right)^{3/2} \dfrac{\rho R S}{\Delta \rho_s d} = \left(\dfrac{0.01}{0.022}\right)^{3/2} \dfrac{R S}{1.65 \times (0.3 \times 10^{-3})}$

$$= 619.10 \, R S$$

Using Eq. (17.5.3),

$$\phi_B = 8.0 (\tau'_* - 0.047)^{3/2}$$

or $\qquad 1.128 = 8.0 \left(619.10 RS - 0.047 \right)^{3/2}$

$\therefore \qquad RS = 5.135 \times 10^{-4}$

Now using the Manning's equation,

$$Q = \frac{1}{n} A R^{2/3} S^{1/2}$$

$$30 = \frac{1}{0.022} \left(26.02 \times R^{5/3} S^{1/2} \right)$$

$\therefore \qquad R^{5/3} S^{1/2} = 0.0254$

or $\qquad R^{7/6} \left(5.135 \times 10^{-4} \right)^{1/2} = 0.0254$

$\therefore \qquad R = 1.103 \, m$

and $\qquad S = 4.66 \times 10^{-4}$

$$A = Bh + \frac{h^2}{2} = 24h + \frac{h^2}{2} = PR = 26.02 \times 1.103$$

or $\qquad h^2 + 48h - 57.4 = 0$

$$h = \frac{-48 \pm \sqrt{(48)^2 + 4 \times 57.4}}{2}$$

$$h = 1.167 \, m$$

$$P = B + \sqrt{5}h$$

$$= 24 + \left(\sqrt{5} \times 1.67 \right)$$

$$= 26.61 \, m$$

which is close to Lacey's perimeter (=26.02 m)
Hence,

$\qquad B = 24.0 \, m, \quad h = 1.167 \, m \qquad$ and $\qquad S = 4.66 \times 10^{-4}$

PROBLEMS

P17.1　What is the meaning of the term 'incipient motion of sediment'? How would you calculate critical tractive stress for given sediment?

P17.2　Describe various bed forms that occur in alluvial channels.

P17.3　Describe various modes of sediment transport in an alluvial channel.

P17.4 Determine the depth of flow which will cause incipient motion condition in a wide channel having a mean sediment size of 5.0 mm and a channel slope of 0.0004. The specific gravity of the material is 2.65. Also determine the corresponding velocity of flow.

P17.5 Find the depth of velocity of flow at which the bed material of average size 5.0 mm will just move in a wide rectangular channel at a slope of 1×10^{-3}. Assume specific gravity of the material tó be 2.65 and neglect viscous effects.

P17.6 Water flows at a depth of 0.20 m in a wide flume having a slope of 1×10^{-4}. The median diameter of the sand placed on the flume bed is 1.0 mm. Determine whether the sand grains are stationary or moving.

P17.7 An irrigation canal has been designed to have hydraulic radius equal to 2.50 m, depth of flow equal to 2.80 m, and bed slope equal to 1.5×10^{-4}. The sediment on the canal bed has median size of 0.25 mm. Find: (i) the bed condition, i.e., flow regime that may be expected, (ii) height and length of bed forms, and (iii) the advance velocity of the bed forms.

P17.8 For an alluvial stream, the average width is 120.0 m and the cross-section may be considered as rectangular. The longitudinal slope of the stream is 0.0002. Prepare a stage-discharge curve up to a depth of 4.0 m. The size and specific gravity of the sediment are 0.30 mm and 2.65, respectively.

P17.9 Determine the bed load transport in a wide alluvial stream for the following conditions:

Depth of flow	= 2.50 m
Velocity of flow	= 1.50 m/s
Average slope of water surface	$= 8 \times 10^{-4}$
Mean size of sediment	= 5.0 mm
Specific gravity of the sediment	= 2.65

P17.10 Determine the rate of bed load transport in a wide alluvial stream for the following data:

Depth of flow	= 4.50 m
Velocity of flow	= 1.30 m/s
Slope	$= 2.0 \times 10^{-4}$

Size distribution of the sediment is as follows:

d (mm)	0.20	0.44	0.78	1.14	1.65	2.6	5.20
% finer	2.0	10.0	30.0	50.0	70.0	80.0	100.0

P17.11 For an alluvial stream having a slope of 0.00015 and depth of flow equal to 2.40 m, the following velocity profile was observed:

y (m)	0.215	0.30	0.425	0.670	0.885	1.035	1.28	1.77	2.07	2.35
Velocity (m/s)	1.31	1.37	1.45	1.56	1.65	1.66	1.68	1.69	1.70	1.65

If the fall velocity for the average size of the suspended load is 8.00 mm/s, plot the distribution of the suspended load in a vertical section. Assume Karman's constant equal to 0.4 and the concentration of sediment at $y = 0.215$ m as equal to 4 N/litre.

P17.12 What are the 'true regime' conditions in an alluvial channel as stipulated by Lacey?

P17.13 A trapezoidal channel is 5.0 km long and has a trapezoidal cross-section with side slopes of $\frac{1}{2}H : 1V$. Up to the first 2 km, the bed-width of the channel is 4.90 m and the depth of flow is 1.05 m. Thereafter, the bed-width and depth are 4.0 m and 0.885 m, respectively, right up to the downstream end of the channel. Estimate the total seepage loss through the channel if the rate of seepage loss is 3 m³/s per million square metres of the water surface area exposed.

P17.14 A stable channel is to be designed for a discharge of 40 m³/s and the silt factor of unity. Calculate the dimensions of the channel using Lacey's regime equations. What would be the bed width of this channel if it were to be designed on the basis of Kennedy's method with critical velocity ratio equal to unity and the ratio of bed-width to depth of flow the same as obtained from Lacey's method.

P17.15 For the design of an irrigation channel through a flat ground it has been decided to avail fully the available slope of 25 cm/km, and adopt Lacey's method of design by restricting the discharge. Design the channel taking the average size of bed material as 0.0005 m and side slope of $0.5H : 1V$.

P17.16 Design an irrigation channel to carry 200 m³/s of flow with a bed load concentration of 100 ppm by weight. The average grain diameter of the bed material is 1.0 mm. Use Lacey's method and a suitable bed load equation for the design. Side slope of the channel is $0.5H : 1V$.

REFERENCES

(1) Arora, AK: *"Suspended sediment transport in rigid open channels"*, Ph.D. Thesis, University of Roorkee, Roorkee, India, 1983.

(2) Einstein, HA: *"Formulas for the transportation of bed load,"* Trans. ASCE, Vol. 107, 1942.

(3) Einstein, HA, and NL Barbarossa: *"River channel roughness,"* Trans. ASCE, Vol. 117, 1952.

(4) Einstein, HA: *"The bed load function for sediment transportation in open channel flows"*, USDA, Tech. Bull. No. 1026, Sep., 1950 [Garde and Ranga Raju (9)].

(5) Engelund, F and E Hansen: *"A Monograph on Sediment Transport in Alluvial Streams"*, Teknisk Forlag, Denmark, 1967 [Garde and Ranga Raju (9)].

(6) Garde, RJ and ML Albertson: *"Sand waves and regimes of flow in alluvial channels"*, Proc. 8th IAHR Congress, Vol. 4, Montreal, 1959.

(7) Garde, RJ, and KG Ranga Raju: *"Regime criteria for alluvial streams"*, J. of Hyd. Div., Proc. ASCE, Nov. 1963.

(8) Garde, RJ and KG Ranga Raju: *"Resistance relationship for alluvial channel flow"*, J. of Hyd. Div., Proc. ASCE, Vol. 92, 1966.

(9) Garde, RJ and KG Ranga Raju: *"Mechanics of Sediment Transportation and Alluvial Stream Problems"*, 2nd Ed., Wiley Easern Limited, New Delhi, 1985.

(10) Kennedy, RG: *"The prevention of silting in irrigation canals"*, Paper No. 2826, Proc. ICE (London), Vol. 119, 1895 [Ranga Raju (22)].

(11) Keulegan, GH: *"Laws of turbulent flow in open channels"*, U.S. Deptt. of Commerce, NBS, Vol. 21, Dec. 1938 [Garde and Ranga Raju (9)].

(12) Kondap, DM and RJ Garde: *"Velocity of bed-forms in alluvial channels"*, Proc. 15th IAHR Congress, Vol. 5, Istanbul, 1973.

(13) Lacey, G.: *"Regime flow in incoherent alluvium"*, CBIP Publication No. 20, 1939 [Garde and Ranga Raju (9)].

(14) Lane, EW: *"Stable channels in erodible material"*, Trans. ASCE, Vol. 102, 1937 [Ranga Raju (22)].

(15) Lane, EW: *"Design of stable channels"*, Trans. ASCE, Vol. 120, 1955, pp. 1234-1279 [Ranga Raju (22)].

(16) Lindley, ES: *"Regime channels"*, Proc. Punjab Engg. Congress, Vol. 7, 1919 [Ranga Raju (22)].

(17) Meyer-Peter, E and R Müller: *"Formulas for bed load transport"*, Proc. 2nd IAHR Congress, Stockholm, 1948 [Garde and Ranga Raju (9)].

(18) Ranga Raju, KG and JP Soni: *"Geometry of ripples and dunes in alluvial channels"*, J. of Hyd. Research, Vol. 14, 1976.

(19) Ranga Raju, KG: *"Resistance relation for alluvial streams"*, La Houille Blanch, No. 1, 1970.

(20) Ranga Raju, KG, KR Dhandapani and DM Kondap: *"Effect of sediment load on stable canal dimensions"*, J. of Waterways and Harbours Division Proc. ASCE, Vol. 103, 1977.

(21) Ranga Raju, KG and RL Misri: *"Simplified canal design procedure using Kennedy's theory"*, CBIP J. of Irrigation and Power (India), April 1979.

(22) Ranga Raju, KG: *"Flow Through Open Channels"*, Tata McGraw-Hill Publishing Company Limited, New Delhi, 1986.

(23) Rouse, H: *"Modern conceptions of mechanics of fluid turbulence"*, Trans., ASCE, Vol. 102, 1937 [Garde and Ranga Raju (9)].

(24) Shields, A. *"Anwendung der Aehnlichkeitsmechanic und der turbulenzforschung auf die Geschiebebewegung"*, Mitteilungen der Pruessischen Versuchsanstalt fur Wasserbau and Schiffbau, Berlin, 1936 [Garde and Ranga Raju (9)].

(25) Swamee, PK and MK Mittal, *"An explicit equation for critical shear stress in alluvial streams"*, CBIP J. of Irrigation and Power, New Delhi, April, 1976.

(26) Yalin, MS and E Karahan, *"Inception of sediment transport"*, J. of Hyd. Div., Proc. ASCE, Vol. 105, Nov. 1979.

Bibliography

Asawa, GL: *Irrigation and Water Resources Engineering*, New Age International (P) Limited, Publishers, New Delhi, India, 2005.

Asawa, GL: *Laboratory Work in Hydraulic Engineering*, New Age International (P) Limited, Publishers, New Delhi, India, 2006.

Chanson, H: *The Hydraulics of Open Channel Flow*: An Introduction, 2nd Edition, Elsevier Butterworth Heinmann, 2004.

Chow, VT: *Open-Channel Hydraulics*, McGraw-Hill Book Company, USA, 1959.

Douglas, JF, Gasiorek, JM, and Swaffield, JA: *Fluid Mechanics*, 4th Edition, Pearsonn Education (Singapore) Pte. Ltd., 2002.

Fox, RW, and McDonald, AT: *Introduction to Fluid Mechanics*, 4th Edition, JohnWiley and Sons, Inc., USA, 1994.

French, RH: *Open-Channel Hydraulics*, McGraw-Hill Book Co., Singapore, 1994.

Garde RJ, Ranga Raju, KG: *Mechanics of Sediment Transportation and Alluvial Stream Problems*, 3rd Edition, New Age International (P) Limited, Publishers, New Delhi, India, 2000.

Garde, RJ and Mirajgaokar, AG: *Engineering Fluid Mechanics*, Scitech Publications (India) Pvt. Ltd., Chennai (India), 2003.

Gupta, V, Gupta, SK: *Fluid Mechanics and its Applications*, New Age International (P) Limited, Publishers, New Delhi, India, 1984.

Henderson, FM, *Open Channel Flow*, The Macmillan Company, New York, 1966.

Massey, BS (revised by John Ward-Smith): *Mechanics of Fluids*, 7th Edition, Nelson Thornes Limited, U.K., 1998.

Munson, BR, Young, DF, and Okiishi, TH: *Fundamentals of Fluid Mechanics*, 4th Edition, JohnWiley and Sons, Inc., Singapore, 2002.

Ranga Raju, KG: *Flow through Open Channels*, 2nd Edition, Tata McGraw-Hill, Publishing Company Ltd., New Delhi, India, 1993.

Rouse, H:, *Elementary Mechanics of Fluids*, Wiley eastern Private Limited, New Delhi, India, 1970.

Schlichting, H, Gersten, K: *Boundary Layer Theory*, 8th Edition, Springer, 2001.

Streeter,VL, Wylie, EB, and Bedford, KW: *Fluid Mechanics*, 9th Edition, McGraw-Hill Companies, Singapore, 1998.

Subramanya, K: *Flow in Open Channels*, 2nd Edition, Tata McGraw-Hill, Publishing Company Ltd., New Delhi, India, 1996.

White, FM: *Fluid Mechanics*, 5th Edition, McGraw-Hill Companies, USA, 2003.

Index